Physical Geography

A Human Perspective

Richard Huggett, Sarah Lindley, Helen Gavin, Kate Richardson

ARNOLD

A MEMBER OF THE HODDER HEADLINE GROUP
LONDON

Distributed in the United States of America
by Oxford University Press Inc., New York

First published in Great Britain in 2004 by
Arnold, a member of the Hodder Headline Group,
338 Euston Road, London NW1 3BH

http://www.arnoldpublishers.com

Distributed in the United States of America by
Oxford University Press Inc.
198 Madison Avenue, New York, NY10016

Hodder Headline's policy is to use papers that are natural, renewable and
recyclable products and made from wood grown in sustainable forests.
The logging and manufacturing processes are expected to conform to the
environmental regulations of the country of origin.

The advice and information in this book are believed to be true and
accurate at the date of going to press, but neither the authors nor the publisher
can accept any legal responsibility or liability for any errors or omissions.

British Library Cataloguing in Publication Data
A catalogue record for this book is available from the British Library

Library of Congress Cataloging-in-Publication Data
A catalog record for this book is available from the Library of Congress

ISBN 0 340 80962 0

1 2 3 4 5 6 7 8 9 10

Typeset in 11 on 13 pt Garamond by Phoenix Photosetting, Chatham, Kent
Printed and bound in the UK by Martins the Printers, Berwick upon Tweed

What do you think about this book? Or any other Arnold title?
Please send your comments to feedback.arnold@hodder.co.uk

For my family (RJH)

For my family and partner (SJL)

For Alistair Graham (HG)

For Adam Holden (KR)

CONTENTS

PREFACE

At the time of writing, the human domination of the planet continues apace. Human impacts on the environment are wide-ranging, affecting weather and climate, soil fertility, the quality and quantity of water in rivers, lakes and oceans, landscapes and landforms and the diversity of living things. The changes being wrought by humans are radical and pervasive, so much so that people–environment interactions now provide a coherent focus for physical geography. This book explores physical geography in a human-dominated world, and in doing so creates a new structure for the subject.

The authorship of the book grew in the writing. The original plan was for a single-authored volume, but that quickly moved to two authors, and later to three and four. Apart from easing the workload of the ageing senior author, the chief purposes of bringing other authors on board were to increase the level of expertise in the various aspects of the environment and to bring youthful minds and ideas into the frame. Only you, the reader, can judge the success of the venture, and we hope that you enjoy wading through the next 200,000 words as much as we enjoyed writing them.

We have several people to thank. Liz Gooster at Arnold first approached Richard Huggett about the project. Abigail Woodman at Arnold took over as editor and has kept us to schedule, more or less. Nick Scarle in the Cartographic Unit at Manchester worked wonders with rough sketches, turning them into first-rate diagrams. Assorted individuals kindly provided us with photographs: Mike Acreman, Andrew Bennett, Tessa Francis, Andrew Derocher, Gary Marrone, Pat Morris, Rich Norby and Ersnt Detlef-Schultze.

Richard Huggett
Sarah Lindley
Helen Gavin
Kate Richardson

Manchester
January 2004

ACKNOWLEDGEMENTS

Figures 1.3 and 1.4 from J. Rotmans (1990) *IMAGE: An Integrated Model to Assess the Greenhouse Effect,* Dordrecht: Kluwer Academic Publishers, reprinted with kind permission of Kluwer Academic Publishers and Jan Rotmans.

Figure 1.5 from R. A. Berner (1994) '3GEOCARB II: a revised model of atmospheric CO_2 over Phanerozoic time' (*American Journal of Science* 294, 56–91), reprinted by permission of *American Journal of Science* and Robert A. Berner.

Figure 1.6 from J. E. Kutzbach and F. A. Street-Perrott (1985) 'Milankovitch forcing in fluctuations in the level of tropical lakes from 18 to 0 kyr BP' (*Nature* 317, 130–4), © 1985 Macmillan Magazines Limited, reprinted with kind permission of Nature Publishing Group, John E. Kutzbach and Alayne Street-Perrott.

Table 1.1 from L. Hannah, D. Lohse, C. Hutchinson, J. L. Carr and A. Lankerani (1994) 'A preliminary inventory of human disturbance of world ecosystems' (*Ambio* 23, 246–50), adapted by kind permission of the Swedish Academy of Sciences.

Table 1.3 Data used with kind permission of the IUCN.

Figures 2.1 and 2.11 from R. J. Huggett (1997) *Environmental Change: The Evolving Ecosphere,* reprinted with permission of Routledge.

Figure 2.6 from T. Whitten, M. Mustafa and G. S. Henderson (2002) *The Ecology of Sulawesi,* reprinted by kind permission of Periplus Editions (HK) Ltd.

Figure 2.7 from *Applied Hydrogeology* (with disk), 3rd edn, by C. W. Fetter © 1994, reprinted by permission of Pearson Education, Inc, Upper Saddle River, NJ.

Figure 2.10 from E. M. Bridges (1997) *World Soils,* 3rd edn, reprinted by kind permission of Cambridge University Press.

Figure 2.12 from F. K. Lutgens and E. J. Tarbuck (1998) *The Atmosphere: An Introduction to Meterology,* reprinted by permission of Pearson Education, USA.

Figure 3.2 from J. Seidensticker, S. Christie and P. Jackson (eds) *Riding the Tiger: Tiger Conservation in Human-dominated Landscapes* (1999) by kind permission of Cambridge University Press.

Table 3.1, Figure 7.17 and Figure 7.18 from M. Claussen *et al.* (1999) 'A new model for climate system analysis: outline of the model and application to palaeoclimate simulations' (*Environmental Modeling and Assessment* 4, 209–16), reprinted with kind permission of Kluwer Academic Publishers and Martin Claussen.

Figure 4.2 from H. P. Huntington 'Beluga Whale TEK: Traditional Ecological Knowledge of the Beluga Whale', available online at http:/www.si.edu/arctic/htnl/tek/html, by kind permission of Henry P. Huntington.

Table 4.2 from J. Studley (1998) *Dominant Knowledge Systems and Local Knowledge,* Mountain-Forum On-line Library Document, http://www.mtnforum.org/resources/library/stud98a2.htm, by kind permission of John Studley.

Figure 5.4 with permission from the Advisory Centre for Education.

Figure 5.5 from S. A. Schumm (1991) *To Interpret the Earth: Ten Ways to be Wrong,* by kind permission of Cambridge University Press.

Figure 5.6 from E. C. Mackey and G. J. Tudor (2000) 'Land cover changes in Scotland over the past 50 years' in R. W. Alexander and A. C. Millington (eds) *Vegetation Mapping: From Patch to Planet,* © John Wiley & Sons Limited. Reproduced with permission.

Table 5.6 from B. K. Wyatt (2000) 'Vegetation mapping from ground, air and space – competitive or complementary techniques' in R. W. Alexander and A. C. Millington (eds) *Vegetation Mapping: From Patch to Planet,* © John Wiley & Sons Limited. Reproduced with permission.

Figures 6.1, 6.3, 6.4, 6.5, 6.6 and 6.14 from I. Heywood, S. Cornelius and S. Carver (2002) *An Introduction to Geographical Information Systems,* 2nd edn (Harlow: Prentice Hall), reprinted by permission of Pearson Education Limited.

Figure 6.9 from J. Gao and Y. Liu (2001) 'Applications of remote sensing, GIS and GPS in glaciology: a review' (*Progress in Physical Geography* 25, 520–40), reprinted by permission of Arnold.

Figures 6.10, 6.11, 6.12, 6.13 and Table 6.1 © John Wiley & Sons Limited. Reproduced with permission.

Figure 7.2 from http://www.bio2.edu/Research/flow.htm with the kind permission of Biosphere 2 Centre.

Figure 7.3 with permission from Oak Ridge National Laboratory.

Figure 7.6 from J. D. Phillips (1991) 'The human role in Earth surface systems: some theoretical considerations' (*Geographical Analysis* 23, 316–31), by kind permission of Jonathan D. Phillips.

Figure 7.8 from H. T. Odum (1971) *Environment, Power, and Society* (New York: Wiley Interscience) by kind permission of Mary Logan.

Figure 7.11 by kind permission of Jerry Meehl and Warren Washington.

Figures 7.12, 7.13, 7.14 and 7.15 and Table 7.5 by kind permission of the Intergovernmental Panel on Climate Change.

Figures 7.16 and 7.17 from M. Claussen *et al.* (1999) 'A new model for climate system analysis: outline of the model and application to palaeoclimate simulations', *Environmental Assessment* 4, 209–16, reprinted with kind permission of Kluwer Academic Publishers.

Figure 7.19 from A. M. Solomon (1986) 'Transient response of forests to CO_2-induced climatic change: simulation modeling experiments in eastern North America' (*Oecologia* 68, 567–79), © 1986 Springer-Verlag, reprinted by kind permission of Springer-Verlag and Allen M. Solomon.

Figure 7.20 from M. J. Sykes and A. Haxeltine (2001) 'Modelling the response of vegetation distribution and biodiversity to climate change' in F. S. Chapin III, O. E. Sala and E. Huber-Sannwald (eds) *Global Biodiversity in a Changing Environment: Scenarios for the 21st Century* (Ecological Studies 152), pp. 5–22. Springer: New York. © 2001

Springer-Verlag, reprinted by kind permission of Springer-Verlag and Martin Sykes.

Table 7.1 from F. H. Sklar, R. Costanza and J. W. Day (1990) 'Model conceptualization' in B. C. Patten (ed.) *Wetlands and Shallow Continental Water Bodies*, Volume 1 (pp. 625–58), The Hague: SPB Academic Publishing, Reprinted with kind permission of Backhuys Publishers.

Figures 8.1, 8.2 and 8.3 from C. D. Ahrens (1998) *Essentials of Meteorology: An Invitation to the Atmosphere* (Belmont, USA: Wadsworth Publishing Ahrens), reprinted by permission of Wadsworth Publishing.

Figure 8.4 from R. G. Barry and R. J. Chorley (2003) *Atmosphere, Weather and Climate*, 8th edn (London: Routledge), reprinted by permission of Routledge.

Figures 8.5, 8.6, 8.7 and 8.8 from *The Atmosphere: An Introduction to Meteorology* by F. K. Lutgens and E. J. Tarbuck © 1998, reprinted by permission of Pearson Education, Inc, Upper Saddle River, NJ.

Figure 8.9 with permission from Oak Ridge National Laboratory.

Figures 8.11, 8.12 and 8.13 by kind permission of the Intergovernmental Panel on Climate Change.

Table 8.1 from R. J. Andres and A. D. Kasgnoc (1998) 'Time averaged inventory of subaerial volcanic sulphur emissions' (*Journal of Geophysical Research* 103, 251–61), reprinted with permission of *Journal of Geophysical Research*.

Table 8.2 from J. W. Erisman, P. Grennfelt and M. Sutton (2003) 'The European perspective on nitrogen deposition and emission' (*Environment International* 29, 311–25), © 2003, with permission from Elsevier.

Figures 9.1, 9.7–9.11, 9.16 and Plates 2, 3, 7, 8, 10 and 11 with permission from Mike Acreman.

Figure 9.2 from H. Riehl (1965) *An Introduction to the Atmosphere*, 1st edn (New York: McGraw-Hill), reprinted by permission of Routledge.

Figure 9.3 from W. Brustaert (1982) *Evaporation into the Atmosphere: Theory, History and Applications*, Dordrecht: D. Reidel Publishing Company, reprinted with kind permission of Kluwer Academic Publishers and Wilfried Brustaert.

Figure 9.4 from M. Newson (1994) *Hydrology and the River Environment*, Oxford University Press, reprinted by permission of Oxford University Press.

Figure 9.6 reprinted by kind permission of Wetlands International.

Figure 9.11 with permission from Nicholas Bailey/Rex Features (1990).

Figures 9.13a, 9.18 and Table 9.1 with permission from European Environment Agency.

Figures 9.13b, 9.14 and 9.15 use data from Table 2 in *Irrigation in Africa in Figures* (Water Report No. 7) with kind permission of the Food and Agriculture Organization of the United Nations.

Figure 9.17 with permission from David Grigg.

Figure 9.20 was kindly provided by Phil Sutherland.

Figure 10.2 from C. P. Burnham (1980) 'The soils of England and Wales' (*Field Studies* 5, 348–63), with kind permission of the Field Studies Council Publications.

Figure 10.7 Data from L. R. Brown and E. C. Wolf (1984) *Soil Erosion: Quiet Crisis in the World Economy* (Worldwatch Paper 60) (Washington, DC: Worldwatch Institute).

Figure 10.8 from F. J. Dent (1980) 'Major production systems and soil related constraints in Southeast Asia' (in M. Drosdoff, H. Zandstra and W. G. Rockwood (eds) *Priorities for Alleviating Soil-related Constraints to Food Production in the Tropics*. Manila: International Rice Research Institute, and Ithaca: Cornell University).

Figure 10.9 from L. R. Oldeman, R. T. A. Hakkeling and W. G. Sombroek (1990) *World Map of the Status of Human-induced Soil Degradation*, revised edn (Nairobi/Wageningen: UNEP/ISRIC GLASOD Project).

Table 10.7 reprinted from A. Lufafa, M. M. Tenywa, M. Isabirye, M. J. G. Majaliwa and P. L. Woomer (2002) 'Prediction of soil erosion in Lake Victoria basin catchment using a GIS-based Universal Soil Loss model' (*Agricultural Systems* 76, 883–94), © 2002, with permission from Elsevier.

Table 10.9 reprinted from D. E. Rocheleau, P. E. Steinberg and P. A. Benjamin (1995) 'Environment, development, crisis, and crusade: Ukambani, Kenya, 1890–1990' (*World Development* 23, 1037–51), © 2002, with permission from Elsevier.

Figure 11.1 from J. Warburton and M. Danks (1998) 'Historical and contemporary channel change, Swinhope Burn' in J. Warburton (ed.) *Geomorphological Studies in the North Pennines: Field Guide* (77–91). Durham: Department of Geography, University of Durham, British Geomorphological Research Group, with permission of Jeff Warburton.

Figure 11.2 from E. D. F. Bird (2000) *Coastal Geomorphology: An Introduction*, © John Wiley & Sons Limited. Reproduced with permission.

Figure 12.4 with permission from Andrew Bennett.

Figure 12.7 from A. J. Kozol, J. F. A. Traniello and F. M. Williams (1994) 'Genetic variation in the endangered burying beetle *Nicrophorus americanus* (Coleoptera: Silphidae)' (*Annals of the Entomological Society of America* 87, 928–35) with kind permission of the Entomological Society of America.

Figure 12.8 from L. D. Harris (1984) *The Fragmented Forest: Island Biogeography Theory and the Preservation of Biotic Diversity,* with a Foreword by Kenton R. Miller. Chicago and London: Chicago University Press. Reprinted with kind permission of The University of Chicago Press.

Figure 12.9 reprinted from A. F. Bennett, K. Henein and G. Merriam (1994) 'Corridor use and the elements of corridor quality: chipmunks and fencerows in a farmland mosaic' (*Biological Conservation* 68, 155–65), © 2002, with permission from Elsevier.

Figure 12.10 from J. Krauss, I. Steffan-Dewenter and T. Tscharntke (2003) 'How does landscape context contribute to effects of habitat fragmentation on diversity and population density of butterflies?' (*Journal of Biogeography* 30, 889–900), reprinted with kind permission of Blackwell Publishing Limited.

Figures 12.11, 12.12, 13.2, 13.3, 13.5, 13.6 and Plates 1, 14 and 15 are © Pat Morris.

Figure 13.1 from C. Parmesan *et al.* (1999) 'Poleward shifts in geographical ranges of butterfly species associated with regional warming' (*Nature* 399, 579–83), © 1999 Macmillan Magazines Limited, reprinted with kind permission of Nature Publishing Group and Camille Parmesan.

Figure 13.7 from F. S. Chapin III and K. Danell (2001) 'Boreal forest' in F. S. Chapin III, O. E. Sala and E. Huber-Sannwald (eds) *Global Biodiversity in a Changing Environment: Scenarios for the 21st Century* (Ecological Studies 152), 101–120. Springer: New York, © 2001 Springer-Verlag, reprinted by kind permission of Springer-Verlag and Terry Chapin.

Figures 13.8 and 13.9 from R. P. Neilson (1993) 'Transient ecotone response to climatic change: some conceptual and modelling approaches' (*Ecological Applications* 3, 385–95), © 1993 Ecological Society of America.

Figure 13.14 with permission from The American Association for the Advancement of Science.

Figure 13.15 reprinted with permission from O. S. Sala *et al.* (2000) 'Global biodiversity scenarios for the year 2100' (*Science* 287, 1770–4). © 2000 American Association for the Advancement of Science.

Table 13.1 and Figure 13.10 from J. N. Galloway and E. B. Cowling, (2002) 'Reactive nitrogen and the world: 200 years of change' (*Ambio* 31, 64–71), reprinted by kind permission of the Royal Swedish Academy of Sciences.

Figure 14.1 from T. R. Oke (1988) 'The urban energy balance' (*Progress in Physical Geography* 12, 471–508), reprinted with permission of Arnold.

Figure 14.2, 14.5 and Table 14.1 are reproduced by permission of Oxford University Press Australia from *Urban Biophysical Environments* by Bridgman, Warner, and Dodson © Oxford University Press, www.oup.com.au.

Figure 14.3 from L. R. Burton (2003) 'The Mersey Basin: a historical assessment of water quality from an anecdotal perspective' (*The Science of the Total Environment* 314–316, 53–66), © 2003, with permission from Elsevier.

Figure 14.4 from E. van Beek (1998) 'Modelling for integrated water management' (in P. Huisman, W. Cramer, G. van Ee, J. C. Hooghart, H. Salz and F. C. Zuidema (ed. Committee) *Water in the Netherlands*, pp. 162–6; Delft: Netherlands

Hydrological Society (NHV)), reprinted with permission.

Table 14.2 from J-P. Savard, P. Clergeau and G. Mennechez (2000) 'Biodiversity concepts and urban ecosystems' (*Landscape and Urban Planning* 48, 131–42), © 2000, with permission from Elsevier.

Table 14.3 Data used with kind permission of the IUCN.

Figure 15.3 from UN Centre for Human Settlements (2001) *The State of the World's Cities Report* (Nairobi: UNCES (Habitat)).

Figure 15.4 from G. V. Iyengar and P. P. Nair (2000) 'Global outlook on nutrition and the environment: meeting the challenges of the next millennium' (*The Science of the Total Environment* 249, 331–46), © 2000, with permission from Elsevier.

Figure 15.5 from WGBU (1998).

Table 15.1 and 15.2 UN Centre for Human Settlements (2001) *The State of the World's Cities Report* (Nairobi: UNCES (Habitat)).

Table 15.3 from H-J. Schellnhuber (1998) *Syndromes of Global Change: An Integrated Analysis of Environmental Development Issues* (Villa Borsig Workshop Series – A policy dialogue on the World Development Report 1999: Development Issues in the 21st Century). Internet reference at http://www.dse.de/ef/vbws98/contents.htm. Last accessed November 2003.

Plate 4 is reproduced by kind permission of Ordnance Survey © crown copyright NC/A7/04/44271.

Plate 9 with permission from Gary Malone.

Plates 12 and 13 with permission from Tessa Francis.

The copyright of photographs remains held by the individuals who kindly supplied them.

Note
Every effort has been made to contact copyright holders for their permission to reprint material in this book. The publishers would be grateful to hear from any copyright holder who is not here acknowledged and will undertake to rectify any errors or omissions in future editions of this book.

FOREWORD

Physical geography is being reinvented. During the 1980s and 1990s, two developments made physical geographers start to feel increasingly uneasy about their discipline (e.g. Stoddart 1987; Slaymaker and Spencer 1998; Gregory 2000). First, the subject was becoming evermore specialized and in imminent danger of falling apart, the fragments being soaked up by related disciplines. Coupled with the fear that the subject would collapse was a worry that physical geography and human geography were moving towards opposite poles. The response to these perceived dangers was to annouce a surprisingly unanimous message: refocus physical geography around a core of global-scale studies that looks at big and pressing environment issues. Such refocusing would at once bring specialists from the branches of physical geography back together and reforge closer links with human geography. *Physical Geography: A Human Perspective* is an attempt to provide a first-year undergraduate textbook structured around the 'new physical geography'.

SUGGESTIONS FOR THE NEW PHYSICAL GEOGRAPHY

What does the 'new physical geography' look like? This question is difficult to answer unequivocally as no fully-fledged structure has yet emerged. So specialized have the sub-fields of physical geography become that few 'physical geographers' take time to step back and consider the wider discipline of which they are part. In truth, the specialization has progressed to such an extent that very few all-round physical geographers are left. Gardner (1996) was probably right in comparing physical geography to a polo mint for, as a discipline, it is all periphery and no core. Gregory (2000, 278) agreed with this analogy, but insisted that each sub-field has its core and that there should be a focal point in the approach adopted by physical geographers. Slaymaker and Spencer (1998) too had a vision of physical geography in the twenty-first century.

Gregory's and Slaymaker and Spencer's views are worth outlining as they both identify the potential shape of the new physical geography.

Slaymaker and Spencer (1998, xi) were worried that 'Contemporary physical geography is increasingly fragmented into the sub-fields of climatology and meteorology, geomorphology, hydrology, biogeography and soils'. Their solution was to place physical geography within 'a modern and timely context' (p. xi) by doing three things:

1   Showing how the pools, fluxes and budgets in established biogeochemical systems help show the complex interlinkages within the ecosphere.

2   Showing how the advances in reconstructing Quaternary environments made over the past fifty years have sharpened and heightened understanding of environmental change over various timescales.

3   Looking at regional case studies of environmental issues to illustrate the complexity of interlinkages.

In addition, they elaborated these ideas 'by considering the human responses to the risks, uncertainties and ethical issues posed by the threat of future global environmental change' and offered 'the principle of sustainability as a focus for twenty-first century physical geography' (Slaymaker and Spencer 1998, xi).

Gregory (2000, 287–88) set down eight 'tenets for the pursuance of physical geography', as guidelines rather than commandments, to 'stimulate thought, debate and reaction'. In our interpretation and words, the tenets are as follows.

1   Build up the spatial perspective.

2   Model the past, present and future.

3   Give due recognition to the history of ideas in physical geography.

4   Tackle big issues at regional and global scales, contributing to Earth system

science and environmental design and management.

5   Promote physical geography as a composite discipline, in the sense of Osterkamp and Hupp (1996), without suggesting that it is some kind of umbrella subject.

6   Reap benefits of a discipline that takes in the totality of the environment and bridges the human and physical sciences.

7   Cultivate multidisciplinary and interdisciplinary research.

8   Make clear the specific objectives of physical geography and get them across to other disciplines, to decision-makers and (through the media) to the general public (and, we would add, greenhorn geography undergraduates).

Gregory submitted a provisory definition of modern physical geography, which, with all the emphases of the original omitted for clarity, ran:

Physical geography/Physiography focuses upon the character of, and processes shaping, the land-surface of the Earth and its envelope, emphasizes the spatial variations that occur and the temporal changes necessary to understand the contemporary environments of the Earth. Its purpose is to understand how Earth's physical environment is the basis for, and is affected by, human activity. Physical geography was conventionally subdivided into geomorphology, hydrology and biogeography, but is now more holistic in systems analysis of recent environmental and Quaternary change. It uses expertise in mathematical and statistical modelling and in remote sensing, develops research to inform environmental management and environmental design, and benefits from collaborative links with many other disciplines such as biology (especially ecology), geology and engineering. In many countries, physical geography is studied and researched in association with human geography. (Gregory 2000, 288–89)

Gregory's and Slaymaker and Spencer's visions have much in common. Inspired by their ideas, we have devised a structure for physical geography that is the basis of our book.

A STRUCTURE FOR THE NEW PHYSICAL GEOGRAPHY

Our starting point in giving shape to the new physical geography is a short definition of the subject that should help in discussing the nature of the subject and in identifying a subject core. The definition is this: physical geography is the study of the form and function of the human sphere (anthroposphere), which is the zone of interaction between the ecosphere and the mental sphere. As with all definitions, this one begs further definitions, and in particular, an explanation of the terms 'ecosphere' and 'human sphere' are required. Although the use of the term ecosphere is disputable, it may be taken as the biosphere, atmosphere, hydrosphere, pedosphere and toposphere. Obviously, defined in this way, the components of the ecosphere are the objects of study of the traditional sub-fields of physical geography. Climatology and meteorology are the study of the atmosphere, biogeography is the study of life, hydrology is the study of the hydrosphere, pedology is the study of the pedosphere and geomorphology is the study of the toposphere. Variations on this theme could be suggested, for example, soils are sometimes considered part of biogeography or even geomorphology, but as a generalization, the connection between terrestrial spheres and sub-fields of physical geography seems acceptable. These variations begin to highlight the critical interconnections that are necessary to provide a deep understanding of the 'big issues' about human–environment interactions. The mental sphere (or noösphere) may be defined as the sphere of influence of the human mind, and may be regarded as the sum of the mental activity behind human impacts and conscious human planetary resource use, management and conservation. It engages cultural, social, ethical, economic and technological structures and processes.

Given these definitions, it may be argued that three spheres are therefore crucial to the subject matter of physical geography. First is the natural ecosphere (i.e. the small remaining portions of the ecosphere that are still largely unaffected by human activities) and its biological and physical structures and processes. Second is the mental sphere (noösphere). Third is the human-influenced ecosphere or human sphere (anthroposphere) that is the product of the other two. In our view, it may be useful to take the position that physical geography explores the interrelations between all the surficial terrestrial spheres as manifest in the form and function of the human sphere. In fact, such a 'holistic' view was taken by many nineteenth and early twentieth-century physical geographers, admittedly without a human focus, so the 'new physical geography' is perhaps not so new as might be supposed.

A problem arises here, because global ecologists and Earth-system scientists also study the form and function of ecosystems and the anthroposphere at the global scale. So, have Earth-system scientists commandeered the role of the physical geographer? Have physical geographers missed that bandwagon and lost the wonderful opportunity of reclaiming their academic heritage? Many physical geographers, quite rightly, seem miffed that their rich and venerable tradition of studying the 'big environmental picture' has passed unnoticed by the Earth-system scientists. Slaymaker and Spencer (1998, ix) believed that the launching of their 'Understanding Global Environmental Change: Themes in Physical Geography Series' was urgent, partly because they were concerned with re-establishing the historical legitimacy of the physical geographic tradition within the academy. The big question then arises of whether the new physical geography has a distinct approach to studying the human sphere that sets it apart from global ecology and Earth-system science. We think that the individuality of physical geography lies in its interconnections with human geography. In particular, we think that physical geography may claim as its core the human sphere and that its individuality lies in an approach boasting two patent geographical foci: the interconnectedness of environmental and human factors and the significance of local, regional and global scales.

Our view of physical geography fits with some recent ideas that emphasize the subject's innate and traditional link with human geography. It seems undeniable that, in physical geography, 'the emphasis on sub-fields has loosened the links with societal application and human geography . . . in spite of a strong intellectual tradition within physical geography that has explicitly engaged with such questions over the last 150 years' (Slaymaker and Spencer 1998, xi). Many big names in British geography have lent their

weight to the case for keeping human geography and physical geography together (e.g. Johnston 1986; Goudie 1986; Douglas 1986). Douglas was one of the most outspoken commentators in arguing that 'Some of us must escape from the narrow research followed to gain PhDs and broaden our horizons to explore the greatest challenges to our discipline, those big, fundamental, important issues in which the unity of geography is obvious'. Likewise, Stoddart (1987) vigorously argued that 'Land and life is what geography has always been about' and 'that it is time we got out again into the great wide world, met its challenges . . .'. Slaymaker and Spencer (1998) saw social connections as important too. In their Series preface, they said that 'in common with many thoughtful persons of goodwill, we are still impressed by the recklessness with which the natural spheres of Earth continue to be exploited and believe passionately that we are accountable to subsequent generations for the way in which we respond to both biogeochemically and societally induced global environmental change' (Slaymaker and Spencer 1998, ix). Newson (1992) argued that, as society is likely to draft geographers to the front line of environmental management, then geography should have an unambiguous disciplinary label. Furthermore, geographical practitioners should 'use a wealth of experience with natural systems analysis and a contemporary research familiarity with some of nature's surprises to develop a more enthusiastic, socially responsible and committed physical geography' (Newson 1992). This feeling is echoed by the tendency for geographers to make significant contributions to such professions as planning and to wider development policy. Indeed, planners are increasingly charged with the responsibility of delivering a framework through which the principles of sustainable development may be promoted and maintained. In view of this, it is important that students aiming for these professions should gain a full appreciation of aspects of the physical and human realm, particularly at their interfaces.

The structure proposed in *Physical Geography: A Human Perspective* has people and environment interactions at its heart. Of course, we are aware that any attempt to reunite human and physical geography is full of potential pitfalls. Johnston (1997, 344) allowed that connections had been made between human and physical geography, but that no integration of the study of physical and societary processes had yet emerged, and he rightly pointed out that, for human geographers, links with other social scientists were still very much more secure than those with environmental scientists. Harman *et al.* (1998, 277) pinpointed the special problems that arise when human geography meets physical geography: '. . . many environmental problems present us with very difficult choices today. Whether the issues be regional (e.g. aquifer depletion, land use zoning, river management) or global (climate change), much of this difficulty stems from the combination of scientific, economic, and ethical uncertainly entangled in the details'. The challenge in reuniting geography is enormous. *Physical Geography: A Human Perspective* is a small step in that direction.

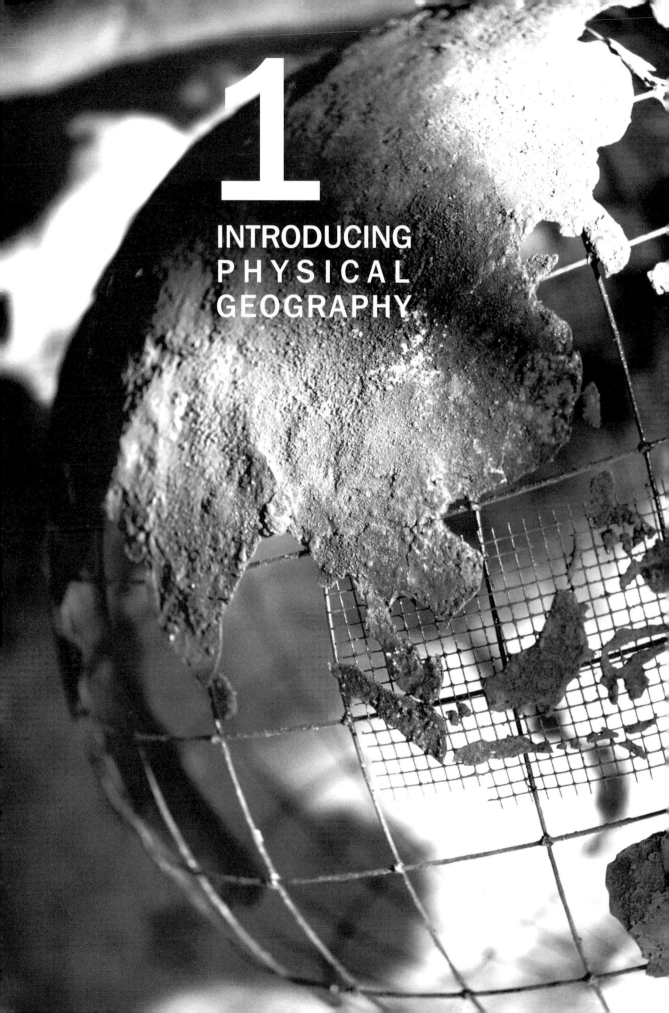

1

INTRODUCING PHYSICAL GEOGRAPHY

CHAPTER 1
HUMANS, THE ENVIRONMENT AND ENVIRONMENTAL CHANGE

OVERVIEW

This chapter sets down the basis of a new physical geography. It starts by taking a novel view of the subject, in which humans and the human-altered environment are set on an equal footing with the natural environment. This viewpoint is justified by the ever-increasing domination of the planetary surface by the human species, which has drastically altered the natural flows of gases, soluble materials and solids, and transformed the land-cover, reducing biodiversity through species extinctions in the process. This new view also sees the Earth as a global ecosystem. The 'Earth system' thus described can be measured, monitored and mapped using remote sensing and other sophisticated techniques (as well as simple ones); and it can be modelled using high-powered computers (and humble PCs too). The current environmental change must be measured against the yardstick of past environmental change and environmental reconstruction involves using the techniques described here to 'read the pages of Earth history'. An alternative method of reconstructing past environments explained here is to 'postdict' (make predictions for past changes) environmental conditions using mathematical models. Finally, based on the arguments presented, a 'blueprint' for a new physical geography is outlined.

LEARNING OUTCOMES

After reading this chapter, you should understand:
- The human domination of the planet
- Human impacts on flows in the Earth system
- Human impacts on biodiversity
- Methods for studying the past, present and future Earth system.

1.1 A FRESH LOOK AT PHYSICAL GEOGRAPHY

REDEFINING PHYSICAL GEOGRAPHY

ecosphere: the global ecosystem; life plus life-support systems (air, water, soil)

biosphere: the totality of life on Earth

hydrosphere: all the water of the Earth, including atmospheric water vapour in some definitions

pedosphere: the totality of soils on the Earth

toposphere: the surface features of the globe, natural and human-made

noösphere: the sphere of influence of the mind

anthroposphere: the portion of the ecosphere influenced by humans

This book explores the possibilities of a new definition of physical geography: *physical geography is the study of the form and function of the human sphere (or anthroposphere), which is the zone of interaction between the mental sphere and ecosphere* (Figure 1.1). The **ecosphere** is the **biosphere**, atmosphere, **hydrosphere**, **pedosphere** and **toposphere**. The mental sphere (or **noösphere**) is the sphere of influence of the human mind. It is the sum of the mental activity behind human impact and the conscious human use, management and conservation of planetary resources. It engages cultural, social, ethical, economic and technological structures and processes. Three 'spheres' are therefore crucial to the subject matter of physical geography:

- the natural ecosphere (i.e. the small remaining portions of the ecosphere that are still largely unaffected by human activities) and its biological and physical structures and processes

- the mental sphere (noösphere)

- the human-influenced ecosphere or human sphere (**anthroposphere**) that is the product of the other two.

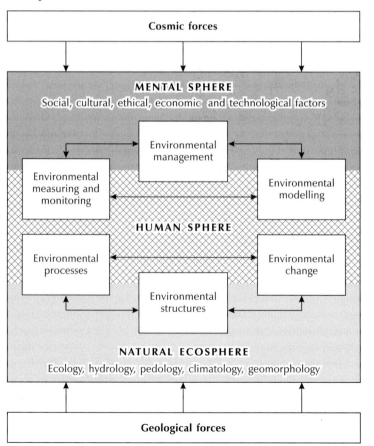

Figure 1.1 Interaction between mental sphere, human sphere and ecosphere

The subject matter of a new physical geography is thus the interrelations between all the surficial terrestrial spheres as manifest in the form and function of the human sphere.

PUTTING HUMANS IN THEIR PLACE

Why put humans at the centre of physical geography? The justification lies in the leading role that humans now play in the shaping and running of environmental systems. Modern humans deeply influence the form and function of the ecosphere. Like all organisms, humans interact with their environment. Unlike other species, they have acquired the ability to alter the environment in novel ways, putting land under crops, building towns and cities, extracting minerals and fuels and constructing factories. The exponential growth of the human population, coupled with advances in agricultural and industrial technology, transport and the desire of western civilization to occupy new lands, led to a dramatic, global change in the land-cover of the planet known as the **Great Transformation**, that started in the late seventeenth century and has since continued at a quickening rate. In the head office lobby of Canada's International Development Research Centre in Ottawa the IDRC Resource Clock, or World Human and Natural Resources Clock, ticks. It shows the current world population and area of productive land, second by second. At 22.30 on 10 July 2003 it read: World Population 6,288,853,300; Productive Land 8,578,375,048 hectares. Half an hour later, it read: World Population 6,288,858,454 and Productive Land 8,578,374,774. That gives an extra 5,154 people in the world and 274 hectares productive land lost in half an hour, or very roughly a gain of 250,000 people and a loss of 12,000 hectares of productive land a day.

Human action on the ecosphere is varied. Humans have transformed about one-half of the land surface (Table 1.1). Regional variations are marked: Europe

> **Great Transformation**: the conversion of the terrestrial land-cover from natural to human-dominated types that started to grow quickly in the late eighteenth century

Table 1.1 Human disturbance of habitats by continent

Continent	Area (millions km²)	Habitat		
		Undisturbed (%)	Partially disturbed (%)	Dominated by humans (%)
Europe	5.8	15.6	19.6	64.9
Asia	53.3	43.5	27.0	29.5
Africa	34.0	48.9	35.8	15.4
North America	26.1	56.3	18.8	24.9
South America	20.1	62.5	22.5	15.1
Australasia	9.5	62.3	25.8	12.0
Antarctica	13.2	100.0	0.0	0.0
World total	162.0	51.9	24.2	23.9
World total less area of bare rock and barren land	134.9	27.0	36.7	36.3

Source: Adapted from Hannah *et al.* (1994)

is the most transformed continent, with only about 16 per cent remaining undisturbed by human action; Antarctica is the least, with practically 100 per cent undisturbed. Since the beginning of the Industrial Revolution, the concentration of carbon dioxide in the atmosphere has increased by nearly 30 per cent. Humanity fixes more atmospheric nitrogen than is fixed by all natural **nitrogen-fixing processes**. Humans use more than one half of all accessible surface fresh water. Human-induced habitat destruction, habitat fragmentation and the introduction of exotic species have driven around a quarter of the bird species on Earth to extinction. Humans affect global **biogeochemical cycles**, including the carbon, nitrogen and sulphur cycles; they mobilize such metals as arsenic and mercury; they alter the water cycle and they reduce biodiversity. They have caused accelerated erosion of soils in many parts of the world and have nearly exhausted some non-renewable resources and placed some renewable resources in jeopardy. So systematic is the global change wrought by the human species that some writers judge it appropriate to designate the last two hundred years as a new geological epoch – the Anthropocene. And that epoch has only just begun. The latest worst-case prediction by the Intergovernmental Panel on Climate Change (IPCC) is for a 1.4–5.8°C average increase in global surface temperature over the period 1990 to 2100. Changes on this scale will have profound effects on the global ecosystem, and a new global science has emerged to study them – global ecology and Earth system science. Global ecologists and Earth system scientists both study the planet as a single, complex and dynamic entity comprising the **geosphere** and biosphere, although their foci are slightly different.

A set of environmental problems are associated with the later stages of the Great Transformation, the prime causes of which are the rapid growth of human populations, their increasing consumption rate and fast-expanding access to technology and their fondness of building ever-bigger cities. The human population has grown from about 2.5 billion in 1970 to just over 6 billion today (2003), will grow to a projected 9 billion or just under by 2070, and may fall after that (cf. p. 424). Just a few urban areas in 1950 had populations over 4 million (Lutz *et al.* 2001). The number of **megacities** grew from two (London and New York) in 1950, to five in 1975, to fourteen in 1995; while predictions suggest there will be twenty-six by 2015 and thirty-six by 2151, of which seventeen will be in Asia (World Resources Institute 1999).

Two broad environmental concerns arise from ever-growing human populations (Vitousek *et al.* 1997). The first is the mounting human impact on the ecosphere and its component ecosystems that are affected by changes in carbon pools (stores), biogeochemical cycling, sediment budgets and climate. The second is the changes in the Earth's biodiversity of species and communities caused by species introductions and extinctions, the fragmentation of habitats and changes of land cover.

nitrogen-fixing process: processes that convert inactive atmospheric nitrogen into a reactive form usable by living things

biogeochemical cycles: the repeated movement of elements essential to life (bioelements) through the ecosphere

geosphere: the non-living spheres of the Earth

megacities: cities with 10 million or more inhabitants

1.2 HUMAN IMPACTS ON FLOWS AND STORES IN THE EARTH SYSTEM

Life is the motive force behind biogeochemical cycles: it takes biogeochemicals from the environment, uses them, and releases them back to the environment. Such exchanges of materials between life and life-support systems are an integral part of biogeochemical cycles. In addition, on geological time-scales, forces in the geosphere produce and consume rocks, so influencing the cycles. Biogeochemical cycles involve the storage and transfer of all terrestrial elements and compounds, except the inert ones. At their grandest scale, biogeochemical cycles involve all the materials of the Earth. However, an exogenic cycle, involving the transport and transformation of materials near the Earth's surface, is normally distinguished from a slower and less well understood endogenic cycle involving the lower crust and mantle. In addition, exogenic cycles may be divided into gaseous and sedimentary varieties. Cycles of carbon, hydrogen, oxygen and nitrogen are gaseous cycles, so called because their component chemical species are gaseous for a leg of the cycle. Other chemical species follow a sedimentary cycle because they do not readily **volatilize** and move between the biosphere and its environment in solution.

> **volatilize**: become volatile, that is, capable of being readily vaporized

Humans make an increasingly material impact on biogeochemical and sedimentary cycles. They do so on a local, regional and global scale. A large number of scientists conduct research into establishing the magnitude of the human impact on biogeochemicals and methods of developing sustainable practices. The next sections will show the flavour of the intense research activity surrounding these critical and urgent issues.

SEDIMENT FLOWS

Locally and regionally, humans transfer solid materials from the natural environment to the urban and industrial built environment. The role of human activity in geomorphic processes was first put on a quantitative footing by Robert Lionel Sherlock who, in his book *Man as a Geological Agent: An Account of His Action on Inanimate Nature* (1922), supplied many illustrations of the quantities of material involved in mining, construction and urban development. Recent work confirms that human activities transform natural landscapes and are a potent agent of Earth surface change. In Britain, such processes as direct excavation, urban development and waste dumping are driving landscape change: humans deliberately shift some 688 to 972 million tonnes of Earth-surface materials each year; the precise figure depends on whether or not the replacement of overburden in opencast mining is taken into account. (Appendix 1 discusses units of measurement.) The astonishing fact is that the deliberate human transfers move nearly fourteen times more material than natural processes. British rivers for example, the next largest transferers, export only 10 million tonnes of solid sediment and 40 million tonnes of solutes to the surrounding seas. The British land surface is changing faster than at any time since the last ice age, and perhaps faster than at any time in the last 60 million years (Douglas and Lawson 2001).

Globally, every year, humans move about 57 billion tonnes of material through mineral extraction processes. Rivers transport around 22 billion tonnes of sediment to the oceans annually, so the human cargo of sediment is nearly three times more than river load. Table 1.2 gives a breakdown of the figures. The data suggest that, in excavating and filling portions of the Earth's surface, humans are at present the most efficient geomorphic agent on the planet. Even where rivers, such as the Mekong, Ganges and Yangtze, bear the sediment from **accelerated erosion** within their catchments, they still discharge a smaller mass of materials than the global production of an individual mineral commodity in a

> **accelerated erosion**: soil erosion exacerbated by human activities

Table 1.2 Natural and mining-induced erosion rates of the continents

Continent	Natural erosion (Mt year^{-1})[a]	Hard Coal, 1885 (Mt)	Brown coal and lignite, 1995 (Mt)	Iron ores, 1995 (Mt)	Copper ores, 1995 (Mt)
North and Central America	2996	4413	1139	348	1314
South America	2201	180	1	712	1337
Europe	967	3467	6771	587	529
Asia	17,966	8990	1287	1097	679
Africa	1789	993	None	156	286
Australia	267	944	505	457	296
Total	**26,156**	**18,987**	**9703**	**3357**	**4442**

Note: [a] Mt = Megatonne = 1 million tonnes

Source: Various, including Douglas and Lawson (2001)

> **fluvial**: of, or pertaining to, flowing water (rivers)

single year. Moreover, **fluvial** (river-borne) sediment discharges to the oceans from the continents are either similar in magnitude, or smaller than, the total movement of materials on continents.

CARBON FLOWS

Given the real threat of climatic change associated with global warming, it is not surprising that there is much research into studying the effects of elevated carbon dioxide levels on ecosystems. Measurements of carbon stores and fluxes, and the understanding of carbon-cycling processes, are becoming ever more refined.

Modelling studies in the early 1970s suggested that, to balance the global carbon budget, carbon must be stored in terrestrial ecosystems. Further research in the 1990s confirmed that terrestrial ecosystems play a vital role in regulating the carbon balance in the Earth system. The global carbon cycling research, so critical to the global warming issue, has stimulated a wealth of studies on several aspects of plant biology, including leaf photosynthesis, plant respiration, root nutrient uptake and **carbon partitioning** (the allocation of carbon to roots, trunks, leaves, etc). The studies range from response to increased carbon dioxide concentrations at the molecular level, through impacts on species diversity in communities, to carbon fluxes in the ecosphere. At the global scale, global biosphere models take in experimental results to predict potential changes in terrestrial ecosystems as the globe warms up (see Chapter 7).

> **carbon partitioning**: the allocation of carbon to different chemical forms (such as sugars, structural carbohydrates and starch) within, or to different parts (such as leaves, stem, roots) of plants

Fascinatingly, some basic plant processes appear to have a varied response to elevated carbon dioxide levels. The apparent respiration rate increases in some species and decreases in others when subjected to short-term and long-term elevated levels of carbon dioxide. Similarly, several processes display mixed responses. These include nitrogen uptake, stomatal conductance (the rate at which gases pass through the stomatae or small openings in leaves), carbon allocation (as reflected in the ratio of roots to shoots) and photosynthetic capacity. The varied response of plant processes to increased carbon dioxide levels is presently inexplicable and makes it difficult to integrate experimental results into global studies. Indeed, some field evidence and theoretical work suggests that experiments that use free-air carbon dioxide enrichment (FACE) actually perturb the ecosystem under investigation, mainly by altering the carbon influx (see pp. 176–77). The question raised by this work is: to what extent is the response to a rather abrupt change in carbon dioxide concentration able to help in predicting the long-term ecosystem responses to the very gradual increase in carbon dioxide level that is actually occurring?

Another interesting recent finding illustrates the complexity of the carbon cycle. Two opposing processes are associated with climatic warming (Figure 1.2). The first process is a **negative feedback** mechanism where plant growth and carbon sequestration increase because of stimulated nutrient mineralization and longer growing seasons. The second process is a **positive feedback** mechanism, where the warming stimulates biological metabolism in terrestrial vegetation, which in turn increases a greater release of heat-trapping gases to the atmosphere and augments human-induced warming. Recognition of the second mechanism partly explains the increased temperature rises predicted by the IPCC. An experiment to test this mechanism in the field, conducted in a tall-grass prairie in the Great Plains of the USA, revealed another complicating factor – acclimatization. The process here seems to be that, rather than soil respiration simply increasing with increasing temperature, the soil tends to 'acclimatize' to the higher temperatures, and does so more fully at high temperatures, so weakening the positive feedback effect. In addition, increased carbon dioxide levels increase microbial activity, which in turn promotes the formation of **soil aggregates**. As soil aggregates tend to protect particles of organic matter in the soil against microbial attack, an increase in soil aggregation resulting from higher carbon dioxide levels may lead to a sequestration of soil carbon. This is another negative feedback mechanism.

Further evidence of the problematic nature of predicting processes in the carbon cycle comes from studies of **litter decomposition**. Litter decomposition is an important component of the global carbon and nitrogen cycles and, as climate

negative feedback: a self-stabilizing change in a system – the change triggers a series of subsequent changes that eventually neutralize the effects of the original change, so bringing the system back to its original state. An example is the regulation of body temperature in warm-blooded animals

positive feedback: a self-reinforcing change in a system – the change triggers further changes that magnify the effects of the original change. Exponential population growth is an example

soil aggregates: basic structural units of soil, formed of individual soil particles (mineral and humus), with planes of weakness between them. Individual aggregates are called peds

litter decomposition: the decay and disintegration of litter lying on the soil surface

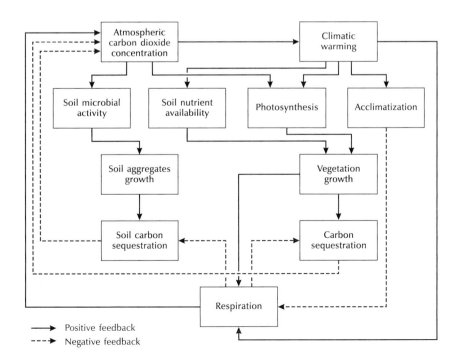

Figure 1.2 Main feedbacks between climate and the carbon cycle

wields a strong control over rates of litter decomposition, climatic change may affect the carbon and nitrogen cycles. Climatic change may directly increase the litter decomposition rates by producing warmer and wetter conditions. It may also have an indirect influence by bringing about changes in litter quality (the chemical status of leaf litter, often measured by the relative proportions of carbon, nitrogen and other nutrients). The importance of direct climatic effects and indirect litter-quality effects on litter decomposition varies from one ecosystem to another. In consequence, it is not easy to predict ecosystem responses to climatic change.

Elevated carbon dioxide levels significantly boost the growth and development of nearly all plants. Studies from field-grown trees suggest a continued and consistent stimulation of photosynthesis of around 60 per cent for a 300 ppm increase in atmospheric carbon dioxide concentration. Such enhanced growth may alter the concentrations of food-plant constituents that are consumed by animals and humans and are therefore related to animal and human health. The situation is complicated because a host of environmental factors and not just carbon dioxide levels affect the nature of plant constituents. However, some generalizations are possible. An increase in atmospheric carbon dioxide concentration normally reduces the nitrogen and protein levels of animal-sustaining forage and human-sustaining cereal grains, but only when soil nitrogen levels are below their optimum. In addition, higher atmospheric carbon dioxide levels tend to reduce oxidative stress in plants, which accordingly manufacture fewer antioxidants for protection, and increase the concentration of vitamin C in some fruits and vegetables. Likewise, the carbon-dioxide enrichment increases the biomass of plants grown for medicinal purposes and at the same time increases the concentrations of disease-fighting substances produced with them, such as heart-helping digoxin (a cardiac glycoside) in the woolly foxglove (*Digitalis lanata*) and a suite of cancer-fighting substances found in the tropical spider lily (*Hymenocallis littoralis*). So, work to date suggests that rising carbon dioxide levels will increase food production while sustaining the nutritional quality of food and heightening the yield of certain disease-inhibiting plant compounds.

Despite the problems of modelling the carbon cycle, considerable success has accompanied efforts to predict the global distribution of vegetation, primary productivity, carbon stored in biomass and soil carbon (see Chapter 7). A recent development of such models is to **postdict** these variables for pre-industrial conditions.

postdict: predict for times past (i.e. after the event)

TOXIC CHEMICAL FLOWS

heavy metals: ions of metallic elements such as copper, zinc and iron with densities greater than 5–6 g cm^{-3}

Humans upset the natural cycles of **heavy metals**, including mercury and arsenic. For example, humans extract some 30,453 tonnes of arsenic globally each year for industrial purposes, mainly for pesticides and wood preservatives. Another 28,400 to 94,000 tonnes of arsenic finds its way back into soils and groundwaters in wastes of various sorts. Once the arsenic-bearing wastes are in the soil, arsenic sulphides cause soil acidification, which in turn increases the mobility of arsenic. Tailings (the refuse produced by, and left after, mining activity) from metal-sulphide mining operations are substantial sources for both heavy metal and arsenic contamination of groundwaters and surface waters.

Several harmful compounds from agricultural and industrial processes are being spread around the globe. Organochlorines (chlorinated hydrocarbons) include the best known of all the synthetic poisons – endrin, dieldrin, lindane, DDT and others. They are widely used as **biocides**. Plastics manufacturers use polychlorinated biphenyls (PCBs), another harmful compound, as flame-retardants and insulating materials. A very worrying development is the high levels of organochlorines recorded in Arctic ecosystems, where they are biomagnified (become increasingly more concentrated as they pass up a food web) and have adverse effects on consumers (p. 384). The organochlorines drift northwards from agricultural and industrial regions, carried by air currents. They then enter Arctic ecosystems through plants and start an upward journey through the trophic levels. PCBs and DDT are the most abundant residues in peregrine falcons (*Falco peregrinus)* and occur in polar bears (*Ursus maritimus*). A wide range of PCBs and organochlorine pesticides have been found in muscle tissue and livers of lake trout (*Salvelinus namaycush*) and Arctic grayling (*Thymallus arcticus*) from Schrader Lake, Alaska. PCBs occur in marine Arctic ecosystems, too. They occur in clams, mussels, sea urchins and some fish found in Cambridge Bay, Hall Beach and Wellington Bay, all in Northwest Territories, Canada.

> *biocides*: a poison or other substance used to kill pests

1.3 HUMANS IMPACTS ON BIODIVERSITY

Biodiversity is a measure of the variety of genes, species, communities and ecosystems that form the biosphere. Several factors drive current biodiversity change and there are four main groups of drivers: land cover change, species exploitation and exchange, climate and environmental chemistry. Land cover change works through habitat destruction and habitat fragmentation. The intercontinental spread of species, often abetted by humans, and the overexploitation of species directly affects communities. Climate affects biodiversity by reorganizing the distribution of bioclimatic space in which species and communities thrive. Chemical changes in the environment, such as a build-up of nutrients in lakes and soil acidification, acts on the biosphere as a whole.

> *biodiversity*: the variety of genetic information, species, communities and ecosystems in a particular place, be it a small pond, a continent, or the entire world

During the history of life on Earth, several episodes of mass extinction caused global biodiversity to nose-dive. The current 'biodiversity crisis' (Table 1.3) differs from the previous ones: it is running at an unprecedented rate, and it is the direct result of human activities. Erosion of biodiversity occurs at various levels, from the genetic diversity of many natural and domesticated species to the diversity of whole ecosystems and landscapes. Current human-induced rates of species extinctions may be about 1000 times greater than past background rates. Biodiversity loss is a matter of concern, not only because of the aesthetic, ethical or cultural values attached to biodiversity, but also because it could have numerous far-reaching, often unanticipated, consequences for the ecosphere as a whole. It may weaken the ability of natural and managed ecosystems to deliver ecological services such as production of food and fibre, carbon storage, nutrient cycling, and the ability to resist climatic and other environmental changes. The assessment of the causes and consequences of biodiversity changes, and the institution of foundations for the conservation and sustainable use of biodiversity, are the foremost scientific challenges to which physical geographers make substantial contributions.

Table 1.3 Evidence of the current biodiversity crisis, 2002

Group of organisms	Number of species in group	Number of threatened species	Percentage of total in group threatened	Percentage of total assessed threatened
Vertebrates				
Mammals	4842	1130	23	24
Birds	9932	1194	12	12
Reptiles	8134	293	4	62
Amphibians	5578	157	3	39
Fishes	28,100	750	3	49
Invertebrates				
Insects	950,000	553	0.06	72
Molluscs	70,000	967	1	46
Crustaceans	40,000	409	1	89
Others	130,200	30	0.02	55
Plants				
Mosses	15,000	80	0.50	86
Gymnosperms	980	304	31	34
Dicotyledons	199,350	5768	3	75
Monocotyledons	56,300	511	1	65
Lichens	10,000	2	0.02	0.02

Source: Adapted from IUCN 2003. *2003 IUCN Red List of Threatened Species* http://www.redlist.org. Last accessed 20 November 2003

1.4 STUDYING THE GLOBAL ECOSYSTEM

Several methods are available to investigate the issues of human impacts on stores and flows in the Earth system, and on biodiversity. These fall into four groups: measuring, monitoring, mapping and modelling (the four ms).

MEASURING, MONITORING AND MAPPING THE ENVIRONMENT

Starting in the second half of the fifteenth century, a picture of the natural and human worlds began to emerge from explorations by the western European nations who sought out new lands and new resources. In the wake of this quest, the framework for a global system of observation stations emerged providing an important basis for understanding the world today. The historical goal of global inventorying has continued through the application of a range of technologies associated with engineering, telecommunications and the computer. Scientists have now mapped and categorized much of the land surface.

New technologies, including those associated with aerial photography, remote sensing and Global Positioning Systems (GPSs), have greatly improved the potential to continuously, or near continuously, monitor the global ecosystem. In conjunction with computer-based interpretation and management tools, such as image processing and geographical information systems (GIS), these new technologies have extended the ability to integrate disparate datasets and re-create aspects of the world virtually. Such models shed further light on a wide range of patterns and processes, such as those associated with past and present human–environment interactions, and open up the opportunity to assess the potential impacts of human actions and activities and create a more sustainable future.

Applications of remote sensing to the ecological and geographical sciences fall into three periods. The first period, which ended in 1950, saw the initial applications of aerial photography. From 1950 to 1972 was a transitional period from photographic applications to unconventional imagery systems (such as thermal infrared scanners and side-looking airborne radars), and from low-altitude aircraft to satellite platforms. A decisive development during this time was the analysis of images by computer software. About the time that the first satellites were launched to observe the Earth, computer technology advanced to the stage where digital processing of remotely sensed data was possible. The first use of a computer in image analysis occurred in 1966. Multispectral measurements collected by an aircraft over an Indiana field in the USA were analysed by computer at the Purdue Laboratory for Agricultural Remote Sensing. Image processing enabled scientists to distinguish wheat from oats, even oats planted within a crop of wheat, and scattered wheat along a stream from seeds transported downstream. Thus arose a new field of scientific and technological research and development – modern remote sensing – that saw a burgeoning of systems using non-photographic techniques to create digital images of reflected, emitted and transmitted energy from objects at and near the Earth's surface. From 1972 to the present, the application of multispectral scanners and radiometer data obtained from operational satellite platforms has predominated, along with refinements in image processing by computers. Since its first use with remotely sensed data, the computer has proved to be an excellent means of storing, retrieving, manipulating and displaying spatial information. Remotely sensed images are a form of spatial information, and their incorporation within geographical data handling systems is an effective way of teasing out the information that they contain. Subsequent developments in remote sensing and computer-based GIS have generated a new potential for conducting investigations of the global ecosystem at all scales. One emphasis lies in global ecology and in studying the Earth as an integrated system. It envisages an Earth Reconnaissance System integrating remote sensing and **ground reference sites**, including Continental Scale Experiments (CSE), the Earth Observing System (EOS), the Global Network of Flux-Measurement Stations (FLUX-NET), Global Climate Observing System (GCOS), the Global Observation of Forest Cover (GOFC), the Global Ocean Observing System (GOOS) and the Global Terrestrial Observing System (GTOS).

> ***ground reference sites**: a phrase used to describe ground data collection points, usually with reference to remote sensing studies*

The speedy development and use of GIS, which allows input, storage and manipulation of digital data representing spatial and non-spatial features of the Earth's surface, have accompanied advances in remote sensing. Other related developments include Electronic Distance Measurement (EDM) in surveying and, more recently, the Global Positioning System (GPS), which made the very time-consuming process of making large-scale maps much quicker and more fun. Interestingly, Arthur C. Clarke, the science-fiction writer, predicted the invention of geopositional satellites in 1945. The US Navy originally developed the forerunner of the modern GPS in 1964 for tracking US submarines using a single satellite. Now, the US GPS, which is maintained by the Department of Defense, uses a constellation of twenty-four satellites and associated control systems that allow a suitable radio receiver to find its location anywhere on Earth, day or night. The Russian Global Navigation Satellite System (GLONASS) positioning system, set up in 1982, theoretically does the same thing, but it has experienced operational problems. The GPS enables latitude, longitude and altitude to be determined, although their accuracy depends on the method used to obtain a 'fix' (absolute, differential or static). Originally, the US GPS was implemented for military purposes, but an increasing number of non-military applications have appeared in, for example, biogeography and geomorphology.

Since 1972, many studies have used remote sensing, often in tandem with GIS. A major advance has been using different wavebands of the electromagnetic spectrum to gather information on various parts of the ecosphere and anthroposphere. Within the hydrosphere and cryosphere, remote sensing aids the study of the land-surface water budget and ice sheets. Within the surface **lithosphere** and pedosphere, it is possible to map soils and surficial deposits, to measure some landscape processes and to sense biogeochemical cycles in drainage basins. Within the biosphere, agricultural areas, wetlands, forest and rangeland, and global vegetation have received close attention. Measurable ecosystem properties include soil-surface moisture and temperature, terrestrial primary productivity, marine photosynthesis, leaf-area index, litter and soil organic matter decomposition, water and energy exchange, canopy biochemistry and trace gas fluxes. Remote sensing may reveal the spatial and temporal dynamics of vegetation. Often used in combination with GIS technology, remote sensing enables the accomplishment of sophisticated land cover mapping, the many and various applications of which include the mapping of regional vegetation, specific vegetation types and vegetation layers; of vegetation and land cover changes; and of specific resources and land cover degradation. It has also provided a basis for some models that predict species distributions from environmental factors.

An enormous stimulus to inventorying was given by the signing of the United Nations' Framework Convention on Climate Change at the Earth Summit in Rio de Janeiro in 1992, and developments at subsequent meetings (e.g. Kyoto in 1997 and Buenos Aires in 1998). Signatory governments are endeavouring to limit atmospheric greenhouse gas levels and circumscribe the human contribution to global warming. One plan is to increase carbon storage by changing land cover and land cover management. However, to monitor the usefulness of this process, it is necessary to measure the current stores of carbon. For this reason scientists are drawing up carbon inventories in many parts of the world. For example, one study used data on the extent of land cover types produced by the CORINE (Co-ordination of Information on the Environment) programme of the European Commission to construct a carbon inventory for Ireland. Another set of studies attempted to assess the live biomass and carbon of Siberian forests and the carbon budget of Russian terrestrial ecosystems at a national level.

MODELLING THE ENVIRONMENT

It is very useful to be able to predict changes in the global ecosystem. One way of doing this is to build **mathematic models** of the whole Earth system or parts of it. Mathematical models have a relatively long history in the physical

lithosphere: the 'sphere of rocks', the outer shell of the solid Earth

mathematical model: a system represented using the formal and symbolic logic of mathematics

geographical and ecological sciences, but mathematical modelling flowered with the arrival of electronic computers in the late 1950s. At this time, analogue and digital computers simulated, for example, the dynamics of relatively simple ecosystems. They looked at the interactions of several species, the passage of **radionuclides** through forests, and a little later DDT through food webs. The International Biological Programme (IBP), started in 1965, created the role of the ecological modeller. Systems models of several IBP biomes were constructed. Similar developments took place in meteorology and climatology, in geomorphology, and to a lesser degree, in soil science.

In the 1990s, mathematical modelling entered a new era that was heralded by three developments: the building of spatial models; the accessibility of time series from remotely sensed images; and the appearance of supercomputers and parallel processing.

Models in global ecology now cover a range of topics, including biogeochemical models, climate models and models that integrate world energy production, emissions of gases, deforestation and other human components of the global ecosystem. An example of the latter is IMAGE (Integrated Model to Assess the Greenhouse Effect) (Rotmans 1990). This model tries to capture the basic aspects of climatic change in a simplified form. It is a multidisciplinary model, involving economics, atmospheric chemistry, marine and terrestrial biogeochemistry, ecology, climatology and glaciology. The full model comprises a set of models that interconnect, although each functions, in large measure, independently. In practice, this means that the output of one model is the input to another model further down the causal chain. The key models are a world energy model, a spreadsheet model for other sources of greenhouse gases, an emissions model, an atmospheric chemistry model, a carbon cycle model, a climate model and a sea-level rise model. Figure 1.3 shows the deforestation

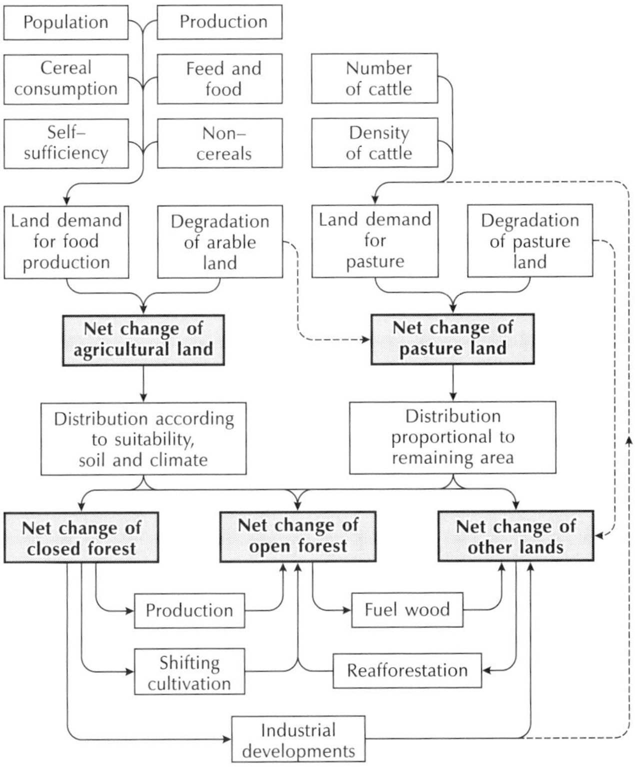

> *radionuclides*: nuclides (nuclei of atoms) that disintegrate spontaneously, releasing radioactivity in the process. Examples are nuclei of uranium and strontium atoms

Figure 1.3 The structure of the IMAGE deforestation model. *Source:* Adapted from Rotmans (1990)

model, which is part of the carbon cycle model. The full model tackles time as discrete steps of six months, and each simulation run covers 200 years of actual time, starting in 1900, roughly corresponding to the close of the pre-industrial era, and ending in 2100. Figure 1.4 gives the flavour of the results obtainable using the deforestation model. Taking Asia as an example, it shows the simulated changes in logging, reforestation by plantations, area of closed tropical forest, area of open tropical forest, area of agricultural land and the degradation of agricultural land, under the differing assumptions of the four scenarios used (Table 1.4).

1.5 THE EARTH SYSTEM IN THE PAST

One problem with measuring states and flows in the global ecosystem is that, although it establishes current operative processes and their rates, it does not provide a dependable guide to processes that were in action a million years ago, ten thousand years ago or even a hundred years ago. In trying to work out the long-term evolution of the global ecosystem, scientists have two main options open to them – environmental reconstruction or environmental modelling.

RECONSTRUCTING PAST ENVIRONMENTS

It is fortunate indeed that the environment records, even if in fragmentary form, its own history. If it did not do so, environmental reconstruction would be

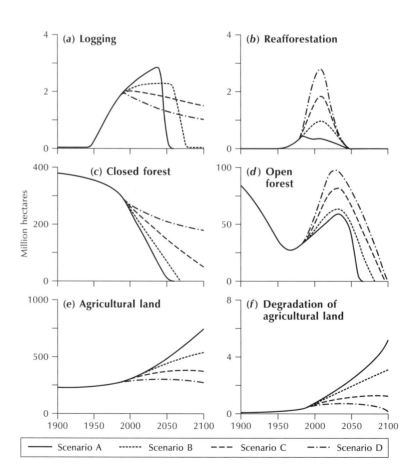

Figure 1.4 Land use changes in Asia predicted by the deforestation model. *Source:* Adapted from Rotmans (1990)

Table 1.4 Scenarios used in the IMAGE model

Scenario	Trends in driving variables	Description
A	Unrestricted trends (business-as-usual)	A major increase of energy consumption based on fossil fuels (coal)
		No incentives to increase the efficiency of energy use, an increasing per capita energy use (especially in the Third World)
		No stimulation in the use of renewable and non-fossil energy, economic growth not curtailed by environmental considerations
		Rapid conversion of available forest resources by shifting cultivation, by the demand for wood as a fuel and timber, and by demand for grazing land
B	Reduced trends	Price increases due to resources becoming scarcer and the imposition of some costs for environmental damage. Consequently:
		Slight shift towards non-fossil fuel energy sources
		Minor increase in per capita energy consumption
		Increasing energy efficiency
		Gradual reduction in the rates of deforestation
C	Changing trends	Concern over the condition of the environment leads to modest changes in policy:
		Incentives for an increased efficiency of energy use and for shifts towards renewable energy sources
		Efforts made across the globe to reduce the rates of deforestation
D	Forced trends	Major policy changes are implemented owing to grave concern over the prospect of greenhouse warming:
		Reduction in the use of fossil fuels
		Vigorous effort to promote greater energy efficiency
		Stagnation of the world economy
		Strenuous effort to halt deforestation by forest management and reforestation programmes

Source: From discussion in Rotmans (1990)

impossible. It is fortunate too that many environmental materials can be dated (Appendix 2). The record of change may be incomplete, but it does hint at the way things were. The evidence of environmental change is varied and falls into four categories – geochemical, physical, biological, and historical and instrumental.

1. Geochemical evidence depends on the fact that environmental factors often determine the mineralogy, elemental composition and **isotopic composition** of sediments and soils. Geochemical data, such as the presence of **evaporite** minerals or particular types of clay, are thus a guide to past environmental states. Isotopes are very helpful in environmental reconstruction; for instance, oxygen-isotope ratios in foraminifers (microscopic, unicellular marine organisms) provide an invaluable thermometer of past climates.

2. Physical evidence of environmental change is highly varied. It includes geomorphological features that point to specific erosional environments

> *isotopic composition*: an isotope is a form of an element with a different number of neutrons, the number of protons being the same in all forms of the same element. Isotopic composition is the make-up of isotopes, for example the amounts of oxygen-16 and oxygen-18
>
> *evaporite*: a water-soluble mineral (such as halite or rock salt), or a rock composed of such minerals, precipitated out of saline water bodies such as salt lakes

glacial troughs: valleys, often U-shaped, eroded by valley glaciers or by ice flowing in ice sheets and ice caps

ice wedges: v-shaped masses of ice that run from the ground surface to the permafrost layer

deep weathering profiles: thick, much-weathered, Earth-surface mantles or crusts, often of great age

till plains: almost flat, slightly rolling and gently sloping plains formed of a thick blanket of till (a mainly unsorted and unlayered deposit laid down beneath a glacier)

glaciofluvial landforms: landforms fashioned by water from melting ice

periglacial landforms: landforms created in the 'periglacial zone' (the region where intense frosts during winter alternate with snow-free ground during summer)

palaeosols: old, fossil or relict soils of soil horizons, usually buried

moraines: various kinds of landforms formed by the deposition of glacial sediments

diatoms: microscopic, single-celled, marine, planktonic (drift in ocean currents or swim a little) algae with a skeleton composed of hydrous opaline silica

desmids: mainly single-celled, freshwater algae with cell walls made of cellulose and pectic materials

Pleistocene: a slice of geological time between 2 million and 10,000 years ago

Holocene: the most recent slice of geological time spanning 10,000 years ago to the present

(e.g. **glacial troughs**), to cold climates (e.g. **ice wedges**) or to tropical weathering (e.g. **deep weathering profiles**). Bedding in some sediments betrays past wind regimes. Other terrestrial sediments offer a general guide to past environments. **Till plains, glaciofluvial landforms** and **periglacial landforms** are signs of frigid climates. **Palaeosols**, which are found in sedimentary rocks and surface deposits, are very useful in reconstructing past environments. Sediment cores supply information on environmental conditions. Measurements from deep-sea cores, especially oxygen–isotope ratios, have revolutionized ideas about global climatic change, providing a detailed time series of surface-water and deep-water temperatures. Ice is an exceptionally rich source of palaeoenvironmental information. Records of glacier advances and retreats, either historical records or dated **moraines**, suggest changes in temperature, the duration of the melting season, sunshine and cloudiness, and snowfall. Ice sheets, which have grown year-by-year, are huge repositories of historical data. Ice cores extracted from the Greenland and Antarctic ice sheets yield up a variety of palaeoclimatic and palaeoenvironmental signals.

3. Biological evidence stems from the sensitivity of most animals and plants to climatic conditions. The more climatically sensitive organisms are, the better suited they are for enlistment as indicators of past climates. Plant microfossils comprise the pollen grains of higher plants, the spores of lower plants, and less commonly leaf hairs and epidermis and the skeletons of **diatoms** and **desmids**. Pollen analysis has been a mainstay of much **Pleistocene** and **Holocene** environmental reconstruction. Pollen grains and spores mostly fall into sediments from the air. This pollen rain derives from local and regional vegetation. A change in the pollen spectrum at a site thus registers local and regional changes in vegetation. Diatoms, desmids and other members of the aquatic microflora originate in the water-body from which the sediment was deposited. These organisms are sensitive to nutrient supply and climatic conditions. Their abundance in sediments thus provides a guide to change in lake environments. Tree-growth rings are excellent palaeoclimatic indicators. Ring width, width of early wood, variations in cell size and year-to-year variability of ring width register water availability, temperature and the length of the growing season. Where the fossil record permits, analysis of plant community structure, vegetation form and leaf form, and growth rings and vascular systems (systems for transporting sap) in wood make it possible to define terrestrial palaeoclimatic parameters, such as mean annual temperature and mean annual range of temperature, with better resolution than most sedimento-logical techniques. Many animal species are good indicators of past climates and habitats. Modern mammal species tend to have specific habitat preferences therefore fossil forms belonging to the same genera as living species give a guide to past habitats. Insects, and especially beetles, are good guides to past environmental conditions. Many have a narrow feeding range and are associated with specific plant communities. *Oodes gracilis*, a marshland beetle, today ranges over central and southern Europe, but not Britain. However, it is found in 120,000-year old river terrace deposits at Trafalgar Square, London, that contain hippopotamus. Many

molluscs, **ostracods**, corals and foraminifers are also palaeoenvironmental indicators.

ostracods: minute, chiefly freshwater crustaceans with a bivalve carapace (hard outer covering)

4. Historical and instrumental evidence covers the last three hundred years or so in good detail, though only in some places. Records go back to about 3000 BC, but they are exceedingly fragmentary. Documentary evidence comprises descriptive weather registers and weather diaries, ships' logs, a miscellany of accounts, annals, books, chronicles and documents, and grain prices. Meteorological instrument records are usually more precise than documentary sources, but data are patchy for most places before about 1850. Documentary information about environmental conditions other than climate also exists. For instance, some historical records contain information on river flood levels and lake levels.

The recent global environmental change agenda gives environmental reconstruction techniques an impetus. Past Global Changes (PAGES), for example, is one of eight core projects of the IGBP (International Geosphere–Biosphere Programme) (p. 89). It concentrates on two slices of time: (1) the last 2000 years of Earth history, with a temporal resolution of decades, years and even months; and (2) the last several hundred thousand years, covering glacial–interglacial cycles, in the hope of providing insights into the processes that induce global change.

MODELLING PAST ENVIRONMENTS

Another way of probing environmental change is to use computer models to postdict former states of the ecosphere. Two areas where this practice is helpful are changes in biogeochemical cycles and in Holocene, Pleistocene and **pre-Quaternary** climates.

pre-Quaternary: all geological time before the start of the Quaternary subera, 2 million years ago

Biogeochemical-cycle models involve a set of biogeochemical reservoirs (stores) connected by biogeochemical fluxes. They comprise a set of storage equations, one equation for each reservoir. Typically, reservoirs would include rocks, air, oceans and organisms. Fluxes between reservoirs are either estimated using empirical evidence, or defined by process equations (equations that determine the amount of a biogeochemical leaving a store in a unit time), which define the rates as functions of temperature, reservoir size and so on. The full model is a set of storage and process equations, which a modeller converts to a computer program. Running the program provides a numerical solution to the model that shows how the system being modelled changes with time. Solutions may simulate present-day conditions, where interest might focus on such problems as the effects of increasing atmospheric carbon dioxide on other parts of the ecosphere, or for geological time-scales. GEOCARB II, for example, investigated the long-term carbon cycle (Berner 1994). It modelled the transfer of carbon between the atmosphere, oceans, biosphere and lithosphere. Processes modelled include the weathering of calcium, magnesium silicates, carbonates and sedimentary organic matter, on continents; the burial of organic matter and calcium–magnesium carbonates in sediments; the thermal breakdown of carbonates and organic matter at depth, with resultant carbon dioxide degassing (release into the atmosphere). Figure 1.5 shows atmospheric carbon dioxide levels over the last 600 million years predicted using GEOCARB II. Notice how much higher the mass of atmospheric carbon dioxide was during much of geological time in comparison to the current level that is causing so much concern.

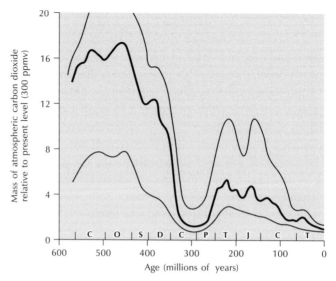

Mesozoic: a geological era lasting from 248 to 65 million years ago and comprising the Triassic, Jurassic and Cretaceous periods.

Cenozoic: the youngest geological era, lasting from 65 million years ago to the present

Figure 1.5 Carbon dioxide levels over the last 600 million years as predicted by GEOCARB II, a store-and-flux simulation of the long-term geochemical carbon cycles. The steady decline in atmospheric carbon dioxide levels is due to the increasing solar constant. Fluctuations about this steady decline reflect geological and topographical factors. The high **Mesozoic** carbon dioxide levels result largely from the low relief prevalent at the time, whilst the low **Cenozoic** levels result mainly from mountain uplift and decreasing metamorphic and volcanic degassing of carbon dioxide. *Source:* Adapted from Berner (1994)

Climate models are usually complex sets of equations that simulate the behaviour of the atmosphere, oceans, and in some cases, the biosphere as well. Simulations of past climates require the specification of appropriate boundary conditions, for instance, the position of the continents and mountain ranges, the solar constant and the tilt of the Earth's rotation axis. Run with appropriate boundary conditions, climate models provide an invaluable benchmark against which to test the past climates reconstructed using geochemical, physical, biological, and historical and instrumental evidence. Equally, the similarity between a predicted climate and the climate at the same period suggested by palaeoclimatic indicators, assists in judging the success of a climate model. Thus, theory and observation work in tandem. However, modellers should take pains to avoid circular arguments: the palaeoclimatic data used to calibrate (fix boundary conditions and process rates) in the model should not be the same data used to test the worth of the model.

Figure 1.6 is an example of a climate model used to postdict environmental conditions. The model tested the idea that variations in the Earth's orbit around the Sun influenced lake levels during the late Pleistocene and Holocene epochs. It involved running an atmospheric general circulation model to simulate the January and July climates at 3000-year intervals, starting 18,000 years ago (Kutzbach and Street-Perrott 1985). Figure 1.6a depicts the orbital changes in the seasonal distribution of solar radiation receipt (the driving variable) and the changing surface conditions assumed to accompany deglaciation (the boundary conditions). Figures 1.6a–d display the simulated changes within the latitude belt 8.9°–26.6°N for the last 18,000 years. From 15,000 years ago onwards, the model predicted a strengthening of the monsoon circulation and increased precipitation and effective precipitation (precipitation less evaporation) in the Northern Hemisphere tropics, which features reached their peak in the period

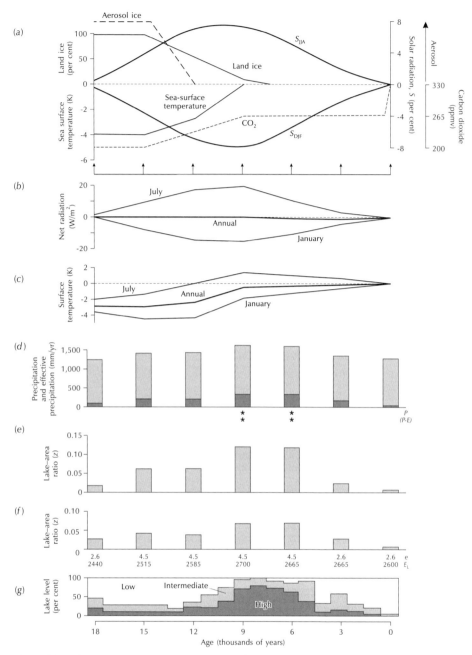

Figure 1.6 (a) Schematic diagram of changing external forcing (Northern Hemisphere solar radiation in June, July and August (S_{JJA}), and December, January and February (S_{DJF}), as a percentage different from present solar radiation. Also shown are internal boundary conditions (land ice as a percentage of the ice volume 18,000 years ago; global mean sea-surface temperature as a departure from the present value; excess glacial aerosol (dust content of the atmosphere), including sea ice on an arbitrary scale; and atmospheric carbon dioxide concentration) over the last 18,000 years. Up-pointing arrows correspond to seven sets of simulation experiments using the climate-model. (b) Simulated net radiation, expressed as departures from the modern case, in the latitude belt 8.9°–26.6° N. (c) Simulated surface temperature, expressed as departures from the present case, in the latitude belt 8.9°–26.6° N. The tops of the columns represent precipitation, the lightly shaded areas evaporation, and the darkly shaded areas effective precipitation (which is available for runoff). The stars indicate that values of P and $P-E$ are significant at $p < 0.05$ in a two-sided t-test. (e) Lake-area ratios, z, assuming a constant ratio of land area to internal drainage, e, and constant lake evaporation, E_L. (f) Lake-area ratios, z, assuming variable values of e and E_L. (g) Temporal variations in the percentage of lakes with low, intermediate or high levels. Maximum sample size = 73. *Source:* After Kutzbach and Street-Perrott (1985) © Macmillan Magazines Ltd

9000–6000 years ago (Figure 1.6d). The water budgets predicted by the model were used to estimate the area of closed lakes located within the latitude belt. Predicted lake levels closely matched actual lake levels as reconstructed from geological evidence (Figures 1.6e, f and g).

1.6 A BLUEPRINT FOR A NEW PHYSICAL GEOGRAPHY

Physical geographers contribute to the multidisciplinary research themes identified in the previous sections. To appreciate the nature of a new physical geography for the twenty-first century, several topics require discussion:

- The form and function of the terrestrial spheres – the natural ecosphere, the human sphere and the mental sphere – and other environmental spheres
- The tools and techniques for tackling environmental systems – measuring and monitoring, mapping and analysis and modelling
- The physical environment and its components – air, water, soils and landforms – and their interaction with humans
- The biological environment (species and communities) and its interaction with humans
- The ecological environment and its interaction with humans.

The rest of the book explores these substantial topics.

SUMMARY

Humans interact with their planetary environment. They affect, and are affected by, environmental change. Therefore, human–environment interactions viewed in the context of the Earth as a system provide a useful focus for physical geography. A novel approach to physical geography stems from regarding humans and the human-altered environment as equal partners with the natural environment. Two chief facts validate this viewpoint. First, the human species has come to dominate much of the land-surface and in doing so, has radically altered the natural flows and storage of gases and soluble materials in biogeochemical cycles, and altered the flow and storage of sediments in the sedimentary cycle. Perhaps even more significantly, it has transformed the land cover, mainly by turning land over to agricultural and urban use, causing the loss and fragmentation of habitats in the process. A consequence of these human-induced environmental changes is a substantial loss of biodiversity, with humans being responsible for much species extinction. Much can be gained by viewing the Earth as a global system and new techniques, such as remote sensing, allow the global ecosystem to be measured, monitored and mapped. High-powered computers enable the global ecosystem to be simulated in a manner scarcely imaginable in the 1970s, while past environmental changes provide a historical perspective for the current changes. Environments in the past can be reconstructed using geochemical, physical, biological, and historical and documentary evidence. They may also be studied using mathematical models of present environments applied to past conditions. A 'blueprint' for a new physical geography is presented that takes a human perspective.

Questions

1 To what extent have humans come to dominate the Earth's surface?

2 How important are humans as agents of erosion?

3 Explain how humans alter flows and stores in biogeochemical cycles.

4 Describe the advances in monitoring and mapping the environment made since the 1940s.

5 Describe the techniques used to reconstruct past environments.

6 Is a global approach to physical geography justified?

Further Reading

Goudie, A. (2000) *The Human Impact on the Environment*, 5th edn. Oxford, Blackwell.
An excellent introduction to environmental problems.

Goudie, A. (2001) *The Nature of the Environment*, 4th edn. Oxford, Blackwell.
A comprehensive and approachable introduction to physical geography with more breadth than depth.

Mannion, A. M. (1999) 'Global change: prospects for the next 25 years'. *World Futures* 54, 211–30.
A useful summary of measures of human impacts on the planet, such as the habitat index, the ecological footprint and the living planet index.

Park, C. C. (2001) *The Environment: Principles and Applications*, 2nd edn. London, Routledge.
A stimulating read.

Slaymaker, O. and Spencer, T. (1998) *Physical Geography and Global Environmental Change*. Harlow, Longman.
A ground-breaking text. Excellent.

Useful Websites

http://www.gesource.ac.uk/home.html
GEsource is the geography and environment hub of the Resource Discovery Network. It is a portal for Internet resources for geography and the environment.

http://www.idrc.ca
International Development Research Centre, Ottawa, Canada.

http://www.redlist.org/
The IUCN Red List of Threatened Species. A mine of useful information with links.

CHAPTER 2
THE NATURAL ECOSPHERE

OVERVIEW

The ecosphere comprises the spheres of life
(biosphere), water (hydrosphere), soil
(pedosphere) and air (atmosphere). The
biosphere consists of all living things that interact
with each other to form communities and with
their surroundings to form ecosystems.
Organisms have specific roles in communities –
some produce organic matter, others consume it,
and yet others feed on the dead remains, so
forming food webs. Interaction with the
environment and movement of materials through
food webs creates grand cycles of chemicals –
biogeochemical cycles. The hydrosphere is the
waters of the Earth, which are linked through the
hydrological or water cycle – evaporation,
condensation, precipitation, runoff. The
pedosphere (soil) sits at the interface of the
atmosphere, hydrosphere, lithosphere and
biosphere. It is a complex entity resulting from
physical, chemical and biological processes.
Climate, organisms, relief and substrate strongly
influence its character. The atmosphere is a shell
of dusty gas. Its composition is unique among
terrestrial planets due to the action of organisms.
Air constantly moves in grand convective
overturnings and great swirling eddies, collectively
called the general circulation of the atmosphere.

LEARNING OUTCOMES

The basic nature, form and function of these
spheres is revealed in this chapter, which should
help you understand:
- How living things interact with each other to
  form communities and with their surroundings
  to form ecosystems and biogeochemical cycles
- The global water cycle and the nature of
  aquatic life
- Soil-forming process and soil biogeochemical
  cycles
- The composition and processes of the
  atmosphere.

2.1 THE BIOSPHERE

All living things form the biosphere. The biosphere interacts with the non-living things in its surroundings (air, water, soils and sediments) to win materials and energy. The interaction creates the ecosphere, which is defined as life plus life-support systems. It consists of **ecosystems** – individuals, populations or communities interacting with their physical environment. Indeed, the ecosphere is the global ecosystem (Huggett 1995, 8–11; 1997). A **community** is the assemblage of interacting organisms within an ecosystem. Ecosystems and communities range in size from cubic centimetres to the entire world. Basics 2.1 summarizes the largest units.

Communities and ecosystems possess properties that 'emerge' from the individual organisms. These **emergent properties** are not measurable in individuals – they are community properties. Four important community properties are biodiversity, production and consumption, nutrient cycles and food webs. These community properties enable the biosphere to perform three important tasks. First, the biosphere harnesses energy to power itself and build up reserves of organic material. Second, it garners elements essential to life from the atmosphere, hydrosphere and pedosphere. Third, it is able to respond to cosmic, geological and biological perturbations by adjusting or reconstructing food webs.

> **ecosystem**: a community of living things together with their life-supporting surroundings
>
> **community**: an assemblage of species living together in a particular place
>
> **emergent properties**: properties, such as the number of species in a community, that can be measured only for a system as a whole, not in its component parts

PRODUCING AND CONSUMING

All organisms have a role in a community and in an ecosystem, with some organisms producing organic compounds, and some organisms breaking them down. The chief roles are:

- **Producers** (autotrophs). These include all **photoautotrophic organisms**: green plants, eukaryotic algae, blue-green algae, and purple and green sulphur bacteria. In some ecosystems, **chemoautotrophs** produce organic materials by oxidation of inorganic compounds and do not require sunlight.

- **Consumers** (heterotrophs). These organisms obtain their food and thus energy from the tissues of other organisms, either plants or animals or both – herbivores, carnivores and top carnivores. Consumers occupy several trophic levels. Consumers that eat living plants are primary consumers or herbivores. Consumers of herbivores are the flesh-eating secondary consumers or carnivores. In some ecosystems, there are carnivore-eating carnivores; these are tertiary consumers or top carnivores.

- **Decomposers** (saprophages) and **detritivores** (saprovores). Decomposers dissolve organic matter, while detritivores break it into smaller pieces and partly digest it.

In short, 'autotrophs produce organic material, the consumers eat it, and the decomposers and detritivores clean up the mess (excrement and organic remains)' (Huggett 1998, 142).

> **producer**: an organism that uses energy from the physical environment to sustain itself
>
> **photoautotrophic organisms**: organisms – green plants and bacteria – that obtain their energy from sunlight
>
> **chemoautotrophs**: organisms that obtain their energy from the oxidation of inorganic compounds, such as hydrogen sulphide
>
> **consumer**: an organism that gets its energy from other organisms
>
> **decomposer**: an organism that lives by dissolving organic matter for nourishment
>
> **detritivore**: an organism that feeds on organic detritus, breaking it up in the process

BASICS 2.1

BIOMES AND ZONOBIOMES

Biomes

Biomes are characteristically regional communities of animals and plants. The deciduous forest biome in temperate western Europe is an example. It consists largely of woodland with areas of heath and moorland. A plant community at the biome scale – all the plants associated with the deciduous woodland biome, for example – is a plant formation. An equivalent animal community has no special name; it is simply an animal community. Smaller communities within biomes are usually based on plant distribution. They are called plant associations. In England, associations within the deciduous forest biome include beech forest, lowland oak forest and ash forest. Between biomes are transitional belts where the climate changes from one type to the next, creating a change in plant communities. These changeover zones are called **ecotones**.

Zonobiomes

Biomes in different regions that experience similar climates group together as zonal biomes or zonobiomes. The world's present zonobiomes correspond to nine basic climatic types:

1. *Polar and subpolar zone.* This zone includes Arctic and Antarctic regions and is associated with tundra biomes with no trees. Arctic tundra regions have low rainfall evenly distributed throughout the year, short, wet and cool summers, and long and cold winters. Antarctica is an icy desert.

2. *Boreal zone.* This is the cold-temperate belt supporting boreal coniferous forest (taiga) biomes. It usually has cool, wet summers and very cold winters lasting at least six months. It is only found in the Northern Hemisphere where it forms a broad circumpolar band.

3. *Humid mid-latitude zone.* This zone is the temperate or nemoral zone. In continental interiors, it has a short, cold winter and a warm, or even hot, summer. Oceanic regions, such as the British Isles, have warmer winters and cooler, wetter summers.

This zone supports broad-leaved deciduous forest biomes.

4. *Arid mid-latitude zone.* This is the cold-temperate (continental) belt. The difference between summer and winter temperatures is marked and rainfall is low. Regions with a dry summer but only a slight drought support temperate grassland biomes (prairies and steppes). Regions with a clearly defined drought period and a short wet season support cold desert and semi-desert biomes.

5. *Tropical and subtropical arid zone.* The climate of this zone is hot desert that supports thorn and scrub savannah biomes and hot desert and semi-desert biomes.

6. *Mediterranean subtropical zone.* This is a belt lying between roughly 35° and 45° latitude in both hemispheres with winter rains and summer drought. It supports Mediterranean-type biomes – sclerophyllous (thick-leaved), woody vegetation adapted to drought and sensitive to prolonged frost.

7. *Seasonal tropical zone.* This zone extends from roughly 25° to 30° north and south. There is a marked seasonal temperature difference. Heavy rain in the warmer summer period alternates with extreme drought in the cooler winter period. The annual rainfall and the drought period increase with distance from the equator. It favours tropical grassland or savannah biomes.

8. *Humid subtropical zone.* This warm temperate climatic zone has almost no cold winter season, and short wet summers. It supports subtropical broad-leaved evergreen forest biomes.

9. *Humid tropical zone.* This torrid zone has rain all year and supports evergreen tropical rainforest biomes. Its climate is said to be diurnal because it varies more by day and night than it does through the seasons.

Between the zonobiomes are transitional belts where the climate changes from one type to the next. These are called zonoecotones.

---

ecotones: transition zones between two plant communities

photosynthesis: the manufacture of carbohydrates from carbon dioxide and water by chloroplasts, using light as an energy source. Oxygen is a by-product

Primary producers

Green plants manufacture organic matter using solar energy, carbon dioxide, minerals and water in the process of **photosynthesis**. The organic matter so made contains chemical energy, so photosynthesis converts radiant energy into chemical energy. **Production** is the quantity of sugars manufactured by photosynthesis in land plants and autotrophic algae. Some of the production runs metabolic processes and some goes to increase the biomass (Basics 2.2). Part of the photosynthetic process requires light, so it occurs only during daylight hours. At night, slow oxidation consumes some of the stored energy, a process called **respiration** in individuals and **consumption** in a community. Respiration is a complex series of chemical reactions that makes energy available for use. The products of respiration are carbon dioxide, water and energy.

Sunlight comes from above, so ecosystems tend to have a vertical structure (Figure 2.1). The upper production zone is rich in oxygen. The lower consumption zone, especially that part in the soil, is rich in carbon dioxide. Oxygen is deficient in the consumption zone, and may be absent. Such gases as hydrogen sulphide, ammonia, methane and hydrogen are liberated where reduced chemical states prevail. The boundary between the production zone and the consumption zone, which is known as the compensation level, lies at the point where there is just enough light for plants to balance organic matter production against organic matter utilization.

> **production**: the sum of sugars created by photosynthesis in land plants and autotrophic algae
>
> **respiration**: a complex series of chemical reactions in all organisms by which energy is made available for use
>
> **consumption**: community respiration

BASICS 2.2

BIOMASS

Biomass is the weight or mass of living tissue in an ecosystem and possesses an energy content, or 'bioenergy'. Biomass consists of phytomass and zoomass. Phytomass is the living material in producers. It normally excludes dead plant material, such as tree bark, dead supporting tissue, and dead branches and roots. Where dead bits of plants are included, the term 'standing crop' is applied. Zoomass is simply the biomass of animals in a community.

Ecospheric zones	Biogeochemical cycles	Spatial structure	
Gaseous storage zone Air		Atmosphere	
Production zone Photosynthetic plants dominant	Production — Carbon dioxide, water, and nutrients	Hydrosphere	
Compensation level			
Regeneration zone Animals and microconsumers dominant	Oxygen and organic food — Consumption	Lithosphere Pedosphere	
Solid storage zone Soils and sediments	— Consumption		
Circulatory mode		Circulation by water movements, migration of organisms, and (in reefs and algal mats) diffusion	Circulation by roots, stems, gravity and moving air
Ecosystem		Aquatic: the hydrobiosphere (lakes, rivers, oceans and seas)	Terrestrial: the geobiosphere (forests, grasslands, deserts and tundra)

Figure 2.1 The vertical structure of production and consumption in ecosystems. *Source:* After Huggett (1997)

Primary production

Photoautotrophs are the green plants that form the production zone. As their name implies, they produce their own food from solar energy and raw materials. The amount of energy (in the form of organic matter) that they make per unit time is **gross primary productivity** (**GPP**). Plant leaves make most of this energy, from where it travels through the phloem (nutrient-conducting tissue in vascular plants) to other parts of plants, and especially to the roots, to drive metabolic and growth processes. **Net primary productivity** (**NPP**) is the gross primary productivity less the chemical energy used in plant respiration (which is lost as heat to the environment). It is usually 80–90 per cent of GPP. Global mean NPP is about 440 g $m^{-2} yr^{-1}$ or 224 billion tonnes of dry organic matter. This is potentially available to primary consumers. On a global scale, productivity is high in the tropics, warm temperate zone and typical temperate zone; it is low in the arid subtropics, continental temperate regions and polar zones.

Productivity consists of primary productivity, the productivity of producers, and secondary productivity, the productivity of consumers. Environmental factors influence primary productivity, some of which may be limiting factors. Light limits productivity in green plants. Less than 2 per cent of the light falling on a tropical rainforest canopy reaches the forest floor and few plants thrive in the deep shade. Water shortage also limits primary productivity where light is not limiting. In the oceans, **phytoplankton** is the powerhouse of primary production. Light, nutrients and the grazing rate of **zooplankton** influence their productivity.

> **gross primary productivity (GPP)**: the amount of energy (in the form of organic matter) made per unit time in a community
>
> **net primary productivity (NPP)**: gross primary productivity less the energy used in respiration
>
> **phytoplankton**: the plant community in marine and freshwater systems which floats free in the water and contains many species of algae and diatoms
>
> **zooplankton**: aquatic invertebrates living in sunlit waters of lakes, rivers, seas and oceans

Consumers

The material produced by photosynthesis serves as a basic larder for entire ecosystems – it serves as an energy-rich reserve of organic substances and nutrients that supply the rest of the system. The potential chemical energy of net primary productivity is available to organisms that eat plants, both living plant tissue and dead plant tissue, and indirectly therefore to animals (and the few plants) that eat other animals. All these organisms depending upon other organisms for their food are heterotrophs or consumers. They are browsers, grazers, predators or scavengers. They include microscopic organisms, such as protozoans, and large forms, such as vertebrates. The majority are chemoheterotrophs, but a few specialized photosynthetic bacteria are photoheterotrophs. The stored chemical energy in consumers is secondary production. There are two broad groups of consumers. **Macroconsumers** or biophages eat living plant tissues. **Microconsumers** or saprophages slowly decompose and disintegrate the waste products and dead organic matter of the biosphere.

The available phytomass in water is small, but the primary productivity is relatively large because aquatic plants multiply fast. Animals eat the plants and incorporate much of the primary production in their bodies. For this reason, the secondary productivity or zoomass (Basics 2.2) in aquatic ecosystems is commonly more than fifteen times larger than the phytomass. The situation is different in terrestrial ecosystems. Much of the phytomass consists of non-photosynthetic tissue, such as buds, roots and trunks. For this reason, the standing phytomass is always very large, especially in woodlands. Primary productivity, on the other hand, is relatively small – a modest amount of dry organic matter is created each year. Animals living above the ground eat a mere

> **macroconsumers**: organisms that feed on living plant tissues
>
> **microconsumers**: organisms that feed on dead and decaying organic materials

1 per cent, or thereabouts, of the phytomass. The zoomass is therefore small, being about 0.1 to 1 per cent of the phytomass.

The human harvest of plants (for food, fuel and shelter) accounts for about 4.5 per cent of global terrestrial net production. Land used in agriculture or converted to other land uses accounts for about 32 per cent of terrestrial net primary production. All human activities have reduced terrestrial net primary production by around 45 per cent. This probably amounts to the largest ever diversion of primary productivity to support a single species (Vitousek *et al.* 1986).

Decomposers and detritivores

Saprophages include decomposers (or saprophytes) and detritivores (or saprovores). Decomposers are organisms that feed on dead organic matter and waste products of an ecosystem. They secrete enzymes to digest organic matter in their surroundings, and soak up the dissolved products. They are mainly **aerobic** and **anaerobic** bacteria, protozoa and fungi. An example is the shelf fungus (*Trametes versicolor*). This grows on rotting trees and is important in decomposing wood. A very few animals, such as tapeworms, and some green plants lacking chlorophyll, such as Indian pipe (*Monotropa uniflora*) and pinesap (*M. hypopithys*), also obtain their food by diffusion from the outside.

> **aerobic**: requiring oxygen
>
> **anaerobic**: not requiring oxygen

Detritivores feed upon detritus. They include beetles, centipedes, earthworms, nematodes and woodlice. They are all microconsumers. Detritivores assist the breakdown of organic matter. By chewing and grinding dead organic matter before ingestion, they comminute it (break it into smaller pieces) and render it more digestible. When egested, the carbon–nitrogen (C/N) ratio is a little lower, and the acidity (pH) a little higher, than in the ingested food. These changes mean that the faeces provide a better substrate for renewed decomposer (and notably bacterial) growth. Successively smaller fragments of dead organic matter pass through successively smaller detritivores, after having been subject to decomposer attack at each stage. The result is a comminution spiral (Figure 2.2).

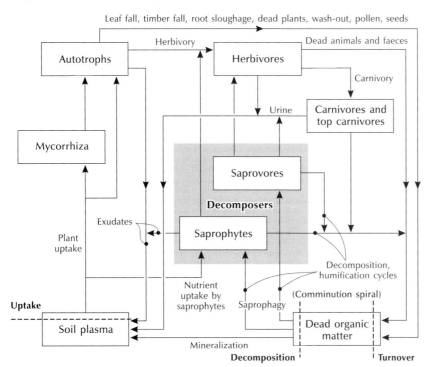

Figure 2.2 Comminution spiral.
Source: After Huggett (1980)

humus: the end-product of organic decomposition

mineralization: the release of mineral components in organic matter through decomposition

Humification accompanies comminution and decomposition to produce a group of organic compounds called humus. **Humus** is, in a sense, the final stage of organic decomposition, but it is also the product of microbial synthesis and is always subject to slow microbial decomposition, the end-product of which is stable humus charcoal. Humus decomposition in the temperate zones is incomplete and humus is brown or black; in the tropics, it is normally more advanced and humus may be colourless. **Mineralization**, the process by which mineral nutrients are released into the soil solution, occurs during decomposition.

ECOSYSTEM TURNOVER

Biogeochemicals

bioelement: an element essential for living things to function

macronutrient: a nutrient required in moderate to large amounts by organisms

micronutrient: a nutrient required in trace amounts by organisms

Biogeochemicals are chemicals involved in living processes, but are not necessarily essential for life. **Bioelements** are elements essential to life. The biosphere is made of three main elements – hydrogen (49.8 per cent by weight), oxygen (24.9 per cent) and carbon (24.9 per cent). Many other elements occur in living things, some of which are essential to biological processes – nitrogen (0.27 per cent), calcium (0.073 per cent), potassium (0.046 per cent), silicon (0.033 per cent), magnesium (0.031 per cent), phosphorus (0.03 per cent), sulphur (0.017 per cent) and aluminium (0.016 per cent). These elements, except aluminium, are the basic ingredients for organic compounds, around which biochemistry revolves. The manufacture of nucleic acids (RNA and DNA), amino acids (proteins), carbohydrates (sugars, starches and cellulose) and lipids (fats and fat-like materials) requires carbon, hydrogen, nitrogen, oxygen, sulphur and phosphorus. Calcium, magnesium and potassium are necessary in moderate amounts. **Macronutrients** are chemical elements required in moderate and large quantities. **Micronutrients** are required in trace amounts. There are more than a dozen micronutrients, including chlorine, chromium, copper, cobalt, iodine, iron, manganese, molybdenum, nickel, selenium, sodium, vanadium and zinc. Functional nutrients play some role in the metabolism of plants but seem dispensable. Other mineral elements cycle through living systems but have no known metabolic role. The biosphere has to obtain all bioelements – macronutrients and micronutrients – from its surroundings.

Biogeochemical cycles

biogeochemical cycle: the repeated movement of a bioelement through the ecosphere

nutrient cycle: the repeated movement of a bioelement through the ecosphere

biotic: pertaining to life

abiotic: devoid of life, that is, non-biotic

Biogeochemical cycles or **nutrient cycles** are the repeated movements of bioelements through living things, air, rocks, soils and water. They all have **biotic** and **abiotic** phases. They involve stores (pools, reservoirs, sinks) of various chemical species in the atmosphere, hydrosphere, pedosphere and lithosphere. The bioelements move between the stores as fluxes. Bioelements such as carbon, that are in a gaseous state for a leg of the cycle, are gaseous biogeochemical cycles; those that do not volatilize readily and that move mainly in solution, magnesium for instance, are sedimentary biogeochemical cycles. The major cycles involve the storage and flux of hydrogen, carbon, nitrogen, oxygen, magnesium, phosphorus, potassium, sulphur and calcium. The micronutrients sodium and chlorine and many other elements suspected of being micro-bioelements (e.g. boron, molybdenum, silica, vanadium and zinc) also have biogeochemical cycles.

Biogeochemicals cycle through the biosphere in three stages: uptake, turnover and decomposition (Figure 2.1). Green plants take up solutes and gases and incorporate them into their phytomass (plant biomass). Photosynthesis releases oxygen. The remaining minerals either pass on to consumers or return to soil and water bodies either when plants die or as spent leaves, flowers and so on. The minerals in the consumers eventually return to the soil, water bodies or the atmosphere.

The carbon cycle is the return movement of carbon through living things, air, rocks, soil and water (Figure 2.3). In brief, the cycle runs as follows: Photosynthesizing plants convert atmospheric carbon, in the form of carbon dioxide (CO_2), to carbohydrates. Producer respiration returns some plant carbon to the atmosphere in the form of carbon dioxide. Animals assimilate and metabolize some plant carbon. A portion of animal carbon returns to the atmosphere as carbon dioxide released through consumer respiration. The balance enters the decomposer food chain and either returns to the atmosphere through decomposer respiration or accumulates as organic sediment (e.g. peat and coal). Combustion during fires and volcanic eruptions releases carbon dioxide into the atmosphere. Agricultural, industrial and domestic practices alter the stores and fluxes in the carbon cycle. Every year, the burning of fuelwood and fossil fuels pumps some 5 billion tonnes of carbon dioxide into the atmosphere. The growing atmospheric carbon dioxide store augments the greenhouse effect and contributes to global warming. Current knowledge of the carbon cycle within the oceans, on land and in the atmosphere is sufficient to conclude that, although natural processes can potentially slow the rate of increase in atmospheric CO_2, they are unlikely to assimilate all the human-produced CO_2 during the twenty-first century. Further understanding requires greater knowledge of the relationship between the carbon cycle, other biogeochemical cycles and the climatic system (Chapter 8).

The nitrogen cycle is the return movement of nitrogen through living things, air, rocks, soil and water (Figure 2.4). Relatively inactive atmospheric nitrogen is a major nitrogen store. It combines with rainwater to form inorganic compounds, or it is fixed by nitrogen-fixing bacteria (nitrifiers) to produce a

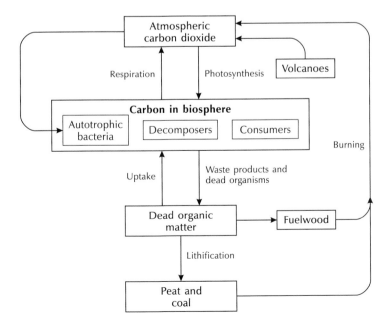

Figure 2.3 The carbon cycle

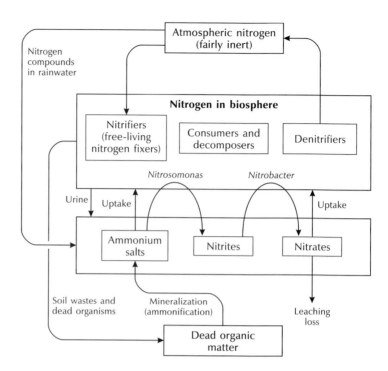

Figure 2.4 The nitrogen cycle

reactive form that organisms can utilize. Animals and plants assimilate and metabolize reactive nitrogen. It then returns to the soil in nitrogenous animal wastes and in dead organisms. Nitrogen in the soil is subject to nitrification (conversion to nitrates and nitrites by nitrifying microorganisms), to mineralization or ammonification (the release of ammonia and ammonium from dead organic matter by decomposers) and to denitrification (the reduction of nitrate to gaseous nitrogen forms that return to the atmosphere). Agricultural and industrial practices modify the stores and fluxes in the nitrogen cycle. Nitrogen fertilizers add to the nitrogen pool in the soil. Growing human and livestock populations increase nitrogenous waste volumes. Industrial activities release nitrogen oxides into the atmosphere. These growing nitrogen pools create environmental problems, including **eutrophication** and ozone depletion.

The sulphur cycle is the return movement of sulphur through living things, air, rocks, soil and water (Figure 2.5). Atmospheric sulphur, occurring as various oxides and sulphides, dissolves in rainwater. Once on land and in the oceans, animals, plants and microorganisms assimilate and metabolize it. It returns to soil and water bodies in dead organisms. Soil sulphates are subject to reduction and soil sulphides to oxidation. Under reducing conditions, some sulphur may return to the atmosphere as liberated hydrogen sulphide, and elemental sulphur may form. Under oxidizing conditions, sulphates may form, including sulphuric acid. Acidification in sulphide-rich soils, as found on spoil-heaps of mines, is a serious problem. The burning of fossil fuels releases sulphur dioxide into the air (as do volcanoes). This quickly oxidizes and falls as acid rain. Steps to remedy the acid-rain problem may reduce atmospheric sulphur concentrations but in consequence, it may become necessary to supplement soil sulphur through fertilizer applications.

A considerable body of evidence indicates that human activities are significantly changing the storage and, to a greater degree, the fluxes of many

eutrophication: the enrichment of water bodies with mineral and organic nutrients. This initially supports a proliferation of plant life but can lead to reduced dissolved oxygen levels and changes in the dominance of plant groups

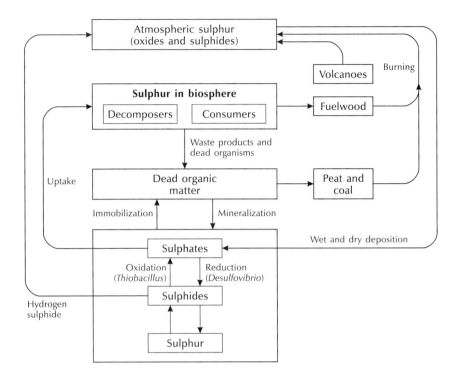

Figure 2.5 The sulphur cycle

bioelements – including carbon, sulphur, nitrogen and phosphorus. The geological effects of these human disturbances of the biogeochemical system are unclear. Some simulations suggest long-delayed, radical shifts in the state of the ecosphere as increased soil erosion transports carbon to the sea floor at an accelerated rate. Of more immediate concern are changes now and in the next century or so. Humans are increasing the atmospheric carbon dioxide store and likely changes in the carbon cycle are under investigation (see Chapters 7 and 13). Extra nitrogen in lakes causes accelerated eutrophication, and sulphur liberated by burning fossil fuels creates acid rain (see Chapter 13).

FOOD CHAINS AND FOOD WEBS

Everything in an ecosystem has either to make its own energy through photosynthesis (or other energy source) or to eat something else. A **food web** is a sort of 'who eats who' in a community. Simple food webs are sometimes called **food chains**. There are two chief types of food web: a grazing food web (plants, herbivores, carnivores, top carnivores) and a decomposer or detritus food web.

> **food web**: a network of feeding relationships within a community
>
> **food chain**: a simple sequence of feeding relationships in a community, starting with plants and ending with top carnivores

Grazing food chains and webs

This simple feeding sequence:

   plant $\Rightarrow$ herbivore $\Rightarrow$ carnivore $\Rightarrow$ top carnivore

is a grazing food chain. An example is:

   leaf $\Rightarrow$ insect $\Rightarrow$ insect-eating bat $\Rightarrow$ bat hawk

Usually though, food and feeding relations in an ecosystem are more complex because there is commonly a wide variety of plants available, different herbivores prefer different plant species, and carnivores are likewise selective about which herbivore they will consume. Complications also arise because some animals, the omnivores, eat both plant and animal tissues. For all these reasons, the energy flow through an ecosystem is better described as a food web (Figure 2.6).

Decomposer food chains and webs

Dead organic matter and other waste products generally lie upon and within the soil. Plants reuse minerals released during their decomposition. Complex food and feeding relations among the decomposers produce a decomposer or detritus food chain, in which the organisms are all microconsumers (decomposers and detritivores).

On land, litter and dead organic matter support a detritus food web. In Sulawesi caves (Figure 2.6), decomposers (mainly bacteria and fungi) slowly digest wet faeces, guano (a substance made chiefly of seabird or bat dung) and dead animals. Detritivores – cockroaches, springtails, beetles and so on – also eat the organic waste. Those that feed on guano – cockroaches and springtails for example – are coprophages; those that feed on dead animals – carpet beetles and other beetles – are necrophages. The coprophages and necrophages themselves fall prey to other detritivores, such as spiders, whip scorpions and assassin bugs. Shrews and toads prey on the some of the detritivores, so forging a link between the grazing and detritus food webs.

In aquatic ecosystems, the producers (mainly phytoplankton) live in the upper illuminated areas of water bodies, where they are prey to animal microplankton and macroplankton. In turn, fishes, aquatic mammals and birds that take their prey out of water, eat the animal plankton. All the organisms in the detrital aquatic food chain are eventually decomposed in the water and in sediments on lake, river or sea bottoms.

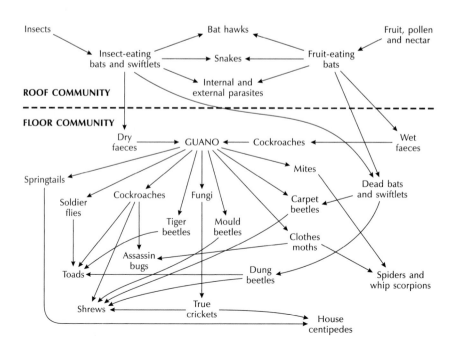

Figure 2.6 Simplified food web of a cave ecosystem in Sulawesi. *Source:* Adapted from Whitten *et al.* (2002)

2.2 THE HYDROSPHERE

Of all the natural resources, water demonstrates the interdependence of the physical and human spheres best. Water is arguably the most important natural resource used by human societies, as it is essential for life and nearly every human activity.

In terrestrial systems, the quantity of available water is a chief determining factor in the location of plant and animal species assemblages. Each species occupies a niche along the spectrum of water availability that ranges from desert ecosystems to wetland environments. Plants and animals have evolved with specialist internal structures or behaviour to enhance or control local water stores and availability. **Xerophytes**, for example, have a thick waxy cuticle that prevents water loss, and wetlands plants, such as mangroves, have specialized tissues that enhance oxygen diffusion in saturated soils. Of all species on Earth, however, only *Homo sapiens* can manipulate the water environment and hydrological cycle on a local to global scale with far reaching impacts. The foundation of the civilized world in ancient times was based on early water management – the redistribution of river flow from its 'natural' pathway for irrigation. Today, the dual need to ensure adequate water resources and minimize the environmental impact of human activities underpins attempts to move towards sustainable water management. The change in water availability in time and space, coupled with the growth of the world's population, means that this is one of the most pressing concerns for the future.

> *xerophyte*: a plant adapted to dry conditions

THE GLOBAL HYDROLOGICAL CYCLE

Water, even in its solid state, seldom stays still for long. It is in a constant state of movement or flux between phases of the **hydrological cycle** (Figure 2.7). Each phase of water moving through the atmosphere, hydrosphere, lithosphere and biosphere in various states – gaseous (water vapour), liquid (streamflow) and solid (ice) – is powered by the Sun. Water covers over 70 per cent of the Earth's surface, but fresh water makes up only 1–3 per cent of the total water of the planet, and only a fraction of this is readily available to life. Concentrations of constituents dissolved in water vary in different parts of the hydrosphere: the obvious example is the concentration of sodium and chlorine in seawater compared to inland waters. The acidity of waters in the hydrosphere is also highly variable, ranging from acid volcanic crater lakes to highly alkaline water, free of carbon dioxide and in contact with ultramafic rocks (dark coloured and base-rich rocks that form deep inside the Earth).

The total volume of water in the hydrological cycle is estimated at 1.5 billion km^3, and at more than 97 per cent, the world's oceans hold the greatest store of water (Figure 2.8). The next largest store is water held in **ice sheets**, **ice caps** and mountain glaciers. These water reserves are inaccessible for human use due to either high levels of salinity or long residence times respectively.

The critical flux for available terrestrial water comes from the evaporation of water from the oceans and land into the atmosphere (estimated at 380,000 km^3 a year: 320,000 km^3 from the oceans and 60,000 km^3 from the land) which then condenses to form clouds and falls as precipitation. About 96,000 km^3 of precipitation falls over land surfaces. Approximately 60,000 km^3 of this

> *hydrological cycle*: the repeated movement of water molecules around the ecosphere by evaporation, condensation and precipitation
>
> *ice sheets*: bodies of ice larger than 50,000 km^2
>
> *ice caps*: bodies of ice smaller than 50,000 km^2

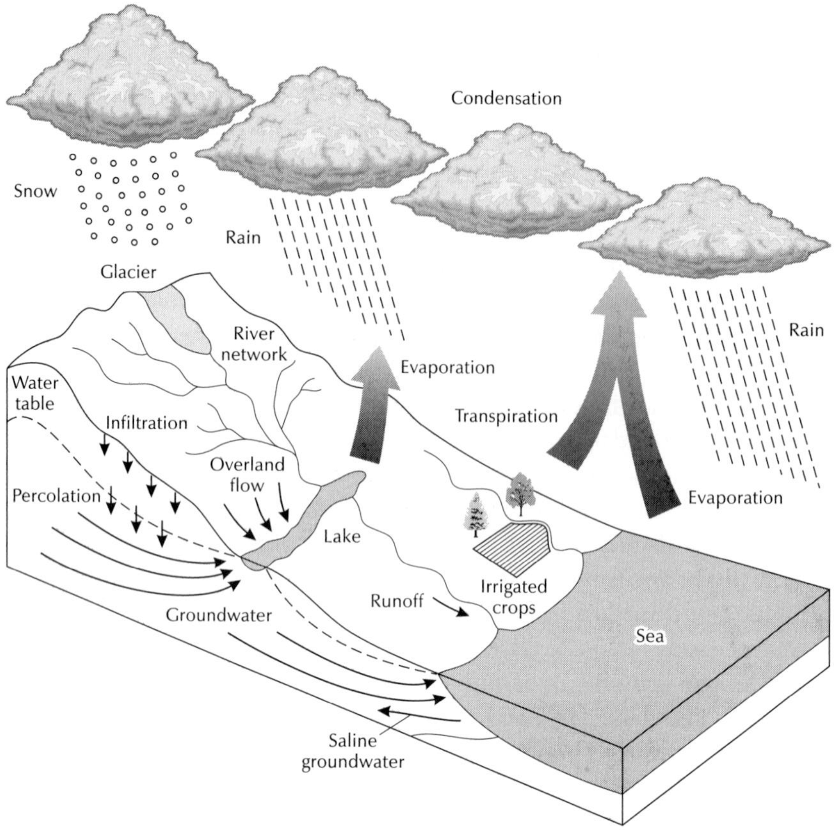

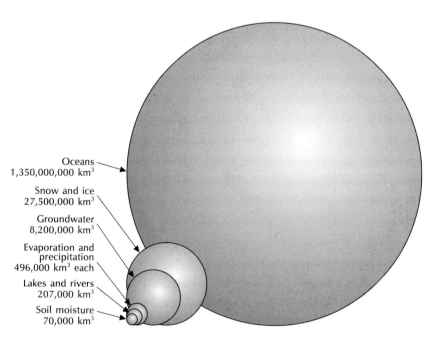

Figure 2.7 The hydrological cycle.
Source: After Fetter (1994)

Figure 2.8 The relative magnitude of water stores

Oceans
1,350,000,000 km³

Snow and ice
27,500,000 km³

Groundwater
8,200,000 km³

Evaporation and precipitation
496,000 km³ each

Lakes and rivers
207,000 km³

Soil moisture
70,000 km³

> *groundwater*: water stored in the pore spaces and joints within rocks

evaporates back into the atmosphere. The 36,000 km³ that remains either flows over the surface into drainage systems (lake or river systems) as overland flow or runoff, or it seeps into the ground. Water that infiltrates into the ground will percolate downwards through the soil to the **groundwater**. Water in the surface drainage systems will eventually flow into the sea where the cycle of evaporation, precipitation and movement over and through the land begins again. Water in groundwater also flows towards the sea, although on a different timescale, and can emerge in springs, streams or wetlands where the land surface is at or below the position of the **water table**.

The time spent by a water molecule in each of the phases of the hydrological cycle determines its influence in affecting the Earth's climate. The time spent by water in the atmosphere (relatively short residence time of 7–10 days), and in flowing over the land surface, influences local and regional weather patterns. This contrasts with the long residence time of groundwater **aquifers**, deep ocean circulation and glaciers (ranging between 3000 and 10,000 years), which influence long-term climate and temperature variability.

The volume of water in each stage of the hydrological cycle also changes over time. These volumetric variations include changing lake levels, sea levels, water table heights, ice cover and, to a lesser extent, average cloudiness. Water fluxes such as evaporation, precipitation, runoff and lake currents fluctuate too.

All these fluctuations of stored and flowing water occur over time-spans ranging from days to aeons. For example, during past ice ages, there were reduced rates of evaporation and surface runoff and an increased store of water in solid state, compared to the current interglacial. Evaporation that did occur from oceans would precipitate as snow onto land and accumulate to transform into glacier ice and ice sheets. Over the duration of the ice age, this could cause sea levels to fall by over 100 m. During the current interglacial, one of the speculated effects of human-induced climate change is increased surface temperature leading to enhanced evaporation and the melting of the existing glaciers and ice sheets, which could raise sea level by over 30 m.

> **water table**: the upper limit of groundwater saturation in permeable rocks; the level at which groundwater pressure is equal to atmospheric pressure
>
> **aquifers**: layers of rock that are capable of storing water

AQUATIC LIFE

Aquatic systems are amongst the most ecologically diverse, with a vast range of habitats ranging from freshwater to saline. The hydrological environment is the dominant feature of such habitats. Biotic influences are used to characterize and differentiate between sites, but the determining features are hydrological: water flow, water levels and water chemistry.

2.3 THE PEDOSPHERE

The pedosphere is a highly organized physical, chemical and biological system. It is related to the wider environment as shown in Figure 2.9. Soil is a facet of the environment that is largely taken for granted by the majority of western, urbanized society. However, soil is one of humanity's most precious assets due to its role in sustaining life. Unfortunately, humans cause local and regional soil degradation through soil use. These issues are discussed in Chapter 10. Soil-forming processes and the role of soil biota are examined below.

SOIL-FORMING PROCESSES

The **weathering** of a rock or mineral mass is the ultimate source of most soils. The creation of soil from the weathering of parent rock is a function of climate, topography, biological activity, substrate type and time. Soils continually evolve

> **weathering**: the chemical, mechanical and biological breakdown of rocks on exposure to the atmosphere, hydrosphere and biosphere

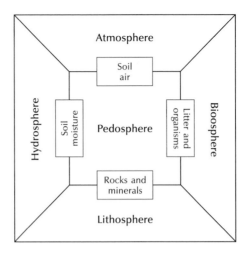

Figure 2.9 The relationship of the pedosphere to other spheres

and change, reacting to feedback mechanisms, external and internal changes, as shown in Figure 2.10. The complexity and range of processes that bring about soil development can be seen in Table 2.1.

Time

regolith: weathered mantle; all the weathered material lying above unaltered or fresh bedrock

pedogenesis: the formation of soils

Soil development occurs over a long time-frame. The first stage in soil development is the transformation of the parent material by weathering into loose material known as **regolith**. The nature of the geology and topography of the parent material influences the chemistry, texture, depth and drainage characteristics of the soil. Note that soil development is a constant but slow process, and can span different climatic regimes and waves of **pedogenesis**.

Climate

Climate is very important, and the global distribution of soil types (see p. 291) is linked to the different climate regions over the world. The temperature of soil

pedon: the smallest volume of material that can be called 'a soil'; it extends the idea of a soil profile, which is a vertical sequence of soil horizons seen in a soil pit, to three dimensions

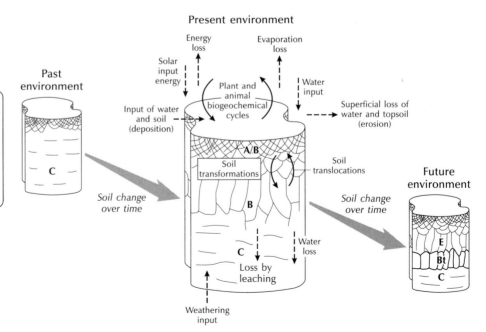

Figure 2.10 A soil **pedon** is shown as a part of a system evolving over time, responding to changing inputs, outputs and transformations. *Source:* After Bridges (1997)

Table 2.1 Soil-forming processes

Process	Description
Alkalization (solonization)	Accumulation of sodium ions on the exchange sites on clay
Brunification (rubification, ferrugination)	Release of iron from primary minerals and the dispersion of iron oxide, the oxidation or hydration of which gives a brown or red hue respectively
Calcification	The accumulation of calcium carbonate
Dealkalization (solodization)	Leaching of sodium ions and salts from natric (clay and sodium rich) horizons
Decalcification	The removal of calcium carbonate by eluviation
Decomposition (weathering)	The breakdown of mineral and organic materials
Desalinization	The removal of soluble salts from salic soil horizons or the entire profile
Desilication (ferrallitization, ferritization, allitization)	The chemical migration of silica from the **solum** leading to a concentration of iron and aluminium **sesquioxides** in the solum
Eluviation	The washing out of material from a horizon or particular location within a soil profile
Enrichment	The addition of material to a soil body, or sometimes horizons
Gleization	The reduction of iron under anaerobic conditions to produce bluish to greenish grey hues, with or without ochre, brown and black mottles and ferric and manganiferous concretions
Hardening	Decrease in the volume of soil voids
Humification	The conversion of raw organic matter into humus
Illuviation	The washing of material into a horizon or particular location within a soil profile
Leaching (depletion)	The eluviation of material from the entire solum (the A and B horizons) and not just a horizon
Lessivage (pervection)	The mechanical eluviation of silts and clays
Leucinization	Pale soil colours due to the loss of organic matter
Littering	Accumulation of organic litter on the soil surface
Loosening	Increase in the volume of soil voids
Melanization	The darkening of light-coloured mineral matter by organic matter (as in dark A1 horizons)
Mineralization	Release of oxide solids through organic matter decomposition
Paludinization	The build-up of organic matter (>30 cm) to form peat
Pedoturbation	Biological, chemical and physical churning of the soil
Podzolization (silication)	The chemical migration of aluminium, iron and/or organic matter resulting in a build-up of silica in the eluviated layer
Ripening	Chemical, biological and physical changes in organic soil when air enters previously waterlogged soil
Salinization	The accumulation of soluble salts (e.g. sulphates and chlorides of calcium and sodium) in salty horizons
Synthesis (neoformation)	The formation of new particles of mineral and organic species

solum: the layers of the soil lying above the weathered parent rock: the A and B horizons

sesquioxides: oxides that contain two metallic atoms to every three of oxygen; for example, Al_2O_3

has a great influence on soil development in a range of activities, from the creation of weathering products to biological activity. Low soil temperatures inhibit or slow down microbial activity in decomposing organic matter, so that thick organic layers are created. Plant growth and development is also inhibited, due to the short growing season.

In saturated soils such as **wetlands**, where the high water content ensures a soil temperature and reduces oxygen levels, the organic content is high due to the inhibition of microbial activity. This is why peat soils and wetland sites are important for archaeological finds or pollen records. In warm moist climates, the higher temperatures allow microbial decomposition of leaf litter and other matter to occur at fast rates so that there is only a small organic layer present if at all.

Slope and aspect

The topography of the land surface, in particular the **slope** and its **aspect**, influences soil development. A slope has a gradient, the steepness of which affects the rate of water runoff and, associated with this, ease of removal of products by erosion. Soil on slopes is subject to a greater erosion potential than flat lands, for example. The slope aspect (the direction it faces), influences soil development by controlling the degree of insolation received at the surface. The amount of insolation received affects other properties such as temperature and evaporation. In the Northern Hemisphere, south-facing slopes receive most insolation as they face towards the Sun. This means that the insolation received at the surface acts to increase soil temperatures and create a greater evaporative demand than on north-facing slopes that receive relatively little insolation. Radcliffe (1982) reports the study of aspect on soils at Coopers Creek, North Canterbury, New Zealand. The pastures on the north-facing slopes were invariably drier than those on the south-facing slope throughout the growing season, sometimes reaching wilting point in summer. In addition, on the north-facing slopes, mean biomass was greater, as was the **carbon–nitrogen ratio**, and soil nitrogen levels were lower.

The sequence of soils on slopes is described as a **catena**. This is where each soil on the slope is different from, but linked to, adjacent soils. The soils differ due to slope processes such as changes in angle, drainage and overland flow and the movement of soil minerals and nutrients, vegetation and biological activity, rather than because of changes in climate or the parent material. Research into catenas in the 1960s promoted the idea that hydrology and the movement of water and solutes link soils on slopes, connecting them and differentiating their properties (Blume 1968).

Biological activity

Soil is teeming with life and is constantly developing in response to the organisms living in it and on it: from microscopic bacteria, earthworms, burrowing animals and, of course, humans.

Soil organisms can be grouped into **macrofauna**, **mesofauna** and **microfauna**. Such vertebrates as rabbits, moles and foxes, burrow into the soil, mixing it and bringing soil to the surface. Grazing animals can strip vegetation from the surface leaving the soil surface vulnerable to erosion. Higher plants extend root systems into the soil, binding it together. The decay of plant matter returns nutrients and minerals to the soil, and former root channels create macropores for aeration and drainage. The macrofauna and mesofauna includes earthworms, mites, springtails, nematodes, and certain insects. They ingest and decompose organic matter, and

---

wetlands: terrain in which the soil is saturated with water, and soil saturation is the chief determinant of soil development and plant communities. Marshes, swamps, bogs and fens are examples

slope: an inclined surface of any part of the Earth's surface

aspect: the compass bearing or exposure of a slope

carbon–nitrogen (C/N) ratio: a useful indicator of the degree of decomposition of organic matter in soils. Well-decomposed soil humus in humid temperate soils has a C/N ratio of around 12; straw has a C/N ratio of about 40

catena: a linked sequence of soils running from hilltop to valley bottom

macrofauna: soil animals at least 1 cm long, including vertebrates, earthworms, and slugs and snails

mesofauna (meiofauna): soil animals of intermediate size, 200 μm–1 cm. Includes most of the nematodes (roundworms), rotifers, springtails and mites, with various small molluscs and arthropods

microfauna: soil animals less than 200 μm long. Includes protozoans, and some nematodes (roundworms), rotifers and mites

their abundance in the soil is greatest in the aerobic litter layer. Microorganisms include bacteria, algae, fungi and **actinomycetes**, and they are concentrated in the region of the largest food supply: the surface layer. Fitzpatrick (1974) estimates that the live weight of bacteria, the most numerous of soil microorganisms, ranges from 1000 to 6000 kg ha^{-1} in the upper 15 cm of soil.

Soil biota play three important roles in soil formation and development:

> **actinomycetes**: fungi-like bacteria that form long, thread-like branched filaments, which are visible in compost

1. The decomposition of detritus and organic matter enables recycling of the nutrients locked within. Macrofaunal detritivores such as woodlice, slugs and earthworms bury and consume detritus, and bacteria and fungi break down organic compounds through the secretion of enzymes.

2. Bacteria play an important role in the nitrogen cycle within the soil. They fix atmospheric nitrogen and convert it to other products that can be dissolved in the soil solution and taken up by plants (pp. 31–32).

3. Burrowing by organisms – ranging from plant roots, through worms, to foxes – creates macropores that aerate and promote drainage.

SOIL BIOGEOCHEMICAL CYCLES

Soil bacteria and other fauna and flora play an immensely important role in biogeochemical cycles, particularly nutrient cycling and nitrogen fixation. Anand *et al.* (2003) comment that the composition of the soil microbial assemblage can affect plant growth and the competitive outcome between plants. The lack or absence of essential microbial processes, such as in degraded or contaminated soils, can result in nutrient deficiencies, which can lead successively to plant death, the inability of further plants to colonize and soil erosion. Very important stages of the nitrogen and carbon cycles occur in soil and are discussed below.

The soil nitrogen cycle

Soils are a major creator of nitrous oxide, a greenhouse gas, through microbial processes. Anthropogenic (human) activities that contribute to atmospheric emissions include the application of nitrogenous fertilizers, conversion of tropical forest to pasture and the burning of biomass (Hillel 1998). Soil nitrogen levels are important as they affect the rate of plant or crop growth and foliage. Soils lacking in nitrogen sustain poorly growing plants with pale leaves that mature prematurely (Davies *et al.* 1993).

Soil microorganisms 'fix' atmospheric nitrogen present in soil pores and convert it to organic nitrogen. This is done by bacteria living freely in the soil and by those in a symbiotic relationship with plants. Plants that form symbiotic relationships with bacteria are called **leguminous**, and include clover and crops such as peas and beans. Once organic nitrogen has been created, it continuously moves from one form to another in the nitrogen cycle (Figure 2.4).

> **leguminous**: refers to plants of the family Leguminosae, which bear pods that split into two halves with the seeds attached to the lower end of one of the halves

Decomposition of organic nitrogen converts it to mineral nitrogen or ammonia nitrogen, NH^{+4}, nitrite, NO^{-2}, or nitrate, NO^{-3}. There are a number of different stages in the cycle (Rowell 1997; Fitzpatrick 1974):

*   Nitrogen fixation. The cycle is started by the conversion of nitrogen gas to organic nitrogen by bacteria, including *Azotobacter*, *Clostrium pasteurianium*, *Beijerinckia* and the *Leguminosae*

- Mineralization. Organic nitrogen is converted to mineral nitrogen by microbial activity

- Nitrification. Ammonium nitrogen is oxidized to nitrite by the microbes *Nitrobacter*, *Nitrosomonas* and *Nitrococcus*, and into nitrate by *Nitrobacter*

- Immobilization. When bacteria cannot get enough nitrogen from organic supplies, they convert mineral nitrogen to organic nitrogen

- Volatization. Conversion of ammonium ion to ammonia gas, which is then lost to the atmosphere

- Denitrification. Under aerobic conditions, bacteria respire by reducing nitrate and nitrite, converting them to nitrous oxide and nitrogen gas.

In addition to atmospheric fixation by bacterial action, ammonia nitrogen and nitrates are added to the soil by rainfall, fertilizers, manures and decaying vegetation, for example.

The soil carbon cycle

Globally, soils are a crucial component of the carbon cycle (Figure 2.3). They are a major holder of carbon, acting as either a source or sink depending on the status of the soils and the pressures upon them. Approximately 1.5×10^{12} tonnes of carbon (Hillel 1998) are present in global soils, second only to the ocean in carbon storage, although this value is destined to fluctuate with potential climate change. Decomposition of organic matter creates carbon dioxide, a major greenhouse gas. Anthropogenic actions that increase levels of carbon dioxide from the soil include the clearance of naturally vegetated areas and ploughing up the soil.

As carbon is a constituent of every living organism, the decomposition of organic matter and recycling of carbon is arguably one of the most important biogeochemical cycles. Photosynthesis converts atmospheric carbon dioxide and water to carbohydrate, which creates new plant growth. When the plants die, the vegetable matter returns to the soil as litter. The soil microfauna and microbial biomass attack the litter, starting the process of decomposition that eventually turns the waste into humus and nutrients. Of the carbonate-based nutrients, the processes of uptake, leaching and respiration eventually return carbon dioxide to the atmosphere.

2.4 THE ATMOSPHERE

The atmosphere can be viewed, to some extent, as a largely self-contained system. There are, however, some important interlinkages with other global systems that cannot be completely ignored, in particular with respect to fresh water through the hydrological cycle and the seas through ocean–atmosphere relationships. External agencies also influence the atmospheric system, most importantly the Sun, shortwave radiation from which drives many atmospheric processes. One of the important 'roles' of the atmosphere is to harness the energy provided by the Sun while shielding the planet from the harmful aspects of this energy. Energy from the Sun drives weather systems that operate on a range of different scales from micro to global. In very basic terms, **climate** refers to the long-term

climate: the long-term characteristics of the atmosphere in a particular geographic zone

characteristics of air in an area, whereas **weather** refers to more short-term characteristics. Climate can also be viewed in terms of a series of spatial scales so it is possible to study global average conditions as well as climates associated with particular zones and even microclimates associated with different land cover types.

Although the composition of air is remarkably constant, it is influenced by inputs from the land surface, from plants, animals, humans, water and land. Some of these inputs, or **emissions**, are entirely natural, such as **sulphur dioxide** from volcanic eruptions or **carbon dioxide** emissions as a result of plant and animal respiration, and are part of the reason that the atmosphere is as it is today. Other inputs are still from essentially natural sources but are modified through the activities of society, such as enhanced soil microbial activity giving rise to increased **nitrous oxide** emissions through the effect of adding fertilizer to agricultural land, or **methane** emissions from intensive animal rearing. Another set of sources relates to the release of emissions because of **fossil fuel** combustion, for example through emissions from power stations or road transport. Still other sources are entirely artificial, such as emissions of **HFCs** from industry, and would not occur if not for the influence of humans.

The contemporary atmosphere is therefore the result of a combination of emissions from all of these types of sources throughout history (with and without human influence), which over time have mixed laterally and vertically through the thin envelope of air surrounding the planet. Over Earth's history the concentrations of different constituents of the atmosphere have undergone some modification with a variety of impacts, such as on weather and climate, and plants, animals and humans have responded to these in different ways. These responses have in turn resulted in changes to the atmosphere so that, in essence, it is most useful to consider the atmosphere as facilitating a number of cycles with a number of interlinked processes, each with inputs, drivers and impacts. These are sometimes viewed in terms of specific chemical elements to help visualize and understand the processes that take place, e.g. the carbon or nitrogen cycles. As well as cycles associated with different chemical elements, the atmosphere has an important role in the hydrological cycle.

The rest of this section will provide an introduction to some of the fundamentals of the atmosphere in terms of its composition, its structure and some of the physical aspects that govern its behaviour.

AN OVERVIEW OF THE COMPOSITION OF THE ATMOSPHERE

The composition of the atmosphere is remarkably constant up to a height of around 80 km or more. The major components by volume are: nitrogen (~78 per cent), oxygen (~21 per cent), and **argon** (~1 per cent). Many more gases occur in very small proportions, including the **greenhouse gases** and **ozone** both of which are essential for sustaining life. The highest concentrations of ozone occur in the upper part of the atmosphere, where it is formed through the action of **ultraviolet radiation** (UV) on oxygen. The highest concentrations of ozone occur in a band between 15 and 35 km from the surface of the Earth and this 'ozone layer' has an important role in shielding the surface from harmful radiation that would otherwise destroy most if not all life on the planet (see p. 47). Greenhouse gases are also critical in that they allow incoming solar shortwave radiation through the atmosphere whilst 'trapping' (i.e. absorbing)

weather: short-term characteristics (temperature, pressure, humidity, clouds, precipitation, visibility and wind) of the atmosphere

emissions: the mass release of a substance (gas or particles) to the atmosphere

sulphur dioxide (SO_2): an acidifying gas most commonly produced by the anthropogenic combustion of materials containing trace amounts of sulphur or as a result of volcanic eruptions

carbon dioxide (CO_2): a greenhouse gas produced from a wide variety of natural (e.g. respiration) and human-made sources (e.g. combustion of carbon-based fuels)

nitrous oxide (N_2O): a greenhouse gas produced through natural processes which can be modified by humans (the activity of soil microbes) and human-made sources (e.g. industry)

methane (CH_4): a greenhouse gas produced largely through natural processes (the decomposition of materials) but which can be influenced by human activity

fossil fuels: carbon based deposits (coal, oil and gas) that have been developed over geological time scales and which produce heat and energy on combustion

HFCs (hydroflurocarbons): man-made compounds containing hydrogen, fluorine and carbon which are used for a variety of industrial processes

argon: one of the inert gases present in the atmosphere. Inert gases are very unreactive

greenhouse gases: gases which absorb outgoing heat from the Earth's atmosphere, thereby acting to warm the atmosphere

ozone (O_3): the highly reactive and unstable tri-atomic (O_3) form of molecular oxygen formed through the action of the Sun's energy on the more common di-atomic (O_2) form of molecular oxygen

ultraviolet radiation: shortwave radiation, around 200 to 400 nanometers wavelength, originating from the Sun

ozone layer: term used to refer to the relatively high concentration of ozone found in parts of the stratosphere. The characteristics of ozone prevent some of the Sun's harmful UV radiation reaching the Earth's surface

reactive: refers to unstable compounds that are prone to change when coming into contact with other constituents and conditions in the atmosphere

particulate matter: any solid particles or liquid droplets that are present in the atmosphere

troposphere: the part of the atmosphere which is in contact with the Earth's surface and which is characterized by a general fall off in temperature with height

stratosphere: the part of the atmosphere above the troposphere and which is characterized by increasing temperature with height

longwave radiation re-radiated from the Earth's surface and using this energy to make the atmosphere warm enough for life to exist. Other components of the atmosphere include **reactive** species, such as those associated with nitrogen and sulphur, and solid **particulate matter** (or aerosols). In order to provide insights into the characteristics of these groups, how they interact with the atmosphere and the impacts that are associated with this interaction, Chapter 8 provides an overview of concentrations and sources of some of the most influential constituents of the atmosphere.

A ROUGH GEOGRAPHY OF THE ATMOSPHERE

Since the atmosphere is largely transparent, it is difficult to visualize the patterns and processes that occur within it. It is also quite difficult to generalize about these patterns because of the atmosphere's incessant movement (see Chapter 8). However, there are some broad generalizations that can be made as a starting point and that help to set the scene for the more detailed discussion to follow. This section will focus on the vertical and horizontal (or more correctly latitudinal) characteristics of the atmosphere.

Vertical characteristics of the atmosphere

Since it rarely exhibits definite boundaries, to talk about the atmosphere as having a particular structure perhaps implies that it is possible to assign rather more precision to the vertical properties of the air than is actually the case. Nevertheless, the atmosphere is often conceptualized into a broad structure in order to communicate basic facts and it is common to see references to various layers or spheres as a function of height from the ground surface of Earth.

This idea of spheres is actually quite useful in terms of understanding the impact of humans on the atmosphere, since the same constituents in one part of the atmosphere may have 'positive' effects, whereas in another part of the atmosphere they may have 'negative' effects. The best example of this is how ozone in the lower atmosphere is considered a pollutant whereas ozone in the upper atmosphere is considered an essential constituent. In the lower atmosphere, or **troposphere**, ozone is considered an atmospheric pollutant because it harms human health and vegetation. However, it plays a critical protective role when in the **stratosphere** in shielding the Earth from harmful rays from space. This also illustrates the idea that 'pollution' is just matter in the wrong place.

Thompson (1998) calculates that the atmosphere technically ends at around 32,000 km from the Earth's surface. This estimate is based on the distance needed to balance the Earth's gravitational forces with forces caused by the Earth spinning. However, since the density of air reduces logarithmically with height, 99 per cent of the atmosphere is actually contained within a band that is only 32 km deep. Gravity ensures that the air is at its most dense at the Earth's surface. Atmospheric pressure at any particular point is dependent on the weight of air in the atmosphere above it, so pressures tend to be highest at the planetary surface. Pressure and density (mass divided by volume) both halve with every 5.6 km of vertical height from the surface. A number of laws have been developed to help in understanding these relationships (see Basics 2.3).

The structure of the atmosphere most commonly used relates to the temperature profile of the atmosphere with height. Figure 2.11 is a diagrammatic representation of this structure that is found in many introductory texts. It is

BASICS 2.3

LAWS OF ATMOSPHERIC PRESSURE, VOLUME AND TEMPERATURE

There are a number of fundamental laws that help the understanding of the relationship between the key parameters of the atmosphere. It is important to appreciate the relationships described here as a foundation for understanding atmospheric processes and some important environmental problems that link back to these, such as urban air pollution.

Boyle's law dictates that for a constant temperature the volume of a mass of gas varies inversely with its pressure. In other words, as pressure increases the space that a parcel of air takes up is reduced. Similarly, as pressure decreases the space that the same parcel of air takes up increases.

Charles's law adds to this by identifying a relationship between temperature and volume. This law states that at a constant pressure, the volume of a given mass of gas varies directly with its pressure. Therefore, as the temperature of a mass of air increases so will its volume, and vice versa.

Since density is mass per unit volume, the density of air can be therefore determined at any point as a function of temperature and pressure (since volume, temperature and pressure are interdependent).

Gravity ensures that pressure and density are related to height above sea level, so we can also relate these parameters to height.

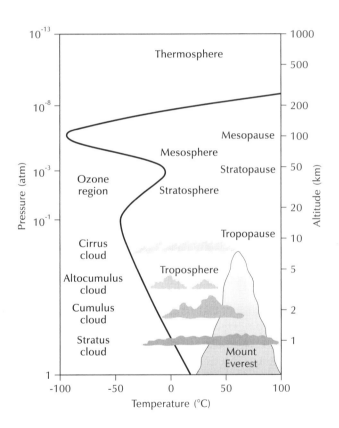

Figure 2.11 Conceptualized structure of the atmosphere. *Source:* After Huggett (1997)

very important to note that the actual profile of the atmosphere at any point of the globe will vary due to localized conditions and as a function of latitude. By referring to this diagram, it is possible to identify four general spheres and each will be examined in turn.

The troposphere

The troposphere is the lowest part of the atmosphere and is characterized by a gradual reduction of temperature with height. You can sometimes observe this during an aeroplane landing where you can see ice crystals that have formed during the trip melting on the outside of your window as you get lower and lower towards the landing strip. This falling temperature with height is the reason that it gets colder the further up a mountain you are and why it is possible to have snow on the top of a mountain even though the surrounding valleys are snow-free.

Many factors (for example, location, season, weather) influence the exact amount by which temperature reduces for a given height – in other words, the **environmental lapse rate** varies considerably from place to place and over time. The average environmental lapse rate is in the region of 6.5°C per km. Using the average environmental lapse rate as a guide, the general height of the troposphere can be taken as around 12 km (Lutgens and Tarbuck 1998). It is deepest over the equator, at around 18 km, and shallowest over the poles, at around 8 km (Thompson 1998). At these points, the normal reduction of temperature with height stops. A band then sees temperatures remaining unchanged with height (i.e. temperatures are constant or **isothermal**). This part of the atmosphere is termed the **tropopause**. Above this point, temperatures actually rise with height and this reversal of the temperature–height relationship is referred to as a **temperature inversion**. It is important to note that within the troposphere many transient temperature inversions can be formed – essentially dividing the troposphere into temporary layers – and some of the processes that cause this layering are discussed in greater detail later in the chapter since they are important in the build up of pollutant concentrations and **air pollution episodes**.

Three quarters of the mass of the atmosphere is contained within the troposphere zone (largely due to the effects of gravity). This is where the movement of the atmosphere is the greatest (turbulence or wind) and where almost all the water vapour occurs (and therefore clouds and precipitation). By extension, this is where most weather happens. It is very difficult for air from the troposphere to get any higher than the tropopause. Indeed, many of the pollutant cycles discussed earlier in this chapter only really operate in this lowest sphere. If this was entirely the case though, there wouldn't be a problem of stratospheric ozone depletion, so obviously some air must in fact be forced upwards through other means than the basic laws of warm air rising.

The stratosphere

Above the tropopause, temperature then begins rising again. This part of the atmosphere is called the stratosphere. It extends from the top of the troposphere for about 50 km (although this does again vary). This band is where the highest concentrations of ozone are found. (See Basics 2.4 for a description of ozone, its characteristics and its role in the atmosphere.) As can be seen in Basics 2.4, the processes that form ozone lead to rising temperatures once again in the stratosphere. Eventually, however, ozone concentrations reduce to a point where the overall temperature of the atmosphere is no longer rising. Here, there is another isothermal band, this time termed the stratopause, before the next sphere is reached.

The upper atmosphere – mesosphere and thermosphere

Above the stratosphere, the atmosphere begins to get very thin. It contains approximately the same proportion of oxygen and nitrogen but there are far fewer molecules of both present in any volume of air. The top of the mesosphere,

---

environmental lapse rate: the rate of temperature change with height observed in the real atmosphere

isothermal: refers to a condition where thermal properties have a constant value

tropopause: the point of the troposphere at which temperatures no longer reduce with increasing height, the 'top' of the troposphere

temperature inversion: where temperatures increase with height rather than reduce with height, a characteristic which inhibits any further upward vertical movement of air

air pollution episodes: a term used to describe a period, usually short-lived, of very high concentrations of pollutants

oxides of nitrogen: a collective term used to group two species of oxidized nitrogen, nitric oxide (NO) and nitrogen dioxide (NO_2), the former is rapidly oxidized to the latter in the atmosphere

ozone 'hole': an area of reduced concentration of ozone in the stratosphere which varies from place to place and time of the year

BASICS 2.4

STRATOSPHERIC OZONE AND ITS CHARACTERISTICS AND ROLE IN THE ATMOSPHERE

This box outlines some of the key aspects of ozone, specifically in relation to its role in the stratosphere. There are other aspects of ozone that are of interest in the troposphere, and these are discussed together with the other constituents of the lower atmosphere that can be considered as pollutants (see Chapter 8). Stratospheric ozone is important because ozone 'shields' the Earth and life on the Earth by preventing the penetration of much of the ultraviolet radiation (UV) that reaches the atmosphere from the Sun. Without the influence of ozone, these rays would cause cell mutations, affect plant photosynthesis and destroy phytoplankton – the basis of the food chain in the oceans.

Ozone formation and destruction

The area of the atmosphere in which most ozone (O_3) is concentrated extends from around 15 km to around 35 km from the Earth's surface (Barry and Chorley 2003) with the highest concentrations occurring 16 km to 25 km (Thompson 1998). Ozone is formed through the action of the Sun's energy on oxygen molecules. This is the case whether the ozone is formed in the troposphere (where less radiation penetrates) or higher up in the atmosphere (where there is more energy available for its formation). Ultraviolet radiation (in particular relatively shortwave UV energy) breaks the bonds between the two atoms in the diatomic form of molecular oxygen (O_2) creating two free oxygen atoms (O) as a result. This can be expressed in the following equation [1]:

$$O_2 + UV \rightarrow O + O \qquad [1]$$

The free (or odd) oxygen (O) atoms then encounter other O_2 molecules and bond with them to create ozone – a process catalysed (speeded up) by the presence of some other molecule (termed M in the equation below [2]).

$$O + O_2 + M \rightarrow O_3 + M \qquad [2]$$

In most cases, this M is nitrogen. The M molecule takes on the extra energy that the free ozone atom had from the previous reaction, and therefore provides the energy balance in the reaction. The extra energy gives the M molecule kinetic energy and it therefore has a warming effect. The energy caused by these reactions is the reason why temperature increases with height in the stratosphere.

However, the ozone that is formed is not stable and there are processes that destroy ozone too. These include photochemical (i.e. involving solar radiation) interactions with: molecular oxygen (above 40 km), **oxides of nitrogen** NO_x (20–40 km) and the hydrogen–oxygen radical (below 20 km), as well as breakdown due to the action of ultraviolet radiation at slightly longer wavelengths than those which cause its formation. The actual cycles that occur between all of these different components are very involved. There is a natural equilibrium reached at around 40 km height although, as has been stated, the maximum concentrations are found some 15 to 25 km lower. Other important mechanisms for ozone destruction involve reactive chlorine and bromine. These processes are problematic as they actually remove the free oxygen atoms (as do the nitrogen oxide processes) so there is a net loss in the atmosphere, and they have severely disrupted the natural cycles.

Trends in stratospheric ozone concentrations

Monitoring of concentrations over the last 15 years suggests an average loss of ozone of around 1 per cent per annum. Thompson (1998) cites estimates in the region of 4 per cent as the anticipated rise in skin cancers in response to this loss of stratospheric ozone. Measurements of CFCs have confirmed the parallel rise in concentrations of these compounds in the atmosphere, during the same period as the observed fall in concentrations of ozone. There are some important natural sources of chlorine (volcanoes and forest fires) but the increases in concentrations are considered too large to be a result of natural sources alone. The presence of **ozone 'holes'** has since been confirmed in both polar regions.

Controls

CFCs have long residence times in the atmosphere (between 75 and 100 years), so even though international agreements banning their use are now well established it will be a long time before the process actually stops. Of course, consumer demands for the products that used these substances has not reduced (in fact, it has increased) so other substances have replaced them. While these substances are not as harmful in terms of ozone depletion, it is very likely that they will have other environmental consequences. For example, the replacement of CFCs by HCFCs is problematic since these compounds still contain chlorine atoms, their replacement with HFCs are then problematic in terms of another major global environmental concern – climate change. Another issue that has received attention of late is the potential influence of aircraft on the stratosphere.

(For more information see Thompson (1998) and Barry and Chorley (2003).)

> **thermosphere**: the outermost layer of the atmosphere, separated from the mesosphere by the mesopause. It sits at about 80 to 550 km above the Earth's surface and extends to around 10,000 km, where it merges into space. Temperatures rise with increasing altitude in the thermosphere to over 1000 °C

at around 80 km altitude is the coldest part of the atmosphere and here average temperatures drop as low as –90°C (Ahrens 1998; Lutgens and Tarbuck 1998). Shortly after 80 km we reach the mesopause, again a section of constant temperature change with height. Here the few remaining molecules are heated through coming into contact with very shortwave solar radiation. Temperatures in this part of the atmosphere, the **thermosphere**, are very high, sometimes in excess of 1000°C. However, since temperature is the measure of the speed of molecules (see Basics 2.5) this is actually misleading when compared to temperatures at the Earth's surface. In such a thin atmosphere there are so few molecules present in the air that objects within the atmosphere would not feel hot. Indeed, orbiting satellites gain far more heat because of absorbing solar radiation than they do from the temperature of the thin atmosphere surrounding them (Lutgens and Tarbuck 1998).

BASICS 2.5

TEMPERATURE AND ENERGY

The temperature of a parcel of air is a measure of the average speed of the molecules contained within it – it is therefore a measure of energy. Molecules in warm air are on average moving more quickly and have more energy than molecules in cold air. The same is true of molecules in water undergoing heating. Under normal environmental temperatures, water exists as a liquid. If we boil water in a pan, the temperature rises, as does the average speed of the molecules. Eventually, some gain enough speed to 'escape' from the rest of the molecules and turn into a vapour (which we see as steam). As the temperature rises to 100°C, the boiling point of water, the average speed of the molecules is such that they can all have enough energy to form steam, and given enough time there would be no molecules left as water. Since the evaporated molecules are able to escape into your kitchen, the surface of the air nearest the water is able to hold further molecules and the process of evaporation continues. This evaporation, in a less extreme form of course, is part of the hydrological cycle which helps move water through the atmosphere from place to place. Evaporation occurs when the Sun, or some

external energy force, heats the surface of water features.

The opposite process is condensation, where the water molecules in cooling vapour lose speed and energy and eventually return to the liquid phase. This occurs when steam from boiling water hits a cool surface like a window or mirror and turns back to water (so the window becomes steamed up). This condensation process is another important part of the hydrological cycle as this enables water from the atmosphere to be returned back to the ground surface through precipitation. If we were to cover the pan of boiling water, we would eventually cause the trapped air to be saturated with water vapour so that for every molecule evaporated (converting from liquid to vapour) another would be condensed (converting from vapour to liquid).

This conversion of water from the liquid to the gas phase transfers energy. The same sort of energy transfer also takes place when water passes from the solid to the liquid phase. The change of the physical state of water, therefore, always involves an exchange of heat.

LATITUDINAL CHARACTERISTICS OF THE ATMOSPHERE

The previous section has given some indication of the differences in the atmosphere that are to be expected with height, such as with temperature through the troposphere or concentrations of ozone at different parts of the stratosphere. It has been mentioned that the vertical profile of the atmosphere is very variable, and the discussion has touched on some of the latitudinal differences to be expected in the vertical profile of the atmosphere. The vertical characteristics are therefore one aspect of this 'rough geography' of the atmosphere.

If attempting a vertical classification is difficult, it is even more difficult to generalize about 'horizontal' patterns. The characteristics of the air at any one point in space on the Earth's surface will be a function of many influences; for example, the type of surface underneath the air, both now and in the days before now. Air behaves differently, has different characteristics over land than over sea, and is influenced by the different types of land cover it has travelled over. These influences are not directly related to latitude or longitude, but if we consider latitude alone (i.e. the distance from the equator to each of the poles) we can come up with some useful generalizations as a starting point.

One particular feature that does exhibit a clear latitudinal influence, and which is very important in appreciating global processes, is solar energy input or the amount of **solar radiation** received. The amount of energy received by the atmosphere at any point depends approximately on latitude. The Sun is the ultimate energy source for most processes on the planet and the atmosphere is no exception. It emits energy in various forms and some of this energy meets the Earth and the Earth's atmosphere. The amount of energy provided by the Sun at the top of the atmosphere (at a point perpendicular to the solar rays and with distance from the Sun averaged) has been measured by satellite to be in the order of 1368 W m^{-2} and this is known as the solar constant. The amount of energy received at any point on the Earth's surface, though, depends on the season, time of day, degree of cloudiness and many other factors. However, all other things being equal, this is at a maximum at the equator and a minimum at the poles (Basics 2.6).

Related to solar energy input, it is also possible to characterize global pressure patterns in terms of latitude. Figure 2.12 presents a schematic representation of the generalized pattern of pressure with latitude. The equator is marked as an area of thermal low pressure, the intertropical convergence zone. Flanked on either side of this zone are two areas of relatively high pressure, approximately centred on 30° latitude north and south. These in turn are bordered by sub-polar low-pressure areas along the polar front which separate high-pressure areas associated with each of the polar regions. The actual positions of all of these zones vary in response to the relative position of the Sun. As Figure 2.12 suggests, the land makes an important contribution to the real movement of air as do the world's oceans. Chapter 9 examines the issue of atmospheric motion in more detail.

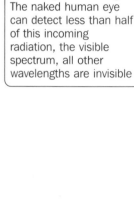

solar radiation: light energy emitted by the Sun, which is by convention grouped according to wavelength. The naked human eye can detect less than half of this incoming radiation, the visible spectrum, all other wavelengths are invisible

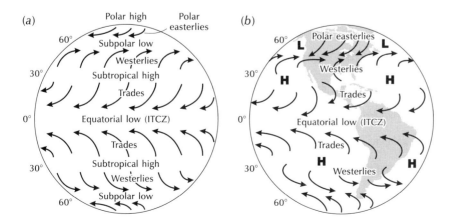

Figure 2.12 Schematic representation of global pressure distribution (a) without the influence of land and (b) with the influence of land. *Source:* After Lutgens and Tarbuck (1998)

BASICS 2.6

SOLAR RADIATION DIFFERENCES AS THE DRIVER OF SEASONAL PATTERNS

In the northern hemisphere (the northern part of the world from 0 to 90° latitude) (see Chapter 7) the maximum solar radiation for the entire year is received on 21 June at noon (the summer solstice). This is the time that the Sun's rays hit the surface at the most direct angle, so covering the least area. The Tropic of Cancer, which lies 23.5° north of the equator, is the location where the Sun is directly overhead. In the southern hemisphere, this time marks the winter solstice or the day of least daylight and solar energy. The Tropic of Capricorn, which is 23.5° south of the Equator, is the location at which the Sun is directly overhead during the summer solstice in the southern hemisphere (21 December) which is then, by extension, the shortest day in the northern hemisphere.

The difference in quantity of daylight hours, the most important driver of seasonality, associated with different parts of the globe on the summer and winter solstice (21 June and 21 December, respectively), can be summarized as below (Ahrens 1998). The examples that are cited as a general guide to locations, represented by 10° increments in latitude, are taken from the northern hemisphere.

- *The Equator – 0° latitude* (for example, northern Indonesia, northern Brazil, Ecuador) – 12 hours on both dates, in fact 12 hours all year round

- *10° latitude* (for example, southern Sudan, southern Vietnam) – 12.6 hours for the summer solstice and 11.4 hours for the winter solstice

- *20° latitude* (for example, southern India, southern Mexico) – 13.9 hours for the summer solstice and 10.8 hours for the winter solstice

- *30° latitude* (for example, Mexico–US border, central China) – 13.2 hours for the summer solstice and 10.1 hours for the winter solstice)

- *40° latitude* (for example, central US, northern China) 14.9 hours for the summer solstice and 9.1 hours for the winter solstice)

- *50° latitude* (for example, central Europe (France, Germany, southern UK), Canada–US border) – 16.3 hours for the summer solstice and 7.7 hours for the winter solstice)

- *60° latitude* (for example, northern Europe (southern Scandinavia), central Russia, central Canada) – 18.4 hours for the summer solstice and 5.6 hours for the winter solstice)

- *70° latitude* (for example, northern Russia, northern Canada) – 24 hours on the summer solstice, in fact 2 months of continuous daylight for the summer solstice and 0 hours for the winter solstice

- *80° latitude* (for example, Greenland and the far Northern reaches of Canada and Russia) – 24 hours on the summer solstice, in fact 4 months of continuous daylight for the summer solstice and 0 hours on the winter solstice

- *The North Pole – 90° latitude* – 24 hours on the summer solstice, in fact 6 months of continuous daylight for the summer solstice and 0 hours for the winter solstice

---

> **air mass**: a large block of air exhibiting similar characteristics over a relatively long time period and which tend to be characteristic of certain geographical areas

The distribution of the globe's main air masses are related to latitude and the amount of solar radiation received at a particular point. **Air masses** are characterized as bodies of air that have specific characteristics that are typical throughout a zone (especially vertically). Although it should be recognized that the usefulness of the idea of air masses has been criticized by some, it is used here as a means of making generalizations about the broad nature of the atmosphere, which the reader can develop in further reading around the subject. This said, air masses can be categorized geographically in terms of their source areas:

- *Polar continental (Pc).* Very cold, dry and stable, exhibiting frequent temperature inversions, sourced from air over the polar regions and high and mid-latitude land masses in winter. Areas influenced by polar continental air masses include Canada and Siberia

- *Polar maritime (Pm).* Cool, moist, unstable air that forms over high latitude oceans such as the North West Atlantic

- *Tropical continental (Tc)*. Hot, dry, unstable air from major subtropical deserts and mid-latitude landmasses in summer and warm, dry, more stable air in winter

- *Tropical maritime (Tm)*. Warm, moist, unstable air in the west and cooler, moist, more stable air in the east because of the influence of warm currents and cold currents respectively in the subtropical oceans.

It is difficult to get a clear idea about how and why these air masses exist and what they mean for key environmental issues without first slotting another piece of the puzzle in place. The missing consideration, which is alluded to in the characterization of air masses, is the role of atmospheric motion. Chapter 8 will describe the key processes and drivers of atmospheric motion. It is only through doing this that a full appreciation of the system can be found.

A DEFINITION OF WEATHER AND CLIMATE

Weather can be summarized as the particular characteristics of seven main parameters: air temperature, air pressure, humidity, clouds, precipitation, visibility and wind, at any point in time or space (Ahrens 1998). These parameters are constantly changing. For example if you stand and observe a weathervane on a breezy day you will see that the direction of the wind fluctuates and the wind can be more gusty at one point than another. Nevertheless, you would be able to generalize and say, for example, that the wind is westerly (from the west) or that it is gentle or strong. Since traditional weather-station measurements can only be made at a finite point of space and time then these too are generalizations. In view of the need to capture large-scale processes over extensive areas, satellite imagery (see Chapter 6) has become an increasingly important tool to help express, understand and predict weather.

In a particular area, the weather, expressed by the parameters identified above, tends to exhibit similar characteristics year on year. People in the UK often refer to 'good' summers (relatively warm and dry) and 'bad' summers (relatively cool and wet) and both 'types' of summers accurately reflect the overall climate of an area since this can be seen as a long-term averaging out of all the characteristics (including extremes). The climate of an area therefore gives generalized estimates for temperature, rainfall and so on for a particular period such as a season. For other parts of the world, seasonal patterns are completely different and the nature (or indeed the existence) of distinct seasonal weather depends upon many factors, such as geographical location, presence of oceans or water bodies, topographic features, elevation and land cover. Some of these factors lead to very localized climatic effects which have a localized yet distinct influence. One type of localized climate is the urban heat island (see Chapter 14). Basics 2.7 provides an overview of the world's broad climatic regions. Climate is not a static phenomenon either; Chapter 8, as well as many other parts of this book touch on the issue of climate change.

BASICS 2.7

CLIMATE CLASSIFICATION – THE KÖPPEN SYSTEM

Since there are so many determinants of climate, creating a simple classification of climate types is not a straightforward task. However, one of the mechanisms for doing this is the well-established and much revised system devised by the German climatologist, Vladimir Peter Köppen. This system categorizes climate according to relationships with vegetation types, recognizing five chief types denoted by the letters A to E:

A. *Tropical moist climates.* Areas that are constantly warm with average temperatures over 18°C and little seasonality

B. *Dry climates.* Areas where potential **evapotranspiration** exceeds precipitation

C. *Moist mid-latitude climates with mild winters.* Areas exhibiting marked cold and warm periods and having winter temperatures generally in the range between −3 and 18°C

D. *Moist mid-latitude climates with severe winters.* Areas exhibiting marked cold and warm periods and having winter temperatures generally below −3°C

E. *Polar climates.* Areas which are constantly cold with average temperatures in the warmest months less than 10°C and little seasonality.

The fact that this can be done with any success at all nicely illustrates the strong interconnections between vegetation and climate.

SUMMARY

> **evapotranspiration**: evaporation plus the water discharged into the atmosphere by plant transpiration

The ecosphere is formed by the interaction of life (the biosphere), water (the hydrosphere), soil (the pedosphere) and air (the atmosphere). The biosphere consists of all life on Earth. Individual organisms interact with each other to form communities. Globally, nine basic community types correspond to nine climatic zones. Individuals and communities interact with their life-supporting environment – air, water and soil – to form ecosystems. Organisms play different roles in ecosystems. Some are producers or autotrophs, others are consumers or heterotrophs. Yet others are decomposers and detritivores that feed on the waste-products of life. The uptake and release of biochemicals by living things creates biogeochemical cycles, examples of which are the carbon and nitrogen cycles. The hydrosphere is the totality of the Earth's waters – the seas and oceans, rivers and lakes, ice caps and ice sheets, glaciers and snowfields and all other stores of water. The hydrological cycles link these water bodies through the process of evaporation, condensation, precipitation and runoff. The soil (pedosphere) is a complex organic and inorganic system formed by the interaction of air, water, rocks and life. Its nature varies from place to place owing to differences in climate, organisms, topography and substrate. The atmosphere is a highly mobile shell of dusty gas. Its composition is unique among terrestrial planets due to the action of organisms. It has both a vertical and horizontal structure. Atmospheric processes produce the weather and climate experienced by other systems of the ecosphere.

Questions

1 What is an ecosystem? To what extent have humans altered the world's natural ecosystems?
2 How cyclical are biogeochemical cycles?
3 Why do evaporation and precipitation vary from place to place?
4 How important is soil within biogeochemical cycles?
5 How important is atmospheric composition in understanding climatic change?

Further Reading

Arnell, N. (2002) *Hydrology and Global Environmental Change*. Harlow: Prentice Hall.
An excellent text on water. Particularly useful because it is up-to-date and takes a global perspective.

Barry, R. G. and Chorley, R. J. (2003) *Atmosphere, Weather and Climate*, 8th edn. London: Routledge.
Deservedly known as a classic and enormously popular. An excellent introduction to the subject.

Bridges, E. M. (1997) *World Soils*, 3rd edn. Cambridge: Cambridge University Press.
A good starting point for studying soil and soil processes.

Davie, T. J. A. (2002) *Fundamentals of Hydrology*. London: Routledge.
An accessible introduction to the study of hydrological processes.

Huggett, R. J. (2004) *Fundamentals of Biogeography*, 2nd edn. London: Routledge.
An introduction to past, present and future individuals, communities and ecosystems.

Useful Websites

http://www.gesource.ac.uk/home.html
GEsource is a good starting point for finding websites dealing with geography and the environment.

Subject-specific websites are mentioned in later chapters.

CHAPTER 3
THE HUMAN SPHERE

OVERVIEW

Human development of the planet has led to a complex of associated problems that may be characterized as 'syndromes'. Syndromes are standard patterns of people–environment interactions that are recognizable around the world. They fall into three groups: utilization syndromes, development syndromes and sink syndromes. Utilization syndromes concern the use or misuse of natural resources. Development syndromes involve the non-sustainable conversion of land cover to agricultural, urban and other human uses. Sink syndromes involve environmental degradation through accidental spillages of natural and human-manufactured substances and the uncontrolled disposal of waste products.

LEARNING OUTCOMES

This chapter will help you understand:
- The nature of a syndrome approach to human-induced environmental problems
- Utilization syndromes resulting from the inappropriate use of resources
- Development syndromes associated with the non-sustainable conversion of land to human uses
- Sink syndromes leading to environmental degradation and involving accidents and the waste products of human activities.

3.1 A SYNDROME APPROACH

The human population is the number one environmental problem: current global change results ultimately from mushrooming population growth (p. 6). Scientists study global change in three chief ways. First, they conduct research within conventional disciplinary boundaries in an effort to identify and explain basic processes within the components of the Earth system. Such research includes studies of processes in ecosystems, studies of human population change and studies of economic development. Second, scientists from several disciplines collaborate to model processes that aid understanding of such areas as climatic change and vegetation change at a global scale. An example is the IMAGE model mentioned in Chapter 1 (p. 15). Third, they undertake field investigations of local or regional human–environment interactions, the results of which sometimes inform work on environmental processes at continental and global scales.

All three approaches add to an ever-growing database of information on aspects of global change. However, they do not reveal much about the underlying structures and mechanisms of global change. Some scientists adopt a fourth approach that helps to reveal functional units that together define the phenomenon of global change. This innovative approach uses syndromes. In this context, syndromes are archetypal patterns of people–environment interactions for describing global change. Table 3.1 lists sixteen syndromes grouped into three categories: utilization syndromes, development syndromes and sink syndromes. Notice that the syndrome names allude either to the regions where they characteristically occur or to their distinguishing features. To illustrate the basic structures and mechanisms in the human sphere that drive global change, this chapter discusses a selection of syndromes in each category. Some other syndromes receive attention elsewhere in the text.

3.2 UTILIZATION SYNDROMES

Utilization syndromes are associated with the unwise use of resources. The Mass Tourism and Scorched Earth syndromes will serve as good exemplars.

MASS TOURISM

The first 'tourists' were rich people with enough time and money to travel. During the last century, innovations in transport and changes in society turned a travelling élite into mass tourists. Busy modern airports like Manchester attest to the enormity of the change. A defining feature of mass tourism is that many people visit small areas, often at beach, coral reef, coastal, mountain and wetland locations. Increasingly, mass tourism seeks exotic locations, such as Bali, where the cultural and natural environments tend to be even more sensitive.

Mass tourism may have substantial impacts on the cultural and natural environments, in some cases altering local areas almost beyond recognition. The changes mainly result from the huge sums of money pumped into an area and the construction of mass tourist facilities. Mass tourism demands extensive accommodation units and services (such as water, power and sanitation services) and associated infrastructures – airports, access roads and the like. Mass tourism has both negative and positive impacts on the local environments (Table 3.2).

Table 3.1 Some syndromes of global change

Syndrome	Description	Example
Utilization syndromes		
Sahel syndrome	Overuse of marginal land	Sahel region, Africa
Overexploitation syndrome	Overexploitation of ecosystem resources	North American bison; sea otter; great auk
Rural Exodus syndrome	Environmental degradation through the abandonment of traditional agricultural practices	Karakoram Mountains, northern Pakistan
Dust Bowl syndrome	Non-sustainable agricultural and industrial use of soils and water bodies	The Dust Bowl in the American mid-west
Katanga syndrome	Environmental degradation through the depletion of non-renewable resources	Mount Etna; central Queensland, Australia
Mass Tourism syndrome	Development and destruction of nature for recreational purposes	Damage caused by cave tourism
Scorched Earth syndrome	Environmental destruction through war and military action	Kosovo; Gulf Wars
Development syndromes		
Aral Sea syndrome	Damage of natural landscapes through large-scale development projects	Shrinking of the Aral Sea
Green Revolution syndrome	Environmental degradation through the use of inappropriate farming methods	DDT application
Asian Tigers syndrome	Disregard for environmental standards in the race for rapid economic growth	Extinction of tiger subspecies and threatened status of remaining tigers in Asia
Favela syndrome	Environmental degradation through uncontrolled urban growth	Karachi; Cairo; São Paulo and Calcutta
Urban Sprawl syndrome	Landscape destruction through the planned expansion of urban infrastructures	Los Angeles
Sink syndromes		
Disaster syndrome	One-off, human-induced environmental disaster with long-term impacts	Minamata disease; Chernobyl
Smokestack syndrome	Environmental degradation through large-scale diffusion of persistent substances	DDT and PCBs in Arctic
Waste dumping syndrome	Environmental degradation through controlled and uncontrolled waste disposal	Disposal of spent nuclear fuel
Contaminated Land syndrome	Local contamination of the environment around industrial sites	Saxony-Anhalt in Germany; Manchester–Liverpool–Birmingham triangle in the UK; Pittsburg in the USA

Source: Adapted from Petschel-Held *et al.* (1999)

Cave tourism started in the late eighteenth and early nineteenth centuries in Europe when tourists used candle lanterns to view the 'wonders of the deep'. Today, cave tourism is a growth industry. Fibre-optic lights illuminate some caves and electric trains transport tourists round the caverns. Tourism has an injurious impact on caves. To combat the problems of cave tourism, cave management has evolved and a body of government and private professionals prosecute it. Several international groups are active in cave management: the International Union of Speleology, the International Speleology Heritage Association, the International Geographical Union, the Commission for

Table 3.2 Good and bad points of mass tourism

Area of impact	Good points	Bad points
Local communities	Employment opportunities, often involving the acquisition of new skills	Loss of local traditions and culture
	Boosts community pride by giving new focus to an area (local scenery, heritage and culture) that was previously taken for granted	
	Possible improvement in quality of life	
Industries	Secondary industries attracted to area providing employment opportunities for local people	May attract unsavoury secondary industries (e.g. sex trade)
	Encourages economic growth	
Services	Provision of new services – waste disposal facilities, sewage farms, clean water supplies and so on – with concomitant health benefits	Overcrowding, littering and other problems may result from services overstretched by huge numbers of tourists
Infrastructure (new airports and roads)	Opens up area to the wider world	Increased habitat loss, increased noise levels, decreased air quality, traffic congestion and an increased risk of accidents
Natural resources and biodiversity	Money may provide means for developing protective measures for areas of outstanding natural beauty to ensure their long-term conservation	Possible contamination of land and water from accidental spills, waste disposal, storage of hazardous materials
		Soil erosion resulting from overexploitation of natural resources
		Loss of biodiversity due to habitat destruction (e.g. loss of coral reefs through overuse) and fragmentation, and contamination of water

National Parks and Protected Areas, and the International Union for the Conservation of Nature and Natural Resources (IUCN) (Focus 3.1).

SCORCHED EARTH

In the past, military activities have viewed the environment as either an obstacle to overcome or a resource to exploit as a weapon to assist the war. More recently, with the increased awareness of human activities upon the Earth, the impact of war and military action upon the environment is coming under scrutiny.

WGBU (1996) term the impact of military action upon the environment as Scorched Earth syndrome, defined as 'environmental destruction through war and military action' and state that the adoption of scorched earth policies by military planners can have 'disastrous consequences for humanity and nature at local level, and may indeed have severe global impacts'. Typical symptoms include the loss of biodiversity due to chemical warfare (e.g. agent orange), permanent soil degradation due to mining, contamination caused by fuels and explosives, health hazards (for example, the inhalation of gaseous pollutants and particulate matter from burning oil wells) and greater flows of refugees.

It is easy to appreciate the devastating effects that war can impart, with the bombing of infrastructure such as chemical and industrial plants, and the use of heavy artillery, missiles and weapons loaded with chemicals for example. Strong visual images and media footage of devastated areas, plumes of smoke

FOCUS 3.1

CAVE TOURISM

Some 20 million people visit caves every year. Mammoth Cave in Kentucky, USA, alone has 2 million visitors annually. Great Britain has some 20 show-caves, with the most-visited receiving over 500,000 visitors every year. About 650 caves around the world have lighting systems, and many others offer 'wild' cave tours where visitors carry their own lamps. Tourists damage caves directly and indirectly through the infrastructure built for the tourists' convenience – car parking areas, entrance structures, paths, kiosks, toilets and hotels. The infrastructure can lead to hydrological changes within the cave systems. Land surfaced with concrete or bitumen is far less permeable than natural karst and the danger is a drastic reduction or even stoppage of the feedwaters for stalactites. Similarly, drains may alter water-flow patterns and lead to changes in **speleothem** deposition. The use of gravel-surfaced car parks and paths, or the inclusion of strips where infiltration may occur, go some way towards alleviating drainage problems. Within caves, paths and stairs may alter the flow of water. Impermeable surfaces made of concrete or steel may divert natural water movement away from flowstones or stream channels, so leading to the drying out of cave formations or increased sediment transport. These problems are in part overcome by the use of permeable steel, wooden or aluminum walkways, frequent drains leading to sediment traps, and small barriers to water movement that approximate the natural flow of water in caves.

Cave tourists alter the cave atmosphere by exhaling carbon dioxide in respiration, by their body heat and by the heat produced by cave lighting. A party of tourists may raise carbon dioxide levels in caves by 200 per cent or more. One person releases between 82 and 116 watts of heat, roughly equivalent to a single incandescent light bulb, which may raise air temperatures by up to 3°C. A party of tourists in Altamira Cave, Spain, increased air temperature by 2°C, trebled the carbon dioxide content from 0.4 per cent to 1.2 per cent, and reduced the relative humidity from 90 per cent to 75 per cent. All these changes led to widespread flaking of the cave walls, which affected the prehistoric wall paintings (Gillieson 1996, 242). A prolonged increase in carbon dioxide levels in caves can upset the equilibria of speleothems and result in solution, especially in poorly ventilated caves with low concentrations of the calcium ion in drip water (Baker and Genty 1998). Other reported effects of cave tourism include the colonization of green plants (mainly algae, mosses and ferns) around continuous lighting, which is known as lampenflora, and a layer of dust on speleothems (lint from clothing, dead skin cells, fungal spores, insects and inorganic material). The cleaning of cave formations removes the dust and lampenflora but also damages the speleothems. A partial solution is to provide plastic mesh walkways at cave entrances and for tourists to wear protective clothing. Recreational cavers may also adversely affect caves (Gillieson 1996, 246–47). They do so in several ways:

- dumping carbide and marking the walls
- compacting sediments with resultant effects on cave hydrology and fauna
- eroding rock surfaces in ladder and rope grooves and direct lowering by foot traffic
- introducing energy sources from mud on clothes and foot residues
- depositing faeces and urine
- widening entrances and passages by traffic or by digging
- performing cave vandalism and writing graffiti.

The best way of limiting the impact of cave users is through education and the development of minimal impact codes, which follow cave management plans drawn up by speleologists, to ensure responsible conduct (see Glasser and Barber 1995).

> **speleothem**: a mineral deposit formed by the precipitation of calcium carbonate from the solution passing through caves: stalactites and stalagmites are examples of speleothems

from bombed factories and oil slicks, to name a few, create the impression that the effects of war exceed previously known levels of devastation on every scale.

Lanier-Graham (1993) classifies the impact of war into three categories:

1. Intentional direct destruction of the environment, e.g. firing of oil wells during the Gulf War in 1991, or destruction of cropland by herbicides

in the Vietnam War. This can be done by any party of war for a variety of reasons, such as the dehabilitatation of the population by destroying food crops and polluting soil and water supplies, draining wetlands, breaking dams to flood land to repel invaders, and to hold environmental resources or technologies (e.g. nuclear power stations) to ransom.

2.   Incidental direct destruction, e.g. digging of trenches, soil disturbance by heavy vehicles and troop movements, landmines or bombardments that could disturb surface soil crusts allowing erosion and increased dust storms.

3.   Indirect or induced destruction as a medium- or long-term consequence of war but still attributable to war. For example, the reduction or extinction of species due to the destruction of habitats, and the migration of people away from the conflict zone may exert environmental stresses elsewhere, or the presence of persistent toxic chemicals at former military sites that contaminate soil and water resources for many years afterwards.

Gulf War, 1991

The impact of the 1991 Gulf War on the environment has been widely studied (Focus 3.2). The effects of repeated military action in the same region in 2003 will undoubtedly compound the impact of the 1991 conflict.

Is war the greatest threat to the environment?

Brauer (2002) comments that it is often difficult to make definite conclusions as to the impact and severity of environmental destruction caused by war compared to other forces. This is due to the general lack of baseline environmental data before the war, and the post-war difficulties of collecting data, such as accessing affected areas and getting funds. Often, early assessments can only provide basic information or vague estimates of, for example, the volume of chemical spilled, the number of birds oiled and so on. Worst-case scenarios are often proposed.

There is debate over the severity and long-term consequences of environmental damage caused by war when compared to peacetime activities. It is clear that war can have a devastating effect upon the environment, but there is also a growing realization within the literature that the most damage occurs, not through war and military actions, but through the normal global activities of industry and business.

WGBU (1996) highlight the fact that the syndromes interlink, with each syndrome having a positive or negative impact on others. For example, the impact of military action upon the environment can cause migration of people away from their homes. Many people migrate to urban settlements potentially aggravating the Urban Sprawl syndrome by increasing population density. Others concentrate in refugee camps causing localized environmental damage, or cultivate new areas of land by ploughing new areas or clearing forests, potentially creating the Dust Bowl syndrome. In addition, the exclusion of human activities

FOCUS 3.2

THE 1991 GULF WAR

The Gulf is particularly sensitive to pollution as it is one of the most productive water bodies in the world, with warm clear water, and an average depth of only 35 metres (Randolph *et al*. 1998). Coral reefs, salt marshes, mudflats and mangroves provide habitats for many organisms, and support a highly valuable fisheries industry as well as thirty-three Ramsar wetland sites. Husain (1998) notes that the movement of troops and vehicles, and the release of potentially toxic materials by ammunitions caused damage to the terrestrial environment and land and sea mines. In addition, the torching of over 700 wells in the Kuwait oil fields and the spill of between 1.0 and 1.5 Tg (teragrams: see Appendix 1) of oil into the Arabian Gulf caused the most obvious environmental problems. This oil spill has been termed 'the world's largest' (Randolph *et al*. 1998).

The release of approximately 819 metric tonnes of oil from tankers and terminals created this spill. About half of the oil evaporated, 136 metric tonnes were contained inland in dug pits, and 123 metric tonnes washed along the shoreline from Kuwait to the Abu Ali peninsula in Saudi Arabia, where damage to habitats seemed irreparable. Randolph *et al*. report that studies performed to assess the level of pollution after the event show that 30 months after the spill in January 1991, levels of petroleum were greater than 1 mg g^{-1} near shore where the slick was positioned. This level of contamination is sufficient to release oil in slicks, re-contaminating surfaces and is toxic to the **benthic** fauna. The main impact was on intertidal habitats, with millions of gallons of oil remaining in the subsurface of the intertidal zone one year after the spill, with an unusually great depth of oil seepage (15–20 cm) into sandy beaches. In addition to the oil spills, bombing of civilian infrastructure included the destruction of sewage treatment plants in Kuwait, resulting in thousands of cubic metres of raw sewage entering Kuwait Bay.

Massoud *et al*. (1998) report the results of sediment analyses from samples taken in 1992 and 1993. They found that although the oil slick contributed trace metals to the sediment, by 1993 the levels of these metals were within the natural background levels of chronic pollution that the Gulf suffers, and that the slick had a minimal effect on the state of pollution overall.

Many organizations conducted air quality monitoring during and after the oil fires using ground and aircraft measurements. Husain (1998) reports that during the burning of the oil wells, levels of gaseous pollutants, all injurious to human health, were increased: sulphur dioxide increased by an average of 71 per cent, nitrous oxide by 98 per cent, nitrogen dioxide by 25 per cent and non-methane hydrocarbon by 200 per cent. The smoke released toxic metals and carcinogenic materials for some months and also caused darkening of the sky in the region, reducing solar radiation by 25–45 per cent, and dropping air and sea water temperatures by up to 10 and 2.5 deg C respectively, which may have severely affected the life cycles of terrestrial and marine organisms. To assess the impact of the war on the environment the World Meteorological Organization (WMO) and United Nations Environment Programme (UNEP) convened two meetings to discuss the findings from monitoring activities. They found that prevailing meteorological conditions ensured that the smoke plume was contained mainly within 2 km altitude thus alleviating fears that the smoke could enter the jet stream and contaminate a wider area of the globe. This meant that pollution levels were higher than normal but less than predicted. The contribution of CO_2 to the atmosphere was only 1.5 per cent of the global annual emission from the burning of fossil fuels and biomass.

Deposition of soot and particulate matter on the ground within 100 km of the source exceeded 500 mg m^{-2}. Particulate matter counts close to the source were high (10^5 per cm^3), decreasing to 50–100 mg m^{-2} at a distance of 200–500 km. Soil samples also showed evidence of contamination by polyaromatic hydrocarbons (PAHs). The deposition of soot affects human health directly and indirectly by covering buildings and water desalinization plants. The desert flora failed to produce seeds and many perished; Malallaha *et al*. (1998) report mutagenic modifications amongst some desert flora in areas close to the blazing oil wells. Particles from gushing oil wells were deposited on the ground, creating oil lakes and endangering groundwater reserves, and coating large areas of the desert surface and vegetation. Ingestion of oil-contaminated vegetation by grazing animals is one mechanism by which contaminants enter the food chain. The heat, smoke or oil from the fires and slick affected birds and the productivity of shrimp fisheries, which fell to 1 per cent of pre-war levels.

---

benthic: a term to describe bottom dwelling organisms or sediment in aquatic systems

from affected areas, or in areas of land controlled by the military, can support the recovery of the environment due to the absence of anthropogenic pressures such as agriculture, tourism or industry.

Brauer (2002) summarizes the environmental effect of war in three statements:

1. Short, major wars, such as the Gulf War of 1991, result in highly visible, but ultimately relatively minor environmental damage when compared to peaceful commercial development.

2. Longer-lasting, major wars, e.g. the Vietnam War, can result in major, long-lasting environmental damage, but largely through deferred effects such as induced population shifts.

3. The environmental effects of long-term wars, such as the long-running sub-Saharan African conflicts are less well studied. Some degree of environmental depletion of major mammals occurs, e.g., rhinos, elephants, primates, either directly through poaching for revenue to buy arms, or indirectly by habitat invasion and destruction.

The terms 'damage', 'destruction' and others are frequently used in the context of environmental war damage. Brauer (2002) attempts to classify environmental effects of war according to severity (Figure 3.1).

Technologies such as remote sensing can detect any impact of war by providing images before and after military action. Changes between dates indicate disturbance to the area, but generally any damage is assessed by measurements or samples taken from the ground. If damage has been sustained, its severity can be determined by reference to the relative amount of damage compared with the overall resource, and the time taken for the environment to rejuvenate. If the damage can be fully remediated (either naturally or assisted) and the environment returned to a pre-war state given time, this is environmental degradation. If only partial restoration is possible this is environmental depletion. Only in the complete absence of possible remediation or restoration over time can the impact of war be termed environmental destruction.

Physical and ecological processes operate over long timescales. In view of this, what should be the timescale that distinguishes between environmental

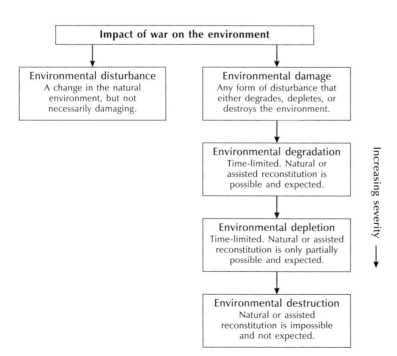

Figure 3.1 The environmental impacts of war

degradation, depletion and destruction? What should be the standard or reference to which judgements are made? Brauer (2002) provides an example of a war-degraded coral reef. The reef may take thousands of years to recover and, in the meantime, the habitat it once provided has changed. This situation could be environmental depletion due to the loss of certain fish and invertebrate species, but could also be environmental disturbance, as other, possibly valued, species would move into the ecological niche. The extent of the damage to the reef is an important factor in assessing the severity of the impact.

3.3 DEVELOPMENT SYNDROMES

Development syndromes are associated with the non-sustainable conversion of land to human uses. The Asian Tiger and Urban Sprawl syndromes are good cases.

ASIAN TIGER

Some of the Asian tiger economies of the Far East are developing extraordinarily fast with little heed for the environment. A consequence of this intensive use of the environment by growing human populations is the destruction and fragmentation of natural habitat that threatens many species. The plight of the Asian tigers – real tigers, not metaphorical ones – typifies this syndrome. The tiger (*Panthera tigris*) is the largest of the cats (Plate 1). It once ranged over huge areas of Asia (Figure 3.2); some 6000 tigers now live in the wild. Over the last century, the landscape in which tigers live has changed radically at the hands of humans and the tiger has suffered:

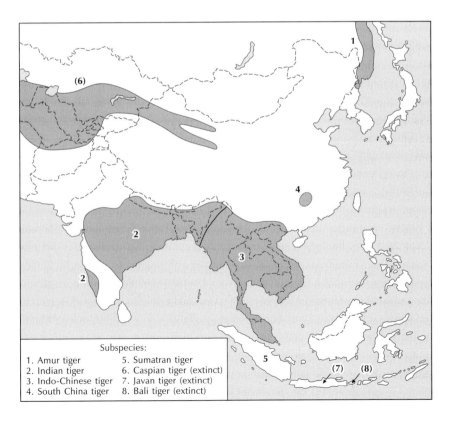

Subspecies:
1. Amur tiger
2. Indian tiger
3. Indo-Chinese tiger
4. South China tiger
5. Sumatran tiger
6. Caspian tiger (extinct)
7. Javan tiger (extinct)
8. Bali tiger (extinct)

Figure 3.2 The distribution of tiger subspecies. *Source:* Adapted from Wentzel *et al.* (1999)

Many small tiger populations are completely isolated, critically endangered and facing a bleak future. Entire subspecies from Bali, Java and areas adjacent to the Caspian Sea have not survived; the Caspian and Javan tigers became extinct after the tiger's endangered status was widely recognized . . . The South China tiger, if not already extirpated in the wild, is down to a few individuals and is slipping away. . . There has been much anxiety for the tiger and its future in the general and conservation press. Concerned, knowledgeable people predict the tiger's demise, with some isolated small populations expected to blink out in the near future. (Seidensticker *et al.* 1999, xv)

Human populations have expanded, putting pressure on the tiger in at least three ways – on the tiger's habitat, on its prey and on the tiger itself (Sunquist *et al.* 1999). Humans have removed, degraded and fragmented forests and grasslands; they have drastically reduced the abundance and distribution of tiger prey species, including the gaur (*Bos gaurus*), banteng (*Bos banteng*) and kouprey (*Bos sauveli*); and they have persecuted the tiger for economic ends. The relative importance of these factors is difficult to assess, but a common assertion is that habitat loss is the main factor in the shrinking range of the tiger. However, some recent studies suggest that the depletion of the tiger's prey is the driving force of the tiger's decline, with habitat loss from agricultural expansion and development projects, poaching for the oriental medicine trade and a loss of genetic variability caused by the isolation of tiger populations taking minor roles. Surprisingly perhaps, recent assessments of forest-cover maps reveal that most countries in the tiger region have extensive stretches of habitat potentially suited to tigers. One piece of evidence for the starring role of a declining prey base comes from comparing leopard and tiger populations in two Indian parks (Focus 3.3). Evidence such as this indicates that tiger conservation in all areas might do well to focus upon strategies to build up and monitor the ungulate prey base that the tiger depends on.

FOCUS 3.3

TIGERS IN TWO INDIAN RESERVES

In India, researchers compared two tiger reserves: the Kalakad-Mundanthurai Tiger Reserve, where leopard (*Panthera pardus*) and tiger populations are in decline, and the Mudumalai Wildlife Sanctuary, where leopard and tiger populations appear to thrive (Ramakrishnan *et al.* 1999). The Kalakad-Mundanthurai sanctuary was designated a tiger reserve in 1988 because tigers and other signs of them (scats for instance) were evident. Since the middle of the 1990s, tiger sightings have declined dramatically. Similarly, sightings of large ungulates, upon which the tigers and leopards feed, have become infrequent, probably because, as revealed by satellite data, grazing land areas have changed to thicket, which is unsuitable as a food source for large ungulates. The change in land cover was due to two major changes in habitat management: a reduction in the frequency and intensity of forest fires; and the exclusion of cattle from most areas of the park. In consequence, unpalatable weeds such as lantana (*Lantana camara*) and eupatorium (*Eupatorium glandulosum*) invaded the grasslands, so reducing the ungulate populations and putting pressure on the tigers. These findings suggest that tiger populations in the Kalakad-Mundanthurai Tiger Reserve might be best boosted by managing the habitat to favour thriving populations of large ungulates.

URBANIZATION

WGBU (1996) identify two specific syndromes associated with urbanization. The first, the Favela syndrome, relates to environmental degradation associated with the unplanned expansion of urban areas. The second, the Urban Sprawl syndrome, refers to the landscape destruction associated with the planned expansion of urban areas. Both of these have far reaching influences over environmental processes and systems as well as on the human sphere itself.

The Favela syndrome

Over the twentieth century there has been an enormous and unprecedented growth in the world's megacities (Figure 3.3). In the developing world, the main drivers of this growth are population expansion and the migration of peoples from the rural hinterland (related to the Rural Exodus syndrome) in pursuit of employment opportunities for regular waged income. In many places, the rate at which this is occurring far outstrips the ability of planners and other authorities to effectively manage for it. As a result there is a lack of essential infrastructure to support this movement (in terms of housing, water supplies, sewage and waste disposal services, medical facilities and so on).

The lack of facilities, however, has not prevented the establishment of extensive informal settlements, often in areas of cities not well suited to

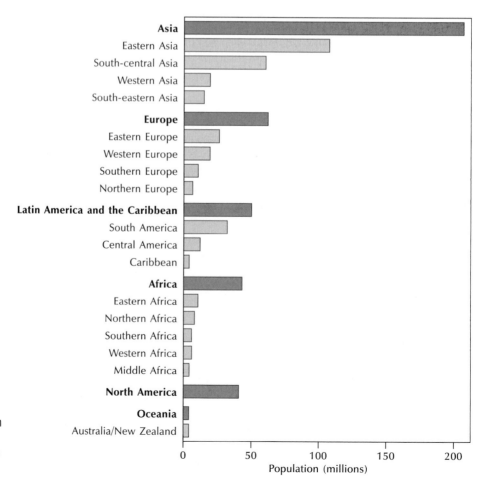

Figure 3.3 The global distribution of megacities. *Source: After United Nations Centre for Human Settlements (UNCHS) (2001)*

habitation, for example through being prone to flood risk or areas where topography prevents the development of permanent construction. The fact that areas may be originally unsafe (in terms of flooding or landslides, for example) and problems caused by the nature of the settlements themselves (e.g. susceptibility to earthquakes, spread of disease and fires), tends to make the residents particularly prone to disasters. Given their low income and few resources, they are also less able to cope with these disasters (Hamsa and Zetter 1998).

Since these settlements still require the basic necessities for life, an informal sector has also grown up with and around these communities providing goods and services to them. The employment opportunities that the city offers for these immigrants, which may still be preferable to the situations many leave behind, are often low paid with minimal health and safety standards. Thus, over time, there is a tendency for social groups within the city to become increasingly polarized and segregated into the 'haves' and 'have nots'. This is a rather simplified picture of the development of shantytowns and urban slums, but is indicative of the key elements of the processes underpinning these developments. The degree to which this is a problem in a specific area depends on a range of factors, from population growth and patterns in wealth and poverty to government and welfare structures. The symptoms of this syndrome include: air pollution and congestion, water pollution, soil erosion and a range of health risks (WGBU 1996). Moore *et al.* (2003) outline the health impacts of global urbanization as related to substandard housing, overcrowding, air pollution, insufficient or contaminated water supplies, poor sanitation and waste disposal systems, the presence of vector-borne diseases and increased vulnerability associated with economic and social stress. A general lack of data hinders the ability of authorities to effectively monitor and control these issues in many informal settlements (Moore *et al.* 2003).

To give a flavour of the intricacies of this issue, it is useful to consider some case study examples in a little more detail. WBGU (1996) identify several examples as illustrations of the Favela syndrome: Karachi (Pakistan), Cairo (Egypt), São Paulo (Brazil) and Calcutta (India). Karachi, for example, has grown from a village to a megacity of over seven million inhabitants in the space of a single century and houses two thirds of all of the country's industry. The country also needs labour to work in its factories in order to facilitate its onward economic development, it is therefore not surprising that the city is a magnet for migrants. In Cairo a similar concentration of demand and supply has developed. Here WGBU (1996) report that there are estimated to be anywhere from 125,000 to 1 million people living in the cemeteries of the city alone.

Although rates of urbanization in Communist China are generally low, some authors have argued that this type of informal development is also evident in some parts of the country, despite its long tradition of strict control over development and its organization (Young and Deng 1998; Zhu 1998). Using the Fuijan province as an example, Zhu notes that 'quasi-urbanization' is clearly evident linked to emerging local economic factors, for example where successful township and village enterprises attract temporary residents. It is noted that in some areas these are becoming a more important influence over growth than the government sponsored enterprises and the related government-recognized urbanization. Also in the context of China, Goodkind and West (2002) highlight the issue of the *liudong renkou* (floating population).

The Urban Sprawl syndrome

Megacities in other parts of the globe are prone to another form of development syndrome, that of the destruction of landscapes through the planned expansion of settlements. Here population pressure may not be as great, although it is still true that cities all over the world are focal points of in-migration (see Chapter 15). The population of cities almost everywhere is still expanding, but the crux of this issue is the ever-expanding amount of space that individuals within the city consume. In view of this, although population growth itself may be slight, the actual extent of the cities themselves still expands (WGBU 1996).

The long-term development of cities, then, has a tendency to subsume the surrounding region, transferring it from a semi-rural hinterland that would have once supplied foodstuffs and other essentials, to an indistinguishable part of the agglomeration. The once isolated towns within this former landscape then become secondary centres to the city itself. The internal distributions of people and employment give rise to more interurban movements supported by ever growing transport infrastructures, usually centred on the private car. In Germany, for example, WGBU (1996) estimate that during the 1990s there was a conversion of natural landscapes to urban uses at a rate approaching some 80–90 hectares per day. They also note the average housing area required by a German citizen from 1950 through to 1981 more than doubled from 15 m^2 to 34 m^2. These patterns are mirrored throughout the developed western world with a reduction in numbers of people per household coupled with increasing sizes of households. Ultimately, this results in displacements of population out towards the periphery of urban zones with a corresponding increase in travel as people purchase more cars and become accustomed to more extensive patterns of travel for both work and leisure activities. Kenworthy and Laube (1999) note that the 'automobile cities' of the USA use eight times more energy per capita than equivalent, more public transport-oriented cities in Asia. Their explanation for this rests on differences between the density of population and the degree of zoning in individual cities.

The symptoms of this syndrome are far-reaching in terms of the scale of the impacts and the prospects for those areas currently at the Favela stage of development. Urban sprawl promotes increasing congestion, loss and degradation of open space, soil compaction and surface sealing, urban air pollution and the impacts of these on the air, water and soil range both from the scale of the cities themselves through to the global scale. These symptoms and their interlinkages are discussed in more detail in Chapter 14.

3.4 SINK SYNDROMES

Sink syndromes lead to environmental degradation. They involve accidental spillages of natural and human-manufactured substances and the uncontrolled disposal of waste products. The Major Accident and Smokestack syndromes serve to illustrate this group of syndromes.

MAJOR ACCIDENT

WGBU (1996, 123) define Major Accident syndrome as 'the mounting threat to the environment in the form of singular localized disasters caused by humans, whereby liability for possible damage is limited or inadequate'. The consequences of major accidents can include the loss of biodiversity and ecosystem degradation, contamination of soil, water and air, and significant impact on health.

As human activities such as production and transport become increasingly global, the risk of major accidents increases. Such accidents include tanker accidents or environmental disasters associated with the transport of hazardous goods. For industrial accidents, the example of Bhopal highlights how developing countries are particularly vulnerable as the regulations are generally less stringent, due to the need for foreign investors.

Other disasters include the introduction of invasive alien species, which can cause the extinction of local species and destroy the structure and functioning of ecosystems and habitats (p. 356). The species can be introduced deliberately (for example, rabbits to Australia or Japanese knotweed (*Fallopia japonica*) in the UK), as a result of animals escaped from farms (e.g. mink (*Mustela vison*) in the UK) via goods (e.g. Dutch elm disease from contaminated wooden packing crates), or planes or ships (e.g. bubonic plague via ship rats).

Chernobyl

A classic example of Major Accident syndrome is the accident at the Chernobyl Nuclear Power Plant, in the Kiev region, Ukraine. The disaster occurred on 26 April 1986 and resulted from a safety experiment to test whether it was possible to shut down reactions in the core in the event of a main power loss. However, the reactor overheated, causing two explosions that destroyed the core, blew off the roof of the reactor, and sent tonnes of radioactive material into the atmosphere. The fire lasted for 10 days. The radioactive cloud reached up to 3 km and the prevailing winds carried it over most of Europe. In the immediate aftermath of the disaster, it is estimated that 31 people died, but many millions across the former USSR and Europe were affected due to the passage of the radioactive clouds.

The first indication of a major nuclear accident came from Sweden on 28 April 1986. Analysis of the trajectory of air movement indicated that the source was Chernobyl. Radioactive fallout was distributed over the Northern Hemisphere, but was mainly concentrated in Ukraine, Belarus and the Russian Federation, the countries closest to the nuclear plant. In these countries, the area of contaminated land (i.e. here defined by soil with a deposition density of ^{137}Cs greater than 37 kBq m^{-2}) is approximately 150,000 km^2 (Rahu 2002).

The European Environment Agency (EEA) (2003) commented that, despite the decay of most deposited **radionuclides**, some tens of thousands of square kilometres in Belarus, the Russian Federation and the Ukraine will remain substantially contaminated for decades by long-lived radionuclides. The long-lived contaminants include caesium-137, strontium-90, plutonium and americium-241. Elevated levels of external exposure to humans will remain over a long period, as will radionuclide concentrations in food products from agriculture and natural sources such as berries, lake fish and mushrooms.

radionuclides: nuclides (nuclei of atoms) that disintegrate spontaneously, releasing radioactivity in the process. Examples are nuclei of uranium and strontium atoms

Health impacts

There is general agreement that the effects of massive doses of radiation are fatal. The longer-term effects of heavy irradiation are unclear, however, and depend on the amount of radiation emitted, how and when it is deposited on the Earth, the number of people affected and the dose they receive (Foley 1992). To assess the effects of radiation, the UN established the Scientific Committee on the Effects of Atomic Radiation (UNSCEAR) in 1955.

Rahu (2002) reviewed the health effects of the Chernobyl disaster, the secrecy surrounding the accident and the difficulties faced by epidemiological researchers. The silence of the Soviet authorities to international questions after the accident caused an atmosphere of fear and distrust and a lack of reliable information. The UNSCEAR report (2002) is considered by some to be the most authoritative review summarizing studies examining health in the wake of the Chernobyl disaster. It reaches the conclusion that reports of health effects caused by radiation have been greatly exaggerated, and that there is currently no internationally accredited evidence of increased leukaemia risk, birth defects over time, overall cancer incidence or mortality that could be associated with radiation exposure. The available evidence points to a sharp increase in childhood thyroid cancer and psychological disorders (non-radiation-related) caused by concerns over radiation, fears of health effects, economic pressures and hardships and so forth. The EEA (2003) recognized that the psychosocial effects are emerging as a major problem, and are similar to those arising from other major disasters such as earthquakes, floods and fires.

Rahu (2002) comments that more research is needed, and the UNSCEAR report does not address some of the challenges facing epidemiological researchers that could obscure the truth of Chernobyl. The most important challenge is overcoming the unreliability of official statistics that would greatly affect the outcome of studies, especially those employing time-trend analyses of before and after the event. The lack of reliable public records have meant that the various studies performed to assess the consequences of the accident have encountered problems assessing the significance of their observations and results. Studies were hampered by a lack of international cooperation, absence of records and exposure data amongst others. In addition, the findings of current studies are difficult to interpret due to the economic and social changes in the affected areas. The low socio-economic status of the people and conditions of poor living quality and reduced economic opportunities following the disaster, result in a wide spectrum of diseases that may mask any obvious impact of radiation exposure.

Environmental impacts

Radioactive contamination can affect flora and fauna in two ways: internally through the uptake of radionuclides into tissues of plants, which are then consumed by animals and humans; and externally by radionuclides deposited in the soil.

Davids and Tyler (2002) examined whether or not radionuclide contamination within the Chernobyl exclusion zone had a measurable effect on the vegetation using remote sensing. The radioactive contamination after the accident caused significant change in the abundance and distribution of tree species in the exclusion zone. Some 400 ha of Scots pine (*Pinus sylvestris*) suffered extensive damage and death and the area was recolonized by silver birch (*Betula pendula*), which appeared to exhibit only minor effects. The study found that the radioactive contamination did cause silver birch and Scots pine to differ in their spectral characteristics and concluded that remote sensing has the potential of providing a valuable monitoring technique for assessing the ecological impact of radionuclide contamination.

Different species have different tolerance levels or sensitivities to radionuclide contamination. Skwarzec *et al.* (2001) investigated differences in the bioaccumulation of Chernobyl-derived plutonium in Gdansk Bay. In the marine environment, the greatest concentration of plutonium was found in sediments, but the biogeochemical cycling of the element resulted in concentrations in all other components. The researchers also examined the levels of plutonium nuclides in the organs and tissues of fish samples using alpha **spectrometry**. Four representative species of Baltic fish were studied: flounder, herring, cod and sprat. Results indicated that bioaccumulation plutonium levels were lowest in flounder stomach, herring skin and cod intestine, while the highest levels were found in cod gills and skin. $^{239+240}$Pu concentrations were measured with flounder 0.94 mBq (a thousandth of a Bequerel) kg^{-1}, herring 2.22 mBq kg^{-1}, cod 2.35 mBq kg^{-1} and sprat 0.33 mBq kg^{-1}.

One of the main processes by which deposited radionuclides migrate within the environment is through the hydrological transport of both dissolved and particulate forms. Contamination can still spread from the heavily affected areas around the nuclear plant by water toward the nearby Pripyat and Dnieper rivers, by soil water movement, groundwater flow and flooding. Amano *et al.* (1999) examined the transfer capability of long-lived Chernobyl radionuclides from surface soil to river water in dissolved forms. Soil samples from the exclusion zone indicated that radioactive caesium (Cs), strontium (Sr) and transuranic isotopes, such as plutonium (Pu) and americium (Am), were dominant in surface soil layers. Sr-90 in the soil was highest in the water soluble and exchangeable fractions, which meant that it could be easily dissolved in river water and transported away. Similarly, transuranic elements were associated with mobile materials such as fulvic acids in **leachates**, again indicating the high potential for the radionuclides to move within the soil and be uptaken by plants.

> ***spectrometry***: techniques for identifying chemicals in a substance
>
> ***leachate***: the product formed by water percolating through soil, dissolving soluble bases

SMOKESTACK

Until relatively recently, much environmental policy in the western world has hinged on the notion of 'dilute and disperse'. Originally a response to the appalling air, water and soil pollution of urban and industrialized areas in and immediately after the industrial revolution, this was, in essence, an attempt to solve what was perceived as a very localized pollution issue. It assumed that the Earth could be treated as an endlessly recyclable purification system for any type of substance release into it. As a consequence, waste was discharged to watercourses and the oceans, buried beneath the ground and emitted from high stacks into the atmosphere (see Chapter 14).

However, towards the end of the twentieth century, it became increasingly apparent that this approach was not working. Instead, the problem was being transferred elsewhere and affecting large areas. A good example of this is acidification, where emissions from fossil-fuelled power stations were leading to the deposition of acid compounds that were causing forest decline in central and northern Europe and impacts on fragile freshwater ecosystems (p. 377–82). The cause and effect processes taking place were difficult to prove and governments therefore resisted taking the required action. However, modelling and empirical studies eventually obtained unequivocal proof and this, together with public pressure, finally led to action to curb emissions and provide more at source technological solutions such as flue-gas de-sulphurization. Nevertheless, the impacts of this will take a long time to overcome, if ever, and the modification of the ecosystems in these modified environments is the subject of ongoing research (p. 377–82). The debate that surrounded the 'acid rain' issue has some similarities with that associated with ozone depletion and global warming. Eventually, the relationship between the emissions of chlorinated compounds and changes in stratospheric ozone patterns was confirmed, and national and international policy put in place to prevent further impacts. With global warming, the debate is still very much in hand, and there is yet to be total consensus about how to tackle this growing problem. Other symptoms of the Smokestack syndrome include the accumulation of toxic substances in ecosystems. Here, activities such as the application of pesticides to agricultural crops can lead to its dispersal into waters, soil and plants and, subsequently, the food chain of both animals and humans. The long-term impacts of these substances in terms of the development of species on Earth cannot be known and the results of the global experiment are yet to be made clear.

The symptoms of the smokestack issue and the influence of this on physical systems at work in the atmosphere are covered in detail in Chapter 8.

SUMMARY

In coming to dominate the environment to create the human sphere (anthroposphere), humans have created a complex set of interconnected problems called 'syndromes', which are basic patterns of people–environment interactions that are seen around the world. Syndromes come under three broad headings: utilization syndromes, development syndromes and sink syndromes. Utilization syndromes concern the use or misuse of natural resources. Examples are the Mass Tourism syndrome and the Scorched Earth syndrome. Development syndromes involve the non-sustainable conversion of land cover to agricultural, industrial, commercial, residential and other human uses. Examples are the Asian Tiger syndrome and the Urban Sprawl syndrome. Sink syndromes involve environmental degradation through accidental spillages of natural and human-manufactured substances and the uncontrolled disposal of waste products. Examples are the Major Accident syndrome and the Smokestack syndrome.

Questions

1 How useful is a syndrome approach to environmental problems?
2 Examine pros and cons of using caves as tourist attractions.
3 What effects does war have on the environment?
4 Explain why tiger populations are struggling to survive.

5 Explore the environmental problems associated with urban sprawl.
6 What are the long-term consequences of the Chernobyl accident?
7 Discuss the environmental problems associated with smokestack emissions.

Further Reading

Blowers, A. and Hinchliffe, S. (eds) (2003) *Environmental Responses*. Chichester: John Wiley & Sons and The Open University.
An informative read.

Hinchliffe, S., Blowers, A. and Freeland, J. (eds) (2003) *Understanding Environmental Issues*. Chichester: John Wiley & Sons and The Open University.
Well worth delving into.

Seidensticker, J., Christie, S. and Jackson, P. (eds) *Riding the Tiger: Tiger Conservation in Human-dominated Landscapes*. London: The Zoological Society of London; Cambridge: Cambridge University Press.
An excellent account of the plight of the tiger, with a mixture of overview and detail. Tackles conservation issues at length.

Useful Websites

http://www.wgbu.de/wbgu_jg1996_engl.html
The online source for the original syndrome approach book: WGBU (Germany Advisory Council on Global Change) (1996) Annual Report World in Transition: The Research Challenge (Berlin: Springer). *Not an easy read as it requires background knowledge.*

http://www.wolfson.ox.ac.uk/~ben/Tiger/Tiger2.htm
Some useful tiger links for those with an interest in bio-geography.

CHAPTER 4
THE MENTAL SPHERE

OVERVIEW

The mental ingenuity and physical handicraft of the human species have combined to create the human sphere. The products of the human mind – society, culture, ethics, economics, religion, science and technology – are factors that help to shape the environment and in part determine its present and future use. Difficult issues can arise when interested groups of people decide the best way of tackling perceived environmental crises, whether at a particular place, a region or globally. These knotty issues have pragmatic and moral dimensions. Three groups of topics illustrate the issues. The first group involves the people–environment themes of environmentalism, sustainability, ecosystem management, traditional ecological knowledge and the 'new conservation'. The last of these is a shift in biodiversity conservation from Nature-centred, exclusive, protected areas toward conservation centred more on people or communities. The second group centres around the international programmes designed to investigate the environment and environmental change, for instance the World Climate Research Programme (WCRP), the International Geosphere–Biosphere Programme (IGBP), the International Human Dimensions Programme (IHDP), and DIVERSITAS, an international programme of biodiversity science. The third group comprises centres that investigate global change, such as the Potsdam Institute for Climate Impact Research, Germany, the Earth System Science Center at Pennsylvania State University, USA, and the Hadley Centre in the UK.

LEARNING OUTCOMES

After reading this chapter, you should be able to understand:
- The chief themes focusing on people–environment interactions
- Views espoused by different kinds of environmentalist
- The idea of sustainability
- The rationale behind ecosystem management as a way of unifying ecological, economic and social needs
- The nature of traditional ecological knowledge and its relation to scientific knowledge
- The reasons for setting up international research programmes to study global environmental change from an interdisciplinary standpoint
- The rationale behind and the scope of centres of research for global change.

The products of human intelligence – society, culture, ethics, economics, religion, science and technology – constitute the mental sphere, which impinges upon the environment to create the human sphere. The mental sphere is thus a key factor in understanding views and policy-making about the environment and environmental change. This chapter will explore its nature.

4.1 HUMAN–ENVIRONMENTAL THEMES

The Age of Ecology opened on 16 July 1945 in the New Mexican desert 'with a dazzling fireball of light and a swelling mushroom cloud of radioactive gases' (Worster 1994, 342). Observing the scene, a phrase from the *Bhagavad-Gita* came into project leader J. Robert Oppenhiemer's mind: 'I am become Death, the shatterer of worlds'. For the first time in human history, a weapon of truly awesome power existed, a weapon capable of destroying life on a planetary scale. From under the chilling black cloud of the atomic age arose a new moral concern – **environmentalism** – that sought to temper the modern science-based power over Nature with ecological insights into the radiation threat to the planet. The publication of Rachel Carson's *Silent Spring* (1962), the first study to bring the insidious effects of DDT application to public notice, was more grist for the environmentalist's mill. The dual threats of radiation and pesticides set environmentalism, which had its roots in the nineteenth century, moving apace.

> **environmentalism**: a movement centring around a moral concern for the environment

ENVIRONMENTALISM

There are several brands of environmentalism, each of which promotes a different attitude towards the environment (see O'Riordan 1988; 1996). These brands tend to polarize around, at one extreme, **ecocentrism**, with its decidedly motherly attitude towards the planet; and, at the other extreme, **technocentrism** with a more *laissez-faire*, exploitative attitude (Table 4.1). The 'green' end of this spectrum was initiated by the atomic bomb and galvanized into action by pesticide abuse. Each strand represents a system of beliefs, although no individual would necessarily believe in just one strand. The chief characteristics of some of the players in these brands of environmentalism are described below. Please note that these descriptions are gross caricatures – they tend to highlight the differences and tone down the similarities between the groups.

> **ecocentrism**: an environmentalist viewpoint that sees humans as part of the ecosphere and subject to its laws
>
> **technocentrism**: an environmentalist viewpoint that recognizes environmental problems and believes that humans can surmount them

Technocentrics

Technocentrics tend to have a human-centred view of the Earth coupled with a managerial approach to developing resources and protecting the environment. They favour maintaining the *status quo* in existing government structures. Technocentrics tend to be conservative politicians of all political parties, leaders of industry, commerce and trades' unions and skilled workers. They aspire to improve wealth for themselves and for society. They enjoy material acquisition for its own sake and for the status it endows. And they are politically and economically very powerful.

There are two brands of technocentric – the cornucopians and the accommodationists. Conservative technocentrics are cornucopians or optimists, named after the legendary goat's horn that overflowed with fruit, flowers and corn, and signified prosperity. They have an unwavering faith in the application

Table 4.1 Brands of environmentalism and their characteristics, as recognized by O'Riordan (1988)

Environmental inclination	Brand	Characteristics
Ecocentrism (Nature-centred)	Deep ecologists or Gaians	Stress the intrinsic importance of Nature for humanity
		Believe that ecological (and other) natural laws should dictate human morality
		Ardently promote bio-rights – the right of endangered species, communities and landscapes to remain unmolested
	Communalists or self-reliance, soft-technologists	Emphasize the smallness of scale ('small is beautiful') and hence community identity in settlement, work and leisure
		Interpret concepts of work and leisure through a process of personal and communal improvement
		Stress the importance of participation in community affairs, and guarantee the rights of minority interests
Technocentrism (human-centred)	Accommodationists or environmental managers	Believe that economic growth and resource exploitation can continue assuming: suitable economic adjustments to taxes, fees, etc.; improvements in the legal rights to a minimum level of environmental quality; compensation arrangements satisfactory to those who experience adverse environmental and social effects
		Accept new project appraisal techniques and decision review arrangements to allow for wider discussion or genuine search for consensus among representatives of interested parties
	Cornucopians or optimists	Believe that humans can always find a way out of any difficulty, be it political, scientific or technological
		Accept that pro-growth goals define the rationality of project appraisal and policy formulation
		Optimistic about the human ability to improve the lot of the world's people
		Faith that scientific and technological expertise provides a firm foundation for advice on matters pertaining to economic growth, public health and public safety
		Suspicious of attempts to widen basis for participation and lengthy discussion in project appraisal and policy review
		Believe that all impediments can be overcome given will, ingenuity and sufficient resources arising out of growth

of science, market forces and managerial ingenuity to sustain growth and survival. Liberal technocentrics are accommodationists and have a more cautious attitude. They have faith in the adaptability of institutions and mechanisms of assessment and decision making to accommodate environmental demands. They believe that adjustments by those in positions of power and significance can meet the environmental challenge. The adjustments involve adapting to and moulding regulation (including environmental impact assessment), and modifying managerial and business practices to reduce resource wastage and economically inconvenient pollution, without any fundamental shift in the distribution of political power.

Pressure groups aligned to technocentrics tend to be NIMBY (not-in-my-back-yard) groups – they may or may not like development, but they certainly do not like it taking place near them – and private interest groups (consumer groups, civil liberties groups and so on). They are very vocal and command much media attention.

Ecocentrics

Ecocentrics have a Nature-centred view of the Earth in which social relations cannot be divorced from people–environment relations. They hold a radical vision of how future society should be organized. They would give far more real power to communities and to confederations of regional interests. Central control and national hegemony, so treasured by technocentrics, are the very antithesis of ecocentrism. Ecocentrics tend to occupy the fringe of modern economies. They are characterized by a willingness to eschew material possessions for their own sake, and to concern themselves more with relationships; and by a lack of faith in large-scale modern technology and associated demands on élitist expertise and central state authority. Pressure groups aligned to ecocentrics tend to be green groups and alternative subcultures (e.g. feminists, the peace movement and religious movements).

There are two brands of ecocentrics. The more conservative ecocentrics are communalists or self-reliance, soft-technologists. They believe in the cooperative capabilities of societies to be collectively self-reliant using appropriate science and technology. That is, they believe in the ability of people to organize their own economies if given the right incentives and freedoms. This view extends a 'bottom up' approach to the development of the Third World based on the application of indigenous customs and appropriate technical and economic assistance from western donors. Liberal ecocentrics are Gaians (named after the Greek goddess Gaia, who was seen as the personification of Mother Earth) or deep ecologists. They believe in the rights of Nature and the essential co-evolution of humans and natural phenomena. Extreme ecocentrics avow that the moral basis for economics development must lie in the interconnections between natural and social rights: there is no purely anthropocentric ethic. Within this group might be included a powerful religious lobby (Focus 4.1).

SUSTAINABILITY

Ever since humans began gathering in large communities, some people have expressed concern about the impact of the large communities on the 'natural' environment. Brimblecombe (1987), for example, traces concern for the quality of air in the UK back as far as the thirteenth century. With the advent of the industrial revolution and the corresponding patterns of population growth and concentration, industrial expansion and resource exploitation, the environmental voice grew stronger as commentators witnessed the transformation of a largely rural, subsistence-based **agrarian** society with local and regional perspectives to a complex, industrial and post-industrial, consumer-driven community with an increasingly global outlook. While this transformation began in earnest in some nations during the eighteenth and nineteenth centuries, it is an ongoing process across the rest of the world. It has given rise to a number of environmental crises, some of which this chapter and the remainder of this book identify and attempt to explain.

Although concern for the state of the environment is not new, the perspectives of commentators in terms of both the problem and the solution have shifted over

> **agrarian**: relating to ownership, cultivation and tenure of the land

FOCUS 4.1

RELIGION, NATURE, THE WORLD WILDLIFE FUND AND ARC

In autumn 1986, conservation, as represented by the World Wildlife Fund, and five of the world's chief religions, forged a unique alliance. The venue was Assisi, in central Italy, home of the patron saint of animals. A pilgrimage, conference, cultural festival, retreat and interfaith ceremony in the Basilica of St Francis formed the basis of a permanent bond between conservation and religion.

For religions, ecological problems pose major challenges in relating theology and belief to the damage to Nature and the human suffering caused by environmental degradation. All too often, scientists and technologists have used or abused religious teaching to justify the destruction of Nature and natural resources. In the declarations issued at Assisi, leading religious figures mapped out the way to re-examine and reverse this process. And in 1987, the Baha'i' Movement, which had its origins in Islam, joined the World Wildlife Fund's New Alliance. At the same time, starting with the conference and retreat held in Assisi, religious philosophers are helping inject some powerful moral perspectives into conservation's ill-defined ethical foundations.

Recently, the WWF and the Alliance of Religions and Conservation (ARC), which was founded in 1995, joined together to encourage, secure and celebrate significant new pledges by the world's major religions for the benefit of the environment and the natural world. In a special ceremony held in Nepal on 15 November 2000, participants from the Baha'i, Buddhist, Christian, Hindu, Jain, Jewish, Muslim, Sikh, Shinto, Taoist and Zoroastrian faiths agreed actions within a Sacred Gifts Programme to combat forest and marine destruction, climate change and a wide range of other environmental issues. These faiths represent billions of people worldwide, all of whom are enjoined to take part in conservation actions locally, nationally and internationally. Such extensive mobilization of communities should make a gigantic contribution toward 'saving the environment'. ARC's website contains many examples of specific actions, such as the Catholic Church in Dili, East Timor, announcing that it would start replanting native trees and plants on parish land, using species that will help the local community. The hope is that people across the island will adopt this initiative, every parish becoming a mini-forestry unit that, in the long term, will help to support the food needs of the community, decrease deforestation (and associated soil erosion) and work as a living and growing investment for the future.

time. These perspectives range from notions of the loss of so-called 'pristine' landscapes or, perhaps, quintessential landscapes, to a more economics-driven viewpoint. Sachs *et al.* (1998) see the development of environmental politics being in step with these changing perspectives. For example, during the 1970s, much environmental debate centred around the idea of the environment as a resource and the recognition of the serious depletion and likelihood of ultimate exhaustion of resources upon which society, and especially western society, was increasingly reliant. During the 1980s, the debate expanded to recognize the **carrying capacity** of the environment, both in terms of direct resource utilization and the ability of the land, air and water to absorb the waste products of production (i.e. pollution). The 1990s moved the debate still further to encompass the idea of 'sustainability' and 'sustainable development' as bringing together the shared needs of society, economy and environment. The term 'sustainability', although initially purported to be something of an oxymoron, has gained significant momentum and acceptance and remains a dominant philosophy for the beginning of the twenty-first century.

> **carrying capacity**: the maximum number of individuals a particular environment can support without causing environmental degradation

Definitions of sustainable development

The most widely recognized definition of **sustainable development** is that given in the **Brundtland Report** *Our Common Future* in 1987. Here, it is given as 'meeting the needs of the present without compromising the ability of future generations to meet their own needs' (WCED 1987, 8). Satterthwaite (1999) notes that, at this point, the term was not new and had in fact been in use for at least ten years before the definition given in the influential Brundtland report. During the development and adoption of the idea of 'sustainable development', its general ambiguity and vagueness has been identified both as a strength and as a weakness. Reid (1995) cites O'Riordan's concerns that this exposes the idea to abuse and that despite the 'beguiling simplicity' of the Brundtland Report definition, there is no easy way to translate theory into practice (O'Riordan 1988). However, others see the ambiguity as a positive attribute, encouraging debate and fostering progress. Indeed, for some, 'sustainability' represents less a framework for problem solving and more a guiding principle on ethical and moral grounds (see Reid 1995 for further discussion). Regardless of the rhetoric surrounding the term itself, its success in focusing environmental efforts is undeniable. This success is due not in small part to the UN General Assembly's follow up of a recommendation in the Brundtland Report, which culminated in the 1992 **Earth Summit** and Global Forum. Hailed as the largest international conference ever held (178 countries were represented), this meeting was reported to have involved 30,000 people, 7000 journalists, and over 1500 official representatives of non-governmental organizations (NGOs) (Reid 1995). Amongst the outcomes of the conference were conventions on climate and biodiversity. Although the attempt to produce a convention on forests was unsuccessful, there was at least recognition of the importance of forests for both biodiversity and climate. The other key outcome was **Agenda 21** (the Rio declaration in Environment and Development), a detailed plan to translate the guiding principle of sustainability into a reality.

The global crises from the perspective of sustainable development

Efforts towards providing a framework and associated plan to assist with implementing sustainable development came in response to a number of key global pressures or crises. Summarized from Reid (1995) and Carley (1994), these pressures included:

- Resource consumption and flows of waste products as pollutants that impact upon the environment and through the environment on human society

- Inequality in consumption and ability to cope with the negative impacts of consumption

- Population growth and changing patterns of population distribution.

Processes associated with these pressures operate on different spatial scales, from local to global, and exhibit high rates of change that make adaptation to negative impacts more difficult for all species on the Earth. The chapter on biodiversity highlighted the different abilities of animals and plants to adapt to rapid climate change and other human-induced change. It is important to note here that there is also a varied ability to adapt within the human species, favouring the rich and powerful of the local and global communities at the expense of the poor and vulnerable members of those societies. At the global scale, there may also be further inhibitions to adaptation as well as perpetration of the global crises through political instability and poor governance (Reid 1995).

---

sustainable development: using natural resources to meet current human needs without harming the biosphere and with an eye to the possible needs of future human generations

Brundtland Report: the final report of the World Commission on Environment and Development, chaired by Gro Harlem Brundtland, that was published in 1987 and proposed a major international conference to deal with environmental issues, which materialized as the United Nations Conference of Environment and Development (UNCED – or Earth Summit) in Rio de Janeiro in 1992

Earth Summit, 1992: the United Nations Conference on Environment and Development (UNCED) held in Rio de Janeiro, Brazil, in June 1992. Its main aim was to discuss ways of safeguarding biodiversity

Agenda 21: an 800-page document coming out of the Earth Summit. It was an innovative attempt to describe the policies needed for environmentally sound development

Concepts and paradigms

There has been considerable progress towards conceptualizing environment and development problems as a means of providing the basis for solutions. The literature on this topic is very large. The following are therefore some of the most commonly used terms, and further information lies in the suggested introductory readings listed at the end of the chapter:

- Natural capital or environmental capital – the stock of resources both renewable and non-renewable and the waste assimilation capacities of the planet (Satterthwaite 1999)

- Ecological footprint – Rees (1995) identifies this as the aggregate land area required (1) to provide the resource inputs to production processes, and (2) to absorb the associated waste outputs. He writes about an exercise he carried out with his students to assess the land area needed to support each individual's consumer lifestyle in North America. They concluded that an average North American requires something in the region of 4–5 hectares. If all people on Earth had the same patterns of consumption as a North American, this would result in a need for around 24 billion hectares of ecologically productive land. He notes that since the Earth has around 8.8 billion hectares of ecologically productive land, given existing technologies, the resultant ecological footprint would demand a minimum of two additional Earths to satisfy it. This idea relates to the pressures driven by resource consumption and population growth

- Equity and justice – Haughton (1999) considers there to be five 'equity concerns': intergenerational, intragenerational, geographical, procedural and interspecies. These ideas relate to the need to understand and plan for pressures associated with inequality and population growth (although in some cases population decline may also lead to some forms of inequality)

- Ecological carrying capacity and limits to growth – it is evident from the discussion on ecological footprints that there are clear limits to growth and 'development' as traditionally seen. This relates to the overall quantity of available resources as well as the finite limits to the ability of the air, land and water to absorb wastes. There has been considerable research into the definition and quantification of limits and capacities that are discussed elsewhere in this book (p 76).

- The acceptance of the notion of 'ecological footprint' and finite limits to development depends on the accepted paradigm or 'worldview'. Rees (1995) discusses two contrasting worldviews: the expansionist paradigm and the steady-state paradigm. The former treats the economy as independent of the environment and the environment as an infinite source and sink to the processes of production. Here, the assumption is that technological solutions will be found to support the status quo and business as usual. The latter sees development and the economy as inextricably linked and with the economy entirely subsumed and dependent on a finite ecosphere.

These ideas provide a framework for understanding the threats facing the global (and local) environment as an essential foundation to the development of workable solutions.

Progress towards solutions

Progress towards sustainable development is slow and painful, but some important milestones are in place. One of these is the rise in the importance of a **holistic view** of processes and impacts together with the development of mechanisms and forms of governance and study that support this view. Another is the development of indicators that help quantify and measure these processes and impacts and provide the basis for assessing progress towards the ultimate goal of sustainability. However, important obstacles still exist, one of which is the issue of uncertainty.

There is still considerable uncertainty about the magnitude and impact of the environmental and developmental pressures identified above. This uncertainty can help promote the expansionist paradigm at the expense of the steady-state paradigm and represents an important obstacle to progress towards sustainability. However, with more research, the range of uncertainty should reduce over time. This will result from the increasing body of scientific evidence that demonstrates the indisputable impacts of pressures on the ecosphere and the highly complex and interrelated nature of human and physical processes on the Earth. Human and physical geographers will play a critical role in this body of research. The need is for physical geographers to view the world through the lens of the human sphere. The work undertaken to date has helped establish a considerable movement towards the adoption of a **precautionary principle** to future development, the importance of which can only increase as humans go through the twenty-first century.

> **holistic view**: an approach that emphasizes the importance of whole systems, such as the global ecosystem, and the interdependence of their parts
>
> **precautionary principle**: the view that a lack of scientific data should not prevent the implementation of measures to avert environmental degradation; a 'better safe than sorry' approach

ECOSYSTEM MANAGEMENT

People are part of Nature. The philosophy behind a new brand of conservation – **ecosystem management** – accepts that simple fact. Ecosystem management emerged in the late 1980s. Many scientists and others interested in the environment are its advocates. Its ultimate aim is to enhance and to ensure the diversity of species, communities, ecosystems and landscapes. It is a fresh idea that covers a spectrum of approaches.

What is ecosystem management? This question is surprisingly difficult to answer. It is easier to say what ecosystem management is not – it is not the traditional model of resource management. Traditional resource management strives to maximize the production of goods and services through a sustained yield from 'balanced' ecosystems. It gives much weight to utilitarian values that regard human consumption as the best use of resources, and that hold a continuous supply of goods for human markets as the purpose of resource management. Ecocentrics feel that this unashamed 'resourcism' is patently flawed, pointing out that it fails to recognize limits to exploitation and, in consequence, a growing number of species, and even entire ecosystems, are currently endangered. Nevertheless, flawed or not, it persists: even now, ecosystems are viewed by some as long-lived, multi-product factories, or, if you prefer, as Nature's superstores.

> **ecosystem management**: in its most extreme form, a system of resource management that tries to balance economic needs, ecological needs, and social and cultural needs

Ecosystem management, though not universally welcomed, is a new model of resource management. Some of its advocates see their endeavours as an extension of multiple-use, sustained-yield policies (e.g. Kessler *et al.* 1992). They promote a stewardship approach, in which an ecosystem is seen merely as a human life-support system. In this view, public demands for habitat protection, recreation and wildlife uses are simply seen as constraints to maximizing resource output. A more radical approach, which has made much headway in discussions of ecosystem management, is to accept Nature on its own terms, even where doing so means controlling incompatible human uses. This extreme and ecocentric form of ecosystem management reflects a willingness to place environmental values, such as biodiversity, and social and cultural values, such as the upholding of human rights, on an equal footing.

Ecosystem management is variously defined, embraces multifarious approaches, and has several facets. It is intimately linked to radical ideas about humanity's place in Nature. Here is a working definition:

> Ecosystem management integrates scientific knowledge of ecological relationships within a complex sociopolitical and values framework toward the general goal of protecting native ecosystem integrity over the long term. (Grumbine 1994, 31)

It is, at root:

> an invitation, a call to restorative action that promises a healthy future for the entire biotic enterprise [and promises a means of bridging the growing gap between people and Nature in] a world of damaged but recoverable ecological integrity. (Grumbine 1994, 35)

Ecosystem management has several dominant themes. These fall into three categories – ecosystem considerations, management practices and human concerns.

Ecosystem considerations

Studying all connections in the ecological hierarchy

A tenet of modern ecology is that everything is connected to everything else, so that nothing in Nature can be understood in isolation. Such ecological interdependence has profound management implications: an ecosystem component cannot be altered without effects being felt throughout the system. This is why a solution to one problem often creates new, unforeseen problems. A 'single problem–single solution' approach is ecologically unsound, and may create more problems than it solves. Ecosystem managers should instead opt for solutions that attempt to embrace the gamut of problems. This will normally mean considering connections between all hierarchical levels of the ecosphere – genes, species, populations, communities, ecosystems and landscapes.

Including the true ecological boundaries

Ecosystems function on their own scales of space and time. Many are larger than a single administrative or management institution. This was recognized by the first exponents of modern ecosystem management, Frank and John Craighead, whose work showed that the needs of the grizzly bear (*Ursus arctos horribilis*) population could not be met within the confines of Yellowstone National Park, USA (e.g. Craighead 1979). The management of most ecosystems requires cooperation of managers from several different administrative and political units. It also requires an integrated ecological approach that often demands international action and cooperation. This is true for managing coastal resources, as well as terrestrial resources. A case in point concerns human impacts acting through terrestrial runoff on coastal marine ecosystems. In managing these impacts, it is useful to adopt the 'marine catchment basin' as a unit of study: such catchments may cross regional or even state boundaries.

Managing ecological integrity

This is the overall goal of ecosystem management. It means protecting the total native diversity (of species, populations, ecosystems and landscapes) and the ecological patterns and processes that maintain them. In practice, this normally includes one or more specific goals (Grumbine 1994, 31):

- Maintaining viable populations of all native species *in situ*, and reintroducing extirpated species where necessary

- Ensuring that, within protected areas, all native ecosystem types are represented. In other words, protecting the full natural range of ecosystem variation

- Maintaining ecological and evolutionary processes. This means guaranteeing the continuance of natural disturbance regimes, hydrological process and ecological processes

- Managing ecosystems for long enough periods to maintain the evolutionary potential of species, communities and ecosystems

- Accommodating human use and occupancy in the light of the above. This acknowledges the vital, if unsure, role that humans must play in all aspects of ecosystem management.

Sound conceptual models of ecosystem structure and function should inform ecosystem management. It is essential to understand natural ecosystem processes before attempting to manage an ecosystem. This means understanding the full gamut of ecological processes and patterns, from feeding relationships between species, to the principles of landscape ecology and the question of disturbance. The ecological role of **disturbance** is now widely recognized and is essential for successful ecosystem management practice. For instance, an ecological framework of natural disturbance, and knowledge of its component processes and effects, provides a sound basis for managing forests as a renewable resource.

> **disturbance**: the perturbation of an ecosystem by abiotic (e.g. fire and wind) and biotic (e.g. pathogens) agents

Environmental monitoring and data collection

Ecosystem management requires continuing research and data collection, as well as better management and use of existing data. This includes data on modern processes and historical data. An example of the use of modern data is the collection of resource inventory data on abiotic, biotic and cultural components of the Great Lakes shoreline environment that has enabled key issues and areas of concern to be identified (Lawrence *et al.* 1993). Historical data on environment, climate and culture in the fragile high Andean puna ecosystems has contributed to their management and preservation (Baied and Wheeler 1993). The behaviour of managed ecosystems must be carefully monitored so that the relative success or failure of actions can be tested and adjusted if necessary. Sometimes the remedial measures are more damaging than the problem. This is sometimes the case with marine pollution, and notably with oil slicks. It is also desirable to know whether environmental assessment objectives have been met. This requires long-term monitoring programmes, as instigated, for example, on streams in Pacific north-west forests of North America (Wismar 1993).

Management Practices

Three facets of management practice are highly pertinent to ecosystem management – adaptive management, interagency cooperation and organizational change.

Adaptive management

Ecosystem management is adaptive in the sense that scientific knowledge is regarded as provisional. New findings, especially from monitoring programmes, may alter advice to managers. Managers must thus remain flexible and adapt to the changeability of 'expert' scientific opinion while scientists must likewise adapt to social changes and demands. Ecosystem management calls for all parties to be adaptable. It is claimed, for instance, that the eventual restoration of depleted fish stocks will require adaptive management procedures (Alexander 1993).

Interagency cooperation

Ecosystem boundaries transgress administrative and political boundaries. National, regional and local management agencies, as well as private parties, must therefore learn to work together and integrate conflicting legal mandates and management goals. An example of this, from the United States, is the National Research Council's *Mandate for Change* report (1990) that urged close collaboration between forest managers and forest-user groups in order to develop a new paradigm of forestry research. Another example is the remedial action plans for Great Lakes water management, where the need for interdisciplinary and intergovernmental participation, and for political and public support, was recognized (MacKenzie 1993).

Organizational change

Ecosystem management demands that land management agencies change the way in which they operate. Forming interagency committees or changing professional norms and altering power relationships may do this. There is little doubt that the collaborative decision making called for by ecosystem management is likely to require extensive revision of traditional management practices and institutions (Cortner and Moote 1994). Adaptability and flexibility by all parties are the twin keys to success in ecosystem management.

Human Concerns

Humans in Nature

Ecosystem management accepts that people are part of Nature. Humans exert a profound influence on ecological patterns and processes, and in turn ecological patterns and process affect humans. The connection between the two is largely uncharted territory: interactions between ecosystems and social systems require far more research. However, policies are tending to move away from the administrator-as-neutral-expert approach, to policies that engender public deliberation and the discovery of shared values. Naturally, such extension of ecological matters to the social and political arena presents difficulties, but difficulties that are worth addressing.

Human Values

Ecosystem management accepts that human values must play a leading role in policy decision-making. Conservation strategies must take account of human needs and aspirations; and they must integrate ecosystem, economic and social needs (Kaufmann *et al.* 1994). The key players in ecosystem management are scientists, policy makers, managers and the public. The public, many of whom have a keen interest in environmental matters, are becoming more involved in ecosystem management as professionals recognize the legitimacy of claims that various groups make on natural resources. In Jervis Bay, Australia, the marine ecosystem is used by many existing and proposed conflicting interests (national park, tourism, urbanization, military training) (Ward and Jacoby 1992). Similarly, in the forests of the south-western United States, ecosystem, economic and social needs are considered in policy decision-making concerning ecology-based, multiple-use forest management (Kaufmann *et al.* 1994) (Figure 4.1).

The notion of sustainable development, as discussed in the previous section, encompasses the human dimension of ecosystem management.

TRADITIONAL ECOLOGICAL KNOWLEDGE

Indigenous peoples worldwide, living in different ecological zones, have accumulated huge bodies of knowledge about the human use of their environment. This repository of knowledge goes by many names: indigenous knowledge, traditional knowledge, indigenous technical knowledge, local knowledge, traditional cultural knowledge, traditional ecological knowledge and traditional environmental knowledge. For lucidity, it is always referred to as '**traditional ecological knowledge**' in this book.

Traditional ecological knowledge contrasts with western scientific or 'modern' knowledge (Table 4.2). To appreciate the differences between these two knowledge systems, it may be useful to consider the nature of knowledge. Knowledge is a body of information that grows as humans pigeonhole, codify, process and give meaning to their experiences. Knowledge includes 'scientific' knowledge and everyday or 'non-scientific' knowledge. It is the sum of human understanding and not just a specialized and sophisticated body of ideas pieced together by scientists. Every human being possesses knowledge that comes out of a complex set of social and cultural processes. These social–cultural processes are such that, as a rule, knowledge is linked to a specific place, culture or society; and it changes with time. Traditional ecological knowledge is a body of empirical knowledge and beliefs – a system for understanding the local environment – built by generations of people living in a particular place, through a process of cultural inheritance involving a mixture of oral-based traditions (stories and legends) and, in some cases, sacred texts. It captures the relationships of living beings with each other and with their environment, taking in religious and spiritual relationships, relationships between people and relationships between

> ***traditional ecological knowledge***: the huge mental database of knowledge about local environments and their human use held by indigenous peoples around the world

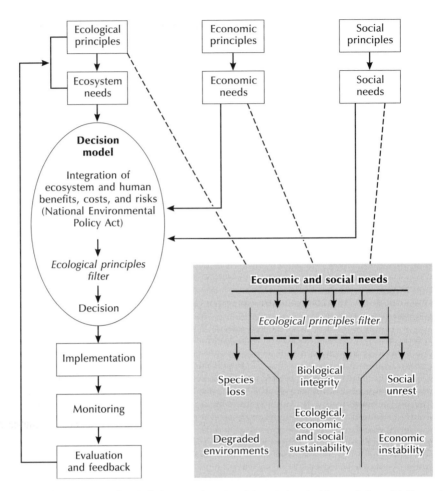

Figure 4.1 The integration of ecological, economic and social needs in a decision-analysis model. Economic and social needs are tested against an 'ecological filter', which is shown in the shaded box. The aim is to determine economic and social actions that will produce the most desirable bridge between biological integrity and ecological, economic and social sustainability. Bowing fully to economic and social needs would lead to species loss and environmental degradation. Bowing fully to ecological needs would lead to social unrest and economic instability. A compromise position allows the maintenance of biological integrity while catering for economic and social needs. The resulting decision model leads to the implementation of an environmental policy, the effects of which are carefully monitored and evaluated. If the policy should fail to work as desired, then amendments can be made and the process started anew, until a satisfactory outcome is achieved. *Source:* After Kaufmann *et al.* (1994)

people and their natural environment, including the use of resources. It is mirrored in language, social organization, values, institutions and laws of various cultural groups. As a system of knowledge, traditional ecological knowledge is based on individual and collective observation and experience, appraised in light of information handed down by elders. Indigenous peoples have relied on traditional ecological knowledge for their survival, staking their lives on its accuracy and repeatability. Human groups studied for their traditional ecological knowledge include circumpolar peoples and other hunter–fisher–gatherer communities from other areas.

Scientists are coming round to the view that it may be beneficial to work with the people who live in a particular place and who have acquired a deep insight into the natural processes at work in their environment. Purveyors of scientific

Table 4.2 Traditional ecological knowledge and scientific knowledge: a comparison

Comparators	Scientific knowledge	Traditional ecological knowledge
Epistemology (the nature of knowledge)		
Means of knowledge acquisition	Learned in an abstract manner, not always linked to application and from the separation of the observer from the object of knowledge	Generated through observations and experiments of uses and by identification with the object of knowledge
Basis of cognition	Analytical and objective	Intuitive and subjective
Process of knowledge transmission	Transmitted deductively through written word	Usually recorded and transmitted orally, sometimes via sacred texts
Integration with worldview and culture	Reductionist, objective, positivist, disembedded and compartmentalized	Holistic, subjective, experiential, embedded and integrated in the social, cultural and moral dimension
Cosmology		
View of life forces	Recognizes only plants and animals as having life force – separation between God and people	Views all matter as having life force, including inanimate forms – Animistic
Perception of Nature and life forms	Hierarchically organized and vertically compartmentalized – the environment is reduced to conceptually discrete components	Ecologically based on worldviews which emphasize social and spiritual relations between life forms
Explanation of environmental phenomena	Explanations derived through testing of hypotheses, using theories and laws of Nature	Spiritual explanations of environmental phenomena, revised and validated over time
Basis of relationship with Nature	Predicated on people's ability to dominate Nature	Shaped by the ecological system in which it is located
Nature of knowledge as a 'good'	Infinite good	A finite good
View of universe	Instrumentalism (views everything as sources of gratification)	Sees all entities in a relational context
Equality between life forms	Sees humans (especially western men) as superior life form, with an inherent right to control and exploit Nature	Stresses interdependency and equality of all life forms
Ontology (self)		
Basis of self worth	Predicated on individualistic values – nothing but the sum of a biological core and behavioural surfaces – the product of random genetic activity – identity and significance are derived from economic production or consumption	Predicated on group values or 'holism'
View of technology	A measure of civilization or backwardness	A phenomenon to be rejected or integrated into worldview
Context		
Dealing with change over time (phenomenological)	Synchronic-based on short time-series over a large area	Diachronic-based on a long time-series in one locality
Time measurement	Time is linear	Time is measured cyclically
Contextual validity	Superior on the basis of universal validity	Bound by time and space, social context and moral factors
Geographic context	Values mobility and weakens local context	Requires a commitment to the local context
Accountability		
Social accountability	Not usually associated with a system of social accountability except theoretical physicists in their role as 'high priests of science'	Associated with a system of social accountability (e.g. a shaman)

Source: Adapted from Studley (1998)

empirical: knowledge based on observation, experience and experiment – evidence that is seen, heard, touched or smelt

knowledge and traditional ecological knowledge have much to offer each other and need not be at loggerheads. Obvious areas of overlap in the two types of knowledge lie in **empirical** observation. Indigenous peoples may know the migrations patterns of bird, fish or mammal species; scientists would presumably be able to accept that information. Views are more likely to diverge over the causes of the migrations patterns. Indigenous peoples may well invoke religious or spiritual explanations that would clash with the scientific paradigm.

Traditional ecological knowledge is often difficult to retrieve. It is seldom written down, and few natural scientists are trained in the techniques of interviewing people and working in a cross-cultural setting. An example of how to overcome this obstacle comes from a study of indigenous communities in Alaska and eastern Russia. The method involved interviewing people to record what they knew about the environment. A pilot project studied beluga whales in three indigenous communities in Alaska and four indigenous communities in Chukotka, Russia, using a semi-directed interview. The researchers interviewed people either individually or in groups, and asked questions that started a discussion or conversation. Then, as the participants talked about what they had seen and learned, they discussed new topics as they cropped up. This method allowed the participants to make connections that the scientist may not have anticipated in a questionnaire, and led to some interesting observations and new ideas (Focus 4.2).

Many scientists, governments and indigenous peoples concur that scientific and indigenous knowledge systems should be integrated, owing to the failure of many conventional development models, the pluralistic nature of society and the ecological interdependence between nations. The role and importance of such integration is at the early stages of assessment and carried out in relatively few human groups. Moreover, the effective use of such integrated knowledge in decision-making and resource management has not yet been fully tested, and a number of problems remain unresolved. These problems include:

- The disappearance of traditional ecological knowledge and the lack of resources to document it before it is lost

- The reconciliation of two very different worldviews

- Translating ideas and concepts from one culture to another

- An acknowledgement of the value of rival knowledge systems

- Differences between social and natural scientists regarding appropriate methods to document and integrate indigenous knowledge

- The link between political power and modern knowledge.

It is questionable if integration alone goes far enough: if western people and western institutions wish to continue having a role managing the developing world, they may need to experience a sea change of opinion and start treating local knowledge as valid and local-knowledge holders as peers. They must be willing to countenance local worldviews and learn to understand and interpret the world from the perspective of indigenous peoples. Failure to do so may extend cultural imperialism and possibly lead to non-sustainable development.

FOCUS 4.2

TRADITIONAL ECOLOGICAL KNOWLEDGE OF BELUGA WHALES

A pilot project studied traditional ecological knowledge of beluga whales (*Delphinapterus leucas*) in three indigenous communities in Alaska and four indigenous communities in Chukotka, Russia. The research documented many details about the timing, location and direction of beluga movements in the areas around each community. These included the annual migrations as well as local movements when the belugas stay in an area for some time. The participants also described details of beluga behaviour, including feeding and calving, and the environmental factors such as ice, fish, wind and killer whales that affect belugas. This information was prepared in text and as maps (Figure 4.2).

In addition to the expected information described above, the semi-directed interviews led to other connections and observations that scientists might not have learned about by other methods. During one group interview session, the conversation suddenly switched to beavers instead of belugas. The connection turned out to be logical. Further conversation revealed that beavers dam streams where salmon and other fish spawn. Since the beaver population is increasing, this may mean loss of spawning habitat, changing the fish populations that the belugas feed on. In consequence, the beavers' activities may affect the belugas. Interestingly, many observations were similar in distant communities. One community in Alaska and one in Russia both described how belugas assist a female when she is giving birth. One beluga swims on either side of the female, pushing against her to help squeeze the calf out. The two communities are separated by hundreds of kilometres, speak different languages and are in different countries. Nonetheless, their knowledge of belugas is precise and consistent.

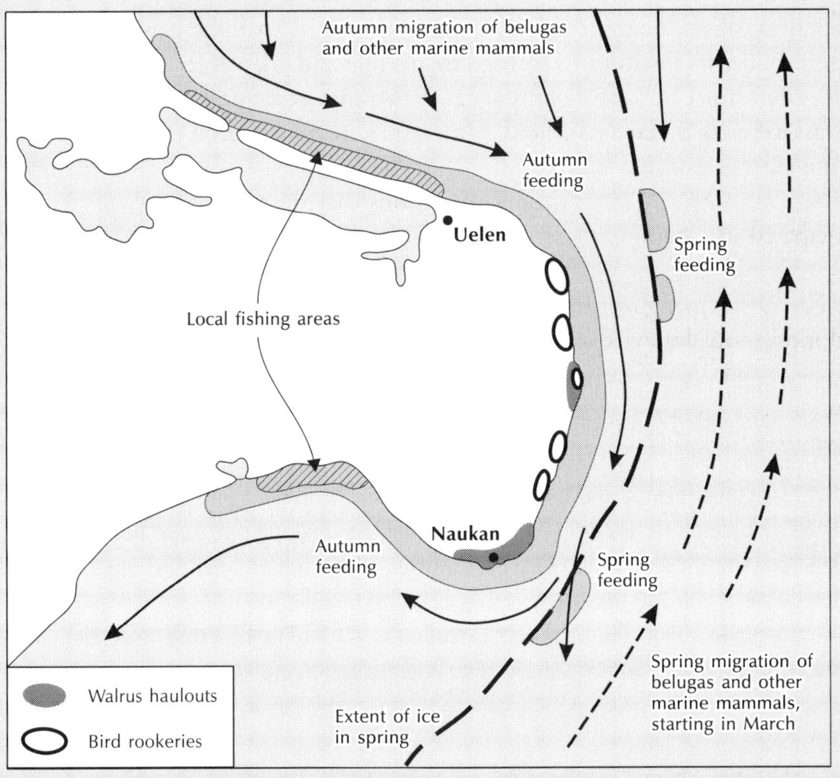

Figure 4.2 Traditional ecological information in the Uelen area, Chukotka, Siberia. *Source:* After Huntington and Mymrin (2002)

THE NEW CONSERVATION

The discussion of ecosystem management and traditional ecological knowledge touches on what is seen as a paradigm shift in biodiversity conservation from Nature-centred, exclusive, protected areas toward a conservation centred more around people or communities. This 'new conservation' involves three huge changes in conservation thinking and practice: (1) a shift of focus from the state to community level; (2) a reconceptualization of conservation using ideas of sustainable development, utilization and ecological dynamics; (3) an incorporation of neo-liberal ideas and market forces to 'make conservation pay' (Hulme and Murphree 1999). Critics of new conservation deride it as an attempt to 'get people on board' by re-labelling and repackaging 'old conservation'. New conservation policy, practice and institutions are, according to the critics, run by experts, undemocratic and dictatorial. To fend off such brickbats and to become a reality, the new conservation must make fundamental changes in priority setting, decision-making and organization (Brown 2003). In particular, it must face up to three challenges (Brown 2003). First, new conservation must adopt a more pluralist approach to the understandings, meanings and values of biodiversity held by different groups of people. Traditional ecological knowledge may be useful in meeting this challenge. Second, the new conservation must find a fair and just means of including greater deliberation and greater inclusion of various stakeholders in decision-making, and letting the stakeholders participate in an active way, with their different positions being recognized and respected. Third, the new conservation must remodel its institutions to become flexible and adaptable. The remodelled institutions must be capable of managing complex ecological systems and accommodating diverse stakeholder interests and values. Brown (2003, 91) sums up the current position nicely:

> ... the policy and institutional landscape of conservation and development is rapidly and quite significantly changed and changing. A wide range of different strategies and approaches will be necessary in the future to reconcile and trade-off the needs and demands of global to local societies. These will include traditional protected areas, but increasing [sic] more integrated and people-centred approaches to biodiversity conservation. I have outlined only three of the challenges for more successful, meaningful people-centred conservation. These are not exclusive and are not the only challenges. Very serious issues such as economic and cultural globalization and global environmental change provide the backdrop against which resilient and adaptive policy will evolve. Furthermore, issues of power inequalities and politics may act as serious constraints to these new approaches. However, integrated and pluralist approaches to new conservation are necessary if it is to bring benefits both to society and to biodiversity conservation. If this new conservation is really to make a difference, then more novel approaches to priority setting and implementation are needed.

4.2 INTERNATIONAL PROGRAMMES IN ENVIRONMENTAL MATTERS

global environmental change: environmental changes that move in roughly the same direction and are observed in most regions at about the same time; enhanced global warming is a case in point. Local effects of deforestation and urbanization may accumulate to produce changes of global extent

The second half of the twentieth century saw the emergence of several international research programmes set up to study **global environmental change** from an interdisciplinary perspective. The chief among these are the International Geosphere–Biosphere Programme (IGBP), the International Human Dimensions Programme on Global Environmental Change (IHDP), DIVERSITAS and the World Climate Research Programme (WCRP). These four programmes have formed the Earth System Science Partnership (ESS-P), and are all sponsored by the International Council of Scientific Unions (ICSU). It is worth exploring the aims of these groups as they represent an international human response to the current environmental crises that receive an airing in this book.

THE INTERNATIONAL GEOSPHERE-BIOSPHERE PROGRAMME (IGBP)

The ICSU General Assembly established the **IGBP** in September 1986. Since then, the IGBP has developed significantly, receiving support from various organizations, including UNEP (United Nations Environment Programme), UNDP (United Nations Development Programme), **UNESCO** and private sponsors. IGBP activities are steered by an ICSU Scientific Committee for the International Geosphere–Biosphere Programme.

The IGBP is an international scientific research programme founded on an interdisciplinary outlook, networking and integration. It addresses scientific questions that demand an international approach to provide an answer. In doing so, it adds value to a large number of individual, national and regional research projects by integrating activities to achieve enhanced scientific understanding. The IGBP's mission is to deliver scientific knowledge that will help human societies to develop in harmony with Earth's environment. Its scientific objective is to describe and to understand the interactive physical, chemical and biological processes that regulate the whole Earth system, the unique environment that the Earth system provides for life, the changes that are occurring in the Earth system and the manner in which they are influenced by human actions. To assist this aim, the IGBP's research programme is built around a family of core projects, each of which carries out integrative research in an area of Earth System Science. At present, there are ten projects, including six established Core Projects, a land use project in development that incorporates the social sciences, and three established framework activities that integrate research efforts and results (Basics 4.1).

> **IGBP**: the International Biosphere–Geosphere Programme
>
> **UNESCO**: United Nations Educational, Scientific and Cultural Organization, established 16 November 1945

THE WORLD CLIMATE RESEARCH PROGRAMME (WCRP)

The WCRP was established in 1980, under the joint sponsorship of the International Council for Science (ICSU) and the World Meteorological Organization (WMO). Since 1993, the Intergovernmental Oceanographic Commission (IOC) of UNESCO has also been a sponsor.

The objectives of the WCRP programme are to develop the fundamental scientific understanding of the physical climate system and the climate processes that is needed to determine to what extent climate can be predicted and the extent of human influence on climate understood. The programme embraces studies of the Earth's physical climate system – the global atmosphere, the oceans, sea and land ice and the land surface. The WCRP studies are specifically directed to provide scientifically founded quantitative answers to the questions being raised on climate and the range of natural climate variability, as well as to establish the basis for predictions of global and regional climatic variations and of changes in the frequency and severity of extreme events.

WCRP activities thus forcefully address outstanding issues of scientific uncertainty in the Earth's climate system, including transport and storage of heat by the ocean, the global energy and water cycle, the formation of clouds and their effects on radiative transfer and the role of the **cryosphere** in climate. These activities match the scientific priorities identified by the Intergovernmental Panel on Climate Change (IPCC), and provide the basis for responding to issues raised in the UN Framework Convention on Climate Change (UNFCCC). WCRP

> **cryosphere**: all the frozen waters of the Earth (snow and ice)

BASICS 4.1

IGBP PROJECTS

Core Projects

Biospheric Aspects of the Hydrological Cycle (BAHC)

Includes local-scale observational and modelling studies of soil–vegetation–atmosphere transfer processes; regional-scale studies of water fluxes and other land–atmosphere interactions; spatial and temporal synthesis of biospheric parameters at the regional to continental scale; and the downscaling of weather information from General Circulation Models (GCMs), for application to ecosystem research.

Global Change and Terrestrial Ecosystems (GCTE)

Aims to develop a predictive understanding of the effects of changes in climate and atmospheric composition and land use on terrestrial ecosystems (both natural and managed), and to determine feedback effects to the physical climate system. Activities are being developed in collaboration with the proposed Global Terrestrial Observing System (GTOS). Modelling studies are focused on the construction of dynamic vegetation and agricultural systems models at a variety of scales, both for compiling biogeochemical models and physically based GCMs and for direct impact studies.

International Global Atmospheric Chemistry Project (IGAC)

Addresses the relationships between atmospheric composition, biospheric processes and climate. Emphasis is on the atmospheric constituents that have a role in global climate control: greenhouses gases (such as carbon dioxide, methane and nitrous oxide); aerosols and cloud condensation nuclei; and ozone and other trace reactive gases and radicals. The project also includes work on global emission inventories, measurement intercalibration and standards, the establishment of global monitoring networks and modelling.

Joint Global Ocean Flux Study (JGOFS)

Designed to improve knowledge of the processes controlling carbon fluxes between the atmosphere, surface ocean, ocean interior and its continental margins, and the sensitivity of these fluxes to climate change. Ocean regions under investigation by JGOFS include the North Atlantic, Equatorial Pacific, the north-west Indian Ocean and the Southern Ocean. A global carbon dioxide survey is being carried out in close collaboration with the WCRP's World Ocean Circulation Experiment (WOCE).

Land–Ocean Interactions in the Coastal Zone (LOICZ)

Directed at an assessment of fluxes of matter between land, sea and atmosphere through the coastal zone; the responses of the land–ocean interface to global change, particularly sea level rise; the biogeochemical responses of coastal systems to global change, with emphasis on the carbon cycle and exchanges of certain atmospheric trace gases; and the socio-economic implications of the degradation of the coastal zone and the need for the development of new policies for the integrated management of coastal environments.

Past Global Changes (PAGES)

Charged with providing a quantitative understanding of the Earth's past environment, and with defining the envelope of natural environmental variability within which to gauge human impacts on the biosphere, geosphere, and atmosphere. PAGES seeks to obtain and interpret a variety of palaeoclimatic records, and to provide the data essential for the validation of predictive climatic models.

Land Use and Cover Change (LUCC)

Jointly planned with the Human Dimensions of Global Environmental Change Programme. It will address how land use, and thus land cover and land surface properties, are affected by socio-economic factors. It aims to integrate the driving forces of land-cover change into a global land use and land cover change model.

Framework Activities

Global Analysis, Interpretation and Modelling (GAIM)

This framework activity aims to develop comprehensive, prognostic models of the global biogeochemical system that can be coupled with GCMs, in collaboration with modelling groups of the WCRP. It has recently started to bring together previously unlinked models of interacting processes, with initial emphasis on the global carbon cycle and the interaction between climate and terrestrial ecosystems at regional scales.

IGBP Data and Information System (IGBP–DIS)

This facility is at the University of Paris, France. Its purpose is to aid IGBP Core Projects, when necessary. Specifically, it helps them develop their individual data system plans. It also helps to provide an overall data system plan for IGBP; carries out activities directly leading to the generation of data sets; ensures the development of effective data management systems and acts, where appropriate, to make certain the data and information needs of IGBP are met through international and national organizations and agencies.

Global Change System for Analysis, Research and Training (START)

This provides a system of networked regional research centres and sites that encourage research into the regional origins and impacts of global environmental changes. In addition, through training and fellowship programmes, it enhances indigenous scientific capacity to engage in focused research on decisive regional environmental issues of global importance.

also lays the scientific foundation for meeting the research challenges posed in Agenda 21. Together with the International Geosphere–Biosphere Programme (IGBP) and the International Human Dimensions of Global Environmental Change Programme (IHDP), WCRP provides the international framework for scientific cooperation in the study of global change. Basics 4.2 lists some WCRP projects.

The development of global climate models is an important unifying component of WCRP, building on scientific and technical advances in the more discipline-oriented activities. These models are the fundamental tool for understanding and predicting natural climate variations and providing reliable estimates of anthropogenic climate change. Models also provide the essential means of exploiting and synthesizing, in a synergistic manner, all relevant atmospheric, oceanographic, cryospheric and land surface data collected in WCRP and other programmes. The Working Group on Numerical Experimentation (WGNE), jointly sponsored by the JSC of WCRP and the World Meterological Organization (WMO) Commission for Atmospheric Sciences (CAS), leads the development of atmospheric circulation models for both climate studies and numerical weather prediction, but each project also has a numerical experimentation group, that supports the respective WCRP components.

In addition, the Working Group on Coupled Modelling (WGCM), jointly sponsored by JSC and CLIVAR, leads the development of coupled ocean–atmosphere–land models used for climate studies on longer time-scales. WGCM is also WCRP's link to the Earth system modelling in IGBP's Global Analysis Integration and Modelling (GAIM) and to the Intergovernmental Panel on Climate Change (IPCC). Activities in this area concentrate on the identification of errors in model climate simulations and exploring the means for their reduction by organizing coordinated model experiments under standard conditions. Under the auspices of WCRP, the Atmospheric Model Intercomparison Project (AMIP) has been set up in which the ten-year period 1979–88 has been simulated by thirty different atmospheric models under specified conditions. The comparison of the results with observations has shown the capability of many models to adequately represent mean seasonal states and large-scale interannual variability for several basic climate parameters. A similar Coupled Model Intercomparison Project (CMIP) is also now being organized.

THE INTERNATIONAL HUMAN DIMENSIONS PROGRAMME ON GLOBAL ENVIRONMENTAL CHANGE (IHDP)

The IHDP is an international, non-governmental, interdisciplinary science programme. The International Social Science Council (ISSC) launched the IHDP in 1990 as the Human Dimensions Programme (HDP). In February 1996, ICSU joined ISSC as co-sponsor of the IHDP, and the Secretariat moved to Bonn, Germany. The restructured IHDP is a full partner with the IGBP,

BASICS 4.2

SOME WCRP PROJECTS

WCRP has formulated a broad-based multi-disciplinary science strategy offering the widest possible scope for investigation of all physical aspects of climate and climate change. This multi-disciplinary strategy is reflected in the ongoing WCRP core projects. Scientific steering or working groups normally meet once a year and lead all projects.

Arctic Climate System Study

The ACSYS is a regional project that studies the climate of the Arctic region, including its atmosphere, ocean, sea ice and hydrological regime. It is being expanded into a global project, Climate and Cryosphere (CliC), investigating the role of the entire cryosphere in global climate.

Climate Variability and Predictability

CLIVAR is the focus in WCRP for studies of climate variability, extending effective predictions of climate variation, and refining the estimates of anthropogenic climate change. CLIVAR is attempting particularly to exploit the 'memory' in the slowly changing oceans and to develop understanding of the coupled behaviour of the rapidly changing atmosphere and slowly varying land surface, oceans and ice masses as they respond to natural processes, human influences and changes in the Earth's chemistry and biota. CLIVAR will advance the findings of the successfully completed Tropical Ocean and Global Atmosphere (TOGA) project, and will expand on work now underway in WCRP's World Ocean Circulation Experiment (see below).

The Global Energy and Water Cycle Experiment

GEWEX is the scientific focus in WCRP for studies of atmospheric and thermodynamic processes that determine the global hydrological cycle and water budget and their adjustment to global changes such as the increase in greenhouse gases. One of the main thrusts of GEWEX is the implementation of a series of atmospheric/hydrological regional process studies such as the GEWEX Continental-scale International Project (GCIP) embracing the whole Mississippi river basin, the GEWEX Asian Monsoon Experiment (GAME) or the Baltic Sea Experiment (BALTEX). Observational projects established in order to satisfy specific and evolving scientific needs include the International Satellite Cloud Climatology Project (ISCCP), the International Satellite Land Surface Climatology Project (ISLSCP), Global Water Vapour Project (GVaP) and the Global Precipitation Climatology Project (GPCP). Progress has also been made in the GEWEX Cloud System Study (GCSS), which aims to develop improved parameterizations and models of cloud systems used in climate and numerical weather prediction studies.

Stratospheric Processes And their Role in Climate

SPARC plays an important role in better understanding the climate system. It concentrates on the interaction of dynamical, radiative and chemical processes. Activities organized by SPARC include the construction of a stratospheric reference climatology and the improvement of understanding of trends in temperature, ozone and water vapour in the stratosphere. Gravity wave processes, their role in stratospheric dynamics and how these may be parameterized in models is another topic taken up.

World Ocean Circulation Experiment

The WOCE is a central element of the WCRP scientific strategy to understand and predict changes in the world ocean circulation, volume and heat storage, resulting from changes in atmospheric climate and net radiation. The project uses a combination of *in situ* oceanographic measurements, observations from space and global ocean modelling.

WCRP and DIVERSITAS (an international global environmental change research programme). It is devoted to promoting and coordinating research that describes, analyses and probes the human dimensions of global environmental change. The human dimensions of global environmental change are the causes and consequences of individual and collective human actions, including the changes that modify the Earth's physical and biological systems and that affect the human quality of life and sustainable development in different parts of the world. Typical topics covered include forest management, material flow analysis and peri-urban ecosystems.

An international scientific committee, comprising scientists from different national and disciplinary backgrounds, guides the IHDP. At present, four science projects receive the full support of the IHDP:

- Land Use and Land Cover Change (LUCC, co-sponsored by the IGBP)

- Global Environmental Change and Human Security (GECHS)

- Institutional Dimensions of Global Environmental Change (IDGEC)

- Industrial Transformation (IT).

These projects are a key mechanism used in recognizing and creating new IHDP research activities in priority areas, in promoting international collaboration and in linking policy-makers and researchers.

National Committees are an essential component of the IHDP's networking and research strategy, which advocates a bottom-up approach. There are currently approximately thirty-five National Human Dimensions Committees and Programmes at various levels of development worldwide. One of the top priorities set by the IHDP is to promote and strengthen existing national committees and programmes, and to support the establishment of new ones.

DIVERSITAS

DIVERSITAS is an international global environmental change research programme that was established 1991 as an overall global cooperation programme for research on biodiversity. Its sponsors are the International Council for Science (ICSU), SCOPE (Scientific Committee on Problems of the Environment), IUBS (International Union of Biological Sciences), IUMS (International Union of Microbiological Societies) and UNESCO–MAB (Man and the Biosphere). It has two avowed missions:

- To promote integrative biodiversity science by linking biological, ecological and social disciplines to create a socially relevant new knowledge

- To provide the scientific basis for an understanding of biodiversity loss, and to draw out the implications for the policies for conservation and sustainable use of biodiversity.

DIVERSITAS works towards its objective in close collaboration with three other international global environmental change research programmes: the IGBP, IHDP and WCRP. These four programmes have formed the Earth System Science Partnership (ESS-P).

DIVERSITAS has established Core Programmes Elements in five areas:

- The effect of biodiversity on ecosystem functioning

- Origins, maintenance and change of biodiversity

- Systematics: inventorying and classification

- Monitoring of biodiversity

- Conservation, restoration and sustainable use of biodiversity.

It has also established six special target areas of research (stars):

- Soil and sediment biodiversity

- Marine biodiversity

- Microbial biodiversity

- Inland water biodiversity

- Human dimensions

- Invasive species and their effect on biodiversity.

The past decade has seen the birth of the Convention on Biological Diversity, of many conservation programmes aimed at protecting biodiversity, as well as many national research programmes dedicated to developing biodiversity science. Scientific efforts, however, need international coordination to address the complex scientific questions posed by the loss and change of biodiversity globally. Many of these questions also require a research framework integrated across disciplines. DIVERSITAS endeavours to establish international, multidisciplinary networks of scientists working on biodiversity that address the scientific priorities presented in its draft science plan, which is articulated around the three following core projects:

- Core Project 1: Understanding, monitoring and predicting biodiversity changes

- Core Project 2: Assessing impacts of biodiversity changes

- Core Project 3: Developing the science of conservation and sustainable use of biodiversity.

The expertise supplied by DIVERSITAS core projects is relevant to the aims of Agenda 21 and to the aims of the Convention on Biological Diversity (CBD) and its Subsidiary Body on Scientific, Technical and Technological Advice (SBSTTA). With its interdisciplinary focus, ranging from the natural to the human dimensions of biodiversity, DIVERSITAS provides key insights into a broad range of important socio-economic policy questions. It also encourages innovations and partnerships between research institutions and universities of developed and developing countries.

4.3 CENTRES FOR GLOBAL CHANGE RESEARCH

The following three examples give the flavour of work done in centres devoted to researching global change.

EARTH SYSTEM SCIENCE CENTER AT PENNSYLVANIA STATE UNIVERSITY

Created in 1985, Pennsylvania State's Earth System Science Center is boldly interdisciplinary. It brings together several departments: geosciences, meteorology, geography, and energy, environmental and mineral economics. Its avowed aim in doing so is to seek out links between the Earth's biophysical processes and global change past and future. It coordinates and conducts extensive research on four big issues: the global water cycle, biogeochemical cycles, Earth system history and human impacts on the Earth system.

Faculty and students cooperate closely in the planning and the execution of research projects, share expertise and insight into research problems, and work together on education and outreach programmes. It plays a foremost role in the development of Earth system science as an academic discipline.

POTSDAM INSTITUTE FOR CLIMATE IMPACT RESEARCH

Taking advantage of German reunification, representatives of ministries, universities and research communities set in motion plans that led to the founding of the Potsdam Institute for Climate Impact Research (PIK) on 1 January 1992. PIK is a member of the Leibniz Gemeinschaft (the 'Leibniz community', which is a group of 80 scientifically, legally and economically independent research institutes and service facilities in Germany) and has annual core funding of around 6 million euros from the German Federal Republic and the State of Brandenburg. It also receives soft funding from national and international sources amounting to several million euros, including major contributions from the European Community. More than 170 members of the Institute make up the five scientific departments – Integrated Systems Analysis, Climate System, Global Change and Natural Systems, Global Change and Social Systems and Data and Computation – together with the small administrative and service units. This departmental structure does not hinder interdisciplinary investigation: small project teams carry out the actual research, the members of each team coming from at least two different departments. The Institute runs a supercomputer (a 200-processor, IBM SP parallel computer), which now runs at teraflop level (a terraflop is a trillion calculations per second). This computer is central to the Potsdam research approach. No institute in the world could single-handedly collect the detailed information or study all the processes relevant to the subject of climate change. Members of the international scientific community, probably numbering tens of thousands, undertake the bulk of such data collection and processes studies. PIK's role is to look at the data in parallel to get a macroscopic view of the problems, the better to inform priorities and possibilities for policy-making.

HADLEY CENTRE, UK

The Hadley Centre for Climate Prediction and Research is part of the UK's Met Office. It is a focal point in the UK for scientific issues connected to climatic change. Its chief aims are:

- To understand the physical, chemical and biological processes within the climate system, and to develop state-of-the-art climate models that represent them

- To use climate models to simulate global and regional climate variability and change over the last century, and to predict changes over the next century

- To monitor global and natural climate variability and change

- To identify the factors responsible for recent climatic changes

- To understand and to predict the year-to-year and decadal variability of climate in the UK.

Most of the funding for the Hadley Centre comes from contracts with the Department for Environment, Food and Rural Affairs (DEFRA), other UK Government departments and the European Commission. The Centre currently employs around 100 staff and uses two Cray T3E supercomputers. An example of its output is the predictions (released on 16 April 2002) of the UK's climate over the next 100 years, made using the Centre's new regional climate model with a spatial resolution of 50 km. The results are more detailed than any before and suggest more extreme variations of climate than in previous scenarios. For the 2080s, the predicted weather changes for the worst-case scenario are:

- winter rainfall increases by up to 30 per cent, particularly in the south, with the potential for a greater incidence of flooding

- days of heavy rain will also increase in winter

- summer rainfall could decrease by up to 50 per cent, again particularly in the south, threatening water supplies.

In addition, rising sea levels will lead to a much greater frequency of high water levels from storm surges on some coasts.

SUMMARY

Human minds have responded to global environmental threats. The response appears in the environmental movement, and in notions of sustainability, ecosystem management, traditional ecological knowledge and the new conservation. A common focus of these topics is the placing of human societies and economies on an equal footing with the natural world, so that social and cultural needs, economic needs and ecological needs 'balance' in a sustainable way. Another response of the human intellect to pressures on the global environment is the setting up of international programmes to tackle environmental problems. The big players are the International Geosphere–Biosphere Programme (IGBP), the World Climate Research Programme (WCRP), the International Human Dimensions Programme (IHDP) and DIVERSITAS, an international programme of biodiversity science. Several research centres also have global change as their prime concern. Examples are the Potsdam Institute for Climate Impact Research, Germany, and the Earth System Science Center at Pennsylvania State University, USA. Others, such as the Hadley Centre in the UK, have an interest in global issues, but also focus on regional changes.

Questions

1 Why does environmentalism cover such a wide range of views?
2 What is sustainability?
3 To what extent does ecosystem management offer a sustainable approach to conservation?
4 Are traditional ecological knowledge and scientific knowledge reconcilable?
5 What are the pros and cons of international programmes for studying the environment?

Further Reading

Berkes, F. (1999) *Sacred Ecology: Traditional Ecological Knowledge and Resource Management*. Philadelphia: Taylor & Francis.
A worthwhile text on indigenous knowledge and its applications in natural resource management.

Brown, K. (2003) 'Three challenges for a real people-centred conservation'. *Global Ecology and Biogeography* 12, 89–92.
A useful appraisal of the 'new conservation'.

Inglis, J. T. (ed.) (1993) *Traditional Ecological Knowledge: Concepts and Studies*. Ottawa, Canada: International Development Research Centre (IDRC).
A good source book.

Miller, K. R. (1996) *Balancing the Scales: Guidelines for Increasing Biodiversity's Chances Through Bioregional Management*. Washington DC: World Resources Institute.
Includes discussion of ecosystem management.

Reid, D (1995) *Sustainable Development: An Introductory Guide*. London: Earthscan.
A good place to start learning about sustainable development.

Sillitoe, P. (2000) *Indigenous Knowledge Development in Bangladesh: Present and Future*. Rugby: Intermediate Technology Development Group (ITDG) Publishing.
A rich collection of examples of traditional ecological knowledge applied to development in Bangladesh.

Sachs, W., Loske, R. and Linz, M. (1998) *Greening the North – A Post-Industrial Blueprint for Ecology and Equity*. London and New York: Zed Books.
Worth a look.

Sillitoe, P., Bicker, A. and Pottier, J. (eds) (2002) *Participating in Development: Approaches to Indigenous Knowledge*. London and New York: Routledge.
An anthropological slant on traditional ecological knowledge.

Satterthwaite, D. (ed.) (1999) *The Earthscan Reader in Sustainable Cities*. London: Earthscan.
Chapters deal with key issues of sustainable cities and sustainable urban development, including environmental justice, health, transport, industry and designing with Nature.

Useful Websites

http://www.ihdp.uni-bonn.de/
The IHDP site. Contains some excellent links.

http://www.diversitas-international.org/
The DIVERSITAS site.

http://www.ipcc.ch/
The IPCC site.

http://www.ankn.uaf.edu/index.html
The Alaska Native Knowledge Network, a good starting place for information and links on traditional ecological knowledge.

http://www.met-office.gov.uk/research/hadleycentre/
The Hadley Centre for Climate Prediction and Research. Part of the Met Office.

http://www.pik-potsdam.de/
The Potsdam Institute for Climate Impact Research.

http://www.essc.psu.edu/
Earth System Science Center at Pennsylvania State University.

2

**TOOLS AND
TECHNIQUES**

CHAPTER 5
MEASURING AND MONITORING

OVERVIEW	LEARNING OUTCOMES
Measurement and monitoring, in combination with theory, are vital in understanding how environmental systems are put together, how they work and how they change. Several questions are crucial to the process of measuring and monitoring and need thinking through before an investigation starts. First, it is essential to establish what will be measured and monitored and why. Second, it must be decided how the measuring and monitoring will be done, and third, when and where. Finally, it is advisable to appreciate the limitations of the techniques being used. This chapter begins with a general discussion of these questions. By taking examples from each of the key environmental media – air, water, soil, sediments and land cover – it then shows how measuring and monitoring help in the understanding of how the Earth system and its components work.	Reading this chapter should help you understand the overall purpose of measuring and monitoring and you will learn how to measure and monitor: • Air, especially urban air • Water • Soil • Earth's surface processes • Land cover and land cover change.

5.1 INTRODUCTION

In the quest for a greater understanding of physical form and physical processes, it is first necessary to find a way of identifying, representing and capturing their fundamental characteristics. One of the key ways this is done is through measurement, whether this is the sediment load of the River Rhine or the concentration of air pollutants in the air around a freeway in Los Angeles. Once there is a measurement of a particular phenomenon, it becomes comparable with a whole series of other measurements of other phenomena, and this comparability is a critical starting-point to a better understanding of why the world is as it is. However, a single measurement, or even a set of measurements made at the same time within a region, can only give a snapshot of the environment at a particular time and at a particular location. To gain a deeper understanding of the physical environment, it is necessary to view these characteristics as the result of one or more ongoing processes of change. Building a series of measurements over time may help to elucidate the behaviour of these dynamic processes. Over human timescales, making systematic measurements over time and space can show and monitor environmental change. Through doing this, and by investigating long-term environmental records, it is possible to begin to appreciate the nature and complexities of countless interrelated processes operating over a range of timescales – from seconds to millions of years – and over a range of spatial scales – from a few square centimetres to the whole of the globe.

In considering the measurement and monitoring of physical characteristics and phenomena, it is useful to consider a number of key questions from the start:

- What is being measured and monitored and why?

- How is it being measured and monitored?

- When and where is the measurement and monitoring taking place?

- What are the limitations to measurement and monitoring?

This chapter begins with a general discussion of these questions. To explore the issues in more detail, it then looks at examples in each of the key environmental media in turn, to determine how measuring and monitoring aids understanding of the processes at work.

Figure 5.1 shows an example of the different ways of measuring and monitoring Earth surface systems, and how these relate to the general questions highlighted above.

THE PURPOSE OF MEASURING AND MONITORING: WHAT AND WHY?

Most of the time, measurement and monitoring work responds to a need to know more about something in particular. In other words, a researcher has a particular problem or issue in mind and this provides the focus for the measurement and monitoring activity to be conducted. The purpose of the measurement activity is important as it helps to determine some of the answers to the other questions in this section, that is, what techniques to use and when

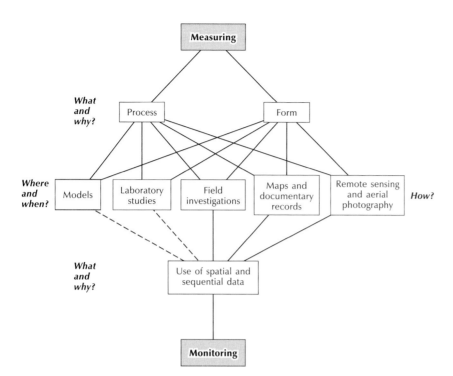

Figure 5.1 The where and when, what and why, and what and how of measuring and monitoring Earth surface systems

and where to use them. It is therefore very important to give careful and considered thought to this question before embarking on any further work. Not doing so could result in a lot of effort and expense going into the generation of an extensive dataset, but this dataset being meaningless or inappropriate for looking at the issues of interest. Since, ultimately, measurement and monitoring data need to be turned into meaningful information, it is also useful to give some thought as to the type of analysis that might be required and the nature of the data needed to carry this out (see Basics 5.1, 5.2, 5.3).

BASICS 5.1

WHAT ARE DATA?

Data are observations made from observing, measuring and monitoring the real world. They are collected as raw facts or evidence, and require processing before turning into useful information about a particular phenomenon. This information is then interpreted in order to further understanding of characteristics and processes. Information is created when a meaning and a context are associated with the raw data that are collected.

Environmental data have three components or dimensions associated with them:

1. A thematic dimension – describing the nature of the data (i.e. the phenomenon or characteristic being measured). This will also include some reference to the units in which the measurement is being made. Measurement data without reference to the corresponding measurement units are completely meaningless.

2. A temporal dimension – describing the time point or time period to which the data relate.

3. A spatial dimension – describing the geographical location at which the measurement is made. More information about spatial data is given towards the end of the chapter.

BASICS 5.2

THE SCIENTIFIC METHOD AND SAMPLING

Most physical geographers use quantitative analysis to measure properties of processes or phenomena and to find relationships between them. Quantitative analysis rests upon a theoretical approach to research that assumes an objective ordered world exists independently of the researcher and the researched. Knowledge is only composed of 'facts' that can be observed, and the aim of the researcher is to establish law-like statements about phenomena and the relationships between them. To do this, researchers use the scientific method, which usually operates as follows:

- Identify a research question
- Formulate a hypothesis to test the research question
- Collect relevant data
- Interpret the results and test the hypothesis
- Accept or reject the hypothesis
- Make generalizations about the phenomena.

Statistical analysis provides a means of testing hypotheses and explaining relationships between observed phenomena. The use of statistics also allows inferences from a sample to a population, where a **population** is a finite or infinite set of observations, and a **sample** is a small subset of these (note that this is a statistical population, not a set of people). Often, samples are taken, as the 'population' under examination is infinite and it is unnecessary and would be too costly and time-consuming to take complete counts or measurements.

A properly taken sample can give valuable information concerning the nature of the population from which it was drawn. However, there is uncertainty about the statistical validity of the sample method, as it concerns only part of the entire data array, i.e. the population, and there is a chance that the sample data may not completely reflect the population. To use samples with confidence, the **total sampling error** must be considered. The total sampling error comprises the **sampling error**, which is the potential difference between the 'true' value from the population and the estimation gained from the sample (i.e. how representative the sample is of the population), and the **non-sampling error**, which is the measurement error (e.g. writing wrong numbers, or the limitations of the equipment).

Sampling error can result from too few measurements being taken to allow precise and accurate measurements or statistics to be made (see Concepts 5.1). It can also arise from the samples not being representative of the subject under study. This can occur for a variety of reasons, such as sampling at the wrong time, season or location, or using the wrong or inadequate equipment. For example, in a study to define the particle-size distribution of a beach, the use of small sample containers may be good for sand samples but do not allow larger particles such as shingle or cobbles to be collected, and so a non-representative sample is collected.

There is no relationship between sampling and non-sampling error but each is related to sample size. Sampling error decreases with increasing sample size; accuracy increases in relation to the square root of the sample size. This means that to double the accuracy, the sample size needs quadrupling. Non-sampling error generally increases with increasing sample size but the relationship with sample size is not as defined as with sampling error.

---

population (in statistics): a set of observations on a particular phenomenon; populations may be finite, as in the number of fields on the Isle of Wight or infinite, as in sand grains on a beach

sample (statistics): a representative subset of a population

total sampling error: the combined potential error of the sample being unrepresentative and human errors taking the sample

sampling error: the potential difference between the 'true' value from the population and the estimation gained from the sample

non-sampling error: the measurement error, e.g. writing wrong numbers

BASICS 5.3

SCALES OF MEASUREMENT

Measurement follows a set of rules that define a hierarchy of measurement scales: the nominal scale, the ordinal scale, the interval scale and the ratio scale (Table 5.1). The scale of measurement will determine what forms of statistical analysis can be applied to a dataset.

Nominal scaling

Scientists mainly use the nominal (or categorical) scale to identify objects and events by numbering them. Land cover types may be designated as 1 = arable, 2 = grassland and 3 = woodland. Nominal scaling simply establishes equivalence of objects according to a measured attribute and the values have no 'numerical' meaning in the usual sense of numbers. It does not allow arithmetic manipulation of the classes recognized: grassland (land cover type 2) cannot be added or subtracted from arable (land cover type 1). The counting of numbers of objects that fall within some particular class is a derivative of nominal scaling. For example, a particular daily rainfall total might be set to decide whether a day was wet or dry. Such count totals – the number of wet and dry days in the example – yield class frequencies that are susceptible to arithmetic manipulation. Nominal scaling is a weak form of measurement, but is applicable to many objects in the physical environment, and includes such subjective qualities as noisy or quiet, beautiful or ugly.

Ordinal scaling

Ordinal scales order objects from most to least, but provide no information about the intervals separating the measured attribute, which are indeterminate. So, a set of objects is ranked from highest to lowest by applying the criterion 'greater than' or 'less than'. For instance, a stream network ordered according to Strahler's system of stream ordering ranks streams as first-order (the smallest), second-order (the next smallest) and so on. Ordinal scales lend themselves to sophisticated analytical procedures.

Interval scaling

In an interval scale, the dimensions of the steps between intervals are known, as well as the equivalence and the rank. Consequently, the ratio of any two intervals, say between 10°C and 20°C, is known. In this example, the ratio would be 10:20 or 1:2. Note that the ratio between two points on an interval scale depends upon the unit of measurement: 0°C to 100°C and 32°F to 212°F are not the same ratio.

Ratio scaling

Ratio scales bear all the properties of interval scales, but additionally, the ratio of any two scales values is known and the same, irrespective of the units of measurement. For instance, a 40-kg wombat has twice the body mass of a 20-kg wombat; and likewise a 40-pound wombat has twice the body mass of a 20-pound wombat. All ratio scales have a real zero point, whereas zero points on interval scale are arbitrary. This is why 20°C is not twice the temperature of 10°C, but 20 K is twice the temperature of 10 K.

Table 5.1 Scales of measurement

Scale	Determinative operation	Arithmetic operations	Examples
Nominal (categorical)	Equality Equivalence	Counting	Land-cover types (woodland, arable, etc); presence–absence data; gender; marital status
Ordinal	Greater or lesser (rankings), as well as equivalence	Greater than or less than operations	Size classes of objects (e.g. hamlet, village, town, city, metropolis); land use capability classes; stream order
Interval	Equality of intervals or differences	Addition and subtraction of scale values	Temperature in °C; time in years AD or BC; IQ scores
Ratio	Equality of ratios	Multiplication and division of scale values	Temperature in degrees Kelvin (K); biomass in grams; spot heights in metres; gross national product; distance in km

Source: After Thomas and Huggett (1980)

TECHNIQUES FOR MEASURING AND MONITORING: HOW?

Physical geographers may measure and monitor aspects of the environment directly and indirectly. They make **direct measurements** in real time and *in situ* (i.e. when and where the processes actually occur) by going into the field armed with eyes, noses, ears, notebooks and pieces of measuring and sampling equipment ('kit'). They may also take samples from the environment back to the laboratory to make such measurements. They make **indirect measurements** by use of remote sensors mounted on airborne or space-borne platforms that do not make direct contact with the air or ground that they 'sense'. They also make indirect measurements by creating an artificial environment in the laboratory and using this to try to simulate real-world processes. This may be necessary, for example, in the case of processes that take place over very long timescales, or which take place in conditions that make it very difficult to carry out direct measurements, or which occur at global scales.

Direct methods

Ground-based measuring and monitoring of the environment can be an arduous and time-consuming business, especially when conducted in inaccessible areas or difficult terrain. It may involve the logging of field observations in notebooks or in digital form. Modern ground-based procedures commonly involve the use of electronic instruments, such as **automated instruments** and **data-loggers**, which record measurements on digital media. In many cases, samples of materials – soils, sediments, air and so on – are suitably stored and taken to a laboratory for measuring using equipment unsuitable for field use.

Indirect methods I: remote sensing

Sensors mounted on airborne platforms – balloons or aircraft – take various kinds of aerial photograph. Aerial photographs are a valuable source of information about the environment. Skilled men and women, trained in aerial photographic interpretation, extract the information in aerial photographs, though some automated techniques are now available. Airborne platforms may also carry digital scanners that are similar to the digital scanners mounted on space-borne platforms (satellites). The chief differences between airborne and space-borne digital scanners lie in the land surface area that they can resolve and the operation characteristic of the carrying platforms. Satellite-borne scanners may cover the globe through continuous and repeated over-flights along preset orbital patterns, but cloud cover and unfavourable atmospheric conditions disrupt their records. Aircraft-borne scanners can take advantage of good weather, but cannot hope to obtain a global coverage. Nonetheless, they do furnish excellent datasets for small areas. Some analogue products from digital remote sensing are subject to photo-interpretation. Digital image processing methods are now becoming the standard.

Indirect methods II: laboratory experiments

Where field conditions hamper measuring and monitoring, or where processes operate too slowly to study directly over human timescales, or where global dynamics are of interest, physical geographers may resort to a range of techniques

---

direct measurements: measurements taken of the object under study

indirect measurements: measurements of a variable that is not the feature of interest but can be used to infer information about the desired item

automated instruments: equipment that can be programmed to record information at set time intervals or over set thresholds without the operator being present

data-loggers: devices for the storage of data from automated instruments

that are conveniently lumped under the heading 'laboratory experiments'. For example, weathering is, in general, a slow process. By recreating weathering environments in the laboratory, natural processes can be speeded up by, for example, increasing the frequency of freeze–thaw or wetting–drying cycles. Similarly, soil evolution is a rather tardy affair, some aspects of which are amenable to study in artificial soil columns set up in laboratories. Where the dynamics of global systems are the focus of attention, large-scale laboratory experiments may help. The biosphere is difficult to tackle directly as a global entity, for example, and human-made miniature biospheres provide a useful substitute for the real thing (see p. 174).

Precision and accuracy

Whether made directly or indirectly, measurements involve assorted degrees of precision and accuracy. Before using any measured data about the physical environment, or any other subject for that matter, it is critical to appreciate the difference between the precision with which a measurement is made and the accuracy of the measurement itself (Concepts 5.1).

SYSTEMS FOR MEASURING AND MONITORING: WHEN AND WHERE?

Physical geographers pose hypotheses about the structure and function of the environment. To test these hypotheses, they collect data by taking measurements of environmental factors according to some pre-designed sampling strategy. The research hypothesis dictates which bits of the environment to measure and monitor, and the sampling strategy dictates when and where to measure and monitor them. Details of what, when and where to measure vary from one environmental medium to another – air, water, land and land cover – but issues common to all media centre around sampling in time (temporal issues) and sampling in space (spatial issues).

Temporal issues

Environmental systems operate over a range of timescales. One-off measurements of system variables at a point in time may be useful in answering some research questions, but it is normally more revealing to take repeated measurements at the same place as part of a monitoring programme. Monitoring programmes are needed to establish such temporal patterns as diurnal changes (occurring over a 24-hour period), seasonal differences, annual patterns and longer-term cycles of change.

Systems with fast turnover times – atmospheric systems for instance – contrast with systems that have slow turnover times – as with some geomorphic systems, for example. Rapidly changing systems, like those in air and water, pose special problems, since the materials of which they are made are often in a never-ending state of flux. Under these conditions, measurements must be averaged over suitable time intervals for useful results to emerge. Furthermore, the averaging periods need standardizing so that temporal trends are assessable in a consistent and vigorous way. Moreover, historical trends are best identified in measurements collected at a specific site during the entire period of analysis.

CONCEPTS 5.1

PRECISION AND ACCURACY

Precision can be thought of as being equivalent to the exactness with which a measurement is made and how repeatable it is (i.e. the degree of agreement between individual measurements in a set of measurements of a particular characteristic or phenomenon). Accuracy, on the other hand, relates to how truthful the measurement is or, in other words, how representative it is of reality. In order to make suggestions as to the degree of precision and accuracy in a particular set of measurements, statistical techniques can be used. The degree of precision can be ascertained using standard error calculations and the degree of accuracy through an assessment of confidence levels.

It is tempting to think that if one has a set of exact measurements given to several decimal places and with little variation between them, then the values represent very 'correct' measurements. Indeed, they could be 'correct', but it is also possible that they are very precise but very inaccurate values. To illustrate this idea, it is useful to consider an example of darts thrown at a dartboard (Figure 5.2). Example (a) shows an accurate but imprecise throwing technique, in that although individual shots do not hit the central target, these are in the right general area, and can be taken as generally on target. Contrast this with Example (b). Here the thrower shoots very precisely because, although he or she repeatedly hits the same area, these are consistently wide of the central target. In Example (c), shots are neither consistent nor on target. In Example (d), a professional thrower can consistently take shots that are on target. Arguably, Example (c) is the worst-case scenario for a set of measurements, although Example (b) could be equally or even more misleading if it is unclear whether the measurements correspond to Example (b) or Example (d).

Relating this to measurements of the physical environment, data with the precision and accuracy characteristics of Example (a) sometimes have to be used because of the nature of equipment available: it may not be possible to obtain exact and repeatable

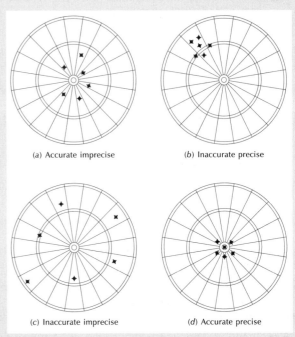

Figure 5.2 Accuracy versus precision in a darts match

measurements even though the results overall are reliable. In Example (b), the tendency for all data to shift by some consistent amount from the 'true' value is sometimes called **bias**. Bias can sometimes be removed from a dataset, for example, if a piece of equipment has previously been measuring over a long time period in an accurate and precise way. As noted above, Example (c) is the worst case for measurement data and would not generally be used, due to the problems in making meaningful interpretations from such a dataset. Example (d) is the aim of any measurement and monitoring programme, but achieving very precise and very accurate measurements can be a very expensive business and is not always practical.

---

bias: the tendency for data to shift by some consistent amount from the 'true' value

Systems with slow-acting processes tend to depend not only on current processes, but also on the history of the system concerned. The typical timescale (and spatial scale) of the operation of a particular system dictates in part how a study is set up and what should be measured or monitored. As a rule of thumb, system history has its maximum impact on systems that operate over long timescales and over large areas. In such systems, measurements of current processes may not be representative of the operation of the system over centuries, millennia or millions of years. Another problem stems from the nature of system dynamics: if a system passes through cycles of change, or if it involves episodes of activity alternating with periods of rest, then measuring and monitoring programmes must be designed accordingly to glean meaningful data. Similarly, some systems display greater variability during the course of a single event (such as a rain storm) than they do averaged out over a year or longer period. Given these points, it is plain that the nature of the system under scrutiny and the use to which results will be put must guide the selecting of a temporal (and a spatial) resolution of a monitoring programme.

Spatial issues

Environmental systems operate over a range of geographical scales, as well as a range of timescales. Before setting up a measuring or monitoring programme, the focus of study must be abundantly clear as it determines the spatial scales that require investigation. Establishing an unnecessarily detailed programme or a programme with too coarse a spatial sampling strategy could waste much time and money. Some problems involve several spatial scales, and monitoring programmes must involve networks of measuring stations that provide information on processes operating at the scales involved.

Deciding on a spatial sampling design has to balance several factors. Ideally, researchers might like to measure or monitor system attributes at all points in an area, but this is not practicable or even strictly necessary. Remote sensing offers some possibilities of acquiring spatially detailed datasets, but current methods are not refined enough to provide all of the information that may be required. In practice, physical geographers design a sampling strategy to give indicative values for sites that capture the range of spatial variability in the system. It is also worth bearing in mind that the understanding of a large system will not necessarily come out of a study using a huge number of data points over a very small area. Conversely, spatial variation in system attributes means that a researcher must use a sufficient density of monitoring points to avoid missing important components of the system in the sampling network, and to reveal the complexity present at a detailed scale.

Limitations of measuring and monitoring

One serious limitation in the dominance of measuring and monitoring as ways of investigating physical geography, is the danger of ignoring the components of environmental change that are not obviously acting in an area during the period of study. This is a particular issue since most measuring and monitoring studies are necessarily restricted to human timescales, and in fact increasingly to the few years which make up the timescale for a particular research project or a PhD study. This means that it is the most quickly changing aspects of the environment that are most likely to be studied, since they offer the opportunity to do so through 'scientific' means. This may be an approach that ignores the episodic event at the expense of regular process, and so gains an incomplete understanding of the system.

Scientific measurement and monitoring of modern processes can offer insights that are not available through other methods, but it is important not to ignore the whole history of environmental systems. This is the case when you simply want to understand why the system looks and behaves the way it does, but also if you are to be able to make predictions about it. In many modern environments, processes may have acted on a system in the past which are no longer doing so, such as in places like the English Lake District, where glaciated valleys can be seen as a dominant feature in a landscape which no longer has ice cover. The form may also still be influenced or even dominated by the effects of past events, even if some of the same forces (such as running water) are still operating in that environment. An extreme example of this is found in the Channeled Scablands of Washington State in the north-west USA, where the channels carved out by outburst floods from pro-glacial Lake Missoula around 15,000 years ago still dominate the landscape, being reshaped only marginally by modern river flow. This historical approach needs to involve a consideration of process if it is to be more than simply descriptive; it relies on our understanding of how the fundamental forces acting on Earth (gravity, air, water, ice, biotic activity and so on) shape it.

Another limitation of measurement and monitoring strategies is the fact that however sophisticated the measuring techniques and however dense the cover of monitoring stations, these results can only ever be a guide to what has happened or what is currently happening: they cannot directly reveal the shape of the future. However, measured relationships between current variables can form the basis of models that predict possible systems states days, weeks and years ahead (see Chapter 7).

5.2 AIR

THE PURPOSE OF MEASURING AND MONITORING AIR: WHAT AND WHY?

The atmosphere, like many aspects of the physical environment, is complex and changeable. Understanding the characteristics and processes of the atmosphere presents a particular problem, though, insofar as air is all around us all of the time but the processes that are operating are generally invisible to the naked eye. Of course, sometimes it is not difficult to be aware of the processes at work in the atmosphere, for example when there is a severe storm or when a clear plume of smoke rises and dissipates from a chimney, but the challenge for measurement and monitoring is especially great in letting us 'see' and understand these processes in detail. Measuring and monitoring many aspects of the atmosphere help in building a better picture of atmospheric structure and function. These aspects relate to climatic and meteorological properties (such as wind direction and speed, temperature, rainfall and pressure) or could focus on atmospheric composition (such as the concentration of a particular pollutant) at a particular point in time and space. Chapter 8 discusses the nature and characteristics of the atmosphere that have been determined through measuring and monitoring programmes, the main purpose of the discussion here is to identify and explore some indicative measurement techniques and monitoring programmes themselves in more detail. Please note that examples are indicative of general approaches only. It has not been possible to cover the full range of measurement techniques for the full range of pollutants – readers are directed to the further reading suggestions to obtain more information.

There are many ways of measuring the characteristics of the atmosphere. Consequently, before carrying out any measurement, it is important to consider its overall purpose, since this will help determine how the measurement should be carried out in terms of equipment, location and duration. A weather station, for example, is a collection of pieces of equipment that measure a whole series of characteristics at the same location to help with monitoring and tracking changes in the weather. Through doing this, general patterns and processes can be discerned which can then be used as the basis of models which are used to predict the weather in the future. These data are also invaluable in looking at related issues such as the build up of areas of poor air quality. The nature, cost and overall precision and accuracy of the equipment used will differ depending on who is doing the monitoring and the purpose of the monitoring. If the data are to be produced by a group of college students carrying out a project to assess patterns of precipitation over a local area, then they may use a large number of relatively crude and inexpensive means of obtaining the information they need. If the data are to be used by a national meteorological office, for example, to inform the public through weather forecasts, then precision and accuracy (see Concepts 5.1) are more important and more sophisticated stations would be used.

Table 5.2 shows the requirements of a climatic station in the UK in order to generate data for use by the UK Meteorological Office. This table highlights

Table 5.2 Minimum requirements of climatological stations in the UK before data can be used as part of the national network

Element	Range	Resolution	Accuracy	Remarks
Main elements				
Maximum temperature (°C)	−30 to +40	0.1	0.5	Daily
Minimum temperature (°C)	−30 to +40	0.1	0.5	Daily
Rainfall amount (mm)	0 to 999	0.2	0.2 mm < 4 mm 5% > 4 mm	Daily
Time (minutes)		1	1	
Additional elements				
Air temperature	−30 to +40	0.1	0.3	1 minute mean
Concrete temperature	−40 to +30	0.1	0.5	
Dew-point temperature	−50 to +40	0.1	0.4	
Grass minimum temperature	−40 to +30	0.1	0.5	
Soil temperature	−30 to +40	0.1	0.3	
Wet-bulb temperature	−30 to +40	0.1	0.3	
Global radiation (megajoules (MJ) m^{-2})	0 to 5 0 to 40	10 kJ m^{-2}	5% or 10 kJ m^{-2} 3% or 10 kJ m^{-2}	Hourly Daily
Net radiation (MJ m^{-2})	−1.5 to 5	10 kJ m^{-2}	10%	Hourly
Wind direction (degrees)	0 to 360	10	10	Hourly modal
Wind speed (knots)	0 to 150	1	1 or 5%	Hourly mean
Gust direction (degrees)	10 to 360	10	10	1-minute mean
Gust speed (knots)	0 to 150	1	1 or 5%	3 seconds
Gust time (minutes)	0 to 59	1	1	
Relative humidity (%)	1 to 100	0.1	5% below 50 2% above 50	
Sunshine duration (hours)	0 to 24	0.1	0.3	Daily

Source: UK Meteorological Office (2002) © Crown copyright 2002

some important considerations in making measurements; namely the characteristic to be measured, the duration and frequency of the measurement and the required precision and accuracy of the measurement. In order to make comparisons between data collected from different locations and different time periods, both the equipment and the way it is used need to be standardized.

Another example relates to the measurement and monitoring of air quality. Colls (1997) identifies three broad purposes:

- The assessment of ambient air quality (associated with concentrations in the general outdoor environment) in terms of potential community health effects;

- The investigation of impacts associated with specific sources (which can include both ambient monitoring in the vicinity of specific sources (such as within urban areas or next to roads), as well as measuring direct releases from exhausts or chimney stacks (emissions monitoring); and

- The validation of model results for research and other purposes.

Measurements of the composition of the atmosphere are also made in order to assess the 'background' trends in pollutant concentrations over recent time. These measurements include carbon dioxide at remote areas, in order to assess global trends in concentrations and their likely implications in terms of global warming. There are also measurements made in terms of the concentrations of particular compounds in different parts of the atmosphere such as ozone or chlorine species. Chapter 8 provides further information on the trends associated with different pollutants in different parts of the atmosphere. Satellites have been very important in assisting with the collection of data about different pollutants around the globe, as have more traditional techniques at or near the ground surface.

In many cases, the overall purpose of the measurements may also be to satisfy legislative requirements. If this is the case, there may be a need to produce data that can be compared to a series of air quality standards that are expressed in terms of pollutant concentrations during a specified period. The specific characteristics of the standards themselves also depend on their purpose; these can range from occupational health to public health to ecosystem protection, such as those produced by the World Health Organization (WHO). Where environmental legislation is involved, the importance of reliability and robustness of data is even more critical. In view of this, government agencies, such as the US Environmental Protection Agency, produce guidance to document appropriate equipment and techniques to use in assessing compliance with standards. In view of the need for data to be reliable for international comparisons, there are also international standards that are issued from time to time to show that a particular piece of equipment or a specific technique is reliable for the purpose it is to be put to. The ISO standards also cover laboratories and laboratory techniques and procedures and help ensure quality assurance and quality control (QA/QC). In this way, governments and other organizations can have some confidence that the data produced is a reliable basis for decision-making.

It is useful to note that this decision-making is based on more information than monitoring alone. Emissions inventories (which are discussed in more detail in Chapter 8) enable the estimation of the mass release of pollutants from particular sources per unit time (e.g. grams per second or tonnes per annum), whether these sources are natural (like volcanoes) or human-made (like power stations). These inventories do not say anything about the concentrations of pollutants in the atmosphere since this requires information about how the pollutants disperse in the air after they have been emitted. Dispersion modelling is required in order to assess ambient concentrations from mass emissions data – these models estimate concentrations through estimating how the pollutants mix in the air and with each other after being released from their respective sources, and what then would be monitored at a specific point in time and space if there was a measurement taken. To do this it is necessary to also input data about the meteorological characteristics of the atmosphere and the nature of the ground over which the air the pollutants mix with is travelling. Chapter 7 gives more information on modelling and Chapter 8 provides more information about the atmosphere. It is useful to introduce the idea of emissions inventories in this chapter in order to point out that emissions monitoring (i.e. of gases directly leaving a car tail-pipe or exhaust or leaving a power station chimney) are very important sources of information for emission factors. These emission factors describe the rate of mass emissions from a particular type of source (for example, a Boeing 747 aircraft) under different operating conditions (e.g. whether the aircraft is landing or taking off) and can be used to estimate the mass emissions from other similar sources without actually measuring emissions from each and every source individually. Some of these emissions factors can even be expressed in terms of mass emissions per person travelling on the plane per kilometre travelled – so each and everyone of us can actually calculate our average contribution to global emissions for a particular holiday or business trip.

Returning to measurements, for more general purposes, several measurement method considerations have been identified (Colls 1997):

- Specificity – does the method involve assessing just the pollutant of interest or is there likely to be some influence from other pollutant species?

- Sensitivity – does the method enable the highest and the lowest possible values to be detected?

- Reliability and stability – can the method be run continuously and generate time series of useable data?

- Response time – what is the timeframe required for obtaining the sample?

- Precision and accuracy – does the method produce results that are precise enough and accurate enough for the specified purpose? (See Concepts 5.1.)

The following sections discuss the specific details of the differences in equipment and networks used in monitoring ambient air quality throughout the UK as an example of the equipment, techniques and issues to be considered in measuring and monitoring an aspect of the atmosphere.

TECHNIQUES FOR MEASURING AND MONITORING AIR: HOW?

Air quality itself is a broad term and depends on the levels, or concentrations, of particular pollutants in the air, which are usually measured and assessed individually. These can be expressed either as a volume mixing ratio – the volume of the pure pollutant against the volume of the air containing it (e.g. parts per million, parts per billion) or as a mass measurement per unit volume (e.g. microgrammes per cubic metre ($\mu g\ m^{-3}$)). Gases can be measured in either unit and data between the two can be easily converted, but particles clearly can only be determined as a mass measure. When broadly categorized, methods are direct or indirect, depending on whether or not they involve exposure to polluted air in the field.

Direct methods: air quality

Direct methods are the most commonly used approaches to the measurement and monitoring of ambient air quality. They can be divided into active methods, automatic methods and passive methods. Although each of these involves exposure to polluted air in the field, there are fundamental differences in the way this exposure occurs and the means by which concentration data are obtained from the air that forms the sample.

The approach with the longest history is **active sampling**. This involves extracting a known volume of air and passing this through a filter or chemical collector over a specific period. Samples collected then go back to a laboratory for analysis using chemical or physical means or both. For example, black smoke, an historically important pollutant, can be measured through assessing the density of large-sized particles trapped on the surface of a coarse filter (since black smoke principally comprises large black carbonaceous particles). This approach relies on a method called **reflectometry**, which cross-references the darkness of a sample against a mass concentration value using standard calibration curves. There are a number of problems associated with this method, not least because of the differing nature of particles in the air in different areas; those associated with cement dust, for example, are much lighter in shade. In view of these problems and the general fall in the importance of black smoke as a pollutant of interest (the technique is not sensitive enough to make effective measurements of the lower concentrations and different characteristics of particles in contemporary UK cities) other methods now dominate. These methods include true **gravimetric methods** involving the collection of relatively large samples from a much larger volume of air enabling mass to be directly determined and a concentration calculated. The use of different sized filters also enables the concentrations of different sized particles in the air sampled to be determined using this technique.

Another example of an active sampling technique is a '**bubbler**' used to measure concentrations of sulphur dioxide. Here the sampled air is literally 'bubbled' through a chemical solution that retains the pollutant, and which can be subsequently measured to determine what concentration it was present at in the air sample. The equipment associated with this method is low cost and portable but considerable skill is required to obtain reliable data (Boubel *et al.* 1994).

Automatic methods are the most resource intensive (in terms of initial cost, maintenance and operational requirements) but have the advantage of generating the most precise and accurate data for short time periods. These are usually the techniques used to test compliance with air quality standards and guidelines, since these may include a need for hourly or even 15-minute data to be

active sampling: air quality monitoring method which involves passing a known volume of air through a filter or chemical collector over a specific period to determine pollutant concentrations in that volume of air

reflectometry: an analysis method for determining concentrations of particulate matter (especially black smoke) through cross-referencing the darkness of a sample against a mass concentration value using standard calibration curves

gravimetric methods: these involve the weighing of the mass of particles collected by drawing a known volume of air through a filter paper

bubbler: air quality monitoring method where sampled air is literally passed through a chemical solution that retains the pollutant and which can be subsequently measured to determine its concentration in the air sample

produced. These methods involve pumping a sample of polluted air into an instrument that determines concentration levels automatically through any of a wide range of methods including, for example, **spectroscopic** or **chromatographic** analysis (for gases or particles) or through **filtration** methods (for particles). Filtration involves passing air through a series of filters; the mass of particles is automatically determined in order to allow an assessment of the mass concentration of different sized particles in the sample. Filters from automatic samplers can also be removed for subsequent laboratory analyses of the physical and chemical properties of the particles on each filter.

Passive techniques result in the most affordable and easy to carry out measurements. Perhaps the most popular device for passive monitoring is the **diffusion tube**, which works by exposing absorbent materials to ambient air without the use of pumping. After a particular exposure period (longer exposure is required for lower concentrations), the equipment is removed and taken to a laboratory for analysis. Diffusion tubes are available for a number of different pollutants. Elsom (1996) identifies seven pollutants for which variants of diffusion tubes are available: ammonia, benzene, carbon monoxide, hydrogen sulphide, nitrogen dioxide, ozone and sulphur dioxide. The reliability of this technique for different pollutants varies, but tubes are in common use to help in the assessment of modern air quality issues. They are particularly helpful for nitrogen dioxide, where the possibility of creating dense networks using inexpensive techniques enables research into the spatial aspects of air quality issues to be carried out in far more detail than could be presently achieved with the more expensive and resource-demanding automatic stations. The cost effectiveness of diffusion tubes means that they are more 'mobile' than some other techniques and can be made to be genuinely portable. Different designs of the equipment (as badges, for example) can even allow the equipment to be worn, so that monitoring can investigate personal exposure in everyday circumstances or the workplace. Some automatic stations are also designed to be mobile, so that they can be moved to different industrial sites or can monitor air quality after a specific event, but they are clearly limited in terms of assessing the personal exposure characteristics that are possible with some of the more portable techniques.

Indirect methods: air quality

There are a number of remote measurement methods that enable concentrations of pollutants to be determined without the need for direct contact with polluted air at all. One important example is the use of satellite-based sensors that can determine patterns of distortion caused by light passing through the atmosphere. Unlike the vast majority of the direct methods, the use of satellite images immediately provides a broad and detailed spatial description of patterns in air quality from sub-national to global scales. Importantly, these can also allow changes in spatial distribution to be identified. For this reason, satellite sensors are also critical to the monitoring of the physical characteristics of the atmosphere (see Chapter 6). **Differential Absorption Lidar** can be used to remotely sense both gases and particles in the atmosphere and can also be used to generate three-dimensional representations of pollutant concentrations in small scale analyses (Boubel *et al.* 1994). Again, these techniques are appropriate for a number of different pollutant species, with an increasing capacity to consider

spectroscopic methods: analysis method where the properties of an object (for example, gases and particles in air) are determined based on its interaction with light – a process called spectroscopy

chromatographic methods: analysis method where the properties of an object (for example gases in air) are determined by the separation of molecules based on differences in their structure or composition – a process called chromatography

filtration: involving the use of filters. For example collecting particulate matter in air through passing a sample of air through a filter that will not allow solid matter of a certain size to pass through

diffusion tube: inexpensive and easy to use equipment for air quality monitoring which works through exposing absorbent materials to ambient air for a certain period of time and then analysing the material to determine a concentration. Available for a number of different pollutants

Differential Absorption Lidar: technique where gases and particle concentrations can be determined through remote sensing

these in detail over large areas of the Earth's surface. Remote sensing and the use of satellites to collect data is commonly thought of as a mapping technique and a more detailed description is given in Chapter 6.

Some ambient air quality measuring techniques are specific to particular pollutants whereas others are suitable for a number of different species. As we have seen above, the suitability of different techniques will depend on the overall purpose of the measurements. Table 5.3 shows some of the advantages and disadvantages of some of the currently available measurement methods for nitrogen dioxide in the UK, as summarized by the UK Air Quality Expert Group (AQEG 2004). For further information about the particular approaches mentioned here, plus numerous others, the reader should consult one of the many texts that include detailed overviews of equipment and techniques for carrying out direct and indirect measurements (e.g. Colls 1997; Boubel *et al.* 1994).

chemiluminescence: analysis technique that assesses the release of light that occurs as a result of certain chemical reactions in order to determine the composition of a sample

electrochemical sensors: analysis method using equipment broadly similar to a battery to generate a slight current in the presence of a specific gas. The amount of current generated is proportional to the concentration of the gas. Sophisticated versions of these instruments have different components that detect different types of substances and have been likened to 'electronic noses'

DOAS: Differential Absorption Spectrometers (see **spectrometric methods** glossary entry)

Table 5.3 Summary of advantages and disadvantages of NO₂ selected measurement methods

Technique	Advantages	Disadvantages
Chemiluminescence	The reference method specified in the EU Daughter Directive (i.e. recognized for policy-making) Lower detection limit of ~1 $\mu g\ m^{-3}$ Provides real-time data with short time resolution (<1 hr) that can be used for public information	Relatively high capital cost High operating costs
Diffusion tubes	Low capital and operating costs Possible to carry out surveys over wide geographical areas to provide information of the spatial distribution of NO₂ concentrations Require no power supply, and minimal training of site staff. Site calibrations are not required	Only provide concentrations averaged over the exposure period (typically 4 weeks) Accuracy of the method, and bias relative to the reference sampler, is dependent upon the method of tube preparation and the laboratory completing the analysis. Results need to be 'bias-corrected' before comparison with limit values and objectives
Electrochemical sensors	Portable samplers that can be easily deployed in the field	Lower detection limit of some samplers (~200 $\mu g\ m^{-3}$) makes them unsuitable for ambient monitoring
DOAS	Concentration integrated over the length of the light path. Gives an 'average' concentration that can be useful to assess public exposure Potential to measure a number of pollutants simultaneously Low running costs	The integrated measurement cannot be directly compared with the EU limit values or the air quality objectives (which are set for a single point measurement) Calibration may be difficult Relatively high capital cost Unfavourable weather conditions, such as fog or snow, can affect the instrument performance

Source: After AQEG (2003) © Crown copyright 2003

SYSTEMS FOR MEASURING AND MONITORING AIR: WHEN AND WHERE?

Air quality changes rapidly over time and space and it is therefore important that measurements are made within this context, with reference to both a relevant time period and the location at which the measurement is made. Clearly, even with a correctly made measurement with this critical information, one measurement is not enough to draw useful conclusions about the 'real world' temporal and spatial dimensions of the characteristic and/or process of interest.

Temporal issues: changing air quality

Since the concentrations of pollutants in the air at a location are constantly in a state of flux (i.e. not static over time), measurements need to be averaged over a particular time period in order to enable them to be interpreted in a meaningful way. As mentioned above, the demands of measurement often also relate to air quality standards, which in turn are determined through an assessment of the level and duration of exposure to specific or combined pollutants that humans can withstand without detrimental effects on their health (usually based on the most susceptible members of society such as the young, ill or elderly). Air Quality Standards are discussed in more detail in Chapter 8.

A second temporal issue relates to the need to standardize averaging periods in order to enable trends over time to be assessed in a consistent and robust way. This need for consistency is also important in relation to spatial aspects, as historical trends in measurements are best made through considering data that have been collected at a specific point during the whole of the analysis period.

A third temporal issue points to one of the limitations of measurement and monitoring strategies, the fact that, however sophisticated the techniques and however dense the cover of stations, these can only ever directly inform us about what has happened or what is currently happening. There is great value in the possibility of forecasting future events, whether that is the likelihood of severe weather conditions or the potential for future poor air quality. The only way to do this is to use models (see Chapter 7) that simulate processes and can project these into the future. It is important to note, however, that measurement and monitoring programmes still have an important role in helping develop and validate models before they, in turn, can be used to reliably inform about future conditions. Since the real world is so complex, the process of model development and validation is an ongoing process, and consequently there will always be an important place for monitoring, regardless of how sophisticated these models become.

Spatial issues: the geography of air quality

Spatial patterns in pollutant concentrations are very important in helping to assess the associated impacts either on humans or on susceptible ecosystems. However, this presents something of a dilemma in terms of measurement and monitoring strategy. Clearly, although policy-makers and researchers would like to know what the levels of pollutant concentrations are at every location over a city or country, it is impossible to directly measure at every possible location to build up a dataset that would allow this to be done. Remote methods, which at least offer some possibility of doing this, are not currently sophisticated enough to be able to provide all of the information that is required. In view of this, a sampling strategy has to be developed to give indicative values for sites with different sets of characteristics. To make data comparable, these different characteristics are categorized and standardized (to a point) so that similar locations in different cities and different countries can be compared. Table 5.4 gives a breakdown of the site characteristics associated with UK automatic monitoring stations.

The spatial component of monitoring also depends on the spatial scales over which particular processes operate. A number of air quality issues operate at the local or regional scale whereas others operate on national or even global scales. Boubel *et al.* (1994) identify five different scales over which air pollution issues operate:

1. Local (up to around 5 km) – usually associated with specific emitters, for example a particular industrial source or a road junction. Elevated carbon monoxide concentrations, for example, can result where buildings, or other structures such as tunnels, restrict the dispersion of exhaust from slow-moving or stationary traffic.

2. Urban (extending to around 50 km) – urban areas are complex sources of air pollutants. Air quality issues can result from the release of primary pollutants (which are directly emitted from particular sources as in the example above), or due to the reactions of primary pollutants to generate secondary pollutants, such as with the oxidation of NO_x (principally primary NO) to NO_2.

3. Regional (extending from 50 to 500 km) – a classic regional scale pollution problem is ozone. Ozone is a secondary pollutant that forms from the interaction of a range of urban pollutants. The more rural areas surrounding large cities, rather than the cities themselves, tend to experience the impacts of elevated ozone concentrations. Another regional scale problem is related to visibility and the haze caused by the presence of fine particles in the atmosphere.

4. Continental (from 500 km to several thousand km) – acidification is a classic and well-publicized 'continental scale' problem. In Europe, this occurred where northern European countries experienced the impacts of secondary pollutants caused by emissions of species such as SO_2 and NO_2 from fossil fuel combustion, principally for power generation, from countries like the UK that tended to be 'downwind'.

5. Global – for some types of pollutants, impacts occur on a global scale. Examples in this category include radiation, chemicals associated with stratospheric ozone depletion and carbon dioxide associated with climate change.

In view of the different scales of different pollution problems, there is a need for monitoring networks to also operate over these different spatial scales. This need relates to monitoring pollutants themselves and also the other aspects of the atmosphere that affect their dispersion, their interaction with other pollutants and their eventual breakdown and/or deposition. Of course, these other aspects of the atmosphere are also monitored for other purposes, some of which have

Table 5.4 Classification of different monitoring stations according to the characteristics of their location (relates to the UK classifications for the UK Automatic Urban Network (AUN) and UK Automatic Rural Network (ARN)

Site Classes	Definition
Kerbside sites	Sites with sample inlets within 1 m of the edge of a busy road. Sampling heights are within 2–3 m
Roadside sites	Sample inlets between 1 m of the kerbside of a busy road and the back of the pavement. Typically, this will be within 5 m of the kerbside. Sampling heights are within 2–3 m
Urban centre	Non-kerbside sites located in an area representative of typical population exposure in town or city centre areas e.g. pedestrian precincts and shopping areas. Sampling heights are typically within 2–3 m
Urban background	Urban locations distanced from sources and broadly representative of citywide background concentrations e.g. elevated locations, parks and urban residential areas
Urban industrial sites	Where industrial emissions make a significant contribution to measured pollution levels
Suburban sites	Typical of residential areas on the outskirts of a town or city
Rural	Open country locations distanced from population centres, roads and industrial areas
Remote	Open country locations within isolated rural areas, experiencing regional background pollution levels for much of the time

Source: After Stanger Science (2002)

been identified above, and which may then have other relationships with the issue of spatial scale. This section ends by giving some examples of networks collecting atmospheric information at each of these different scales. Here, the scales are grouped in terms of local (incorporating the local to regional scales), national and global. Chapter 8 describes and explains the characteristics of air quality problems and the results of monitoring networks in more detail than the brief examples offered here.

Local networks

Local networks for assessing ambient air quality tend to be developed using affordable equipment, such as diffusion tubes, and may look at the spatial characteristics of pollutant concentrations or levels associated with a particular source. There are numerous local networks in operation, some of which feed data into a national monitoring network. An example of such a network is the UK NO_2-diffusion tube network (e.g. Loader 2001), where, although UK Local Authorities run their own networks, the data they produce feed into an extensive national network that makes an important contribution to understanding the distribution and behaviour of this pollutant. Local networks can also look at other aspects of pollution, such as acid deposition and toxic compounds, as well as considering climatological and meteorological variables.

National networks

There are presently some 1500 sites across the UK that monitor air quality, and these are organized into different component networks in recognition of the range of pollutants, information types and method involved (NETCEN 2002).

Examples include acid deposition (e.g. Hayman 2000), hazardous toxic pollutants (e.g. Coleman 2001) and nitrogen dioxide from automatic or passive stations (e.g. Loader 2001).

Some of the sites that form part of the UK's automatic networks have been in operation since the 1970s. They provide valuable information to help with policy and research work. These include sites that are part of the Automatic Urban Network (AUN) and the Automatic Rural Network (ARN) and which are classified according the siting characteristics referred to in Table 5.4. More information and sample datasets for the AUN and ARN can be found on the UK's public and local government information website NETCEN (2002).

International networks

An excellent example of an international network is that operated through the World Meteorological Organization (WMO 2003). During 2002, the WMO organized and synthesized data from around 14,000 land-based stations (around 3000 of which are part of the Regional Basic Synoptic Network which forms the main global network), as well as information reported from around 7000 ships at sea and 700 buoys. The network also includes data collected from 30 or so satellites (both polar orbiting and geostationary) that constantly collect and relay information about weather systems for the World Weather Watch initiative. Vast amounts of up-to-date information are passed to the member states and territories, which in 1996 totalled 185. The network facilitates global climate and meteorological research and helps in understanding the interactions of climate and weather with the environmental system and human society. This information is also useful in facilitating a better understanding of air pollution and air quality in different parts of the globe and complements the information gained from national and sub-national networks.

A final example of an international monitoring network is the WHO Global Environment Monitoring System which has been collecting data about urban air pollution in cities across the globe since 1974 (UNEP 1991). Most participating cities include at least three stations (one site in an industrial, a residential and a commercial area). Over fifty nations contribute to the initiative, which allows international comparisons to be made in terms of concentrations of sulphur dioxide, suspended particulate matter, lead, nitrogen oxides and carbon monoxide. Chapter 8 offers more information about trends in these pollutants and the composition of the atmosphere more generally.

5.3 WATER

THE PURPOSE OF MEASURING AND MONITORING WATER: WHAT AND WHY?

Measuring and monitoring water fluxes in time and space aids knowledge and understanding of the processes of precipitation, evaporation, groundwater fluxes, river flows and the relationships between them. The quality of water is as important as the quantity, as the increased use and pollution of water sources by anthropogenic activities means that virtually no waters are free from human influence. As pollution becomes more common, so the potential for environment degradation becomes greater, with an ever-increasing need for global sustainable management of this precious resource. Such knowledge is important in managing the Earth's limited water resources.

TECHNIQUES FOR MEASURING AND MONITORING WATER: HOW?

Techniques for the measurement of water can be broken down into the main constituents of the water cycle: precipitation, evaporation, groundwater, river flow and water quality. These can further be broken down into direct and indirect methods.

Precipitation

Rain gauges are a direct method of measuring precipitation. A rain gauge is a storage device designed to funnel rain falling over the gauge into a container (Plate 2). The amount of rain collected is measured (units of millimetres) either manually at regular intervals or automatically by the gauge. Manual readings involve measuring the water stored within the container, emptying it and replacing it inside the gauge. Automatic gauges, such as tipping-bucket designs (Plate 3), record when rain falls and for how long, so the intensity of rain can be measured.

Recent technological developments have allowed the use of indirect techniques such as ground-based radar and weather satellites for rainfall monitoring. Radar is used to map the distribution and intensity of rainfall, and is also used as a forecasting tool to detect and give warning of severe storms. Many meteorological satellites now orbit the world, positioned at regular intervals around the equator; these include MSG (successor to Meteosat) positioned above Europe, GOES-EAST over the Americas, GOES-WEST over the Pacific, and GMS over Japan and Australia. These **meteorological satellites** carry sophisticated instruments that collect meteorological data for use in atmospheric computer models forming the basis of modern weather forecasting (Lillesand and Kiefer 2000).

> **rain gauge**: a device for measuring rainfall at a particular point on the ground. Many different designs exist with a typical design of a funnel to capture rainfall over a certain area, which then funnels water either into a storage bottle or tipping device with known volume
>
> **meteorological satellites**: a group of Earth observation satellites designed to monitor synoptic weather systems

Evaporation and evapotranspiration

A direct method of measuring water loss through evaporation is by using an **evaporation pan**. Many different designs of evaporation pan are available, but the standard is the US Class A pan, which is circular and set above the ground on a platform to allow water to circulate around it. The amount of water evaporated (units of millimetres) is found by measuring the water level in the tank at the same time every day and accounting for any water added (to keep it full or from rainfall). A series of empirical coefficients enable the conversion of pan evaporation to water loss or evapotranspiration from other surfaces, such as bare soil or field crops. Specialized investigations use instruments such as the **Bowen Ratio Energy Balance** and **Hydra** to measure water loss from the surface on which they are placed. These instruments are not routinely used, however, due to their cost and maintenance demands.

Indirect methods of computing water loss are numerous (Brutsaert 1982). The water budget method quantifies the other components of the water balance over a unit area, such as the catchment, and computes evaporation as the balance. This method is simplistic but not very accurate over short time periods. It is most useful for large areas over long time periods, such as a year, when any

> **evaporation pan**: a chamber of known dimensions, usually circular but sometimes other shapes, from which the rate of evaporation can be recorded by noting the water level over time
>
> **Bowen Ratio Energy Balance**: an instrument to measure evaporation loss based on the Bowen ratio: the ratio of sensible heat to latent heat lost to the atmosphere by the surface under investigation
>
> **Hydra**: an instrument to measure evaporation using the eddy flux approach

change in storage is usually negligible. The water balance for evaporation can be computed as follows (Shaw 1994):

$$E = P - Q - G \pm \Delta S$$

where E is the water loss from the water surface, vegetation and soil, P is the precipitation (rainfall, snow, dew, etc.), Q is the river and other surface water discharge, G is spring flow and other groundwater discharge, and ΔS is the change in soil moisture.

Lysimeters consist of a container or large block of undisturbed soil and vegetation wrapped in an impermeable membrane. Water loss from evapotranspiration is found either by noting the change in weight for weighing lysimeters, or by using a water balance approach and recording the change in soil moisture and addition of precipitation and irrigation water over regular intervals. The water balance can be calculated for different time periods according to the data available and required accuracy. As the lysimeter is a block of soil isolated from lateral and groundwater movement, the water balance above can be simplified to:

$$E = P \pm \Delta S$$

When installed and maintained properly, lysimeters are a very useful tool. Their disadvantages lie in that they are small scale relative to the area under investigation and it is difficult to construct the lysimeter without disturbing the soil structure and vegetation roots. In addition, the impermeable membrane prevents lateral and groundwater movement (both into and out of the lysimeter), which may be important processes regulating evaporative loss.

The most common approach to computing evaporation loss is through modelling using meteorological data. Many empirical models exist based on **turbulent transfer processes** and/or the energy balance of the surface under investigation, and employ data obtainable from standard weather observation stations. The Food and Agriculture Organization (FAO) (Smith *et al.* 1992) recommend the use of the **Penman–Monteith model** to quantify the water loss from vegetated surfaces. This approach is now the optimal model for computing evaporation, although its accuracy depends on the quality of the data used to characterize the surface under investigation. Weather stations (p. 111) are used to obtain the data required, which include radiation, air temperature, saturated and actual vapour pressures and wind speed.

The use of evaporation formula using meteorological data yield values of the potential rate of evapotranspiration, i.e. the amount that would evaporate with a constant supply of water. Actual evapotranspiration can be calculated by accounting for the soil moisture available, i.e. the **soil moisture deficit** approach. Here, actual evaporation occurs at the potential rate until the maximum soil moisture deficit for that particular vegetation cover has been reached. Thereafter actual evaporation is less than the potential until the soil moisture has been replenished. This approach is the basis of the UK Met Office's MORECS (Met Office Rainfall and Evaporation Calculation System) approach (Hough *et al.* 1997).

lysimeter: a device (of varying size) for measuring the rate of water movement and storage in a soil block

turbulent transfer processes: the movement of water or air in an irregular, eddy flow, in contrast to smooth laminar flow

Penman–Monteith model: the most widely accepted model for computing evaporation from a vegetated surface, based on Monteith's modification of Penman's formula for open water

soil moisture deficit: the amount by which moisture held within the soil falls below the level of field capacity

Groundwater

A basic groundwater metric is its position below the ground surface (units of metres). Direct methods of monitoring groundwater involve the use of shallow

dipwells, which can be dug by hand in unconsolidated sediments, **piezometers** that tap into a particular layer of rock or sediment and **boreholes** that are created by drilling deep in the rock (Fetter 1994). Groundwater enters the well and rests at the same position as within the rock. The water level within the well can be measured manually using a dipper or by automatic methods such as pressure transducers. Indirect methods to determine the position of groundwater include the use of groundwater maps. The position of monitoring wells and the water level data is marked on a scale plan and contours drawn by joining points of equal water level, or by interpolating between points (Brassington 1988). Groundwater flow (units of metres cubed over time, such as per second, hour or day) can be determined using models such as **Darcy's law**, which uses information on the hydraulic conductivity of the rock or sediment and the hydraulic gradient to calculate the speed of groundwater movement. This is of great importance when determining the spread of pollutants within groundwater and the potential abstraction rate, for example.

> **dipwell**: a slotted well used for measuring the position of the water table or phreatic surface
>
> **piezometer**: a well slotted at the base, used to determine the groundwater head for particular strata
>
> **borehole**: a well of particular depth used to tap into a strata of interest e.g. aquifers
>
> **Darcy's law**: an equation describing the flow of water through a porous medium

River flow

If the cross-section of the river has been measured, a **stage–discharge relationship** or **rating curve** (Figure 5.3) can be derived using a series of readings of the stage and velocity (Jones 1997). With enough measurements, a graph can be drawn, quantifying the relationship between the water level and flow. Using this relationship, the discharge can be measured from stage readings alone. The stage of a river is its water level (units of metres), and is measured manually using a stage board (also known as staff gauge) or automatic devices such as pressure transducers.

Where repeated measurements are required over a long period of time, the most accurate flow measurements (units of cubic metres per second or cumecs) are made using flumes and weirs, where the river reach is stabilized by the installation of an indestructible cross-channel structure with standardized shape and characteristics (ISO 1980). **Flumes** are useful to measure flow in small streams as they operate by narrowing the channel, causing an increased velocity and a decrease in depth. **Weirs** operate by forcing water to pile up behind them before gushing over the top. The rate at which water passes through the flume or over the weir depends upon the height of the water on the upstream side, and this relationship is known for standardized structures. Weirs and flumes have an upper limit in the range of flows they can measure and are well suited to low to medium flows; flood flows are not accurately measured as the relationship between water level and discharge no longer exists.

> **stage–discharge relationship (rating curve)**: a chart plotting the river stage against discharge so that flow can be estimated from water level data
>
> **flume**: the artificial channel of known dimensions built within a stream to measure flow
>
> **weir**: a dam across a stream over which the water flows. It is used for flow measurements and water level control

Flow can be measured indirectly by measuring the river velocity (units of metres per second) and multiplying this by the cross-sectional area (units of metres squared) of the river. The simplest direct method of estimating flow velocity is to use a float and time its progression over a known distance to obtain velocity. However, these

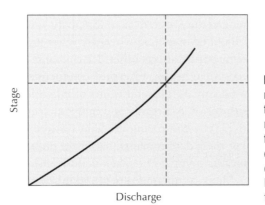

Figure 5.3 A stage-discharge relationship or rating curve built from many coupled measurements of water level and flow. Once the relationship is established, discharge is determined by converting the known stage to a discharge using the chart

measurements give only the surface velocity and correct factors must be applied to estimate the average flow within the channel. Flow velocities are measured more accurately using **current meters**, which use an impeller design or wheel of small cups. When placed in the river, the current turns the impeller, which rotates in proportion to the speed of the water, and the velocity of the river is given directly or read from a calibration chart for the instrument. The **electromagnetic method** is useful in rivers with in-stream vegetation or no stable stage–discharge relationship. These flow meters pass an electric current through a coil within the flow meter that creates an electromagnetic field. Water moving across this field produces a proportional current that is converted into a velocity measurement. The **ultrasonic method** passes acoustic pulses from one side of the river at a set depth; the time taken for the pulse to be recorded by a sensor on the other bank is recorded and converted into velocity. **Dilution gauging** is an indirect method of determining velocity and discharge, and is particularly suitable for turbulent upland streams where the reach characteristics (e.g. low flow, rocky bed, steep gradient) are unsuitable for other types of measurement. Dilution gauging involves the addition of a known volume and concentration of a tracer (dyes, chemicals or mechanical tracers e.g. fine particulate matter) into the stream. The background or natural concentration of the tracer must be known. The concentration of the added solution is measured at a suitable point downstream.

Another approach is the widely used **Manning's equation**, which is based on hydraulic principles. Discharge is computed from the relationship between the channel cross-section, slope and roughness:

$$Q = \frac{A^3 \sqrt{R^2} \times \sqrt{S}}{n}$$

where Q is discharge (m³ s⁻¹), A is cross-sectional area (m²), R is the hydraulic radius (2 × depth plus width, m), S is the slope of water surface (decimal) and n is the Manning's roughness number. The uncertainty with this formula is the choice of the Manning's number, as roughness can change with flow, within the channel and along the river. Typical values for natural channels range from 0.033 to 0.07 (Newson 1994).

Water quality

Direct measurement of water quality involves the determination of the physical and chemical constituents of the water using handheld probes or laboratory-based techniques. Physical features of water that can be measured directly include temperature, electrical conductivity and turbidity. The temperature of water (°C) should be measured whenever water samples are taken, as a change in temperature affects other determinants and is important when assessing ecosystem health. **Electrical conductivity** is a useful measure of the dissolved solid content of water (units of microsiemens per centimetre). **Turbidity**, the fine suspension of organic and inorganic particles, is measured in turbidity units (FTU) by comparing the light scattering of the sample to a standard suspension. For these determinants, handheld probes are available allowing direct measurements to be taken rapidly onsite.

Many compounds are found in solution in natural waters and chemical analyses can be very extensive (Trudgill *et al.* 1999). The principal categories are

---

current meter: a device for measuring the speed or velocity of water

electromagnetic method: technique of measuring stream velocity by creating an electromagnetic force by an electric current. The emf is directly proportional to the average velocity of the river cross-section

ultrasonic method: measuring stream velocity at a certain depth by recording the time taken for sound pulses to pass through the water

dilution gauging: measuring stream discharge by adding a tracer or chemical solution and measuring its dilution downstream. There are two types: constant injection and gulp injection which relate to the addition of the chemical into the water

Manning's equation: an equation relating stream velocity and channel geometry

electrical conductivity: the dissolved solids content of water measured through its ability to conduct an electrical current

turbidity: the reduced transparency of the atmosphere or water caused by dust or particles in suspension

pH, dissolved oxygen (DO), biochemical oxygen demand (BOD) and nutrient (e.g. nitrate and phosphate) content. There are four main methods for the determination of chemical water quality: titrimetric, colorimetric, spectral and potentiometric techniques. **Titrimetric techniques** involve the chemical balancing of reactions using coloured indictors, while **colorimetric analysis** uses the density of colour of the compound under scrutiny as a guide to its concentration. For metal analyses, **spectral methods** and **atomic absorption spectrophotometry (AAS)** compare the light intensity of metal ions in a flame to a known standard solution, for example the white flame of magnesium and yellow colour of sodium. The **potentiometric approach**, used for pH meters, creates an electronic potential by inserting a metal electrode into a salt solution. The concentration is related to the difference in potential between the solution and a known standard.

Indirect methods of assessment are used to assess the aesthetic and olfactory properties of water such as taste, smell and colour. These result partly from impurities, which are either human pollutants or natural impurities such as the brown staining of water from tannins in organically rich soils and sediments. No definite test exists with which to quantify these characteristics and they are judged subjectively using rating scales.

Physical and chemical analyses of water give a snapshot of the water quality of a sample taken at a particular point in time. Water quality can also be defined in terms of ecosystem health by **biological monitoring** (Hellawell 1989; Chapman 1996). The presence of a particular species – plant or animal (Figure 5.4) – indicates that the particular physiochemical characteristics of the habitat are within the tolerance range of that species over a period of time. Some species prefer cleaner waters than others, and so an assemblage of particular species points to a certain level of ecological 'health' and therefore water quality over

titrimetric techniques: balancing of reactions using coloured indicators

colorimetric analysis: a technique that relates the concentration or strength of a solution by its colour intensity

spectral methods: a technique that relates the spectral intensity of ions in a flame to known concentrations of the same ion to determine concentration

atomic absorption spectrophotometry: a method of determining concentration by comparing the absorption of light by atoms at various wavelengths in a solution to known concentrations

potentiometric approach: concentration is measured by the difference in electrical potential between a metal electrode in a salt solution and the unknown concentration of metallic ions in the solution

biological monitoring: monitoring flora and fauna to provide an indirect assessment of water quality of ecosystem 'health'

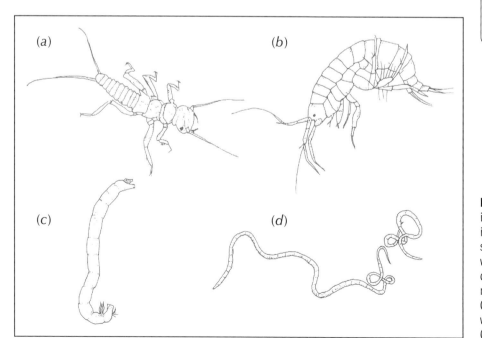

Figure 5.4 Stream macro-invertebrates can be used as indicators of pollution. The four species illustrated characterize water quality from (a) to (d) in declining quality; (a) Stonefly nymph; (b) freshwater shrimp; (c) Chironomid larva; and (d) sludge worm. *Source:* After Advisory Centre for Education (1971)

longer timescales than chemical analysis. Benthic (bottom-dwelling) macro-invertebrates, for example, are valuable biological indicators as they are readily identifiable, have a cosmopolitan distribution, and are easy to sample using kick sampling. Kick sampling involves standing in the river, facing upstream, and disturbing the riverbed by gently kicking at the substrate. The disturbed invertebrates are collected in a net held immediately upstream. Another way to capture benthic macro-invertebrates is to place nets at positions along the river to catch invertebrates as they naturally drift with the water.

Methods of data analysis such as **diversity indices or metrics** (e.g. the Trent Biotic Index and the BMWP Index) determine the status of the water quality from the species sampled. Biological analyses of water can also reveal the presence of microorganisms, which is important when assessing the quality of water for domestic supply. Many aquatic microorganisms transport disease, and it is important to monitor continuously the occurrence of particular organisms, for example the bacterium *Vibrio cholerae* transmits cholera and the presence of *Escherichia coli* is an indication of sewage pollution.

> **diversity indices or metrics**: techniques to measure species number and species evenness

SYSTEMS FOR MONITORING WATER: WHEN AND WHERE?

Temporal issues of water monitoring

One-off measurements are used to determine the status of a particular variable at a point in time, but it is more useful to have repeated measurements at the same place as part of a monitoring programme. One of the characteristics of the water cycle is that it is in a constant state of flux. Monitoring programmes are needed to determine relationships over varying timescales – daily (e.g. precipitation), seasonally (e.g. evaporation loss), annually (e.g. position of groundwater) – and to determine what the natural or background level is for a particular variable at a particular place (e.g. water quality).

Technology has advanced greatly over recent years with the development of numerous new sensors and automatic measuring techniques that are increasingly reliable and affordable. Automatic methods allow for near continual measurements to be taken, allowing observation of changes on the smallest of timescales. Most modern automatic devices are digital and store data on loggers that can either be downloaded onsite using portable computers at fairly long time intervals, or be telemetered remotely back to base using radio frequencies or satellites. Data can also be telemetered in real time, for example, when precipitation data from a network of gauges is needed during heavy storm events for input into models to assess the real-time possibility of flooding.

Spatial issues of water monitoring

Scale is a key issue in hydrology and the catchment is considered to be the basic unit of measurement. The techniques discussed above relate to point-based measurements of water fluxes. One limitation of point-based sampling is that the measurement (e.g. precipitation, water quality) is only a sample of that occurring over the surrounding area. The placement of sampling or instrument sites is therefore very important in order to avoid even small errors that can result in substantial values if generalized over larger areas.

Monitoring networks are used to overcome some of these difficulties and to allow monitoring to occur over regional spatial scales. For example, the amount of rainfall falling over a catchment will be more accurately calculated based on a network of gauges distributed over the area rather than a single monitoring station. The number of sites in monitoring networks is based on the level of accuracy required and the spatial variability of the area or variable under investigation. For example, on a global scale, guidelines issued by the World Meteorological Organization (WMO 1994) recommend a minimum density of gauging stations (Table 5.5) depending upon climate and topography.

Owing to the cost of maintaining monitoring networks, the majority are found in developed countries in the Northern Hemisphere, with large parts of the global landmass not covered by any monitoring network. Therefore, monitoring water fluxes to examine trends on a global scale is problematic. Satellites and remote-sensing techniques are increasingly being used to measure and monitor processes over large spatial areas. For example, measurement of evaporation using ground-based methods is limited in its spatial extent, but large areas can be covered using satellites with sensors that measure surface thermal characteristics.

Table 5.5 Minimum densities (km^2 per gauge) for gauging stations, as recommended by the World Meteorological Organization (WMO 1994)

	Humid climates			Arid climates
	Mountainous regions	Plains	Coastal regions	
Precipitation	250	575	900	10,000
River flow	1000	1875	2750	20,000

5.4 SOIL

THE PURPOSE OF MEASURING AND MONITORING SOIL: WHAT AND WHY?

To use and manage soil in an optimal way, it is necessary to measure and monitor its properties. Poor soil management often results from a lack of understanding of the basic soil characteristics that underpin the success or failure of crops. Some techniques to examine these basic soil properties are presented here (see Chapter 10 for more details). The measurement of soil chemical properties is not covered in this chapter, as it is an immense subject worthy of entire textbooks in its own right.

TECHNIQUES FOR MEASURING AND MONITORING SOIL: HOW?

Techniques for measuring soil properties can be either field or laboratory based. Generally, field-testing kits or procedures give rapid results that are sufficient for reconnaissance surveys or broad-scale studies. Laboratory methods require collection of soil samples and are required for more rigorous research. There are many different techniques; those presented here comprise only a few of those available to measure the basic soil properties.

Soil texture

Soil texture can be determined in the field by assessing touch and malleability (see p. 284). Laboratory assessment of texture is more accurate. Two methods are available. The first method passes a soil sample through a series of sieves to separate the soil into clay, silt and sand grades. The weight of soil collected in the sieves with different mesh sizes gives the proportion of each grade. Alternatively, the proportion of particle sizes can be determined using sedimentation and **Stokes's law**. The procedure is to mix a gravel-free soil sample with water, place it in a glass column, and allow the suspension to settle for a given period. Samples extracted from the column at certain depths, in conjunction with Stokes's law of settling velocities, enable particle size to be determined.

> **Stokes's law**: an expression expressing the rate of settling of spherical particles in a fluid of known density and viscosity

Soil organic matter content

The traditional method to determine soil organic matter content is to dry a soil sample in an oven at 105°C for 48 hours, which removes any moisture and leaves only the soil matrix and organic content. The samples are then weighed before burning in a furnace at 550°C for six hours, a process that destroys the organic matter in the sample. When re-weighed, the weight loss constitutes the organic fraction of the soil. Taking this as a proportion of the weight of the dry soil, the percentage organic content of a soil can be calculated. This important measure relates to the water holding capacity of a soil and the method described is known as 'loss on ignition'.

Bulk density, particle density and soil porosity

Bulk density is the mass of a unit volume of dry soil (i.e. both solids and pore spaces). A value of bulk density is often required to calculate volumetric soil moisture content. To measure bulk density, a ring is pushed into the soil to collect a core of known volume. Great care should be taken with the samples to prevent any compression, particularly in clay soils. Bulk density is calculated by dividing the weight of dry soil by the ring volume. Typical, published values of bulk density fall between 1 and 1.8 (for clays and sands respectively) and are expressed as unit weight per unit volume, for example g cm^{-3}. In a heavily agricultural area, the bulk density will vary spatially, and more importantly, with depth, due to soil compaction caused by heavy machinery, for example.

Particle density is the mass of the soil particles alone. It is found by calculating the volume of a known mass of water. This is done by dispersing a known mass of soil with water in a beaker and dispelling air through boiling the suspension. The cooled suspension is decanted into a flask to a known volume. The weight of the flask and solution are noted. The mass of water in the flask is found by subtracting the soil mass from the suspension. From this, the volume of water is determined and thus the volume of the soil. Dividing the soil mass by volume gives the particle density.

Once the bulk and particle densities of a soil are known, these data are used to work out soil porosity. Porosity is determined by the equation:

$$Porosity = 1 - \left(\frac{bulk\ density}{particle\ density} \right)$$

Soil moisture

Many methods are used to measure soil moisture. This section discusses some of them.

One method is to dry soil samples in the laboratory. The sample is weighed, and then reweighed after drying in an oven at 105°C for 48 hours. The amount of moisture is calculated as the difference in weight between the wet and dry samples. This is the **gravimetric soil moisture measurement**, that is, the mass of water per unit mass of soil.

Calculation of the amount of moisture in the soil by volume is a much more useful measure, as the result can be scaled to the field. This is done by multiplying the gravimetric soil moisture content by the soil bulk density.

Other methods of calculating soil moisture involve more specialized pieces of equipment, such as **gypsum blocks**, **neutron scattering**, **Time Domain Reflectometry** (TDR) and remote sensing. The equipment is calibrated to the soil by taking a reading and then taking a soil sample and determining the volumetric soil moisture by drying, for example. Multiple measurements are taken in this way in order to encounter the likely range of soil moisture. The probe output and soil moisture values are plotted together to highlight the relationship between the two. Thereafter, soil moisture measurements can be taken using the probe and the calibration relationship.

Hydraulic conductivity and matric potential

The hydraulic conductivity of a soil can be measured either in the laboratory or in the field. Laboratory methods include both falling and constant head design **permeameters**, saturating an undisturbed core sample of the soil, or a sieved and packed soil sample and measuring the steady outflow of water. This is hardly representative of field conditions, however, due to the very small sample used, problems of compaction or smearing and the difficulty of relating small-scale results to field processes.

Traditionally, **tensiometers** have been used to generate suction values *in situ* (Curtis and Trudgill 1975), while equipment such as the pressure-plate apparatus is used on soil samples in the laboratory. Published empirical formulae derive the soil moisture characteristic curve from basic soil properties such as particle size distribution, organic content and bulk density (e.g. Gupta and Larson 1979).

SYSTEMS FOR MONITORING SOIL: WHEN AND WHERE?

Chapter 10 discusses how different soils form in different locations as a function of climate, slope, aspect and other factors. In addition, soil properties change down the profile because of surface processes and leaching, for example. Thus, soil properties change spatially and soil-sampling programmes must be large enough to characterize any natural changes within the area of interest. It is important to ensure that enough measurements are taken to be representative of the area.

For sampling widely over an area, an **auger** is used to the depth(s) of interest. When sampling from particular horizons using soil pits, it is generally assumed that, provided enough samples are taken, a bulked sample can be created which is representative.

Soil properties naturally develop and change over long timescales. However,

---

gravimetric soil moisture measurement: the measurement of soil moisture by computing the difference in mass before and after the soil has been dried. The soil moisture is expressed as the amount of water per weight of soil

gypsum blocks: a block of gypsum buried in the soil connected to an electrical resistance meter. Soil moisture is measured on the relationship between moisture content and electrical current

neutron scattering: a technique of soil moisture measurement based on the relationship between water content and speed of scattering of neutrons from a neutron source

time domain reflectometry: an accurate method to determine moisture content and electric conductivity by dielectric permittivity and signal attenuation respectively

---

permeameter: device to record the soil permeability by recording the flow of water through a soil column. There are two types: falling and constant head permeameter

tensiometer: piece of equipment comprising a porous cup sealed to a tube which is filled with water and inserted into the ground to record the soil wetness by measuring the suction or water tension

---

auger: a tool for obtaining a small soil sample by boring into the ground

some properties can change relatively quickly depending on the purpose for which the land is used, for example soil structure and mineral content. Studies that examine temporal changes in soil properties include those examining the impact of particular agricultural or cropping systems, for instance looking at the reduction of fertility over time.

5.5 THE SURFACE OF THE EARTH

THE PURPOSES OF MEASUREMENT AND MONITORING EARTH-SURFACE SYSTEMS: WHAT AND WHY?

The components of Earth-surface systems to be measured or monitored can be broken down into weathering, erosion, transport and deposition. Weathering, the initial breakdown of bedrock, makes it more susceptible to further processes. Erosion, transport and deposition form a closely linked system. Erosion is the most commonly thought about, but transport is one of the system components that is most amenable to measurement, and deposition is frequently what provides a long-term sedimentary record and thus an insight into the history of the system. The systems discussed below (mainly weathering, slope processes and aeolian systems) need considering in conjunction with hydrological systems (Section 5.3), since water (and ice) are the dominant land-forming agents in so many environments.

TECHNIQUES FOR MEASURING AND MONITORING EARTH-SURFACE SYSTEMS: HOW?

Investigation of the systems that make up the surface of the Earth is carried out through a balance of field work, laboratory studies, modelling, the use of maps and records and Earth observation (Figure 5.1). Particularly important is fieldwork, which bridges the gap between the studies of process and form, and allows links to be made between different forms of investigation. In recent years, the development of data-loggers and of Earth observation techniques has proved a huge breakthrough in many areas of monitoring, allowing systems to be studied in depth without constant attendance.

Direct methods for assessing geomorphic processes

Weathering

Goudie (1990) distinguishes four ways in which rates of weathering are measured or monitored in the field:

- Hydrologically (looking at the chemistry of waters discharged from a system undergoing weathering)

- Instrumentally (using micro-erosion meters or rock tablets)

- Archaeologically (using human systems of known age, such as gravestones, buildings or archaeological sites)

- Geomorphologically (using natural systems of known age, such as lava flows, erratic boulders or dunes).

The degree of weathering can also be determined more qualitatively by the use of weathering indices (Gardiner and Dackombe 1980; Bland and Rolls 1998), and the state and mineralogy of weathered surfaces can also be used either qualitatively or quantitatively to indicate the degree to which they have been weathered.

The simplest means of monitoring weathering are descriptive and qualitative. Indices of weathering have been developed for a variety of purposes (including engineering purposes and for monitoring building stone) that describe the state of a surface and categorize the degree of weathering.

A large-scale quantitative method for determining the rate of weathering in a hydrological system, such as a drainage basin, is to investigate the geochemistry of the water discharged from that system to measure the dissolved ions resulting from weathering (Goudie 1990). This method has been used in particular, but not exclusively, for limestone areas. The contributions of other inputs, such as from the atmosphere, from biological activity and due to the chemistry of the water as it enters the system under investigation, need to be taken into account if a sensible rate is to be determined (Goudie 1990).

Instrumental measurements of weathering rates are generally at a much smaller scale, but give more resolution within the area investigated. The **micro-erosion meter** (MEM) will measure surface lowering on a rock due to both weathering and erosion (Goudie 1990). Permanent studs are fixed to a rock surface and these act as a datum. The MEM is mounted on these, and consists of a spring-loaded probe and calibrated dial, which can measure the height of the surface relative to the datum to a precision of 0.01 μm (Thomas and Goudie 2000). Sequential measurements will show how the surface height is changing. This can be carried out in detail over even quite small areas if a large number of studs are mounted. This will allow the degree of precision for individual mineral crystals to be monitored. It can be hard to stabilize the studs when they are mounted on the rock, and this may affect the areas chosen for study. There can also be problems with the growth of lichens, or with changes in temperature. Possibly the most difficult issue with using MEMs to monitor weathering is that they measure surface lowering, which can result from a variety of processes, not least erosion (Goudie 1990).

A further quantitative means of measuring weathering rates is the use of **rock tablets**: pre-weighed pieces of rock of a standard size and shape (Goudie 1990). These have been used in a variety of locations, including urban environments as well as natural ones. When used at ground level, the tablets can be enclosed in a mesh bag in order to protect them from abrasion, but this may interfere with the soil matrix in which they are placed. Tablets can be removed and weighed to monitor the disintegration of the rock; it is also possible to monitor the qualitative change in the condition of the rock.

Human environments have the great advantage in weathering studies that they contain many surfaces of known age, which allows the rate of weathering to be determined more effectively when the difference between the original and current states of the surface is measured. There are also important social and economic interests in the way in which building stone and concretes are affected by weathering.

Gravestones provide a particular case for the monitoring of weathering. One way in which this has been particularly successful is by using lead insets in the writing carved onto the stones (Cooke *et al.* 1995). The difference in the surface

micro-erosion meter (MEM): a device to measure surface height at a number of predetermined points, relative to datum studs set into the surface under observation

rock tablets: pieces of rock of known weight used to assess the rate of weathering in a particular place or environment. The rate of weathering is determined by weighing the rock over time and noting any change caused by disintegration

level when the lead was inserted and its current level can be used together with the date of the stone to give a rate of weathering. Graveyards exist in many different environments, from the remote rural to the urban, and this can give a good idea of the different rates of weathering in such environments.

Archaeological remains can give an even longer record of weathering (Goudie 1990). Features such as **barrows** will protect the surface beneath them while the surface around remains exposed. If such features can be dated then they will allow a rate of weathering to be determined.

Geomorphic features may also be dated (see Appendix 2) and from this the rates of weathering can be determined through a variety of means by comparing the current state with the estimated original state, in a similar way to that discussed above for the human environment (Goudie 1990). Although geomorphic monitoring gives the scope for much longer records of weathering than do human environments, there is a limit to very long records. Weathering will not remain constant over time, but will tend to slow down as surfaces alter. Temperature and precipitation also affect it considerably, so that very long records may effectively be meaningless.

Weathering can be monitored qualitatively, using indices of weathering. The condition of the surface will also allow more quantitative measurements to be made, as weathering will alter its resistance (White *et al.* 1998). Usually, this is likely to make the surface more friable, but it is also possible that case hardening will occur and the weathered surface will actually be more resistant (Goudie 1990). The state of the surface can be tested quantitatively using a **Schmidt hammer** (simply hitting it with an ordinary hammer will give an easier, but less calibrated, approach). The Schmidt hammer was originally developed to test the compressive strength of concrete, and consists of a spring-loaded rod that hits the surface. Its rebound is measured and an 'R value' determined which can be compared with other surfaces (Gardiner and Dackombe 1983).

The alteration of minerals in the **regolith** can be used to determine the degree of weathering which has occurred, by consideration of the way in which different minerals weather and the products which result when they do (Rolls and Bland 1998). Not all minerals will weather at the same rate, and the presence of very stable minerals such as quartz or zircon can be compared against the levels of more easily weathered minerals, such as hornblende, and the composition of the parent material.

Erosion

An important natural erosive agent is the wind, and the first area for measuring erosion is the wind itself, as this will determine, together with the condition of the surface, what erosion can occur.

In field or **wind-tunnel** studies, wind speed close to ground level is usually measured with anemometers (Goudie 1990). The simplest form of these is the familiar **cup anemometer** in which the speed is derived from the speed at which the cups are driven round by the wind. These are relatively cheap and easy to use, but unfortunately not very efficient at translating the motion of the wind into the motion of the cups; they are also quite sizable and therefore hard to use close to the surface. However, they are relatively cheap and sufficiently sensitive for many purposes, so that arrays of many instruments can be used to produce wind profiles in natural environments (Nickling and McKenna Neuman 1999). More sensitive are **hot wire anemometers** (Livingstone and Warren 1996). These

---

barrow: a communal burial mound used from the Stone age to Saxon times

Schmidt hammer: an instrument used for the non-destructive testing of concrete, or rock based on the relationship of the rebound of the hammer and the compressive strength of the surface under investigation

regolith: a loose layer of rocky material overlying undecomposed bedrock e.g. alluvium, peat and glacial drift

---

wind-tunnel: a structure shaped like a tunnel through which air moves and its speed recorded. Obstacles can be placed within the tunnel to measure their effects on wind speed

cup anemometer: a device for measuring wind speed based on the ability of the wind to rotate a design of three or four hemispherical cups pivoting around a spindle

hot wire anemometer: a device to measure wind speed indirectly using the cooling power of the wind and electrical current required to maintain the resistance and thus temperature of an exposed fine wire

work through the relationship between the temperature of the hot wire and the conductivity of the wire. As the wind blows faster across the wire, the conductivity must be altered to maintain a constant temperature. This relationship can be calibrated for wind speed (Goudie *et al.* 1999). Hot wire anemometers are extremely useful in wind tunnel studies and armoured versions have been produced for measurements where sediment movement is involved. Cost is a major consideration, however, particularly where large arrays are used, as has been the case with cup anemometers, and the more sophisticated versions, furthermore, are vulnerable to destruction by **saltating grains** (Nickling and McKenna Neuman 1999).

> **saltating grains**: sediment grains being transported by bouncing along a surface. Most commonly seen in aeolian transport

Transport

In the case of aeolian systems, only the finest grades – sand, silt and clay – can be moved (see p. 283). The size of sediment will determine how the material is transported and over what distance.

The main way in which sediment in transport is measured and monitored for aeolian systems is by the use of **sand** and **dust traps**, which catch the sediment in transport. There are three major constraints on trap design (Livingstone and Warren 1996): that wind direction is variable through 360°; that intrusive traps can significantly alter the profile of the wind and so of erosion rates and sediment transport; and that sediment flux alters considerably with height. The simplest form of sand trap is a box sunk into a pit along the path of transport (Livingstone and Warren 1996). This has the advantage of being cheap and easy, but must be large enough that sand cannot simply saltate over the hole. It also does not allow for changes in wind direction (Goudie *et al.* 1999).

> **sand/dust traps**: devices designed to capture sand or dust over time. Regular measurement of the build up in the trap can be used to estimate transport rates
>
> **aspirator**: a device attached to an instrument to provide ventilation such as a suction fan

A variety of vertical trapping systems allow sand or dust to be collected at heights above ground level (which is very important to the understanding of aeolian systems). In general, these consist of pillars or towers with vertical slots or vertically mounted traps (e.g. WEWC 2003). These are often divided horizontally so that the sediment flux at different heights can be measured. The simplest versions are just pointed into the wind and are therefore only suitable for short-term measurements, or use in wind tunnels. Other (more expensive) versions have been produced which will rotate so that they always face into the wind (Livingstone and Warren 1996). Electronic measurement techniques may be used so that the sediment flux can be measured continually.

Dust flux may also be measured using **aspirators**: these suck dust in and trap it gravitationally, electrostatically or through filters (Livingstone and Warren 1996). Lower-tech ways of measuring dust deposition include the use of sticky surfaces, simple bucket traps or of placing marbles in a tray or using a water trap so that dust which lands in the trap is not able to be remobilized (Goudie 1990). These techniques can lack rigour, but some similar techniques that have been developed for air pollution testing are also applicable and have been tested to higher specifications (McTainsh 1999).

Measuring the rate of slope processes is of most use for slower, more regular movements such as soil creep, which occur in soft rocks and unconsolidated sediments. For the larger, more episodic and dramatic forms of slope movement, including most that occur in hard rock, the rate of process is more or less irrelevant except over extremely long timescales.

Some traditional qualitative measures of rates of creep, such as observation of the way in which tree trunks bend down the slope, may not give quite the information one wishes, since other factors may be at work in causing their shape (Selby 1993). One way of looking at gradual movement of material on slopes is to tag stones in some way, such as by painting them, and observe their movement over reasonably extended periods (e.g. Kirkby and Kirkby 1974). For reasonably fast movements in soft rocks and sediments, stakes and pegs may be hammered into areas prone to movement and their displacement measured in relation to some fixed reference point (Selby 1993; Goudie 1990). To investigate below surface level, columns of blocks or some similar device may be dug into the ground and their movement traced over time (Goudie 1990).

Rates of rockfall can be determined by observing the accumulation of debris. Other methods of visual monitoring are of use in monitoring harder rocks in order to determine if failure is likely to occur. In road cuttings, for example, bolts may be inserted into the rock and their position monitored; high levels of movement may indicate that failure is imminent and allow mitigation or avoidance measures to be put in place (Goudie 1990).

Deposition

The deposition of wind-blown sand and dust may be measured directly through artificial traps similar to those described above, but for longer scale monitoring, natural traps may give more insight into the systems. Dust, in particular, will be preserved in lake sediments, in ice and snow and in soils where it is deposited (Lowe and Walker 1997). Ocean records also provide records of long-term dust flux from the world's deserts: the North Atlantic, for example, preserves a record of fine aeolian quartz from the Sahara (Pye 1987). The continental equivalent record is **loess**: units of aeolian silt that can be hundreds of metres deep (for example on the Chinese Loess Plateau). These natural traps are important environmental records. Characteristic landforms such as dunes will also provide long-term records of sand mobility.

The shape created by slope processes is widely used as a way of interpreting what form of slope failure has occurred. Processes can be classified into flows, slides and heaves (Selby 1993), and the different modes of movement, together with the type of material and the depth to which movement occurred, will help to determine the shape which is left after movement. Thus, investigation of the morphometry of the slope, by observation, surveying, or use of remotely sensed imagery or maps, will allow interpretation of the way in which the slope has developed. Changes over time may be monitored by using sequential datasets; aerial photography and maps are particularly useful for this, and digital terrain modelling allows quantification of such datasets (see below). Although they only provide a fairly short-term monitoring method, given the nature of the system under consideration, they can give particular insights into systems which are developing fast. They also give the geomorphologist an insight into the way in which form changes according to process, and so help to refine not just our understanding but also this technique.

In order to extend data sets back beyond the time over which slopes have been mapped or photographed, the deposits from mass movements may be dated, using one of the techniques explored in Appendix 2. **Radiocarbon dating** is of particular use where organic material has formed part of the movement and has been preserved as part of the deposits (e.g. Johnson and Vaughan 1983).

loess: silt transported by aeolian processes

radiocarbon dating: a technique of determining the age of organic material based on measuring the proportion of ^{14}C isotope within the carbon

Indirect methods of assessing geomorphic process

Laboratory simulations have proved very useful in weathering studies since weathering is a relatively slow process in much of the world, which will inhibit direct monitoring techniques. Blocks of rock can be kept in climatic cabinets that will control the temperature and humidity of the environment; more basically, such experiments may also be carried out using ovens, freezers and heaters (Goudie, 1990). This technique has been used to carry out experiments into heating and cooling (including freezing), the effects of wetting and drying rock and the effects of different kinds of salt solution on weathering (e.g. Sperling and Cooke 1985; Williams and Robinson 1981).

Chemical weathering can also be investigated through laboratory simulation, with water, fluids of different pHs or even solvents being percolated through rocks or rock fragments (Goudie 1990).

A great deal of insight into the operation of aeolian systems has been gained by using wind tunnels to simulate environments (see discussion in Nickling and McKenna Neuman 1999). Wind tunnels allow the speed of the wind and the grade of the sediment to be controlled in a way that is not possible when making measurements of the natural environment. It is also much easier to instrument a laboratory simulation than a field site, so that more detailed measurements can be made.

Earth observation and related techniques have given great insights into the behaviour of aeolian systems in recent years. Light penetration techniques have been used to measure dust transport rates (McTainsh 1999). This can be carried out from above, using satellite data, or from below, using photometers (Livingstone and Warren 1996). Essentially, dust in the atmosphere will change the way in which it transmits light, and thus spectra and even simple light levels can be examined and calibrated for other factors which will change transmission characteristics (such as gas contents or weather conditions) to give the levels of dust mobility.

Use of aerial photography and satellite imagery has allowed the study of the landforms of aeolian deposition, such as sand dunes, in a way which previously involved a great deal of field mapping, often in remote locations (Goudie 1990). The whole of a dune system can be monitored over time in this way, and insights into process gained by observing the relationships between dune form, direction etc and the behaviour of the wind and the supply of sediment, which often relates to vegetation cover.

Slope processes are one of the points at which geography most obviously meets engineering. Measurements of shear strength of materials, using a shear box or a shear ring, will help to show whether or not the material making up a slope is stable (Goudie 1990). This needs to be combined, in particular, with measurement of **pore water pressure**. Water is critical in many slope failures, since it increases the weight of the sediment, may reduce its coherence and may also lubricate planes along which failure can happen more easily, such as bedding planes in rock (Selby 1993).

Stability analysis is a predictive technique, allowing the likelihood of slope movement to be determined (Selby 1993); it is easier to carry out for soil slopes than for rocks. A factor of safety, F, is calculated:

$$F = \frac{\sum \text{resisting forces}}{\sum \text{driving forces}}$$

> **pore water pressure**: the pressure of water in the pores and voids of soil and rock. Under saturated conditions, this pressure may force particles apart and cause slope failure

The forces are calculated using factors such as slope angle, water pressure, the stresses placed on the material and its strength. In theory, F is <1 for a slope likely to fail, and F >1 where the slope is likely to be stable. In practice, values of 1 to 1.3 also indicate natural slopes that are not completely stable (Selby 1993).

Digital terrain models (DTMs) allow the relationship between process and form in the landscape to be quantified and monitored (Lane *et al.*, 1998). Discussed in more detail in Chapter 7, DTMs can be described briefly as

> ... digital representations of variables relating to a topographic surface, such as digital elevation models (DEMs) and digital models of gradient, aspect, horizontal curvature and other topographic attributes. (Florinsky 1998, 34)

Data from other studies, such as process measurements or vegetation indices, can be added to DTMs. Since DTMs can be derived from a variety of sources, including ground surveys, GPS measurements, photogrammetry and satellite images, or a combination of these, they allow change to be monitored over timescales which are inconvenient for other forms of monitoring change (Lane *et al.* 1998). They also allow digitization and quantification of existing data such as aerial photographs, so archival data can be studied using modern techniques (Dixon *et al.* 1998). Use of temporally sequential DTMs has proved very useful, for example, in the monitoring of glacier mass balance, form and flow (e.g. Theakstone and Jacobsen 1997; Willis *et al.* 1998).

> **digital terrain models (DTM):** a model based on a digital topographical map over which layers of other information can be draped to highlight relationships

SYSTEMS FOR MONITORING EARTH-SURFACE SYSTEMS: WHEN AND WHERE?

Temporal issues in geomorphic systems

Earth-surface systems, perhaps even more than those of air, water or vegetation, are dependent not only on the immediate processes that are happening but also on the history of the system. As can be seen from Figure 5.5, the scales of space and time to be investigated will have an important effect on how an investigation is set up, and just what is to be measured or monitored. In general, the larger the scale of space or time under investigation, the greater the effect that a system's history will have on its current form. By contrast, if one wishes to understand links between form and process in more detail, the modern, process part of that system will be more useful. Similarly, selecting the correct timescale is vital to the way that a system is investigated and understood. Much geomorphic process is episodic: landslides, for example, even when extremely active, do not move constantly down a slope, but will become active at particular times, such as after heavy rain. Understanding in detail how a system develops over a few weeks, or even a few years, may not really explain how it has behaved over the whole of its history or why it has the current form that it does. It may also give only a partial insight when information is required to predict future behaviour. Similarly, a wholly historical or morphometric study is less likely to be of use in understanding the behaviour of the system over human timescales. Variability is also crucial: some systems may show more variability through the course of a single event than they do averaged out over a year or longer. Consideration of these influences will help to determine how an investigation is set up. This is particularly important when monitoring a system: the spatial and temporal resolution of the monitoring programme must be selected with a view to the system under consideration and to the use which is to be made of the results.

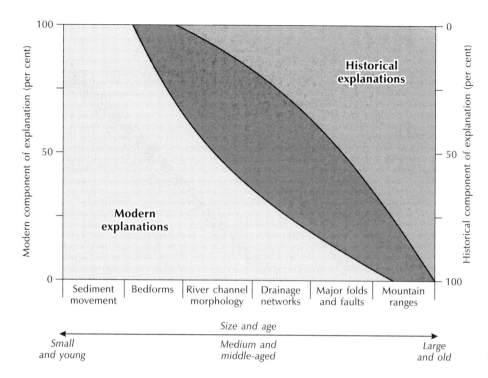

Figure 5.5 The components of historical explanation needed to account for geomorphic events of increasing size and age. The top right of the diagram contains purely historical explanations, while the bottom left contains purely modern explanations. The two explanations overlap in the middle zone, the top curve showing the maximum extent of modern explanations, and the lower curve showing the maximum extent of historical explanations. *Source:* After Schumm (1985, 1991, 53)

Spatial issues in geomorphic systems

> There is no obviously preferred scale that can be used for the study of a natural system. However, we have the best understanding of things that are within the 'human-scale' . . . roughly from 1/10 of a millimeter to a few kilometers in space. (Schumm 1991, 47)

As for temporal issues, it is vital when considering the spatial scales at which measurements and/or monitoring is to be set up for Earth-surface systems, first to know what is the focus of study. As a system is examined in more detail, more complexity will emerge. This means that studies carried out at too detailed a scale for the problem in hand may suffer problems of extrapolation. Earth-surface systems are inherently variable across space and may not multiply up with area in any simple way.

Thus, the project that has the most data points over a very small area is not always the most suitable for understanding a large system. Similarly, spatial variation means that a sufficient density of monitoring points must be set up so that important components of the system are not simply missed in the network, and so that the complexity that is revealed at a detailed scale may be perceived.

Digital terrain modelling is an area in which scale needs consideration before data are obtained. Thus, microscale studies will obtain their data from ground-based studies, such as the use of total stations; mesoscale data is usefully obtained using GPS; and remote sensing is appropriate for macroscale DTMs (Lane *et al.* 1998).

5.6 LAND COVER

THE PURPOSES OF MEASURING AND MONITORING VEGETATION: WHAT AND WHY?

Why bother to map vegetation and other aspects of land cover? A seemingly trivial answer is that it is useful to know the makeup of the land surface. Up until the early 1980s, vegetation and land cover maps were just media of communication, providing a wider audience with information about 'what grows where'. Many vegetation and land cover maps still fulfil that function. An example are the land cover maps produced by CORINE (Co-ordination of Information on the Environment) programme of the European Commission. Over the last few decades, a range of users have pressed vegetation and land-cover maps into service as analytical tools in studying the environmental complex. Examples of such analytical applications are legion, and include applications in vegetation conservation policy and planning, forestry, geology (geobotanical prospecting), environmental impact assessment, military terrain assessment, studies of vegetation change, resource inventorying, and biogeochemical cycling and carbon sequestration. Increasingly, hydrological models, wind and water erosion models and global climate change models utilize digital versions of vegetation and land cover maps.

Vegetation maps show the geographical distribution of such biological units as species, communities, ecosystems and biomes. Land cover maps include all types of land cover, including natural and semi-natural vegetation, agricultural crops, forest plantations and non-vegetated land cover types (e.g. roads, buildings, ponds and lakes).

TECHNIQUES FOR MEASURING AND MONITORING VEGETATION: HOW?

Today, four survey techniques are available for the mapping of vegetation and land cover: ground-based surveys, aerial photography, digital airborne remote sensing and space-borne remote sensing. Table 5.6 lists the pros and cons of each of these techniques. Ground-based surveys are direct, in the sense that humans go into the field and see the features they wish to map. The other three methods are indirect, in that they involve the recording of an image of the land surface (analogue or digital) that human observers or machines later interpret.

Direct methods: ground surveys

Ground-based surveys of vegetation and land cover demand skilled labour (for vegetation at least) and a deal of time-consuming exploration in the field. They also have to contend with inaccessible areas and difficult terrain. Ground surveys involve the logging of field observations in notebooks or in digital form. Modern ground-based procedures, involving the use of electronic instruments that record measurements on digital media, are not generally useful in vegetation and land cover surveys.

Table 5.6 Techniques for mapping vegetation and land cover: pros and cons

	Ground-based surveys	Airborne methods		Space-borne methods
		Aerial photography	Digital airborne remote sensing	
Resolution limits (m)	Less than 1	1–10	1–20	5–5000+
Coverage (km²)	Constrained by resources	100	100	Local to global
Features mapped	Vegetation, land cover, land use	Vegetation, land cover, land use inferred from context	Vegetation and land cover	Vegetation and land cover
Smallest mappable unit	Species, subspecies	Structural vegetation types	Structural vegetation types	Broad land cover classes
Revisit frequency	Constrained by resources	Depends on resources and weather	Depends on resources and weather	Daily to about 20 days, but depends on weather
Recording media	Usually analogue (pen and paper; data recorders)	Analogue (photograph)	Digital	Digital
Interpretation methods	Human observation	Usually human interpretation	Automated image-processing or photo-interpretation	Automated image-processing or photo-interpretation
Financial investment	Low	Medium	High	Very high
Cost per unit area	High	Medium	Medium	Low

Source: Adapted from Wyatt (2000)

Several classic ground-based surveys attest to the value of a large and well-organized labour force making field observations. A case in point are the Land Utilization Surveys carried out in the UK. The First Land Utilization Survey (1931 to 1939) showed seven categories of land, the maps being published at a scale of six inches to the mile. The Second Land Utilization Survey (1961 to 1969) used base maps of the same scale, but recognized 256 categories. Maps from this survey were published at 1:25,000, using 70 categories, but cover just 15 per cent of England and Wales. More recently, a series of ecological surveys conducted in Britain by the Institute of Terrestrial Ecology from 1978, culminating in the Countryside Survey in 1990, involved intensive field observations within a stratified sample of 1-km squares within the National Grid. In Kuwait, a reconnaissance survey, made between 1996 and 1998 at a scale of 1:100,000, used 8351 observation points at which information on vegetation and soils was recorded on field-site cards and later keyed into a database (Omar 2001).

Vegetation mapped by ground survey provides a useful yardstick or baseline against which to measure the power of models predicting vegetation from digital elevation models and remotely sensed data. A ground-based survey of forest types in Croatia served as a test bed for a model predicting Croatian vegetation from several climatic and topographic variables (Antonić *et al.* 2000). Similarly, a field-based map of vegetation in the Podyjí National Park in the Czech Republic provided a reference against which to test a model predicting potential natural vegetation from several environment factors (Tichý 1999).

Indirect methods: surveys from airborne and space-borne platforms

aerial photograph: a vertical or oblique photograph of the Earth's surface taken from the air

multivariate analysis: a statistical analysis considering multiple (three or more) variables simultaneously to assess their interrelationships

Landsat Thematic Mapper: a non-photographic imaging system sensing electromagnetic radiation in seven different bands carried by Landsat satellites to record information about the land surface

neural network methods: techniques of analysis using pattern recognition software that is 'trained' by an input dataset and corresponding output

Aerial photographs are a valuable source of information about vegetation and land cover. The information in aerial photographs is normally extracted by men and women trained in aerial photographic interpretation, though some automated techniques are available. In many aerial photographs, it is feasible to infer land use within a particular land cover category; for instance, grassland in an urban area is likely to have an amenity or recreational use as a park or recreation ground. Such inference is difficult to accommodate in automated digital-image processing.

Digital scanners housed on airborne platforms (aircraft) are similar to digital scanners mounted on space-borne platforms (satellites). The chief differences between them lie in the land surface area that they can resolve and the operational characteristics of the carrying platforms. Satellite-borne scanners may cover the globe through continuous and repeated over-flights along preset orbital patterns, but cloud cover and unfavourable atmospheric conditions disrupt their records. Aircraft-borne scanners can take advantage of good weather, but cannot hope to obtain a global coverage. They do, however, furnish excellent datasets for small areas. Colour infrared (CIR) aerial photography and digital imagery collected by the Airborne Terrestrial Applications Sensor (ATLAS) from the NASA-Stennis Learjet 23 platform, enabled the mapping of riparian (riverbank) vegetation of the Blue Canyon reach of Moenkopi Wash on the Hopi Indian Reservation, Arizona, USA (Weber and Dunno 2001). Some analogue products from digital remote sensing are subject to photo-interpretation. Several land surveys, including the CORINE Land Cover Project, use such an approach. Digital image processing methods are becoming the standard. The usual technique is to carry out a **multivariate analysis** of the radiation spectra or changes in the spectra through time.

Technical developments in image processing are growing apace. It is now possible to produce multi-attribute vegetation maps. For instance, vegetation databases (digital maps) for the United States Department of Agriculture (USDA) Forest Service lands in California, which are created by a mixture of remote sensing (30-m resolution **Landsat Thematic Mapper** data) and GIS methods (digital elevation models), yield several map attributes, including vegetation life-form class, vegetation type and canopy cover. Such multi-attribute information is invaluable in supporting national and regional land cover inventory and monitoring, conservation, fire-risk assessment, wildlife habitat evaluation and land management planning and forest inventory work (e.g. Franklin *et al.* 2000). **Neural network methods** are improving the efficiency of automatic vegetation mapping from Landsat Thematic Mapper and terrain data (e.g. Carpenter *et al.* 1999). Other advances are occurring in the ability to map multiple vegetation types at large scales. For instance, techniques for finding an appropriate plot size and spatial resolution are improving. Researchers are also getting to grips with the equally challenging problem of scaling-up of land cover classification from site-specific maps to global maps, the latter forming the basis of many global change models.

SYSTEMS FOR MONITORING VEGETATION: WHERE AND WHEN?

Spatial issues: local, regional and global monitoring

Table 5.6 shows that scientists monitor vegetation and land cover at local, regional and global scales. Ground-based surveys cover small areas, and larger areas if time and a plentiful skilled labour supply permit. Airborne scanner systems provide analogue and digital products of areas up to about 100 km² with a resolution of some 1 to 20 m. Space-borne scanner systems may give global coverage, but the resolution drops to between 5 m and 5 km. These scale differences affect the features that scientists can map. Ground-based surveys identify individual species or subspecies in small areas. Airborne surveys identify structural vegetation types. Space-borne surveys are limited to broad land cover classes. Financial investment is lowest in ground-based surveys and highest in space-based surveys. However, costs per unit area surveyed are highest using ground-based methods and lowest using space-based methods.

Technological advances in remote sensing make previously 'hidden' features visible. Laser altimetry, known as **lidar** (light detection and ranging), improves the accuracy of biophysical measurement and extends spatial analysis of vegetation to the third dimension, revealing the three-dimensional distribution of plant canopies and subcanopies (e.g. Lefsky *et al.* 2002).

> **lidar (light detection and ranging)**: an active remote sensing system that measures vertical distance using a laser light beam

Temporal issues: vegetation and land cover change

The current debate over global change has given a big boost to studies of past changes in vegetation and land cover. Repeated ground surveys along well-established monitoring routes provide a vehicle for studying such change. Such a system was set up in the Dutch Province of Noord-Brabant in 1995, where results to 2000 show an increase in some species associated with a reduction in the influence of salt water in the delta area and an increase in rainfall (van der Linden and Verbeek 2001). Old maps and aerial photographs provide an invaluable source of information for investigating the nature and rate of such change. In the USA, maps from the US Public Land Survey records are widely used to recreate presettlement vegetation maps (e.g. Manies and Mladenoff 2000). In the UK, in late 1983, the Nature Conservancy Council began the National Countryside Monitoring Scheme, which since 1986 has focused on Scotland under the administration of the Scottish National Heritage. This scheme uses current and historical aerial photographs to observe and measure change in the rural landscape. It interprets aerial photographs, with field checking, within a stratified random sampling framework designed according to the information derived from Landsat data. A classification of Landsat MSS imagery for each Scottish local authority region, as mapped in 1984, yielded up to five broad land classes (strata), which approximate to upland, lowland, 'intermediate' and urban. A stratified random sampling of 5 × 5 km² or 2.5 × 2.5 km² squares within each local authority district, detected changes of 10 per cent or more in the extent of land cover features (Figure 5.6). In north-eastern Puerto Rico, vegetation and land cover changes between 1978 and 1995 were assessed from 'old' and 'new' remote sensing and GIS data (González 2001). The study showed that over those years, pasture and agricultural areas shrank by about 50 per cent. At the Happy Creek study site, a scrubland landscape located within

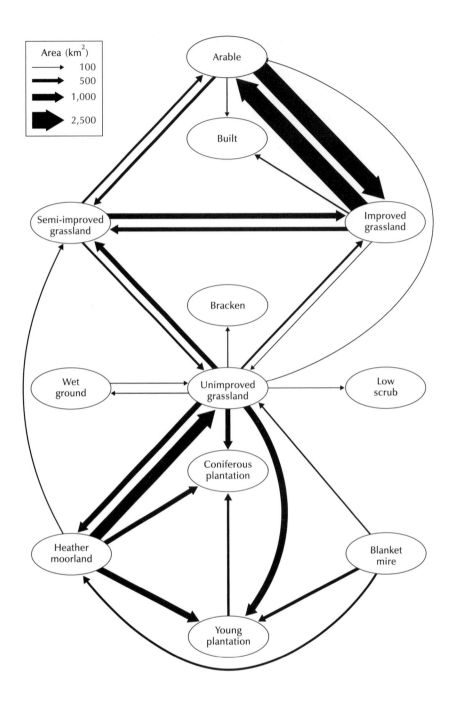

Figure 5.6 Land cover changes in Scotland from the 1940s to 1970s. Only changes exceeding 100 km² are shown. *Source:* Adapted from Mackey and Tudor (2000)

the John F. Kennedy Space Center, Florida, USA, aerial photographs from 1943, 1951, 1958, 1969, 1979 and 1989 were used to map land cover changes (Duncan *et al.* 1999). In Tanzania, long-term trends and effects of conservation on vegetation were examined using a 13-year dataset of satellite imagery (Pelkey *et al.* 2000). An interesting finding was that game-controlled areas, which permit settlement, cattle grazing and hunting, suffered more habitat degradation than areas with no legal protection. In southern Portugal, a 10-year series of Landsat-5 Thematic Mapper images allowed an assessment of the progress of desertification in the region (Seixas 2000).

SUMMARY

Measuring and monitoring are essential tools for understanding the environment and environmental change. Before undertaking measurements or setting up a monitoring programme, it is necessary to give thought to several questions – what, why, how, when and where. All these matters require an appreciation of several ideas, including the scientific method, the nature of data, sampling techniques, scales of measurement, precision and accuracy and a host of field and laboratory techniques. In addition, it is prudent to recognize the limitations of the techniques employed. A wide range of equipment measures and monitors the chief atmospheric variables (temperature, wind speed and so on) and atmospheric composition at climatic stations and other sites. In the hydrosphere, measurements focus on flows in the water cycle – evaporation, precipitation, runoff and so on – as well as water quality. Soil scientists measure nutrient status, acidity and other soil variables, especially those relevant to agriculture. A host of measurement techniques facilitate the study of geomorphic processes, including weathering, erosion, transport and deposition. Measurement techniques allow land cover to be mapped and land cover change to be monitored.

Questions

1 What are the problems of measuring air quality?
2 Compare and contrast the problems of measuring evaporation and precipitation.
3 Why is it useful to measure soil properties?
4 Discuss the issues to consider when setting up a study of a geomorphic process.
5 Compare and contrast the use of ground-based surveys, aerial photography, digital airborne remote sensing and space-borne remote sensing in land-cover mapping.

Further Reading

Good texts that cover the topics introduced in this chapter include:

Arnell, N. (2002) *Hydrology and Global Environmental Change*. Harlow: Prentice Hall.
Full of useful information and ideas.

Brutsaert, W. (1982) *Evaporation into the Atmosphere: Theory, History and Applications*. Dordrecht: D. Reidel Publishing Co.
Rather old, but still worth reading.

Fetter, C. W. (1994) *Applied Hydrogeology*, 3rd edn. Englewood Cliffs, New Jersey: Prentice Hall.
As the title suggests, it covers applied aspects of groundwater.

Goudie, A. (1995) *The Changing Earth: Rates of Geomorphological Process*. Oxford, UK and Cambridge, USA: Blackwell.
A valuable source on geomorphic processes.

Useful Websites

http://www.wmo.ch/index-en.html
World Meteorological Organization (WMO) website.

http://new.filter.ac.uk/database/insightrecord.php?id=48
Useful for fluvial geomorphology and hydrology.

http://www.smuc.ac.uk/RiversWEB/fieldwork/fieldguide.htm
Useful for field hydrology.

http://cru.cahe.wsu.edu/CEPublications/pnw0475/pnw0475.html
Information on the tools and techniques of soil water monitoring and measurement.

http://www.meto.gov.uk/
The UK Meteorological Office website.

Jones, J. A. A. (1997) *Global Hydrology: Processes, Resources and Environmental Management.* Harlow: Longman.
A useful global perspective on the world's waters.

Newson, M. (1994) *Hydrology and the River Environment.* Oxford: Clarendon Press.
Focuses on fluvial environments.

http://www.lbk.ars.usda.gov/wewc/tooltech.htm
A site for Wind Erosion and Water Conservation (WEWC), USDA Agricultural Research Service, with selected wind and water erosion techniques and tools.

http:www.airquality.co.uk
The UK National Air Quality Information Archive.

CHAPTER 6
MAPPING AND ANALYSIS

OVERVIEW

To understanding the nature and complexities of the environment in all of the spheres examined in this book, it is not enough simply to measure and monitor at one location (Chapter 5). It is also imperative to have an idea of the spatial distribution of features and processes. Just as with measurement techniques, there have been great leaps in human ability to map the world around us and to use this to further understanding about what happens under certain circumstances and why. Much of the globe is now mapped and categorized at different spatial scales of investigation as a starting point for further analysis. Unlike in the past, where maps were often the product of many years of toil and involved large teams of people, maps are now often the beginning of projects. Technological developments have also given rise to the development of increasingly sophisticated tools to help in the analysis of spatial and other data. The possibilities for analysis in the twenty-first century are enormous and this will hopefully pave the way for geographers, environmental scientists and managers to help find sustainable solutions to the key threats to the Earth, its ecosystems and its peoples.

This chapter begins by investigating the nature and characteristics of spatial data and shows how it is used to map the globe. The spatial dimension is one of the dimensions of data that were introduced in Chapter 5. The ways of collecting spatial data have changed radically over the last century so that now we now hold vast stores of data describing every sphere of the planet from humans, plants and animals to physical parameters such as temperature and rainfall. In order to appreciate the different forms of spatial data, key spatial data sources are examined, including maps, aerial photographs and satellite imagery. Illustrative examples of each are provided, together with an indication of how these data are used to help understand environmental issues. As well as being collected, data can also be created through numerous spatial analysis techniques. The chapter ends by introducing some of the most important analysis tools, in particular Geographical Information Systems, and shows how GIS-based techniques can be used to turn environmental data into environmental information. Since environmental analysis using computer-based tools is increasingly becoming a norm for research centres, governments and environmental agencies, it is considered appropriate for this discussion on mapping and analysis to pay frequent attention to the implications of issues, concepts and ideas within this context.

LEARNING OUTCOMES

After reading this chapter, you should have knowledge and better appreciation of:
- The nature and characteristics of spatial data and how these are used to map the world around us
- Where spatial data come from, how they are collected and how data are represented within a computer system
- How the development of computer-based tools and techniques for the analysis and visualization of spatial data has helped to improve knowledge about key environmental processes.

6.1 INTRODUCTION

Many of you will have heard of the saying 'location, location, location' as the most important consideration when you are deciding on somewhere to live. Consider your own position, looking for somewhere to live while at University. It is no use finding a fantastic flat for a cheap rent if you find yourself somewhere miles away from college and friends with no means of getting about. Obviously, this is one of the reasons that in many cities there are distinct areas where students all tend to live. Looking at this in another way, patterns of student term-time addresses are not very well spatially distributed. If we map the locations of student addresses, using postcodes or zipcodes, we are likely to find that they exhibit concentrations or clusters in certain areas. We could analyse these patterns through comparing the spatial distribution of students with the spatial distributions of other things such as, bus routes, fast food outlets, colleges, bars or pubs, to see how well they compare. Carrying out such an analysis may show that patterns in student numbers appear related to patterns in some of these other parameters, so where student numbers are high there may be a high density of bus routes or bars. This type of analysis can be carried out just as effectively for helping to understand the natural environment. We could equally examine patterns of soil pH and relate this to geology, waste emission points or some other phenomenon, to try and find out whether there are any spatial relationships that might suggest they are connected in some way. Importantly though, this will only suggest possible relationships, not prove or explain the relationships. For example, do students live in an area because there are lots of bars or are there lots of bars because students live in an area? Despite this limitation, it is clear that the analysis of spatial patterns and processes is essential for understanding the environment and, by extension, understanding environmental issues.

> Almost everything that happens, happens somewhere. Knowing where something happens is critically important (Longley *et al.* 2001, 2)

6.2 GEOSPATIAL DATA AND THEIR CHARACTERISTICS

spatial data: any data that have an explicit locational reference, whether direct (e.g. coordinates) or indirect (e.g. postcodes). Compares to aspatial data that do not have any such reference

temporal: relating to the attribute of time. In this context, the time period a dataset refers to

thematic: relating to subject matter. In this context, the subject or theme of a dataset

Spatial data can be defined as 'data that have some form of spatial or geographic reference that enables them to be located in two- or three-dimensional space' (Heywood *et al.* 2002, 289). Given Longley *et al.*'s opinion that everything that happens must happen *somewhere*, you could view all data as being inherently spatial. However, since not all data that are produced have an explicit geographical or spatial reference associated with them, not all data are actually considered to be spatial data. You could extend this same argument to the **temporal** dimension of data in that all data must happen at some *time* but, once again, the time period may not be explicitly noted. Since all data must be about *something* – it would be meaningless otherwise – the **thematic** dimension of the data is usually well recognized. Recording this and other information about data (the information itself is termed **metadata**) is very important and makes the data more valuable than it otherwise would be, for example, it may allow other people to firstly know about and secondly (and crucially) accurately use data that you had produced for an entirely different purpose.

In the context of this book, spatial data is perhaps more correctly termed **geospatial** data, since we are only interested in geographical data. Data that do have a spatial reference explicitly recognized and recorded are also more likely to have several other characteristics too. Longley *et al.* (2001) cite seven reasons for considering spatial data to be different to non-spatial (aspatial) data (after Longley *et al.* 2001):

1. They are multidimensional – at least two **coordinates** are required to express location, for example latitude and longitude, eastings and northings, *x* and *y*. A third coordinate (*z*) may be used for elevation information.

2. They are voluminous – spatial databases can be very large. Consider global inventories that contain estimates for each area of the globe on a 1° latitude by 1° longitude **spatial resolution**.

3. They must be **projected**. The Earth does not lend itself to a direct representation on a flat medium, such as a piece of paper, and therefore geospatial data must undergo transformation in order that they can be represented.

4. They are associated with special methods, techniques and tools for analysis and representation. Examples include **spatial analysis** techniques through tools such as **Geographical Information Systems (GIS)**, which are discussed later in this chapter.

5. They come from many sources and require considerable effort to integrate into one **geospatial database**, that is the paper or virtual representation of the Earth for the particular purpose to which the data are being put.

6. Data management and update can be complex and expensive. For example, the production of new paper maps may require surveying the entire area, recording changes, redrafting maps and producing output. To some extent spatial data handling tools such as GIS and automated mapping help with this task.

7. Data display requires accessing large volumes of data. Computer-based graphical representations can take up a lot of computer memory. There are many different mapping and representation techniques that can be used to represent geospatial data, and this can also make them prone to misrepresentation.

These features have led researchers to consider spatial data as 'special' data and, certainly, this would seem to be borne out by the amount of intellectual effort that has been invested into developing methods and systems of handling them. These features are highly relevant to a discussion of environmental datasets since the majority of data about the environment are also spatial data.

Burrough (1986) sees spatial data as having three components. First, the data must have a **positional** component that describes the location of the feature(s) of interest. Second, they must have a non-spatial (**attribute**) component (which

---

metadata: data about data – used as a means of describing and cataloguing spatial and aspatial datasets

geospatial: spatial data where the locational reference is given in geographical (real world) coordinates

coordinates: numerical locational references which may or may not be expressed using a real world referencing system

spatial resolution: the level of spatial detail or precision given through a locational reference and its associated data. In the context of remotely sensed images, this refers to the size of individual pixels and, by extension, the area on the ground that is covered by one data value

projected: geographical data converted to a two dimensional form following a certain set of rules and procedures

spatial analysis: the intelligent combination and processing of diverse locational and attribute data to produce new information

Geographical Information System (GIS): a computerized system for handling and analysing geospatial data

geospatial database: a collection of spatial data referenced to the Earth's surface using the same referencing system and usually held in thematic layers

positional characteristic: locational properties that describe where a feature is located (using coordinates)

attribute characteristic: thematic properties that describe what a feature is (using numbers or text)

topological characteristic: relational properties that describe how features are related in space

map: a stylized representation of the real world, usually in a two-dimensional form

describes what the features are and what they represent). Third, a **topological** component describes spatial relationships and patterns in the data. When we read a map or give directions we automatically create these topological relationships so that we know that taking one particular road will lead to another and eventually to where we want to be, or that taking a left turn at the supermarket will take us to the park.

The traditional means of representing geospatial data is a **map** (Basics 6.1). Various characteristics of maps can be used more generally as a means of looking at the characteristics of spatial data (Heywood *et al.* 2002). These characteristics help further expand the three components of spatial data that were considered above. The following sections look at each of these in turn and provide examples to illustrate their relevance in terms of mapping the environment (after Heywood *et al.* 2002). Specific characteristics are also discussed in detail in Kraak and Ormeling (2003) and Longley *et al.* (2001).

BASICS 6.1

WHAT IS A MAP?

Kraak and Ormeling (2003) present an interesting discussion about what constitutes a map and what the discipline of **cartography** therefore is (i.e. 'the conveying of geospatial information by means of maps' (Kraak and Ormeling, 2003, 35)). They give a possible definition as 'a graphical model of the geospatial aspects of reality' (2003, 35). Maps are models because they conform to a set of rules about their production. It is these rules that mean maps have a transferable language that enables them to be read. This set of rules, or map language is, largely, international. If it were not, you would have almost as much difficulty making sense of a Vietnamese map of Vietnam, for example, as you would trying to understand a written description of Vietnam written in Vietnamese. You may not understand what all the features on the map are but you can still get a lot of information about those features and about how they relate to one another.

Although it is most common to find maps that describe the physical characteristics of areas: whether a country, city or neighbourhood, they can also be produced in relation to more abstract phenomena, such as voting behaviour, environmental awareness, recycling rates and so on. Topographic maps, like the ones we are used to purchasing from national mapping agencies, that show the natural and cultural aspects of an area and which are used as way-finders, are composites of many different thematic maps. Thematic maps, as the name suggests, are maps that focus on

one particular theme or subject. A road map is a thematic map, as is a map of historical buildings or places of worship. Another thematic map may be produced to show patterns of elevation change in an area. Plates 4a and 4b illustrate the differences between topographic and thematic maps.

Heywood *et al.* (2002) note some of the rules of map-making in terms of a process that cartographers follow. The process has the following elements:

*   Establishing the purpose of the map
*   Defining the scale of the map at which to represent features
*   Selecting the features to include
*   Selecting a method to represent the features
*   Selecting an appropriate level of generalization for the features
*   Selecting a map projection
*   Selecting and using an appropriate spatial referencing system
*   Deciding on appropriate annotation to help the reader determine the meaning of the map (titles, legends, etc.).

Considering this process is a useful way of getting at the key characteristics of geospatial data as a whole (Heywood *et al.* 2002) and these are further elaborated below.

cartography: the discipline of map-making

PURPOSE

Although not strictly a spatial characteristic itself, the purpose for which geospatial data have been produced is very influential, both in terms of the other characteristics of the data and their representation in map form. Monmonier (1991) explains how these different purposes can influence the message and meaning of mapped data through many examples, such as within development and environment planning, marketing, advertising and political propaganda. In environmental research, spatial data must be objective (or free from conscious or unconscious interpretation) and considerable efforts are made to avoid introducing bias in data collection, analysis and representation.

SCALE

All mapped representations are scaled down versions of what they represent in the real world. A map with a 1:1 scale would be of little use, this would mean that distance representations on the map were the same as those on the ground. The map scale denotes this relationship between real world distance and map distance for a particular map. It is useful to remember that small-scale maps are used to cover large areas (large spatial extent) and large-scale maps are used to cover small areas (small spatial extent). It is common to find global spatial data or mapped representations at a 1:1,000,000 scale, so that every map unit represents a million real world units. This would be very little use for mapping a specific neighbourhood, so here a scale of 1:5000 or 1:10,000 might be more appropriate. Table 6.1 suggests the effect of reading positional data from maps of different scales.

GENERALIZATION

Since all mapped representations are scaled down versions they must also be generalizations, that is, they simplify the features being represented to a greater or

Table 6.1 The effect of map scale on determining positional references

Map Scale	Ground distance
1:1250	62.5 cm
1:2500	1.25 m
1:5000	2.5 m
1:10,000	5 m
1:24,000	12 m
1:50,000	25 m
1:100,000	50 m
1:250,000	125 m
1:1,000,000	500 m
1:10,000,000	5 km

Note: Positions on a map can generally be determined to about 0.5 mm on the map itself. The estimation of the impact of this on reading positional data from a particular map can therefore be determined by multiplying this number with the map scale.

Source: After Longley *et al.* (2001)

lesser extent. The level of generalization of particular features will, in part, depend on the purpose of the map and whether the feature is the focus or just included for contextual purposes. A national road atlas will have enough spatial detail to enable navigation from one city to another, but not necessarily between specific locations within a city. If the atlas had every single road in the whole country it would be very difficult to determine which were the main roads and which the minor roads. The scale and purpose gives a guide as to what should be shown and how. The road atlas contains only roads that are significant in terms of national navigation. Furthermore, the roads themselves are generalized. For national navigation, it is not necessary to show every bend and turn in the road. So, although the mapped version will show that a particular road goes through a particular town, it will not show precisely how it does so. Figure 6.1 shows some of the influences of scale and generalization on a set of features.

SPATIAL ENTITIES

Since a map is a scaled down generalization of the real world, it is necessary to use some sort of stylized means of representing features of interest. In the national road map example, roads are represented as linear features. The most common spatial entities are points, lines and areas. Figure 6.2 gives some examples of the relationship between scale and spatial entities used to represent features.

Figure 6.2 only includes the basic spatial entities contained in traditional mapped representations. However, it is important to note that GIS and spatial data handling systems, such as visualization software, have literally added new

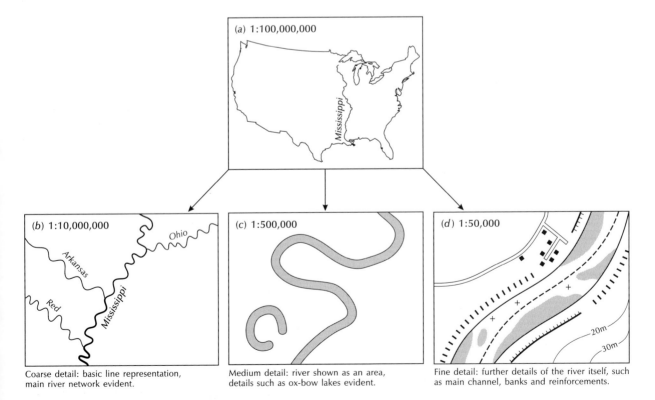

Coarse detail: basic line representation, main river network evident.

Medium detail: river shown as an area, details such as ox-bow lakes evident.

Fine detail: further details of the river itself, such as main channel, banks and reinforcements.

Figure 6.1 The relationship between map scale and generalization level. *Source:* After Heywood *et al.* (2002)

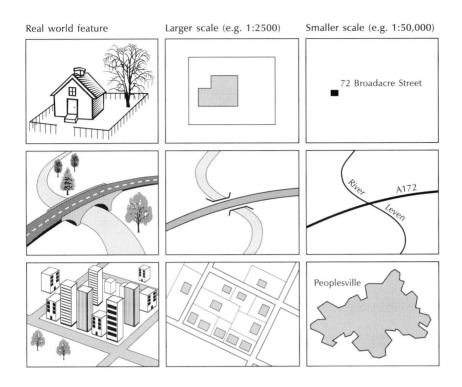

| Real world feature | Larger scale (e.g. 1:2500) | Smaller scale (e.g. 1:50,000) |

Figure 6.2 The relationship between scale and spatial entities

dimensions to data and how they are represented. Computer-based representations increasingly use **surface** representations (i.e. *x*, *y* and *z* coordinates) for features such as topography. This is useful in terms of analysis since this enables the calculation of three-dimensional aspects of the data such as volume. Another extension to the traditional use of spatial entities is to use **network** representations so that a map is not made of a number of discrete and separate lines but in fact a series of interconnected lines, which are associated with flows, directions and other attributes.

> ***surface***: a continuous, pseudo three dimensional representation of a feature comprising both locational data (*x*,*y* coordinates) and some form of elevation attribute (*z* coordinate)
>
> ***network***: a collection of connected lines (with explicit topological information)

PROJECTIONS

The process of transforming spatial data referring to a three dimensional globe (an imperfect geometric sphere) to a two dimensional medium requires the use of a map projection. This process is not perfect and the selection of a projection method can drastically change the nature and look of mapped output. Distortion in the final product is inevitable; it is impossible to faithfully represent all aspects of the three dimensional globe. Monmonier (1991) discusses a wide range of different projections, demonstrating how each performs in relation to five geographic properties: areas, angles, distances, directions and shapes.

There has been an enormous degree of intellectual effort invested in the development of map projections. Indeed, this was traditionally the backbone of any cartographic education and training (Kraak and Ormeling 2003), which is not surprising given the historical importance of maps as navigation tools and the complexities involved in selecting the best projection for a particular task in a particular area. Projections can either mainly preserve shapes (**conformal**) or mainly preserve area properties (equal area or **equivalent**) but not both shape and area.

> ***conformal***: describes map projections which maximize the preservation of the real world shape characteristics of features
>
> ***equivalent***: describes map projections which maximize the preservation of the real world area characteristics of features

> **projection plane**: the characteristics of the two dimensional surface upon which the map will be represented

A broad categorization of map projection types is possible according to the shape of the **projection plane**: cylindrical, azimuthal or conical (Figure 6.3).

- Cylindrical group – This is easily visualized as a piece of paper wrapped around the globe, either encompassing the Equator (normal) or with the paper encompassing the area along a meridian (transverse[1]). Features can then be projected onto this surface and the paper unwrapped. This process can be likened to seeing the globe as a lampshade with a light bulb at the centre. Lines on the lampshade (which could represent land masses on the globe) would be projected onto the paper that is wrapped around the lampshade in the same way as the projection transformation process. This type of projection provides a full representation of the entire surface of the Earth but has some problems. Although positions along the equator are preserved and areas are generally representative, there is increasing distortion towards the top and bottom edges of the paper.

- Azimuthal group – With this family of projections the paper only touches the surface of the globe at a single point. If this point is at the

[1] In actual fact the paper could be wrapped around any part of the globe – this would then be termed oblique. Experimentation with wrapping pieces of paper around a golf ball will reveal how many permutations are possible!

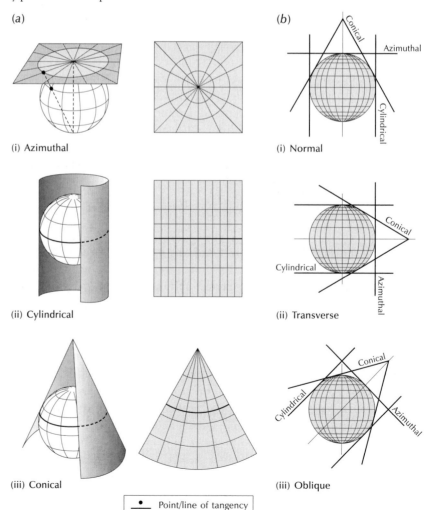

(a)

(i) Azimuthal

(ii) Cylindrical

(iii) Conical

(b)

(i) Normal

(ii) Transverse

(iii) Oblique

Point/line of tangency

Figure 6.3 Map projection groups according to projection approach. *Source:* After Kraak and Ormeling (2003)

poles, it is normal; if it is on the equator, it is transverse. Take the lampshade idea again. It is possible to think through what the consequences of this might be in terms of the associated view of the surface of the Earth on the piece of paper. This enables only part of the world to be viewed and this view is the best in the middle of the paper. Distortion increases in all directions with distance from the centre of the page. Distances are preserved for the most part.

- Conical group – Here, the projection involves making the paper into a conical (cone) shape and placing this cone on the globe. Again, this can be done in various positions depending on where least distortion is needed. This projection does not preserve areas, and distances are especially distorted towards the bottom of the page. This does, however, generally preserve scale.

SPATIAL REFERENCING SYSTEMS

Spatial data that are expressed in terms of the three dimensional Earth are the only true geographic coordinates. In order to be able to make sense of many descriptions of global environmental processes (such as those in Chapter 8) it is important to understand the global coordinate system of latitude and longitude. This system is elaborated in Basics 6.2.

> **spheroid**: a three dimensional, sphere-like shape
>
> **ellipsoid**: a three dimensional, ellipse-like shape with dimensions longer in one plane than the other

BASICS 6.2

THE GEOGRAPHIC COORDINATE SYSTEM OF LATITUDE AND LONGITUDE

The Earth axis is an imaginary line that runs through the centre of the globe from one pole to another, around which the Earth rotates. Another imaginary line can be drawn across the widest point of the Earth (perpendicular to the axis) and this represents the Equator. The Earth is not a perfect sphere, rather a **spheroid**. Since the Earth is flattened at each pole and bulges at the equator (Figure 6.4), an **ellipsoid** is the most reasonable approximation of its shape, where the minor axis (axis of rotation) is shorter than the major axis (equator). The minor axis and major axis are generally referred to as semi-axes since the values used to measure them refer to half of their respective lengths, that is equivalent to the radius of the Earth from the centre to the poles along the axis, and from the centre out to the equator along the equatorial plane. Throughout history there have been many different ellipsoids in use, but now there is growing international consensus associated with the use of the WGS84 (World Geodetic System 1984) ellipsoid. This ellipsoid is more correctly termed a

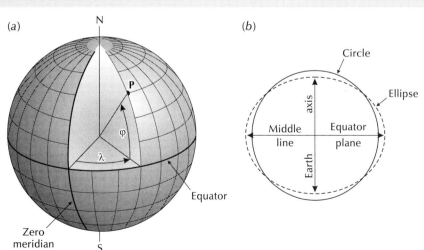

Figure 6.4 The generalized dimensions of the Earth. *Source:* After Kraak and Ormeling (2003)

BASICS 6.2

THE GEOGRAPHIC COORDINATE SYSTEM OF LATITUDE AND LONGITUDE – *continued*

geoid since it is a more precise estimation of the Earth's shape than an ellipsoid. Differences in the means of representing the shape of the Earth are not only influential in terms of location but also in terms of elevation (a height reference point being referred to as a datum). The importance of these measures is demonstrated through the fact that there is a branch of science devoted to **geodesy**, which investigates the precise nature of all local variations from the reference ellipsoid in order to determine the Earth's size, shape and gravitational fields (Clarke 2001). One of the reasons for the need for consensus is the growing importance of global systems such as Global Positioning Systems (which generate geographic coordinates in the first instance).

Lines of longitude (meridians) are drawn from pole to pole (Figure 6.5). They 'begin' at the prime meridian (also called the zero or Greenwich meridian) at 0 degrees, although this starting point is just convention. Convention also dictates that locations are expressed in terms of degrees east or west of that line so the globe is represented between 180° east (or positive values) to 180° west (or negative values). The Greek letter lambda (λ) is a shorthand symbol used to represent longitude values.

Lines of latitude (parallels) are perpendicular to lines of longitude (Figure 6.5). The widest line of latitude is located at the point at which the Earth has the largest circumference; in other words, at the equator. In contrast to lines of longitude that are the same length everywhere, these circular lines of latitude get shorter and shorter from the equator towards the poles. Latitude is expressed in terms of degrees North or south of the equator. The North Pole is at 90° north (or positive values) of the equator and the South Pole is at 90° south (or negative values) of the equator. The area from the equator to the North Pole is the Northern Hemisphere and the area from the equator to the South Pole is the Southern Hemisphere. The Greek symbol phi (φ) is a shorthand symbol used to represent latitude values.

Locations are defined by specifying degrees latitude as the angle from the Greenwich meridian and the meridian of the point in question, and degrees longitude from the equator to the parallel of the point in

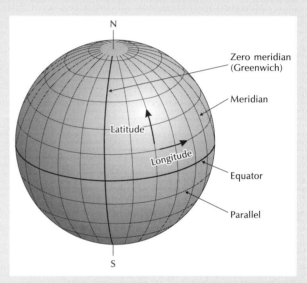

Figure 6.5 The global graticular network. *Source:* After Kraak and Ormeling (2003)

question. Lines of equal longitude (meridians) and latitude (parallels) around the globe can be drawn to make up a reference graticular network (Figure 6.5). Assuming that the Earth has a circumference of 40,000 km, the distance east–west along the equator one degree is 40,000 km divided by 360°, i.e. 111.11 km. Since the circumferences of parallels north or south of the equator get progressively smaller, the east–west distance covered by each degree gets smaller and smaller. At 45° latitude north or south of the equator, for example, a parallel has a circumference of 28,301 km, so one degree east–west along this line is only 78.61 km. The distance *between* one degree parallels though, i.e. along meridians, does not change since the distance along all meridians is equal at our assumed 40,000 km circumference. These therefore remain at 111.11 km per degree longitude. This distinction is useful in interpreting global inventory data since this is often expressed in terms of a spatial resolution of one degree latitude and one degree longitude.

A more detailed explanation of latitude and longitude, from which this summary is drawn, can be found in Longley *et al.* (2001) or Kraak and Ormeling (2003).

---

geoid: a detailed approximation of the shape of the Earth

geodesy: the science of characterizing and modelling the true shape of the Earth

Despite the importance of geographic coordinates, it is also fair to say that most spatial data are actually handled in terms of a rectangular coordinate system. Rectangular or grid-based coordinate systems are generated after the application of a particular map projection to transfer geographic co-ordinates to a two dimensional version. This is sometimes referred to as a Cartesian grid and locations are then denoted through giving values along an *x*-axis and along a *y*-axis (Figure 6.6). Different parts of the world have different rectangular grid systems that are developed specifically for that area. The British Ordnance Survey, for example, uses the transverse Mercator projection as the basis for creating their rectangular grid system. The grid covers a spatial extent of 700,000 m east–west and 1,300,000 m north–south. The grid is given an origin of 0,0 meters as a reference point (a false origin) and references can then be given for any part of England, Scotland or Wales relative to this point.

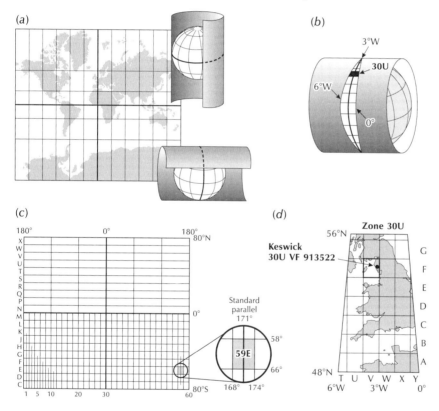

Figure 6.6 The basis of rectangular grid systems. *Source:* After Kraak and Ormeling (2003)

TOPOLOGY

Many of the characteristics of mapped data that have been covered in this chapter are affected by scale. However, topology (or the spatial relationships between features) is independent of scale. It is also independent of map projections. The main geometric properties (Figure 6.7) are: adjacency (what lies next to a particular feature); containment (what lies within a particular feature); and connectivity (what is connected to a particular feature).

Determining these features of topology are part of the map-reader's skills in terms of standard paper maps, although a good cartographer will help the reader determine these relationships easily, through careful use of colour, labelling and keys. Geographical Information Systems require that these topological relationships be encoded into the system before any analysis can be undertaken using them.

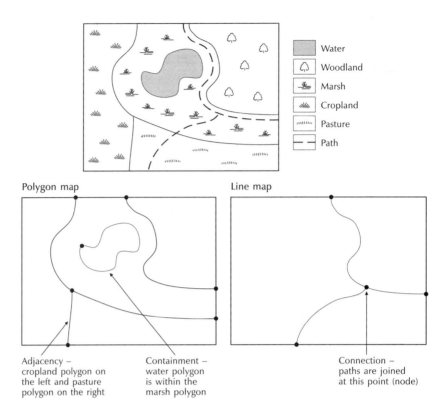

Figure 6.7 Examples of geometric properties associated with topology

Polygon map

Line map

Adjacency – cropland polygon on the left and pasture polygon on the right

Containment – water polygon is within the marsh polygon

Connection – paths are joined at this point (node)

6.3 GEOSPATIAL DATA SOURCES

The previous section showed the various procedures of abstraction of reality that are required to transform spatial data into a mapped representation. These are useful in considering any mapping project and, in the context of this book, they help provide some points for consideration in interpreting (and using) environmental and other geospatial datasets.

Although many environmental datasets exist as maps, they are not the only source of spatial data, as will be explored in this section. Indeed, many maps are composites of spatial data collected through other means. Increasingly, geospatial data are collected as **digital** data and so they can be readily input into a computer-based tool for further manipulation and analysis. In doing this, further levels of abstraction are required (see Basics 6.3). Some key geospatial data sources are discussed below in terms of their utility for computer-based handling and analysis systems.

> **digital data**: data which are held and stored in a computer-readable format

MAPS AS A DIGITAL GEOSPATIAL DATA SOURCE

Much of the discussion so far in this chapter has centred on the characteristics of spatial data in terms of traditional paper maps. The links between the cartographic process and computer-based forms of geospatial data have been made in Basics 6.3, and the point has also been made that traditional maps are often the form in which data that have been collected in different ways are presented. However, not all maps are in digital form and it is useful to complete

> **vector data model**: conceptual means of representing spatial data using coordinate pairs
>
> **raster data model**: conceptual means of representing spatial data using grid cells

BASICS 6.3

SPATIAL DATA MODELS – MODELS OF SPATIAL FORM

Computers require specific, unambiguous instructions about how to interpret what the user inputs. This is just as true for spatial data handling as for handling email addresses, website addresses or any other sort of data that the user might enter. The production of paper maps traditionally uses a process of abstraction for the representation of geographical data, and before data can be encoded in such a way as to enable their use in computers a similar process of abstraction needs to be determined for digital data. In fact, there are two commonly used means of abstraction – the vector spatial data model and the raster spatial data model. Both of these models are of spatial form rather than spatial processes (the latter are considered in detail in Chapter 8).

- **The vector data model** – this model is largely based on the traditional cartographic process that has been discussed in Section 6.1. Features are represented as discrete points (e.g. individual trees), lines (e.g. hedgerows or paths) or areas (e.g. forests or lakes) and the locations and dimensions of these features are determined through plotting locations as *x, y* coordinates in an otherwise 'empty' space (Figure 6.8). Since the method of abstraction is strongly related to the field of cartography, it is common for digital maps to be stored in vector format. The vector data model conforms to the 'discrete object' view of the world (Longley *et al.* 2001) and this makes it particularly useful as a means of representing features that have clear demarcations.

- **The raster data model** – this model is more related to a 'field-based' view of the world (Longley *et al.* 2001) where there are no definite boundaries. Instead, the world is seen in terms of gradations of values of parameters of interest. Elevation, for example, has a value at every location and so it varies continuously over space. The raster model can incorporate this continuum of values (or individual features) using a grid-based model. Row and column references are used to represent locations, and for each cell or pixel in the grid, a single value representing what is at that location. This may be a single cell representing, for instance, a single tree; a set of adjacent cells with the same value representing a hedgerow; or a group of cells with the same value that together might represent an area of forest. In terms of the 'field view' of the

world, a raster representation of forest cover might show tree density across an area, with different cell values representing relative tree numbers across the grid. This data model is linked to those data sources that are also pixel based, such as remotely sensed imagery. The raster data model is very well suited to many environmental datasets.

The selection of a data model will depend on various considerations. Some of these include the source of data and its format, the nature of the computer-based system to be used and the types of analyses to be undertaken.

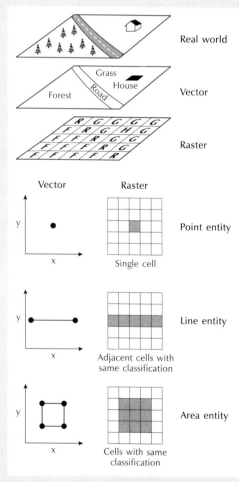

Figure 6.8 Representation of features in the vector and raster worlds

this discussion through a brief consideration of how maps can be converted to digital forms.

There are two main ways that paper maps can be converted to a digital form. The first is through the process of manual **digitizing**. This involves tracing the pattern of features of interest on the paper map with a pen or **puck** while the map is mounted on a special board that can pick up the locations and draw these on the computer screen. As can be imagined, this process is very time-consuming and involved and this method of spatial data input is now becoming less important (not least since most paper-based map data has now been digitized). However, on-screen digitizing using, for example, an aerial photograph as a base, is still a fairly common means of obtaining base data or certain types of environmental data such as areas of greenspace within cities and specific types of vegetations stands. This process is not usually as difficult. The second method is through **scanning** paper maps. This method has grown in popularity with the advent of better scanners and the development of computer procedures to help draw out features of interest from the scanned output. Importantly, however, more and more mapping and environmental agencies are providing data already in digital format. Together with better data management and reporting procedures, this means that there is now more digital map data available than there has ever been.

Undoubtedly, the first impetus for mapping activity related to the need to record locational information for navigation purposes. With increasingly far-flung exploratory expeditions of the globe, the need for accurate mapping grew in importance. The British Library Map Collection was founded in 1753 and houses maps from the fifteenth century to the present day – there are now some 4.25 million records of maps, atlases, globes and books on cartography (British Library 2003). It is the second largest map collection in the world after the Library of Congress. An excellent source of on-line historical maps (for reference purposes) is available through the University of Texas (www.lib.utexas.edu/maps/index.html).

Over time, maps have developed to be more than simple way-finding tools and now record a great breadth and depth of information about the human and physical world. This is particularly true in the case of environmental data, the increasing availability of which has been linked to the wider development of environmental politics (Haklay 2000). In a brief review of key milestones, Haklay (2000) notes the influence of the US National Environmental Policy Act (NEPA) (1969), as well as developments in other countries in response to the growing environmental agenda of the 1960s. In the UK, for example, there was the Royal Commission on Environmental Pollution (1969) and the Department of the Environment (1970) and, through Europe as a whole, the European Economic Commission was considering broader policy horizons (Haklay 2000). In terms of global initiatives, the influence of the United Nations is perhaps the first and most important. The United Nations Environment Programme (UNEP) which is still an important focus for the study of the environment (now embracing the philosophy and ideology of sustainability) forged new ground with the development of GEMS (see Chapter 6) and the related initiative INFOTERRA, an International Environmental Information System that Haklay (2000) suggests was the first truly worthy of the name. The influences of information systems themselves are discussed later in the chapter.

digitizing: the process of converting data from an analogue (non-digital) to a computer readable form. Creates vector data

puck: a piece of equipment rather like a computer mouse which is traced around features on a map in order to capture the features digitally

scanning: automated digitizing. Creates raster data

Increasing volume and accuracy of all types of environmental and other types of geospatial data can be linked to the ongoing developments in remote sensing techniques. Two forms of remote sensing, aerial photography and satellite remote sensing, are discussed below, but it is important to recognize that others are not covered. The reader is directed to the suggested introductory texts for further information.

REMOTELY SENSED GEOSPATIAL DATA SOURCES

One of the earliest **remote sensing** methods of data collection was aerial photography. This involved literally taking photographs of the Earth's surface from an elevated position, whether stationary (e.g. a platform attached to a building or crane) or moving (e.g. mounted on a plane). These photographs could be taken from a range of angles or from directly overhead, with the latter being particularly important as a data source for computer-based spatial data handling tools like GIS (Heywood *et al.* 2002). Individual photographs can be patched together to make an **orthophoto** and therefore cover a large area. This is never a simple task, though, as there are many corrections that need to be made before the patching is satisfactory. For example, corrections are needed to the scale of the photography, since this gets increasingly distorted with distance from the centre of the image. On moving platforms, particularly aircraft, there may also be distortions due to the movement itself. Fortunately, the increasing availability of commercial products that have had all the necessary pre-processing carried out are becoming increasingly common, such as with the Cities Revealed datasets.

It has been shown that maps are already interpretations of the Earth's surface that use a set of recognized rules and the skills of trained cartographers (in most cases). Photographs, on the other hand, have no such interpretation to assist in making sense of them. Instead, they are simply a record of everything below the camera at the particular point in time and space that the photograph was taken. In the words of Curran (2001, 332), 'remotely sensed data are nothing more than a measure of electromagnetic radiation after it has interacted with the environment'. This is true whether the remotely sensed data (imagery) are aerial photographs or satellite images (see later in this section).

Chapter 8 discusses the nature of **electromagnetic radiation** and how it interacts with the Earth in more detail, but there are a number of points that can be made here to demonstrate the characteristics of remotely sensed images. As a starting point, it is useful to note that the **electromagnetic spectrum** contains both visible and invisible energy originating from the Sun that is in part reflected and in part absorbed by features on the Earth's surface. Imagery picks up the reflected radiation for the particular patch of ground that the camera or sensor is trained on and this itself is a function of the properties of the features that the light or energy strikes. So, if this is, for example, a patch of forest, the reflected radiation will be a function of (after Curran 2001):

- The *atmosphere* through which the light has travelled. How dusty is the atmosphere and how effectively does this scatter light between the ground surface and the camera or sensor? Are there clouds in the atmosphere that obscure the surface?

- The *vegetation* of the forest. How many trees are present? What type of vegetation grows under the trees? What type of trees are they and what

---

remote sensing: observation of an object of interest without contact. In this context the collection of data about the Earth's surface without being in direct contact with the Earth

orthophoto: a mosaic of aerial photographs

electromagnetic radiation: energy associated with the Sun. This energy can be in many forms

electromagnetic spectrum: shows the full range of electromagnetic energy, grouped and ordered in terms of relative wavelength

are the shape, density and angles of its leaves? How healthy is the vegetation, e.g. is it well watered or very dry?

- The type of *soil* on which the forest grows. What are the chemical, physical and biological properties of the soil? What depth is the soil? Is it well watered or very dry? What sort of geology underlies the soil?

All this makes the interpretation of images very complex. However, it is precisely because the reflected light is a function of all of these properties, that it is possible to interpret them to be able to give information about all sorts of aspects of the environment.

Aerial photographs usually just pick up a small part of the electromagnetic spectrum. This is commonly just the **visible spectrum**, so the information is limited to what can actually be determined using the naked eye (i.e. what would have been seen by looking out of a window in the undercarriage of the plane on which the camera was mounted). This is, however, also very useful as a source of environmental information. For example, it has been extensively used as a means of determining detailed vegetation patterns. Sometimes, however, vegetation cover can be a problem in using imagery, since it can almost totally obscure what lies beneath it. It is important to note here that many aerial photographs are black and white only (i.e. they are monochromatic) but this does not detract from their utility as a means of mapping some of the key environmental aspects of the Earth's surface. Colour photographs pick up the entire visible spectrum and give further detail to assist interpretation.

However, the most important feature of modern imagery is the ability to pick up reflected light in terms of the other parts of the spectrum (i.e. that are invisible to the naked eye). Such images are termed multi-spectral because for the same area on the ground they record reflectance levels in more than one part of the spectrum. The different physical, chemical and biological properties of a feature result in different patterns of absorption and reflection in different parts of the electromagnetic spectrum. The particular **spectral signature** of a feature allows a huge amount of information to be generated about it. Figure 6.9 shows the different spectral characteristics of snow and ice. This example is explained in more detail in Gao and Liu (2001) who also give an indication of the different types of remote sensing sources useful for glaciology applications. These characteristics can be compared to the generalized reflectance pattern to be expected for water, green vegetation and dry soil shown in Figure 6.10.

Although some traditional aerial photographs can pick up reflected lights in different parts of the spectrum using special photographic film, remotely sensed satellite imagery (using either digital photography or scanning) is the primary source of multi-spectral data.

There are now far too many satellites orbiting the Earth to give details of each and every one, suffice to say that there are satellites associated with media uses (e.g. satellite TV), the military and the Global Positioning System (Basics 6.4, p. 165) as well as satellites associated with Earth Observation.

Earth Observation satellites themselves fall into some general categories. Geostationary satellites 'watch' the same part of the Earth's surface over time and move with the Earth, whereas other satellites move relative to the Earth and collect data for different parts of the surface. Until relatively recently, there has been quite a clear trade off between the frequency with which data are collected and the spatial detail (resolution) of those data. This means that coarse resolution

---

visible spectrum: the part of the electromagnetic spectrum that can be detected with the naked eye

spectral signature: the specific electromagnetic radiation reflectance and absorption characteristics associated with a feature or set of features

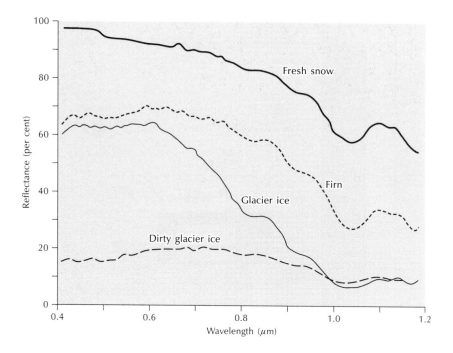

Figure 6.9 Spectral reflectance characteristics of snow and ice. *Source:* After Gao and Liu (2001)

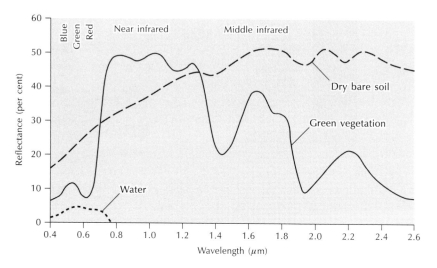

Figure 6.10 Generalized reflectance characteristics for water, green vegetation and dry soil. *Source:* After Longley *et al.* (2001)

data from a high orbiting geostationary satellite like Meteosat (2.5×2.5 km² per pixel, i.e. one reflectance measure per 2.5×2.5 km cell in the image) could be collected for a large area every 30 minutes. Very fine resolution data from a lower-orbiting satellite such as Landsat Thematic Mapper (TM) (30×30 m² resolution for multi-spectral data) could only be collected for the same area on a much less frequent basis, in this case 16 days (Longley *et al.* 2001). This relationship between frequency and resolution is becoming increasingly blurred with data from satellites like IKONOS and Quickbird, which have very fine resolution data and are also collected more frequently (Figure 6.11). The resolution of the data received is a function of the satellite (for example, its orbit height and nature) and the sensor on board the satellite (for example its age and method of data collection).

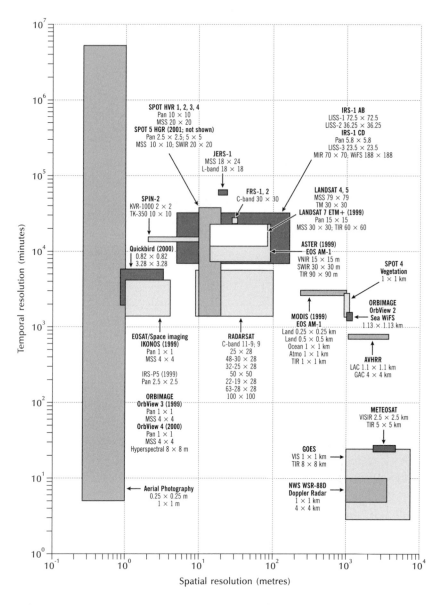

Figure 6.11 Spatio-temporal characteristics of some remote sensing systems and their sensors in operation in the early twenty-first century. *Source:* After Longley *et al.* (2001)

These spatio-temporal characteristics have important implications for the use of the data that are collected by particular systems. The data collected by Meteosat and AVHRR[1] are very well suited to meteorological applications since it is not necessary to have a very detailed set of data, but the frequency of the data is essential in order to track the movement and development of weather systems. Chapman and Thornes (2003) also note some examples of the utility of such imagery to develop climate datasets, such as the land surface temperature maps produced through a combination of **Digital Terrain Model** (DTM) data and Meteosat's **thermal** imaging band (Schadlich *et al.* 2001). In another example of applications in terms of potential hazard detection, Plate 5 shows an oblique image of Hurricane Fabian on 4 September 2003, shortly before it hit the island of Bermuda. The image, taken from the International Space Station,

> **Digital Terrain Model (DTM)**: digital representation of the Earth's surface using positional (*x*, *y*) and elevation (*z*) data. Also referred to as Digital Elevation Model (DEM). Can be used to plot pseudo three dimensional (surface) representations of any physical or human parameter

---

[1] AVHRR stands for Advanced Very High Resolution Radiometer. It should be noted that technology has moved on since this sensor was put into operation and it is no longer considered to be high resolution compared to some of the newer sensors

shows the hurricane at a time when winds were reaching some 120 miles per hour and it was travelling at 12 miles per hour (NASA 2003a). There is also some very interesting imagery depicting the nature and extent of the 2003 European heat wave. This heat wave was responsible for some 3000 excess deaths in France alone, where some areas experienced July temperatures that were 10°C hotter than in 2001. Plate 6 was generated from NASA's Terra satellite using the Moderate Resolution Imaging Spectroradiometer (NASA 2003b). The NASA website contains a wealth of examples of other mapping applications to which satellite remote sensing can be put (http://earthobservatory.nasa.gov). These include drought mapping in Africa, wildfire, flooding and volcanic activity. A short review of some of the other recent applications of remote sensing for base mapping can be found in Aplin (2003).

Finer-scale imagery, with some analytical work, can be used to map other parameters. For example, Voogt and Oke (2003) review work into mapping and characterizing urban climates and Sutton (2003) presents a method for developing an aggregate indicator of urban sprawl using satellite collected night-time imagery. Increasingly these data can be used to investigate urban environmental issues (Carlson 2003) as well as those associated with rural areas, which have perhaps tended to be more commonplace due to the reduced complexity of these areas in terms of interpreting reflectance values.

Both aerial photographs and satellite images, although the data they collect are inherently spatial, require further processing in order to add a recognized real world coordinate system to them (a process called **georeferencing**). Without this, it would not be possible to link the data with other geospatial data sources or to make calculations involving real-world measuring units like area. This process is not always straightforward because, rather like with aerial photographs, scale can vary across an image due to a variety of causes. With satellite images, other corrections are also needed, for example to take account of the effect of the atmosphere and the curvature of the Earth. Sometimes, of course, things in the atmosphere (clouds in particular) obscure features on the surface to such an extent that no useful data can be obtained from the image. This is a particular issue in the tropics.

In view of all these necessary pre-processing tasks and the not-inconsiderable task of then actually interpreting the images themselves, a whole field of study has grown up which is dedicated to the science of remote sensing and digital image processing. Again, it is useful to note that it is not now essential to be able to carry out all of the pre-processing corrections oneself since imagery is increasingly available in a form that can be directly used, or even already processed into thematic maps that are suitable for analytical work with other data types. However, a sound understanding of the science itself is essential to ensure that accurate interpretations of the data are made.

In view of the ability to take images of the same patch of the Earth's surface time and time again, these sources are very useful for detecting changes and giving information that can help determine patterns and processes in operation in the environment. As well as a data source for further analysis, this is also very helpful in terms of environmental modelling (Chapter 7). Images also help through enabling large areas to be 'surveyed' in a short space of time, especially areas where their remoteness or inhospitableness would make generating the same quantity and quality of data using traditional surveying techniques extremely difficult, if not impossible.

> **thermal**: wavelengths associated with emitted heat energy. Can be derived from natural sources (e.g. after being absorbed by materials on the surface, from animals or natural fires) or from artificial energy sources (e.g. central heating or cars)
>
> **georeferencing**: the process of transforming image coordinates into geospatial coordinates through the application of a real-world coordinate system

OTHER SOURCES OF GEOSPATIAL DATA

Traditional geospatial data collection techniques are still important means of getting base data for environmental mapping, analysis and modelling. These and other useful sources of geospatial data are:

1. Ground surveying using traditional equipment. These techniques rely on trigonometry (i.e. relationships between angles and distances) to map unknown locations in terms of x, y and z (height) coordinates from known locations. Traditionally, this was carried out by skilled personnel using equipment like theodolites (for measuring angles) and measuring tapes and chains (for measuring distances), but there are now many electronic tools to assist with the process (Longley *et al.* 2001).

2. Photogrammetry – pairs of aerial photographs can be used to determine relative elevation either using manual or, increasingly, digital stereoplotters (Kraak and Ormeling 2003).

3. Ground Surveying using the Global Positioning System. Initially a navigation tool, this is being increasing applied to the generation of spatial data for environmental and other applications, such as tracking animal species and detailed mapping of vegetation. See Basics 6.4 for a more detailed overview.

IMPLICATIONS OF THE SOURCE OF GEOSPATIAL DATA

Certain data sources are more aligned to one of the two common data models (refer back to Basics 6.3 in the light of the discussion of geospatial data sources). Imagery of all forms that provide a single value for an individual cell or pixel located in a regular grid, are inherently raster data by nature and must therefore, initially at least, be handled as such in terms of computer-based tools. Digital maps and GPS data are much more clearly associated with the vector data model for their representation. It is still true that these distinctions are relevant in terms of the type of spatial analysis that particular data sources are best suited to, but it should be noted that it is possible to convert data from one model to another. This process of vector-to-raster and raster-to-vector conversion does have implications on resultant data, and the choice of data model influences both the nature and results of analysis techniques. However, a detailed discussion of this is beyond the scope of this book and the reader is referred to the various suggested introductory texts at the end of this chapter for a more thorough explanation. This chapter ends with an overview of some of the key spatial analysis techniques and the computer-based tools that are available with which to carry out this analysis.

BASICS 6.4

THE GLOBAL POSITIONING SYSTEM

The Global Positioning System was initially developed as an aid to navigation, but its potential as a means of collecting geospatial data was quickly recognized. With the advent of increasingly sophisticated GPS **receivers**, it became possible to record and store not just locational information about current positions but also past positions, all in an inherently digital form (Kennedy 2002).

There are three components of the GPS. The *space component* comprises the GPS constellation of satellites, a minimum of 24 with 3 spares, the newest of which orbit at around 20,000 km above the Earth's surface (Kennedy 2002). The orbit of the satellites is designed for maximum coverage and both the orbit (emphemeris data) and the internal workings of the satellites themselves are constantly monitored by a network of ground stations (the *ground component* of the GPS). The master control facility for the US NAVSTAR system is at Schriever Air Force Base in Colorado and it, together with other globally distributed stations, are in regular contact with satellites in order to monitor their orbit and the status of on board equipment, such as the clock. Following incorporation of the received data in models, corrections can be computed and sent back to the satellite. The satellite then, in turn, sends some of this data on to the GPS receiver (Dana 1994). The *user component* refers to the users and their individual GPS receivers. A minimal GPS receiver comprises: an antenna and associated electronics to pick up the signals transmitted by each 'visible' satellite in the constellation (i.e. those that are not obscured by buildings, vegetation or the Earth itself); a microcomputer to process the signals into a locational reference; controls by which the user can interact with the receiver and a screen to display the processed information.

The accuracy of the positional information varies because of many factors. For example, these may be associated with the satellites, the atmosphere through which the signal has to pass, the type of GPS receiver, or the users themselves. All GPS-sourced locational data is much more accurate in the twenty-first century than in the twentieth, due to the removal of a deliberate error 'selective availability' that used to be included with the satellite signals for all non-US military uses. This has reduced the error in estimating positions from ± 100 m to nearer ± 10 m using standard receivers (Kennedy 2002). Differential GPS, where positional data collected in the field are compared to data collected at a local base station, can provide locational references, and hence geospatial data, to sub-metre accuracy. This works on the principle that errors in data received across a particular area at a particular time will be similar.

Given these developments, the uses of GPS data have grown enormously and they are now used to map a range of environmental data, from the locations of forest stands to nesting sites and the locations of landslides. GPS equipment can even be discreetly attached to individual animals to track and map their movements in the quest for a greater understanding of their behaviour. Another important use is to provide reference points in the field through which other data sources such as satellite imagery can be georeferenced. This is particularly useful for more remote areas where traditional surveying and mapping might be difficult.

> **receivers**: devices that pick up transmissions from GPS satellites and interpret these signals as a spatial coordinate

6.4 SPATIAL ANALYSIS AND COMPUTER-BASED TOOLS FOR ENVIRONMENTAL ANALYSIS

Curran (2001) highlights eight components of the field of **geoinformatics** (Figure 6.12). All of these components, with the exception of geostatistics, geocomputation and Geographical Information Systems, have already been covered in this chapter. These three remaining components relate to the primary analytical components of the field and each of these will now be considered.

> **geoinformatics**: a term used to denote the collective disciplines associated with the collection, collation and analysis of geospatial data

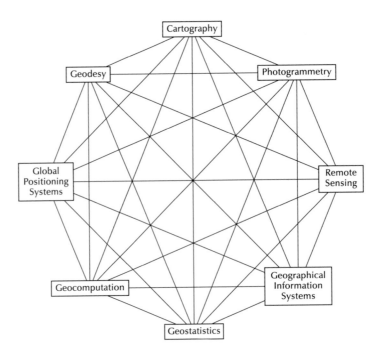

Figure 6.12 The eight components of the field of geoinformatics. *Source:* After Curran (2001)

However, before this is done, it is useful to first consider what is meant by spatial analysis.

Spatial analysis (the intelligent combination and processing of diverse locational and attribute data to produce new information) is not, by any means, a new idea; but there can be no doubt that the computer has revolutionized both its present-day reality and its future potential. Tasks such as distance measurement and the comparison of data from different maps have now been automated and expanded. The notion of sieve mapping, introduced by Ian McHarg in the late 1960s (McHarg 1969), forms the basis of one of the most important analytical functions within GIS. His approach involved creating thematic layers of mapped data on separate sheets of transparent paper or film that could be overlaid upon one another in order to identify areas conforming to particular criteria. The reality of carrying this out for lots of different layers limited the degree to which this could be successfully applied, but it is in essence very similar to the layer-based approach and overlay analysis tools still used in most GIS. Although the original form of overlay analysis is perhaps little used in modern government departments and other similar organizations, some scientists still value the immediacy of the analogue technique. Indeed Weier (2003), discussing recently revised maps of global tectonic activity, notes that these can be easily downloaded from the internet; that 'when overlaid on transparencies, the maps all match up point for point'; and that they have considerable power in an educational setting (Weier 2003; Lowman *et al.* 1999), perhaps where the use of GIS may be impractical or otherwise undesirable.

Several authors trace the short but rapid development of GIS from its roots in the 1960s to the present day (e.g. Heywood *et al.* 2002; Longley *et al.* 2001) (see Focus 6.1) and some also relate this development more specifically to geostatistics and geocomputation (e.g. Burrough 2001; Couclelis 1998). Both of these latter fields have helped to provide some of the intellectual grounding for the development of analytical tools within GIS, although they have each developed

FOCUS 6.1

MAJOR EVENTS IN THE HISTORY OF GIS

1963 The Canadian Geographic Information System is developed, led by Tomlinson, launching the term GIS

1964 Harvard Laboratory is established, directed by Fisher. SYMAP is created in 1966, the first raster-based GIS

1969 ESRI Inc., makers of the ArcGIS family of packages, is established by Dangermond and Dangermond as an independent, commercial spin off from the Harvard laboratory

1969 Intergraph, another influential GIS company, is established by Meadlock and colleagues

1969 McHarg published *Design with Nature* and introduced the idea of the overlay process

1972 Landsat 1 is launched, the first dedicated Earth Observation satellite

1981 ArcInfo is launched, the first major commercial GIS package (ESRI Inc)

1984 Marble *et al*. published *Basic Readings on Geographical Information Systems* the first truly accessible compendium of contemporary ideas

1985 Global Positioning System became operational

1986 MapInfo Corp is established, who later become responsible for the first major desktop package

1987 *International Journal of Geographical Information Systems* is launched by Coppock

1987 UK Chorley Report is published, called 'Handling Geographic Information', it is one of the first government committees to recognize the growing importance of spatial data and GIS (DoE 1987)

1988 UK and US Research Centres are established

1991 Maguire, Goodchild and Rhind collaborated on a state-of-the-art summary of principles and applications of GIS, which became an important reference for key developments (also known as the Big Book or GIS Bible)

1992 Digital Chart of the World is launched

1994 Open GIS consortium is established

1996 Internet-based products are developed and released

1999 GIS day is established

2000 Industry analysts reported that the GIS industry was worth almost seven billion dollars and was experiencing a growth of 10 per cent per annum

2000 Estimated core user community reached over 1 million

Source: After Longley *et al*. (2001)

somewhat in isolation from one another. Helen Couclelis defines geocomputation as the application of computer methods and techniques 'to portray spatial properties, to explain geographical phenomena, and solve geographical problems' (Couclelis 1998, 17). She argues that geographers 'have been "doing geocomputation" for many years without knowing it' and even goes on to suggest that 'GIS is in many ways the antagonist big brother who is robbing geocomputation of the recognition it deserves' (Couclelis 1998, 19). Geostatistics, which links traditional statistical analysis with the spatial dimension of geographical data, has developed over a similar time period to more mainstream GIS, but from different foundations and following different paths (Burrough 2001). Whilst GIS is seen as inherently deterministic and exact (i.e. generating answers with no (quantitative) consideration of the robustness of those answers), geostatistics provides a means to incorporate statistical uncertainty and try and improve answers in the light of that knowledge (Burrough 2001). Given the contributions of these related fields it is now usual to see reference to Geographical Information Science as well as Geographical Information Systems.

The Geographical Information System is the major computer-based analytical tool that is probably the primary catalyst for uptake of advanced geospatial data handling by researchers, governments and other agencies. In 1987, the influential UK Chorley committee defined GIS as 'a system for capturing, storing, checking, integrating, manipulating, analysing and displaying data which are spatially referenced to the Earth' (DoE 1987). There are many other definitions associated with the different viewpoints on the technology. Clarke (2001) for example, identifies the following broad groups as:

- *GIS is a toolbox* – relating to the idea that these are a collection of methods and techniques to facilitate spatial and geographical analysis for different purposes and goals

- *GIS is an information system* – relating to the idea that these are systems for delivering answers and generating information for a particular set of purposes

- *GIS is an approach to science* – relating to the idea that developments relate to an entirely new and internally coherent approach to research

- *GIS is a multibillion-dollar business* – relating to the rapid increase in use and the huge business that GIS has become. The data given in Focus 6.1 illustrates the extent to which this is the case

- *GIS plays a role in society* – relating to changes to the way people work (since GIS has revolutionized some working environments and required a whole new set of skills to be acquired for some) and the way people live (through, hopefully, facilitating better decisions about transport, communications the environment and its planning).

For many, the true worth and potential of GIS is associated with spatial analysis and the ability to generate new 'value added' data products and useful information from existing datasets. So, what are the key analysis functions that make GIS so special and so popular? Some of the more important of these can be summarized as follows:

Buffering functions

The ability to generate a zone of interest or catchment area around a particular feature, which might be represented as a point, a line or an area (Figure 6.13). Examples might be: an evacuation zone around a volcano (point buffer); a zone of potential noise pollution around a proposed new highway (line buffer); or the determination of a species' potential forage zone around an area of its habitat for conservation purposes (polygon/area buffer).

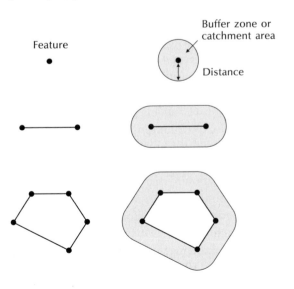

Figure 6.13 Representation of buffer zones around point, line and area features (vector model). *Source:* After Longley *et al.* (2001)

Overlay functions

The ability to compare and integrate one or more layers of geospatial data in order to create a third layer of information. In terms of the vector data model this is like a digital version of McHarg's sieve mapping (p. 165), but rather than just facilitating visual overlay it actually enables data from different layers to be integrated in order to create a new layer. This might be carried out in order to determine whether the locations of nesting sites are associated with certain types of habitat, or whether a pipeline runs through an area that is particularly vulnerable to movement during an earthquake. In the raster (grid-based) data model, layers of data with data at the same cell resolution can be overlaid very easily. In this case, mathematical operators are used to combine data into a third layer. For this reason, raster overlay techniques are sometimes referred to a 'map algebra' or 'mapematics'. Complex equations can be built up from very simple building blocks (Figure 6.14).

Interpolation functions

Estimating the value of a parameter at an unsampled location based on the values of samples at known locations. By extension, this is creating a map or layer of continuous values from a small number of points for which it has been possible to make measurements. There are a wide range of interpolation techniques available in the latest GIS packages and their development is due in large part to developments in the fields of geostatistics and geocomputation, particularly the former. Applications might be to determine patterns of rainfall over an area from a set of rain gauge points, or the pattern of contaminants around a landfill site from soil sample data.

Network analysis functions

Modelling the characteristics and flows of materials along a network of connected linear features. For example, this could be the transport of sediment through a river system.

Terrain modelling functions

The generation of surface models from *x*, *y* and *z* datasets (i.e. location and height). These are called **Digital Terrain Models** or **Digital Elevation Models** and there are different techniques associated with the raster and vector data models. Once produced they can be used to produce in turn a range of surface attributes such as slope and aspect as well as just elevation. Examples of applications might be runoff patterns in a watershed, or visibility analysis to determine where wind-farms could be located to minimize their visual impact in a highly valued amenity landscape.

> *Digital Terrain Model/Digital Elevation Model (DTM/DEM)*: both terms refer to a computer-based representation of a surface using location (*x*, *y*) and height (*z*) data. Most commonly used for representing topography but can be used to visualize any continuous data

As with the examples given for the use of remotely sensed imagery, the literature on GIS applications is enormous. The reader is directed to the suggested introductory texts for further application examples. Some examples related to environmental spheres and issues discussed elsewhere in the book include:

- Snow and ice mapping and analysis applications – e.g. Gao and Liu (2001)

- Soil erosion and soil mapping applications – e.g. Millward and Mersey (2001); Scull *et al.* (2003)

- Climate, meteorology and air quality related applications – e.g. Fedra (1993); Moragues and Alcaide (1996); Koussoulakou (1994) and Briggs (1997)

- Vegetation mapping and analysis applications – e.g. Wilson *et al.* (2003)

- Noise mapping applications – e.g. Kluijver and Stoter (2003)

- Earthquake mapping applications – e.g. Lowman (1999)

- Wildfire and fire-scar mapping and analysis applications – e.g. Ferreruela (2002)

- Hydrological and water management applications – e.g. Maidment (2002); Lyon (2003).

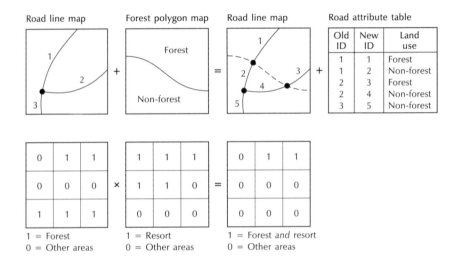

Figure 6.14 Representation of overlay techniques in the vector and raster models. Adapted from Heywood *et al.* (2002)

SUMMARY

GIS is a powerful tool for the integration and analysis of mapped data from a variety of sources. This chapter has provided an overview of the fundamental nature and characteristics of geospatial data as a starting point for an appreciation of much of the environmental data now used by physical geographers and environmental scientists. It then gave a flavour of some of the specific issues associated with key data sources, especially maps, aerial photographs and satellite imagery. The interrelated nature of data collection and analysis techniques was stressed and selected examples illustrated the range of potential applications to which they can be put to enhance understanding of both the physical aspects of the planet and the interaction of humans with their environment. The chapter ended with a look at how spatial analysis techniques can be used with geospatial data and then applied to problem solving and decision making. This analysis can be linked to environmental models or can be used to further analyse the results of environmental models with other datasets. Chapter 7 looks at environmental modelling in more detail.

Questions

1 Using the US Environmental Protection Agency website as a guide, produce a summary of environmental data sources available for one application area of your choice.

2 How are spatial data different to other types of data?

3 What is remote sensing and how does it work?

4 Visit the NASA Earth Observation website; what environmental hazards has imagery helped inform decisions makers about during the last month?

5 What are Geographical Information Systems and how do they relate to the eight other components of the field of geoinformatics?

6 Describe and explain the two common spatial data models used by GIS.

7 Visit the websites of at least two different GIS vendors (e.g. ESRI and MapInfo). What evidence of applications and spatial analysis functions can you find?

Further Reading

The text in this chapter has drawn on a number of introductory texts covering cartography and Geographical Information Systems and Science. These, and some additional texts, are recommended to provide further grounding in some of the ideas that have been introduced. Although all are introductory texts they differ in terms of overall focus and level of detail.

Burrough, P. A. and McDonnell, R. (1998) *Principles of Geographical Information Systems.* Oxford: Oxford University Press.
A detailed and thorough coverage of some of the major spatial concepts and procedures underpinning modern GIS, particularly interpolation procedures and error analysis.

Heywood, I., Cornelius, S. and Carver, S. (2002) *An Introduction to Geographical Information Systems*, 2nd edn. Harlow: Prentice Hall.
An introductory text aimed at absolute beginners that provides a readable overview of the key concepts associated with spatial data and GIS.

Kennedy, M. (2002) *The Global Positioning System and GIS: An Introduction*, 2nd edn. New York: Taylor & Francis.
A 'how to' text which gives a grounding in the practicalities of using GPS and handling GPS data within a GIS environment.

Kraak, M.-J. and Ormeling, F. (2003) *Cartography Visualisation of Geospatial Data.* Harlow: Pearson Education.
A nicely illustrated and interesting introduction to spatial data handling and visualization from a cartographic perspective.

Longley, P., Goodchild, M. and Rhind, D. (2001) *Geographic Information Systems and Science.* Chichester: John Wiley & Sons.
A comprehensive introductory text covering a wide range of topics, perspectives and ideas in the field of Geographical Information Science.

Useful Websites

Some suggested websites for further information and examples of geospatial datasets are:

http://www.agi.org.uk/
The Association of Geographic Information (AGI) – includes GIS dictionary, GIS bibliography, newsletters, frequently asked questions (FAQ) and information about recent/upcoming conferences.

http://www.geo.ed.ac.uk/home/gishome.html
The University of Edinburgh, Department of Geography – a compilation of GIS resources and links to other GIS sites.

http://www.ordsvy.gov.uk/home/index.html
Ordnance Survey – UK National Mapping Agency with introductory information on Geographical Information Systems (The GIS files).

http://www.esri.com/
ESRI – developers of ArcView and ArcInfo, the site gives a range of applications and other resources such as educational resources.

http://www.geoplace.com/
GeoPlace.com – news, software information, articles and links to free geospatial data sources.

http://www.usgs.gov
United States Geological Survey (USGS) site. A wealth of examples of the application of GIS to environmental topics. The site also has a wide range of freely available data resources for download.

CHAPTER 7
ENVIRONMENTAL SYSTEMS MODELLING

OVERVIEW

Being able to judge the potential environmental and social impacts of human actions and activities is crucially important. One way of doing this is to build models of environmental systems. Starting by introducing the chief kinds of model, this chapter explains the workings of a selection of global, regional and local models. Discussion of global models covers tutorial models, comprehensive models and models of intermediate complexity. Tutorial or conceptual models, such as Lovelock's 'Daisyworld' model, are mechanically simple, do not consider spatial differences and are designed to demonstrate plausible processes in the Earth system. Comprehensive models are three-dimensional, use explicit geography and have a high spatial and time resolution. Examples include the atmospheric general circulation models, dynamic global vegetation models and Earth Simulator, a gigantic computer designed to simulate just about everything environmental. Some of the more recent examples of comprehensive climate models attempt to couple atmospheric and oceanic circulations with biological and geochemical processes. Earth System Models of Intermediate Complexity (EMICs) simulate interactions between several components of the natural Earth systems and form a useful bridge between comprehensive models and tutorial models, allowing climate simulations incorporating vegetation dynamics over timescales such as 10,000 years, or even glacial cycles. An example is the Climate and Biosphere Model (CLIMBER), applications of which include predicting the onset of the next ice age, exploring Holocene climatic changes and examining the threat of abrupt climatic change. Many models have local or regional components or are specifically developed to operate at a particular spatial and temporal scale. This chapter also provides examples of models that operate at these different scales.

LEARNING OUTCOMES

After reading this chapter, you may not be expert modellers but you should better appreciate:

- The several types of model – hardware, analogue, conceptual and mathematical – and how they are used in general terms
- The use of mathematical models in studying environmental systems from a global perspective
- Global climate models and their role in investigating climatic change during the last 10,000 years and through the present century (global warming)
- Dynamic global vegetation models and their value in predicting large-scale patterns and changes in world vegetation
- The impact of vast computing power on the potential of modelling projects
- The use of models in probing issues concerning regional and local environmental systems, such as changes in forest composition in a warmer world.

Mimicking the real world using models can provide insights into how environmental systems work, why they changed in the past and how they are likely to change in the future.

7.1 MAKING MODELS

The human species, like all species, interacts with its environment. So far as is known, humans, unlike all other species, strive to understand their interaction with the natural world. In endeavouring to make sense of this interaction, humans seek patterns and processes in the world around them. To this end, they pick out structures that are stable in form and meaningful in function. Especially useful is the conception of structures as relatively stable forms, storing energy and matter, and existing in, and maintained by, fluxes of energy and matter. Basic structures of this type are referred to as systems.

A **system** is a structure presumed to exist in the real world. It is thought to possess characteristic properties, and to consist of interconnected components. The components are meaningfully arranged in that they function together as a whole. A forest is a system because it consists of components (trees and other plants, litter, soils and so on) arranged in a particular way. The arrangement of a forest's components is meaningful because it is explicable in terms of physical processes and because it acts as a whole. Moreover, forests have characteristic properties that serve to distinguish them from other systems, such as meadow, mire and moorland. All natural structures with which humans interact may be termed environmental systems.

In trying to single out the components and interrelations of environmental systems, some degree of abstraction or simplification is necessary: the real world is too rich a mix of objects and interactions for all components and relations to be considered. The process by which scientists condense reality to manageable proportions is termed **model building**. As a general definition, a model is a simplified representation of some aspect of the real world that happens to interest an investigator.

> **system**: a meaningful arrangement of components, as in the solar system and ecological system
>
> **model building**: the process by which humans simplify the 'real world' to help understand and explain it

HARDWARE AND ANALOGUE MODELS

Hardware models

Models can be built at different levels of abstraction (Figure 7.1). The simplest level of abstraction involves a change of scale. In this case, a **hardware model** represents the system. There are two chief kinds of hardware model: scale models and analogue models. **Scale models**, because they closely resemble physically the system they represent, are also termed iconic models. They are miniature, or sometimes gigantic, copies of systems. They differ from the systems they represent only in size. A child's toy train and an architect's model of a building are miniature copies of systems. Three-dimensional structures, constructed out of coloured plastic balls and metal or wooden rods, are gigantic copies of molecular systems. Hardware models of molecules have practical value as well as being useful in teaching. A crude hardware model gave a clue to the structure of the third stable form of carbon, carbon-60 (also called fullerene), for example; a molecular 'ball' with hexagonal and pentagonal faces arranged exactly as in a traditional sewn-leather football. Static scale models like these have been used to replicate certain aspects of environmental systems. In geomorphology, relief

> **hardware model**: a model of a system in which the size of the system is made bigger or smaller
>
> **scale model (iconic model)**: a smaller or larger replica of a system

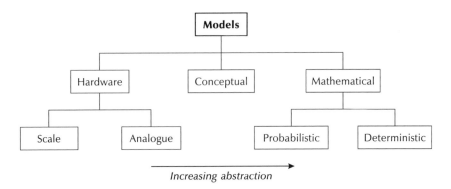

Figure 7.1 Types of model.
Source: After Huggett (1993, 4)

Increasing abstraction

models, fashioned out of a suitable material such as plaster of Paris, have been used to represent topography as a three-dimensional surface. In climatology, models made of glass can depict the distribution of upper air temperatures, and models made of wire describe the structure of a cyclone.

Scale models need not be static: models made using materials identical to those found in Nature, but with the dimensions of the system scaled down, can be used to simulate dynamic behaviour. In practice, scale models of this kind imitate a portion of the real world so closely that they are, in effect, a 'controlled' natural system. An example of this is Stanley A. Schumm's (1956) use of the badlands at Perth Amboy, New Jersey, to study the evolution of slopes and drainage basins. The great advantage of this type of scale model, in which the geometry and dynamics of the model and system are virtually identical, is that a high degree of control can be exerted over the simplified experimental conditions. In other scale models, natural materials are employed but the geometry of the model is dissimilar to the geometry of the system it imitates, the system being scaled down in size. The process of reducing the size of a system creates a number of tricky, but not insuperable, problems associated with scaling. For instance, a model of the Severn Estuary made at a scale of 1:10,000 can fairly easily preserve geometrical and topographical relationships. But, when water is added, an actual depth of water of, say, 7 m is represented in the model by a layer of water less than 0.7 mm deep. In such a thin layer of water, surface tensions will cause enormous problems, and it will be impossible to simulate tidal range and currents. Equally, material scaled down to represent sand in the real system would be so tiny that most of it would float. These problems of scaling can be overcome, to a certain extent at least, and scale models are used to mimic the behaviour of a variety of environmental systems. For example, since the latter half of the nineteenth century, scale models have assisted in the design of civil engineering projects, the dynamics of rivers and river systems have been investigated in simple waterproof troughs and more sophisticated flumes, and the patterns and effects of airflow on sediments have been observed in wind-tunnel experiments. Indeed, it is surprising how many ingenious attempts have been made to replicate the behaviour of environmental systems using scale models. A discussion of this fascinating field is beyond the scope of the present book, but the interested reader should consult Michael Morgan's (1967) paper on the Heath Robinsons of physical geography.

In the 1980s, several enterprising researchers decided to build an artificial biosphere – Biosphere-II – in the Sonora Desert, Arizona. Biosphere-II is a self-contained structure designed as a sort of scale model to replicate the workings of the actual biosphere – Biosphere-I. The structure covers 1.275 hectares, stands 28 metres at its highest point, and contains five biomes (communities of animals and plants) – rain forest, desert, savannah, marsh and ocean (7.5-metres deep) and a human habitat (Figure 7.2). An eight-person crew began a two-year mission in 1991, but this mission failed owing to problems with food and oxygen production, species loss, pests and others factors. A crew of six tried again in 1994 but fared little better. In 1996, Columbia University took over the management of the project and now uses the facility for educational and visitor programmes, scientific research and conferences. Some 200,000 tourists visit Biosphere-II each year.

The Biosphere-II project, although a magnificent failure, at least sparked renewed interest in attitudes towards life and its sustainability. This ambitious nanotechnology (which rebuilds the Earth system or parts of it at a smaller scale) has products that are more modest, if more informative, such as the Ecotron and the free-air experiments currently used to assess the effects of atmospheric carbon-dioxide enrichment on agroecosystems (agricultural ecosystems involved in the extensive or intensive production of crops or stock or both). The Ecotron is the controlled environment facility of the Natural Environment Research Council's Centre for Population Biology and is located at Silwood Park, Berkshire, England. The purpose of the Ecotron is to establish simplified communities of terrestrial plants, animals and microbes as models of the real world. There are sixteen small communities, or microcosms, each housed in separate walk-in chambers with computer-controlled climatic conditions. The facility bridges the gap between the complexity of real field communities and the

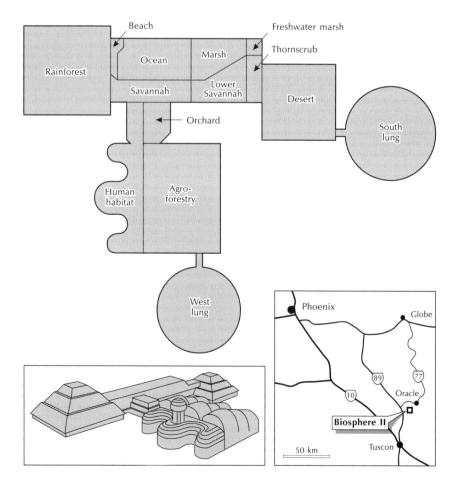

Figure 7.2 The layout of Biosphere-II. *Source*: http://www.bio2.edu/Research/flow.htm

simplicity of laboratory or greenhouse experiments. An Ecotron experiment carried out in the early 1990s studied the effects of varying biodiversity on the functioning of the whole ecosystem, and found that reduced biodiversity reduces community productivity and impairs ecosystem performance. A study completed in 1997 investigated the effects of enhanced temperature and carbon dioxide on populations and communities. The latest experiments examine the relationships between below-ground species diversity and ecosystem processes.

The Global Change and Terrestrial Ecosystem core project of the International Geosphere–Biosphere Programme launched the Network of Ecosystem Warming Studies in the mid 1990s that aimed to foster and integrate studies on the ecosystem-level effects of rising temperatures. In these studies, the application of a range of techniques (electrical heat-resistance ground cables, greenhouses, vented and unvented fields chambers, overheat infrared lamps and passive night-time warming), artificially warm natural ecosystems. Lindsey E. Rustad and her colleagues found that a synthesis of data on the response of soil respiration, net nitrogen mineralization and above-ground plant productivity at thirty-two ecosystem-warming research sites in the latitude band 35–79°N, and one site at 45°S, produced several conclusions (Rustad 2001). Despite considerable variation in response at individual sites, across all sites and years in 2–9 years of experiment, 0.3–0.6°C warming increased soil respiration rates by 20 per cent, net nitrogen mineralization rates by 46 per cent and plant productivity by 19 per cent. Soil respiration increased more in forested ecosystems than in low tundra and grasslands, while plant productivity generally increased more in low tundra than in forests and grassland. Increases in soil respiration and net nitrogen mineralization appeared to be unrelated to the geographical, climatic or environmental factors considered. These findings underline the need to elucidate the relative importance of site-specific factors (such as temperature, moisture, site quality, vegetation, successional stage, land-use history and so forth) at local and regional scales, and show that the scaling-up of responses from plots and sites to landscapes and biomes should be dealt with cautiously.

Other experiments, known as free-air carbon dioxide enrichment (FACE), artificially increase the level of atmospheric carbon dioxide at a study site. FACE technology modifies the microclimate around growing plants to simulate climatic change (Figure 7.3). Usually, air enriched in carbon dioxide issues from a circle of vertical pipes into plots up to 30 metres in diameter and up to 20 metres tall. By pre-diluting the gas and by using fast feedback controls, carbon dioxide levels remain stable. The natural airflow is virtually unaltered, as the equipment does not require containment. Larger plots reduce 'edge effects' and capture fully functioning, integrated processes in ecosystems. The field data suggest plant and ecosystem responses to levels of atmospheric carbon dioxide that are predicted to occur around 2050.

Analogue models

analogue model: a sophisticated scale model, which includes maps and remotely sensed images

Analogue models are like scale models, but more refined. The most commonly used analogue models are maps and remotely sensed images. On a map, the surface features of a landscape are reduced in scale and represented in symbolic form: roads by lines, relief by contours and buildings by point symbols, for instance. Remotely sensed images represent, at a reduced scale, certain properties of the Earth's surface. The level of spatial resolution of an image corresponds to

Figure 7.3 A FACE experiment in a planted sweetgum (*Liquidambar styraciflua*) monoculture located on the Oak Ridge National Environmental Research Park in Tennessee, USA. Photograph by Rich Norby

the actual size of a pixel (picture element) on the ground. Maps and remotely sensed images are, except where a series is available for different times, static analogue models.

Dynamic analogue models are hardware models in which the system size is changed, and in which the materials used are analogous to, but not the same as, the natural materials of the system. The analogous materials simulate the dynamics of the real system. So, in a laboratory, the clay kaolin can be used in place of ice to model the behaviour of a valley glacier. Under carefully controlled conditions, many features of valley glaciers, including crevasses and step faults, develop in the clay. Difficulties arise in this kind of analogue model, not the least of which is the problem of finding a material that has mechanical properties comparable to the material in the natural system. Electrical analogue models of environmental systems, which use assorted electrical components to represent stores and flows of energy or matter, are not uncommon. Groundwater systems and ecosystems in particular lend themselves to electrical analogue modelling. In simple ecological applications, batteries supply energy input, electrical currents simulate energy flow, and amperage and voltage changers simulate energy dissipation.

CONCEPTUAL MODELS

Maps and remotely sensed data are invaluable in monitoring the state of the ecosphere, and can provide the information needed to calibrate and test models of environmental systems. However, to gain a deep understanding of the dynamics of environmental systems, further abstraction is required. This will normally involve the use of a **conceptual model**. In brief, a conceptual model expresses ideas about components and processes deemed to be important in a system, as well as some preliminary thoughts on how the components and processes are connected. In other words, it is a statement about system form and

conceptual model: an abstract model of a system that can be expressed as pictures, box-and-arrow diagrams, matrix models, flowcharts and various symbolic languages

system function. Conceptual models can be expressed in several ways: as pictures, as box-and-arrow diagrams, as matrix models, as computer flowcharts and in various symbolic languages (Table 7.1).

Words, pictures, boxes and arrows

As far as conceptual models are concerned, the old saying is generally true: a picture is worth more than a thousand words. Verbal descriptions of most environmental systems, where the components and relations are multifarious and complex, become cumbersome and unintelligible. Pictures of environmental systems, in contrast, can convey large amounts of information about composition and spatial structure at a glance (Figure 7.4).

More abstract than pure pictures of systems are **box-and-arrow diagrams**. In box-and-arrow diagrams, the boxes stand for the system components and the arrows depict supposed important links and relations between the components. Because it is the flow of matter and energy through the boxes that is of prime concern, the inner workings of the boxes are not explored in detail. For this reason, box-and-arrow diagrams are sometimes referred to as **black-box models**. An example is the model of the terrestrial carbon cycle portrayed in Figure 7.5.

> **box-and-arrow diagram**: a model of a system where boxes represent stores and arrows the flows or relationships between them
>
> **black-box model**: a box-and-arrow diagram. The inputs and outputs to the boxes are known but process inside the boxes are not – hence black box

Table 7.1 Conceptual designs of systems

Model designs[a]	Attributes	Advantages	Disadvantages
Words	Verbal descriptions	Supplement all kinds of conceptualizations	Complexity difficult to convey
Pictures	Illustrations using natural elements	Convey spatial characteristics	Lack temporal and mathematical inferences
Black-box diagrams	Components and relations shown as abstract symbols (boxes and arrows)	Emphasize throughputs of matter and energy	Lack mathematical inferences
Input–output matrices	Inputs and outputs of black-box models shown as a linear matrix	Concise list of the size of interactions	Lack temporal dynamics and assume linearity
Signed digraphs	Black-box models with logic gates	Qualitative interactions	Lack temporal dynamics and assume linearity
Computer flow charts	Sequential ordering of processes	Components can change in space and time	Interactions not obvious
		Use logic gates	Rudimentary symbolic language
Forrester diagrams	Computer flowcharts with feedbacks	Interactions more obvious	
		Rate equations, components, sources and sinks	
Energy circuits	Complex symbolic language	Computational inferences	
		Thermodynamically constrained	
		Components classed by function	
Unique designs	Combinations and extensions of any of the above		

Note: [a] Fuller explanations of these designs, both verbal and mathematical, follow in the text

Source: Adapted from Sklar *et al.* (1990)

Plate 1 The threatened tiger (*Panthera tigris*).
Photograph by Pat Morris

Plate 2 A storage rain gauge.
Photograph by Mike Acreman

Plate 3 A tipping-bucket gauge with the upper section removed to reveal the tipping mechanism within. Each reservoir of the tipping bucket is designed to hold a certain amount of rain (e.g. 0.5 mm). When rain falls, water collects in one of the buckets. Once full, the bucket tips and the other reservoir collects rain. Each tip is recorded by the data logger that totals the rain over a certain time period (e.g. hour). Photograph by Mike Acreman

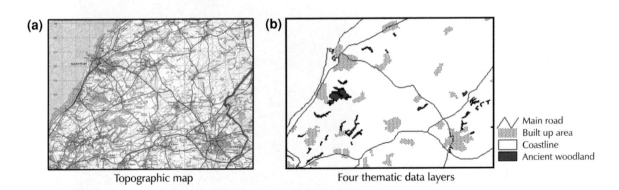

(a)

Topographic map

(b)

Main road
Built up area
Coastline
Ancient woodland

Four thematic data layers

Plate 4 Differences between (a) topographic maps and (b) thematic maps. *Sources*: (a) Ordnance Survey; (b) English Nature

Plate 5 Hurricane Fabian Photo ISS007-E-14419 Earth Sciences and Image Analysis Lab, Johnson Space Center. *Source:* After NASA (2003a)

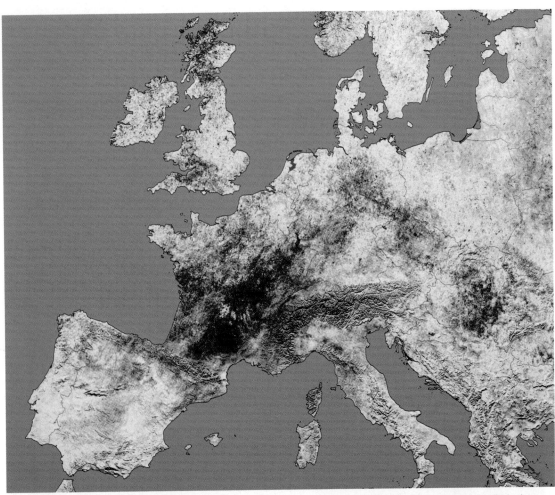

Plate 6 Temperature anomaly map of western Europe showing 2003 July temperatures against July 2001 temperatures. Reds show a positive temperature anomaly, blues a negative anomaly. *Source:* After NASA (2003b)

Plate 7 Harvesting fish is one way in which people depend upon wetlands. In the picture, fish caught in the wetland is drying and being sold in a village on Logone floodplain, Cameroon. Photograph by Mike Acreman

Plate 8 A wetland bird: the comb-crested jacana (*Irediparra gallinacea*) on a lily pad, Kakadu National Park, Australia. Photograph by Mike Acreman

Plate 9 American burying beetle (*Nicrophorus americanus*). Photograph by Gary Marrone

Plate 10 Soil profile on St Lucia. Photograph by Mike Acreman

Plate 11 Soil profile exposed by road, St Lucia. Photograph by Mike Acreman

Plate 12 Undeveloped shoreline along Lake Armstrong, Washington, USA. Photograph by Tessa Francis

Plate 13 Typical developed shoreline on Shady Lake, Washington, USA. Photograph by Tessa Francis

Plate 14 Passenger pigeon (*Ectopistes migratorius*). Photograph by Pat Morris

Plate 15 Sea otter (*Enhydra lutris*). Photograph by Pat Morris

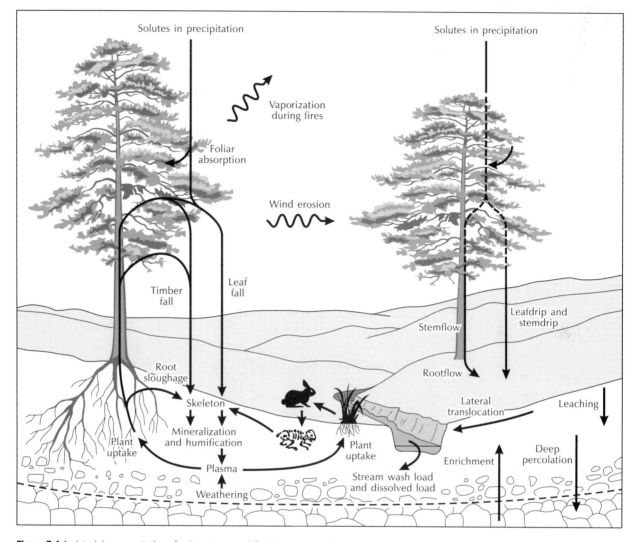

Figure 7.4 A pictorial representation of solute stores and flows in an ecosystem

Matrix representations

Black-box conceptual models usually convert to **input–output models**, which may take the form of mathematical matrices. The inputs and outputs cascading through the components of an environmental system will translate into a system of rows and columns that look similar to the income and expenditure tallies in an accountant's spreadsheet. The simplest form of matrix is an **adjacency matrix**. In an adjacency matrix, the system components are the row and column headings. Each element in the matrix represents a flow between the corresponding components. The matrix elements are either 1s or 0s. The 1s in the matrix correspond to an arrow in a box-and-arrow diagram, and the 0s correspond to no arrows. As the arrows are directed (they go from one component to another), so are the entries in the matrix: flows are either from rows to columns or from columns to rows. As an example, the adjacency matrix shown in Table 7.2 represents the flows in the model of the terrestrial carbon cycle shown in Figure 7.5. Notice that 0 represents zero flow, as from leaves to branches, and a 1 represents a flow, as from leaves to litter.

input–output model: a record of inputs and output links in a box-and-arrow diagram, often represented as a matrix

adjacency matrix: a system represented as flows (non-zero entries) and non-flows (zero entries) in a matrix with system components as row and column headings

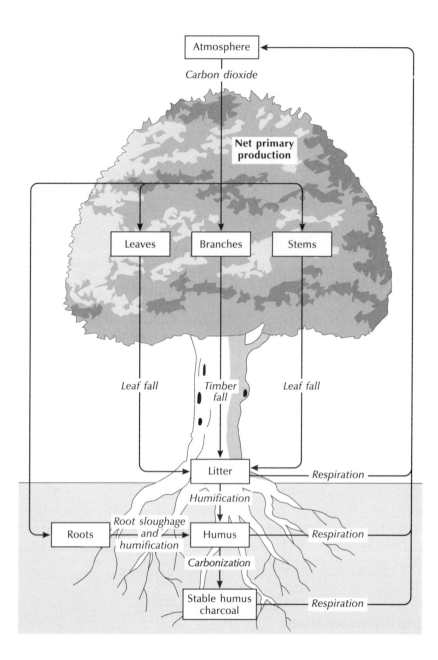

Figure 7.5 Representation of the terrestrial carbon cycle. *Source:* After Huggett (1993)

Converting the non-zero entries in adjacency matrices to actual flows creates input–output matrices. They provide a concise listing of the interactions among system components over a particular time. Table 7.3, for instance, is an input–output matrix of the terrestrial carbon cycle in a tropical forest. Notice that row and column totals provide various measures of the system's functioning. The cycling efficiency is the outflow from a system component passed on to other system components expressed as a ratio of the total outflow, which includes flows that leave the system. The mean path length is the total system throughflow divided by the sum of the inflows (or outflows) in the matrix. Mean path length tends to increase with the number of system components, greater throughflow and more feedback (flow within the system).

Table 7.2 Adjacency matrix of terrestrial carbon cycle as depicted in Figure 7.5

To	From						
	Leaves x_1	Branches x_2	Stems x_3	Roots x_4	Litter x_5	Humus x_6	Charcoal[a] x_7
Leaves x_1	0	0	0	0	0	0	0
Branches x_2	0	0	0	0	0	0	0
Stems x_3	0	0	0	0	0	0	0
Roots x_4	0	0	0	0	0	0	0
Litter x_5	1	1	1	0	0	0	0
Humus x_6	0	0	0	1	1	0	0
Charcoal x_7	0	0	0	0	0	1	0

Note: [a] Stable humus charcoal

Table 7.3 Input–output matrix of the terrestrial carbon flows (Gt C yr^{-1}) (see Appendix 1) cycle in a tropical forest

To	From							Input	Output	Row total
	x_1	x_2	x_3	x_4	x_5	x_6	x_7			
x_1	0	0	0	0	0	0	0	8.34	0	8.34
x_2	0	0	0	0	0	0	0	5.56	0	5.56
x_3	0	0	0	0	0	0	0	8.34	0	8.34
x_4	0	0	0	0	0	0	0	5.56	0	5.56
x_5	8.34	5.56	8.34	0	0	0	0	0	0	22.24
x_6	0	0	0	5.56	8.92	0	0	0	0	14.48
x_7	0	0	0	0	0	0.55595	0	0	0	0.55595
Input	0	0	0	0	0	0	0	0	0	0
Output	8.34	5.56	8.34	5.56	22.23	11.119	0.55594	0	0	61.70494
Column total	16.68	11.12	16.68	11.12	31.15	11.67495	0.55594	27.8	0	126.78089
Cycling efficiency	0.5	0.5	0.5	0.5	0.286	0.04761	0	Mean path length: 126.78089/27.8 = 4.56		

Note: The stores (x_1, x_2 and so forth) are identified in Table 7.2

Signed digraphs and interaction matrices

These are a kind of conceptual model. They extend the information given in adjacency matrices and box-and-arrow diagrams by 'signing' the interactions. In an adjacency matrix, non-zero entries are assigned minuses or pluses to indicate positive or negative interactions between the system components. This is an **interaction matrix**. In a box-and-arrow diagram, lines ending in a small arrowhead represent positive interactions, and lines ending in a small circle stand for negative interactions; this is called a signed directed loop diagram, or **signed digraph**.

Figure 7.6 is a conceptual model of human modifications to the storage of water in lowland soils in eastern North Carolina, USA, expressed as a signed

interaction matrix: an adjacency matrix with non-zero entries signed − or +

signed digraph: a box-and-arrow diagram with positive and negative links indicated by small arrows and circles, respectively

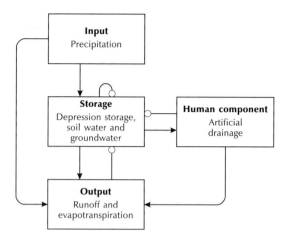

Figure 7.6 A signed digraph.
Source: Adapted from Phillips (1991)

digraph. Water enters the system as precipitation, is stored in surface depressions and the soil body (as surface depression storage, soil water and groundwater), and exits the system as runoff and evapotranspiration. Each link in the system is 'signed', plus or minus, depending on whether the link is positive or negative. Precipitation, the input, has a positive influence on both storage and output. Lines ending in small arrowheads denote its links with these variables. Runoff and evapotranspiration, the output, has a negative influence on storage, being derived from stored water released between precipitation events. A line terminating in a small circle thus denotes its links with storage. Water storage is self-limiting: there is an upper limit to the amount of water soil can store. This property is indicated by a closed loop. To an extent, the store regulates the output: the more water stored, the greater the outflow (especially via groundwater flow); the less water stored, the smaller the output since water is held at higher tensions: hence the positive link from storage to output. The human component, which may seek to maintain water storage at the level most favourable to crop growth, is connected to the rest of the system because storage has a positive influence on humans who strive to manipulate the system, using canals, ditches and so forth, to increase withdrawals. In turn, the withdrawals have a positive link with output, because they increase it.

The signed digraph shown in Figure 7.6 readily converts to an interaction matrix (Table 7.4). The connections are expressed by the letter *a*, with subscripts signifying the variables involved. So, a positive connection between precipitation, variable 1, and soil water storage, variable 2, is denoted by a_{12}. All other connection labels use the same logic.

Table 7.4 The interaction matrix of the soil water system in eastern North Carolina as depicted in Figure 7.6

	Water input x_1	Water storage x_2	Water output x_3	Artificial drainage x_4
Water input x_1	0	a_{12}	a_{13}	0
Water storage x_2	0	$-a_{22}$	a_{23}	a_{24}
Water output x_3	0	$-a_{32}$	0	0
Artificial drainage x_4	0	$-a_{42}$	a_{43}	0

Flow charts

Computer programmers commonly use **flow charts** to design their programs. Given a specific problem, flow charts are used to work out a logical sequence of computations for finding a solution. It is possible to apply the same method in the solution of problems concerned with environmental systems. The sequence of events in a flow chart will then represent an assumed ordering of environmental processes. Once the flow chart is translated into a programming language, the model can be run and, if all goes well, a solution to the problem suggested. A drawback with using flow charts as conceptual models is that the type of interaction occurring within a compartment of the flow chart is not evident. Normally, the name of a subroutine, but little else more informative than that, is mentioned. Look at Figure 7.7. This is a flow chart for a simple spatial model. The model is designed to predict the changing storage of nitrogen within a landscape. Although the chart gives an idea of the overall structure of the computational procedure, it is uninformative on several counts. For example, it fails to indicate the precise nature of the equations used to calculate change in nitrogen storage. It does not make obvious, therefore, the fact that topographic slope gradients determine the rate constants for nitrogen transfer between grid-cells.

> ***flow chart***: a model showing an assumed ordering of processes within an environmental system

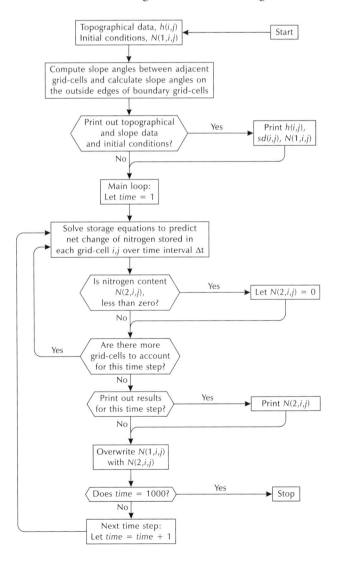

Figure 7.7 Computer flow chart.
Source: After Huggett (1993, 9)

Analogue computer diagrams are another kind of computer flow chart model. Analogue symbols are used to represent storages and flows of energy or matter as a system of electrical voltages and currents. Popular in the early 1960s, before digital computers became widely available, analogue computers are little used today in the study of environmental systems.

Symbolic languages

Some writers have created their own sophisticated versions of box-and-arrow diagrams in which the roles that components play in a system are indicated pictorially. Canonical structures (that is, basic functional units whose operation need not be resolved at a lower level of resolution) represent standard 'roles'. The creators of symbolic systems of canonical structures do not lack imagination and many different schemes have been thought up. Indeed, there are so many that the chances ever of agreeing on a standard form are slim. However, two symbolic languages have caught on, and are commonly employed. The first is the creation of Jay W. Forrester. It is basically a computer simulation language written, initially at least, to tackle industrial and urban problems. Later, it was extended to the so-called world simulation described in *World Dynamics* (1971). The chief canonical structures in Forrester's systems dynamics language are state variables, shown by valves; auxiliary variables that influence the rate of processes, drawn as circles; flows of people, goods, money, energy and so forth, represented by solid arrows; causal relationships, denoted by broken arrows; and sources and sinks of energy and mass, portrayed as clouds.

Howard T. Odum (1971; 1983) devised another symbolic language that enjoyed widespread currency for some years. Odum's energy circuit language, called **energese**, comprises several modules, each a canonical structure, joined by lines representing flows of energy. In isolation, the modules in this language look like the doodlings of an eccentric electrical engineer (Figure 7.8). Energese is, however, a succinct way of capturing in pictorial form an enormous amount of information about the components and relations within systems. In energese, solid lines show flows. Circles stand for potential energy sources, such as the Sun and wind (Figure 7.8a). The symbol that looks like a house with a rounded base stands for passive stores of energy (Figure 7.8b). Arrows going to ground are heat sinks and depict that portion of energy dissipated as heat while work is being done (Figure 7.8c). Combined with arrows going to ground, the passive energy store symbol represents the storage of new potential energy against some storage force, a process requiring work to be done and hence energy dissipation (Figure 7.8d). The short bullet-shaped symbol stands for the reception of pure wave energy (such as sound, light, water waves and wind) and its interaction with a combined cycling receptor and self-maintaining module (Figure 7.8e). In this module, energy interacts with a material cycling round the system producing an energy-activated state, which then returns to its deactivated state passing energy on to the next step in a chain of processes. The kinetics of this module was first discovered in the reaction of an enzyme with its substrate, and is named the Michaelis–Menton reaction after its discoverers. The pointed block (a sort of stubby arrow) is a work gate and represents the work done by a system component in procuring energy from sources outside the system; it acts as a kind of control valve (Figure 7.8f). The hexagon represents a self-maintaining subsystem (Figure 7.8g). It is a combination of two modules – a work gate and

energese: a symbolic language for representing energy and material flows in systems

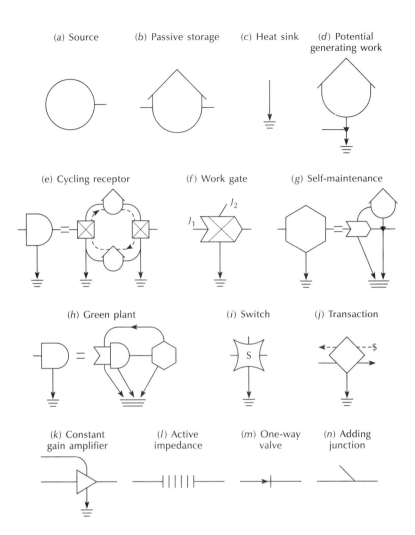

(a) Source   (b) Passive storage   (c) Heat sink   (d) Potential generating work

(e) Cycling receptor   (f) Work gate   (g) Self-maintenance

(h) Green plant   (i) Switch   (j) Transaction

(k) Constant gain amplifier   (l) Active impedance   (m) One-way valve   (n) Adding junction

Figure 7.8 Modules in Howard T. Odum's electrical circuit language called energese. *Source:* After Odum (1971, 38)

potential energy store – that act in concert to ensure that the energy stored is fed back to control the work done by the whole unit. An example is a trophic level of an ecosystem. The long bullet-shaped symbol is a combination of a self-maintaining module and a cycling receptor (Figure 7.8h). Energy captured by the cycling receptor is fed to the self-maintaining unit that in turn keeps the cycling machinery working, and returns necessary materials to it. A green plant is an example. The small square with concave edges represents flows which have 'on' and 'off' states, and which control other flows by switching actions (Figure 7.8i). The small lozenge-shaped symbol denotes transactions in systems involving flows of money (Figure 7.8j). Four other symbols (Figure 7.8k, l, m and n) represent constant gain amplifiers, active impedance (which allow a backforce to develop in the system), one-way valves and additive junctions. All the aforementioned units may be assembled to represent actual systems. By way of example, look at the stores and flows of materials in a Third-World village expressed in energese (Figure 7.9).

Richard J. Chorley and Barbara A. Kennedy in their influential *Physical Geography: A Systems Approach* (1971) adopted yet another symbolic language, this one originating in hydrological literature, to depict physical process–response systems. Arthur N. Strahler and Alan H. Strahler (1973; 1974), father and son, modified Odum's energese, so devising their own method of portraying the

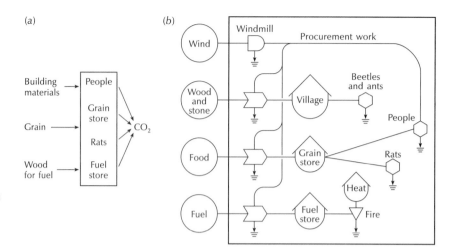

Figure 7.9 Stores and flows in a Third-World village expressed in energese. *Source:* After Huggett (1980, 33)

energetics of environmental systems. To meet the needs of the wide range of systems encountered in physical geography, Arthur N. Strahler (1980) made further revisions to the language. Other schemes have been suggested but we need not describe them here. The point to make is that there are several commonly used symbolical languages for presenting conceptual models of systems. Each has advantages and disadvantages. Any attempt to come up with a definitive set of symbols is probably futile. It is far, far better to spend time thinking about the components and relations in the system depicted, than it is to agonize over their method of depiction.

Conceptual models as hypotheses

It is crucially important to be aware that any system can be conceived of in a great many ways: there are potentially an astronomically huge number of conceptual models of any system. This is because no model can ever be fully correct and achieve identity with the system it represents. The corollary of this fact is that the nature of a system can never be truly or thoroughly known, but will always remain, to some extent, a matter of conjecture. For this reason, a system is itself a kind of conceptual model, the existence of which in reality often rests upon nothing more than shared intuition. A system is a concept that, like beauty, lies in the eyes of the beholder; it is an idea, a hypothesis about how some portion of reality is assembled and how it works (Huggett 1985, 8). This point is absolutely fundamental but commonly neglected. The risk is that shared intuition may lead to an aspect of the nature or dynamics of a system being accepted without question, the necessarily hypothetical nature of the system being overlooked. To take an example: for decades, dogma had it that, whereas materials circulate around ecosystems, energy passes through as a one-way flow. The danger of this doctrine was that the hypothetical nature of the view that energy transfer is a one-way flow might be forgotten. Some years ago, Bernard C. Patten (1985) persuasively demonstrated that, although it differs from matter in its dissipation, energy *does* cycle in ecosystems: the dogma has collapsed and the textbooks need rewriting!

Conceptual models help to clarify loose thoughts about how a system is put together and how it operates, and often provide a foundation for the construction of mathematical models. Yet, the process for devising conceptual models is seldom deliberated fully enough. The dynamics predicted by a model of a system depend very much on the system components selected and the relationships assumed to exist between them. Early astronomers grouped stars into constellations that, it is now believed, have no modern, scientific meaning. Today, stars are grouped into galaxies and galaxy clusters, a conceptual arrangement that seems to make far more sense to twenty-first century astronomers. This is an extreme example, but it serves to stress that the building of conceptual models is not something to be done lightly: it is the most important step in the entire process of mathematical modelling. As Fred H. Sklar and his colleagues (1990, 625) put it, the conceptualization of a system is the backbone of the entire process of mathematical model building.

MATHEMATICAL MODELS

Mathematical model building involves translating the ideas encapsulated in a conceptual model into the formal, symbolic logic of mathematics. The language of mathematics offers a powerful tool of investigation limited only by the creativity of the human mind. Of all modes of argument, mathematics is the most rigorous. It furnishes a means of describing a system in a symbolism that is universally understood. **Mathematical models** seem capable of giving the deepest insight into how environmental systems work, of providing the best means of predicting change in environmental systems, and of affording the most trustworthy guide as to how best to manage or control environmental systems. This is not to say that mathematics can replace the intuition and inspired guesses of environmental scientists. Rather, it offers a standardized way of formalizing thoughts and ideas, and a potent means of analysing problems. Of course, mathematical models merely provide a guide to the credible environmental repercussions of specific human actions. They are uninformative on matters political, philosophical and moral. Nonetheless, mathematical models allow experiments to be run on environmental systems and generate realistic output that can be used as the basis for rational and informed environmental management policies. That, at least, is the hope. In practice, the irrational side of human nature seems often to come to the fore, as in the reluctance of some governments to take steps to reduce 'greenhouse' gas emissions, despite the repeated warnings uttered by the climate modellers about the potential dire consequences of global warming.

> *mathematical model*: a formal, logical and highly abstract expression of a system

The act of quantification, of translating ideas and observations into symbols and numbers, is in itself of no value; it must be validated by explanation and prediction. The art and science of using mathematics to study environmental systems, is to discover expressions with explanatory and predictive powers. Mathematical models are pregnant with power. It is this power that sets mathematical models apart from conceptual models. An unquantified conceptual model is not susceptible of formal proof; it is simply a body of ideas. A mathematical model, on the other hand, is testable: model predictions stand to be matched against the yardstick of observation. By a continual process of mathematical model building, model testing and model redesign, the understanding of the form and function of environmental systems, and the role of human activity in them, should improve.

Stochastic and statistical models

Three chief classes of mathematical model assist the study of environmental systems: stochastic models, statistical models and deterministic models. The first two classes are both probabilistic models.

Stochastic models have a random component built into them that describes a system, or some facet of it, based on probability.

Statistical models, like stochastic models, have random components. In statistical models, the random components represent unpredictable fluctuations in laboratory or field data that may arise from measurement error, equation error or the inherent variability of the objects being measured. A body of inferential statistical theory exists that determines the manner in which the data should be collected and how relationships between the data should be managed. Statistical models are, in a sense, second best to deductive models: they can be applied only under strictly controlled conditions, suffer from a number of deficiencies, and are perhaps most profitably employed only when the 'laws' determining system form and process are poorly understood.

stochastic model: a mathematical model that includes a random component

statistical model: a mathematical model dealing with inherent variability in measured phenomena

Deterministic models

Deterministic models are conceptual models expressed mathematically and containing no random components. They can be derived from physical and chemical principles without recourse to experiment. It is sound practice, therefore, to test the validity of a deterministic model by comparing its predictions with independent observations made in the field or laboratory. Deterministic models come in a variety of forms, the most useful of which in studying the human impact on environmental systems are dynamical system models. Evolved from open system theory, dynamical system models involve a set of system components acting as a whole. They are concerned with change and susceptibility to change in systems, and are used to predict the transfer and transformation of energy and matter in environmental systems. In addition, they provide a theoretical base for the new generation of mathematical models dealing with dissipative structures, multiple equilibria, bifurcations and catastrophes. These fascinating models generate probabilistic-type elements from deterministic relations.

deterministic model: a mathematical model with no random components

The rest of the chapter will illustrate modelling, without going into the workings of the modelling process, by looking at examples of (1) global models, (2) regional models and (3) local models.

7.2 GLOBAL MODELS

Being able to predict changes in the global ecosystem is extremely useful. One way of doing this is to build mathematic models of the Earth system or parts of it. Deterministic mathematical models have a relatively long history in the ecological sciences. Alfred James Lotka and Vito Volterra developed mathematical models of predator–prey systems (for instance, a vole–weasel system) in the 1920s. Later experimental work built on these ideas, but mathematical ecological modelling flowered with the arrival of electronic computers in the late 1950s. At this time, analogue and digital computers simulated the dynamics of relatively simple ecosystems, for example, looking at the interactions of several species, the passage of radionuclides through forests and, a little later, DDT through food webs. The International Biological Programme (IBP), started in 1965, created the role of the ecological modeller. Systems models of several IBP biomes were constructed. The builders of these models saw them as fair facsimiles of real ecosystems. Not all ecologists were so favourably disposed towards systems models, however, and they said so in plain terms. To an extent, the ecological system-modellers were probably misunderstood. Detractors claimed that the charm of turning ecology into a space-age science lured them into using system analysis, with its attendant jargon borrowed from engineering and physics. It does not matter whether that is the case, for mathematical modelling in ecology entered a new era in the 1990s. Three developments heralded this era: the building of spatial models; the accessibility of time series from remotely sensed images; and the appearance of supercomputers and parallel processing.

Models in global ecology now cover a range of topics, include biogeochemical models, climate models and models that integrate world energy production, emissions of gases, deforestation and other human components of the global system. These models fall into three broad categories: tutorial models, comprehensive models and models of intermediate complexity.

TUTORIAL MODELS

Tutorial models are mechanically simple, do not consider spatial differences and are designed to demonstrate plausible processes in the Earth system. Critics claim that such models tend to oversimplify the Earth system to such an extent that they fail to simulate such decisive elements of planetary dynamics as ENSO (El Niño–Southern Oscillation) events (p. 223). James Lovelock's 'Daisyworld' family of models, for example, focus on two or three basic mechanisms, such as feedback between sunlight, vegetation type and the reflectivity (albedo) of the vegetation (Focus 7.1). Although highly simplified, tutorial models may greatly assist understanding of non-linear processes that help to shape the dynamics of the global ecosystem (e.g. Watson 1999; Robertson and Robinson 1998).

> **tutorial models**: simple, non-spatial models of environmental systems

COMPREHENSIVE MODELS

Comprehensive models are three-dimensional, use explicit geography, and have a high spatial and time resolution. They arose from a fast-dawning awareness of the significance of the spatial dimension in comprehending global ecosystem dynamics. Spatial models emerged, virtually independently, in geography, ecology, landscape ecology and fluid dynamics. Developments in computing, GIS and remote sensing allowed the building of large-scale spatial models of the global ecosystem. These models require the grid-based data provided by remote sensing for input and for testing. They also need the power of the latest generation of computers to cope with the sheer number of calculations involved within a reasonable time.

Comprehensive models are expressly spatial and data-hungry. To calibrate a spatial model, modellers need information on several variables for each of the many grid-cells. Obtaining this information from ground surveys is nigh on impossible, especially where the model covers inaccessible terrain. In some cases, existing maps of vegetation, soils, climatic variables and so on have been used to

> **comprehensive model**: a complex mathematical model that resolves space and time in rich detail

FOCUS 7.1

DAISYWORLD

Daisyworld (Watson and Lovelock 1983) is a mathematical model of a hypothetical, Earth-like planet with no ocean. It was constructed by Watson and Lovelock (1983) to show that, without there being any designed purpose to life, a self-regulating biosphere can emerge from interactions between life and its physical environment. The model considers what happens to the biosphere as the Sun grows more luminous – brighter – over billions of years and temperatures rise (Figure 7.10). Two species of daisy – black daisies and white daisies – populate the surface of Daisyworld and form its biosphere. The black daisies warm their local environment because, having low albedos (reflectivities), they absorb sunlight, while the white daises cool their local environment because they have high albedos and reflect sunlight. Once daisies become established, they regulate the planetary surface temperature by competing for space. At the outset, when the Sun is relatively dim and conditions are cool, the seeds of black and white daisies occur equally over the planetary surface. Conditions near the equator are warm enough to stimulate germination, but black daisies have the edge over the white daisies because, with a lower albedo, they absorb more heat (sunlight). As the Sun grows brighter, temperatures rise, the black daisies spread polewards, and the white daisies start to thrive in the warmer temperatures and keep the surface temperature as much as 60°C lower than would be the case on a bare planetary surface. The ever-brightening Sun raises temperatures even more, and eventually the black daisies suffer stress from overheating and can survive only at the poles. The white daisies, with higher albedos reflecting more solar radiation, now occupy the rest of the planet. Eventually, it becomes too hot for even the white daisies and the biosphere is destroyed. However, the model demonstrates that, for a long time, the biosphere stabilizes the planetary temperature while the Sun grows brighter. At the outset, black daisies increase surface temperatures, but as the heat-reflecting white daises come to outcompete their black cousins, the surface temperatures remain roughly constant.

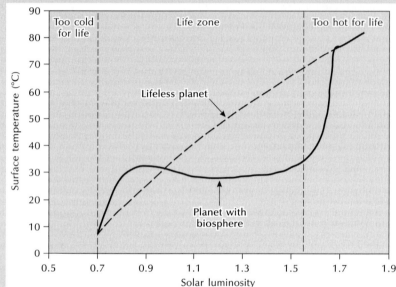

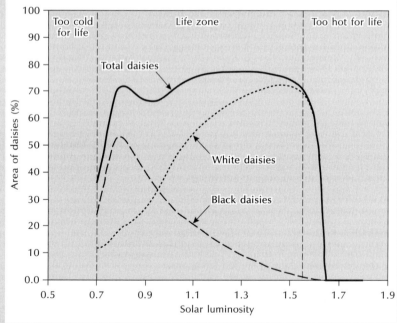

Figure 7.10 Predictions of planetary temperature in the Daisyworld model
More sophisticated versions of the basic Daisyworld model include grey daisies, daisies of several hues, herbivores and carnivores. These more realistic additions to the biosphere seem to increase the stability of the system.

calibrate spatial models. The process involves superimposing grids of the required density on the maps, and measuring the dominant 'state' of particular variables in each one where possible. A disadvantage with this practice is the need to interpolate data. For instance, climate is measured at points (meteorological stations), whereas the climatic variables needed in the calibration of a model must be fitted to all grid-cells, whether or not they contain a meteorological station. All mapped data are to a degree interpolated. They also tend to provide a composite 'snapshot', a picture of the varying conditions within the region being modelled, sketched from scraps of information gathered in various surveys carried out over several years or even decades. Data gleaned from remotely sensed images are an invaluable data source for this type of modelling. These data are, normally, readily fitted to any grid size above a minimum level of spatial resolution, and each pixel in an image can be made to correspond to a grid-cell. Remotely sensed images provide complete coverage of the system's spatial domain at a particular time. In some cases, images may be available for different times of year and for different years. This source of environmental data is a boon to environmental modellers and is very useful in calibrating climate models. The use of GIS in large-scale modelling has become commonplace, and has made possible the calibration and testing of complex dynamic spatial models. GIS also facilitates the efficient handling and processing of large, spatially referenced datasets from a variety of different sources for use in all types of models at a range of scales.

Climate models

The models used to simulate the chemical reactions and physical processes in the atmosphere are exceedingly sophisticated. There is little point in delving too deeply into their workings here, since the novice modeller would be hard pressed to digest and assimilate the complexities of atmospheric physics and chemistry. It is sufficient to outline the chief types of model, describe their basic components and summarize the ways in which they are used.

A range of climate models have been built in recent decades, varying from simple zero-dimensional models to far more elaborate three-dimensional models. Four chief types of climate model now exist:

1.   Energy balance models (EBMs for short) predict the change in temperature at the surface of the Earth in response to a change in heating, with the proviso that the net flux of energy should remain unchanged. The biggest drawback with energy balance models is that they do not include a physically based model of the atmosphere.

2.   Next are radiative–convective models (RCMs for short). These compute the vertical (one-dimensional) temperature profile of a single column of air. Normally, the predicted temperatures are global averages. In general, because they deal with 'average' conditions and the assumptions made about cloud feedback, surface albedo and so forth strongly influence their results, radiative–convective models are not particularly useful.

3.   Statistical dynamical models (SDs for short) normally deal with a two-dimensional slice of air along a line, commonly a line of latitude. They combine the latitudinal dimension of energy balance models with the vertical dimension of radiative–convective models. A set of statistical

equations specifies wind speeds and directions (hence the designation 'statistical' models). Though they can be useful, statistical dynamical models fail to resolve zonal differences within the climate system.

4.   General circulation models (GCMs), the most sophisticated climate models currently available, tackle three-dimensional parcels of air as they move horizontally and vertically through the atmosphere. They are not limited by explicit spatial averaging. Their chief restrictions are the long computing times and the paucity of adequate data with which to calibrate them. There are three types of general circulation model: (1) atmospheric general circulation models (AGCMs), which predict the state of the atmosphere; (2) oceanic general circulation models (OGCMs), which predict the state of the oceans and range from simple, mixed-layer oceans to full ocean general circulation models; and (3) atmosphere–ocean general circulation models (AOGCMs), which, as their name hints, consider the atmosphere and ocean simultaneously.

In coupled atmosphere and ocean models, the grid size is, perforce, coarse (even the fastest and largest computers cannot at present handle the amount of information in coupled models at a very high level of spatial and temporal resolution), and they are very costly to run. The current resolution for an atmospheric section of a typical model is about 250 km horizontally and about 1 km vertically above the boundary layer; and for the oceanic section it is about 200 to 400 m vertically and 125 to 250 km horizontally. The chief physical processes and interactions, and the spatial structure, of a coupled atmosphere–ocean model is set down in Figure 7.11. Basics 7.1 explains the structure of a GCM. Focus 7.2 gives the example of modelling global warming.

Dynamic global vegetation models

Terrestrial Biosphere Models: models designed to simulate global patterns and dynamics of vegetation

Complex dynamic vegetation models are those built to study the global patterns and dynamics of vegetation. They explicitly simulate the growth of plants and competition between different plant types by incorporating biological and geochemical processes. The first generation models, known as **Terrestrial Biosphere Models** (TBMs), included stores (pools) of live and dead biomass (expressed as mass of carbon) and the processes affecting fluxes into, among and between the stores. They also included vegetation structure as a dynamic feature of the biosphere. As a minimum, most TBMs contained modules for photosynthesis, the main powerhouse for carbon assimilation; for autotrophic respiration, the process by which plants consume the carbohydrates they produce; and for heterotrophic respiration, the short-term loss of carbon from biological activity of other organisms in the ecosystem.

Dynamic Global Vegetation Model: a complex model used to simulate the global functioning of vegetation

Dynamic Global Vegetation Models (DGVMs) are the latest generation of models and evolved from TBMs. They have all the mechanisms of TBMs, but, in addition, they include the biological processes that affect longer-term behaviour and are concerned with the plants' life cycles (Figure 7.16 on p. 199). Examples are establishment in a new location, growth, competition with neighbours and responses to natural or human-made disturbances from natural events (fires and storms) or human activity. Vegetation succession models simulate such plant populations, but the direct linking of different succession

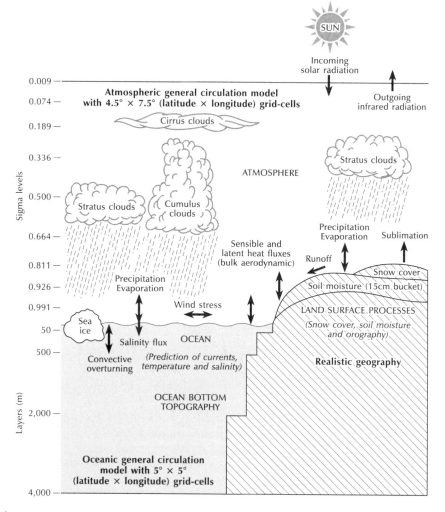

Figure 7.11 A schematic diagram of a coupled atmosphere–ocean general circulation model, showing the vertical structure of atmospheric and oceanic layers and the chief physical processes and interactions modelled. *Source:* After Washington and Meehl (1991)

BASICS 7.1

A BASIC GENERAL CIRCULATION MODEL

A general circulation model consists of a set of equations describing the physical and dynamical processes that determine climate. These equations are prognostic, that is, they enable predictions about the state of the atmosphere (or ocean) to be made. Additionally, many general circulation models include a heat and water balance model of the land surface, and a mixed-layer model of the ocean. At the heart of a general circulation model lie the governing or primitive equations. These are the equations of:

- Motion (conservation of momentum)
- Continuity (conservation of mass or hydrodynamic equation)
- Continuity for atmospheric water vapour (conservation of water vapour)

- Energy (thermodynamic equation derived from the first law of thermodynamics).

To these, storage equations are added of:

- State (hydrostatic equation)
- Surface pressure tendency (in some models).

Additionally, to use a general circulation model, it is necessary to specify parameters such as the solar constant and orbital parameters, and boundary conditions such as the distribution of land and sea, topography and total atmospheric mass and composition.

General circulation models include diagnostic variables – clouds, surface albedo, vertical velocity and the like. Diagnostic variables, such as those related to clouds or ocean convection, result from processes operating at a smaller scale than the model grid. Processes going on within a grid-cell can be included in the model indirectly by representing the state of the grid-cell as parameter values. For instance, individual clouds cannot be included directly in general circulation models because no computer can yet handle small enough cell sizes to take account of cloud-forming processes, at least when the entire globe is being modelled. However, cloudiness can be represented by a few parameters derived empirically from cell-averaged values of temperature, winds and humidity. This process – parameterization – uses physically based relationships between the larger-scale variables. Biological and chemical processes operating within a grid-cell, which clearly influence the atmospheric system as a whole, may also be represented parametrically.

For prescribed boundary conditions and parameters,

the full set of equations is solved to determine the rates of change in prognostic variables such as temperature, surface pressure, horizontal velocity, water vapour and soil moisture. The equations are usually solved for every half hour of model integration.

There is no doubt that state-of-the-art atmospheric general circulation models and coupled atmosphere–ocean general circulation models are complicated, but they are powerful and sophisticated tools in modelling the human impact on several environmental systems. They allow, for example, predictions to be made of future atmospheric states given inputs of several constituents, each of which change with time, and have, therefore, proved invaluable in assessing the credible degree and pattern of warming associated with the release of greenhouse gases by human activities.

FOCUS 7.2

MODELLING GLOBAL WARMING

There is good observational evidence that the concentrations of some atmospheric gases are on the increase owing to emissions from human activity, in particular fossil fuel consumption, agricultural practices and industrial processes. Influential gases include carbon dioxide (CO_2), carbon monoxide (CO), methane (CH_4), nitrous oxide (N_2O), nitrogen oxides (NO_x) and chlorofluorocarbons (CFCs). In 1990, the annual rate of increase of carbon dioxide was 0.5 per cent, methane 0.9 per cent, nitrous oxide 0.25 per cent and CFCs 4 per cent (Watson et al. 1990). There is a strong suspicion that the increasing atmospheric burden of these gases is causing global warming (p. 224).

Evidence that the Earth is warming is mounting. Here are some observational pointers:

- The global average surface temperature (the average of near-surface air temperature over land, and sea-surface temperature) has increased since 1861 (Figure 7.12a). During the twentieth century, the increase was about 0.6°C, although the warming was not even, most of it taking place in the periods 1910 to 1940, and 1976 to 2000

- Globally, it is likely that the 1990s were the warmest decade and 1998 the warmest year in the instrumental record since 1861

- Twentieth-century warming in the Northern Hemisphere is likely to have been the fastest for any time over the last 1000 years (Figure 7.12b).

A warmer, more humid world

Almost without exception, models that simulate the effect of much higher burdens of carbon dioxide and trace gases in the atmosphere predict that the Earth will be a warmer and more humid planet. Predictions suggest a globally averaged surface temperature rise in the range 1.4 to 5.8°C over the period 1990 to 2100. The basic reasons for the increased temperature and humidity are not difficult to grasp. The higher the concentration of greenhouse gases in the atmosphere, the greater the amount of infrared radiation emitted from the Earth's surface absorbed by the atmosphere, and so the hotter the atmosphere. With a warmer atmosphere, evaporation of water from the world's oceans increases, so leading to a more vigorous pumping of water round the hydrological cycle. This results in an increased occurrence of droughts and of very wet conditions, the last created by deeper thunderstorms with greater rainfall. Tropical cyclones also become more destructive. The increased humidity of the air may itself boost greenhouse warming since water vapour absorbs infrared radiation.

Climate models predict an uneven warming of the atmosphere, the land areas warming faster than the seas, and particularly the land areas at northern high latitudes during the cold season. Predictions indicate that the northern regions of North America and northern and central Asia will warm 40 per cent more than the global average. Conversely, south and south-

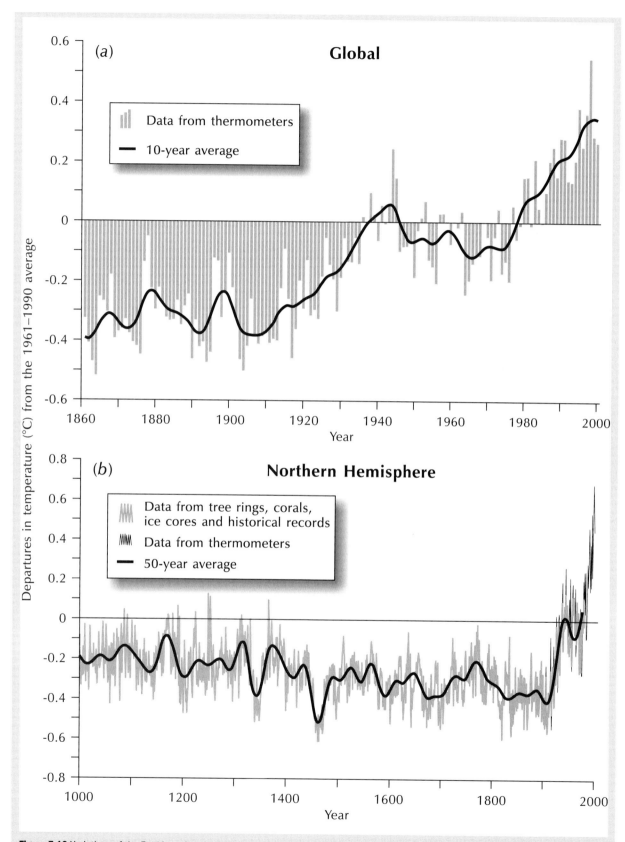

Figure 7.12 Variations of the Earth's surface temperature over (a) the last 140 years and (b) the last millennium. *Source:* After Houghton *et al.* (2001)

east Asia in summer and southern South America in winter will warm less than the global mean. It also seems likely that surface temperatures will become more El Niño-like in the tropical Pacific Ocean, with the eastern tropical Pacific warming more than the western tropical Pacific.

The differential warming should substantially alter the global pattern of evaporation and precipitation and cause radical changes of climate in most regions outside the tropical zone. The models suggest that the global average water-vapour concentration and precipitation will rise over the twenty-first century. After 2050, northern mid- to high-latitudes and Antarctica will have wetter winters. At low latitudes, some regions will become wetter and some drier. Large year-to-year variations in precipitation are likely in most regions with higher predicted precipitation (see Table 7.5).

Scenarios of possible futures

A major effort in evaluating the effects of rising concentrations of greenhouse gases is being undertaken by the Intergovernmental Panel on Climate Change (IPCC). As part of this project, the IPCC devised scenarios to cover possible long-term emissions of carbon dioxide, carbon monoxide, methane, nitrous oxide, nitrogen oxides and chlorofluorocarbons from the present to the year 2100. The scenarios detailed in the Special Report on Emissions Scenarios (Nakićenović and Swart 2000), predicted future greenhouse-gas emissions as the outcome of very complex dynamic systems driven by demographic development, socio-economic development and technological change. A range of different scenarios offered alternative images of how the future might unfold to assist in analysing climatic impacts. Four different 'storylines' – unimaginatively styled A1, A2, B1 and B2 – described the relationships between emission driving forces and their evolution, so giving a context for quantifying the scenarios (Figure 7.13). Each storyline characterized different combinations of demographic, technological and environmental developments. Specific quantitative interpretations of each storyline provided a set of scenarios or scenario family all based on the same story.

The A1 storyline and scenario family

This storyline is one of very rapid economic growth, a global population peaking around 2050 and then declining, and rapidly introduced new and more efficient technologies. It envisages convergence between regions, capacity building, increased cultural and social interactions and a substantial reduction in regional disparities in per capita income. The A1 scenario family develops further scenarios here relating to different technological changes in the energy system: fossil-fuel intensive (A1F1), mainly non-fossil energy sources (A1T) or a balance across all sources (A1B). 'Balance' means non-reliance on any one particular energy source.

Table 7.5 Estimated confidence in observed and predicted extreme weather and climate events

Phenomenon	Confidence in observed changes (second half of twentieth century) (%)	Confidence in predicted changes (twenty-first century) (%)
Higher maximum temperatures and more hot days in nearly all land areas	66–90	90–99
Higher minimum temperatures, fewer cold days and frost days over nearly all land areas	90–99	90–99
Reduced daily temperature range over most land areas	90–99	90–99
Increase of heat index[1] over land areas	66–90, over many areas	90–99, over most areas
More intense precipitation events	66–99, over many Northern Hemisphere mid- and high-latitude land areas	90–99, over most areas
Increased summer continental drying and associated risk of drought	66–99, in a few areas	66–90, over most mid-latitude continental interiors (other areas lack consistent predictions)
Increase in tropical cyclone peak-wind intensities	Not observed in the few analyses available	66–90, over some areas
Increase in tropical cyclone mean and peak precipitation intensities	Insufficient data for assessment	66–90, over some areas

Note: [1] The heat index combines temperature and humidity as a measure of their effects on human comfort

Source: Adapted from Houghton et al. (2001)

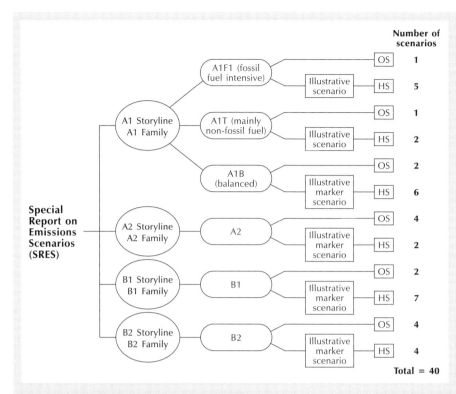

Figure 7.13 The SRES scenarios. *Source*: After Nakićenović and Swart (2000)

The A2 storyline and scenario family

This storyline describes a very heterogeneous world characterized by self-reliance and the preservation of local identities. Important features are the very slow convergence of fertility patterns across regions, which produces a slowly rising global population, regionally orientated economic development, and more fragmented and slower per capita economic growth and technological change than in other storylines.

The B1 storyline and scenario group

This storyline tells of a convergent world with a population peaking mid-century and then declining (as in storyline A1), but with fast changes in economic structures towards a service and information economy associated with reductions in material intensity and the introduction of clean, resource-efficient technologies. It envisages a world characterized by global solutions to economic, social and environmental sustainability, including improved equity, but without additional climatic initiatives.

The B2 storyline and scenario group

This storyline is one of a world dominated by local solutions to economic, social and environmental sustainability. The population continues to grow but at a slower rate than in storyline A2, economic development runs are at intermediate levels, and technological change is slower but more diverse than in storylines B1 and A1. It leans towards environmental protection and social equity and focuses at local and regional levels.

Altogether, there are forty scenarios developed by six

modelling teams (Figure 7.13). Within each scenario family are harmonized scenarios, which share assumptions on global population, gross world product and final energy; and scenarios that explore the uncertainties in driving forces beyond those of the harmonized scenarios. For each of the six scenario groups, an illustrative scenario is given, which is always a harmonized scenario.

Scenario predictions

The future atmospheric concentrations of greenhouse gases and other human-related emissions should have an impact on global climates. Figure 7.14 shows the projected emissions and concentrations of carbon dioxide, and the emissions of sulphur dioxide, for different scenarios. Figure 7.15 shows the predicted temperature rise and sea level rise associated with the illustrative scenarios. Notice that predicted temperatures rise by 1.4 to 5.8°C, depending on the scenario. Scenario A1F1 (fossil-fuel intensive) is the worst-case scenario; scenario B1 (the 'green scenario') is the best-case scenario. Projected global mean sea level will rise by 0.09 to 0.88 m between 1990 and 2100, depending on the scenario. Most of the sea level rise will occur as the increasingly warmer oceans expand and as glaciers and ice caps melt. The fossil-fuel intensive and the 'green' scenarios again spawn the worst and best cases, respectively.

Natural causes of climatic change

It cannot be overstressed that all current observations of global change must be set in the context of natural

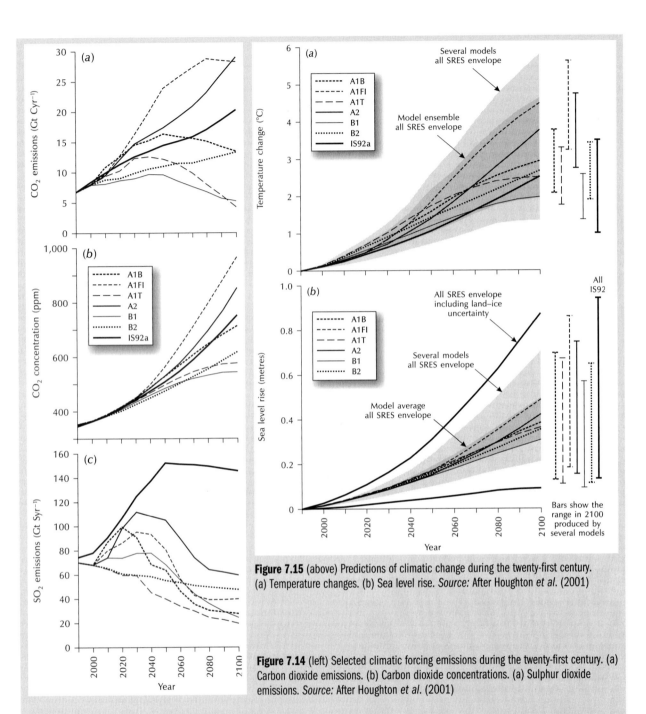

Figure 7.15 (above) Predictions of climatic change during the twenty-first century. (a) Temperature changes. (b) Sea level rise. *Source:* After Houghton *et al.* (2001)

Figure 7.14 (left) Selected climatic forcing emissions during the twenty-first century. (a) Carbon dioxide emissions. (b) Carbon dioxide concentrations. (a) Sulphur dioxide emissions. *Source:* After Houghton *et al.* (2001)

cycles of change in the biosphere and geosphere. The present increase of greenhouse gases in the atmosphere is, almost certainly, an unusual event produced by human activity. Indeed, future greenhouse warming could well create a climate unique in Earth's history (Crowley 1991, 42). A consequence of more greenhouse gases in the atmosphere is a general warming of the globe. But natural cycles of climatic change may also lead to fluctuations in atmospheric carbon dioxide levels and in the levels of some other greenhouse gases, which may bring about a warming and cooling of the atmosphere. Natural climatic cycles must provide a backdrop against which to assess human-induced global warming.

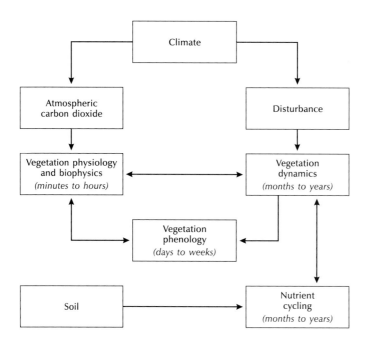

Figure 7.16 General structure of a Dynamic Global Vegetation Model (DGVM)

models to cover the global biosphere is not possible at present because the necessary parameters are known for only a few plant species. Moreover, interactions between different kinds of plants, such as trees and grasses, demand new methods for simulating the processes involved including, for instance, competition for soil water. DGVMs also demand the definition of an appropriate set of Plant Functional Types (PFTs), for example, trees (broadleaf evergreen, broadleaf deciduous, needleleaf evergreen, needleleaf deciduous), shrubs, grasses and forbs (broad-leaved flowering plants other than grass, usually growing in a meadow or prairie). These must be few enough to make parameterization in a global model possible, but complex enough to capture at least some of the range in functional behaviour among plants. DGVMs aim to simulate vegetation response under changing environmental conditions, usually a physical climate alteration, changing atmospheric carbon dioxide levels and nitrogen deposition, or an alteration in disturbance regimes resulting from land use change (or all three at the same time). Models that incorporate land use changes are in the early stages of development. Such models aim eventually to predict interlinked changes of vegetation at global, regional and local scales.

Earth Simulator

Huge sums of time and money are spent trying to increase the sophistication and resolution of comprehensive models. Strenuous efforts are made to construct models of the full climate system that include biological and geochemical processes. The hope is that these comprehensive models of the geosphere–biosphere system will allow a deeper appreciation of the human impact on the environment and provide effective tools for global and regional environmental management. A team of researchers in Japan have taken a colossal step in this direction. They have built a computer called Earth Simulator that occupies the space of about four tennis courts and, since 11 March 2002, runs on the campus of the Yokohama Institute for Earth Sciences. This hyper-computer cost about $350 million, uses 10 megawatts of electric power under standard operational conditions and has 5104 high-speed central processing units connected by 2800 km of cables. It has a peak computing performance of about 40 teraflops (40 trillion calculations per second), has a central storage unit with a 10-terabyte memory, 700-terabyte of disc space and a mass store of 1.6 petabytes. It is hoped to use this beast of a computer to run a climate model with a spatial resolution of 10 kilometres, some five to ten times better than previously possible, that will allow regional-verging-on-local climate changes to be simulated with some confidence. The model is chiefly designed to predict the impacts of global warming. Its core is a climate model comprising coupled atmosphere, ocean and land surface models. It includes an emission model that uses different scenarios, such as rapid-growth society, a differentiated world, global sustainability and local stewardship. And it includes an impact model that predicts the effects of global warming under given scenarios upon land use, water resources, social and economic factors, crop productivity and food supply and demand.

INTERMEDIATE COMPLEXITY MODELS

Earth System Model of Intermediate Complexity (EMIC): a model designed to combine the advantages of simple models with the benefits of complex models, without the drawbacks of either

Critics of comprehensive models point to their over-sophistication, high cost and results that seem no less daunting than the reality they try to simulate. **Earth System Models of Intermediate Complexity** (EMICs) simulate interactions between several components of the natural Earth system and form a useful bridge between comprehensive models and tutorial models, allowing climate simulations incorporating vegetation dynamics over timescales such as 10,000 years, or even glacial cycles (e.g. Claussen *et al.* 1999). They have been used to predict the onset of the next ice age, to explore Holocene climatic changes and to examine the threat of abrupt climatic change.

The Climate and Biosphere Model (CLIMBER) is an example of an EMIC. Developed at the Potsdam Institute for Climate Impact Research in Germany, CLIMBER computes many processes and feedbacks in the climate system, as its comprehensive model cousins do, but at a fairly low spatial resolution (the planetary surface is divided into 126 grid-cells with a latitudinal resolution of 10 degrees) (Figure 7.17). The result is a much quicker computing time: 4000 simulated years take about a day on a workstation (late 1990s top models) and 30 minutes on a supercomputer. CLIMBER is an evolving model and different versions have their own specifications. CLIMBER-2 contains six modules: an atmosphere module, an ocean and sea ice module, a vegetation module, an inland ice module and modules of marine biota and oceanic biogeochemistry. Figure 7.18 shows some sample output. One application of CLIMBER-2 was to investigate the change of the world climate system from a warmer and wetter state to a colder and drier state over the last 8000 years. The increase in aridity led to the expansion of the subtropical deserts. An interesting question is how this climatic change occurred: was it a gradual, long-term trend or did it occur in short steps? Simulations made using CLIMBER-2 suggested a gradual change at high northern latitudes in the abundance of taiga and tundra. At lower latitudes, and especially in the North African subtropics, it suggested an abrupt change in precipitation about 5500 years ago, followed by a further gradual drift (Claussen *et al.* 1999).

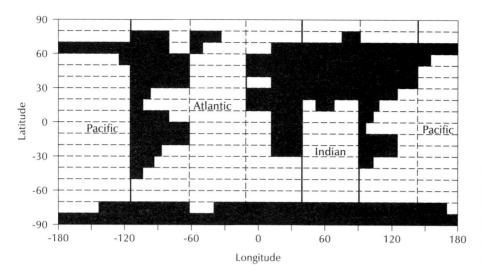

Figure 7.17 Schematic representation of Earth's topography in the CLIMBER-2 model. Dashed lines show the atmospheric grid; solid lines separate ocean basins. *Source:* After Claussen *et al.* (1999)

7.3 REGIONAL AND LOCAL MODELS

MODELLING THE CHANGING COMPOSITION OF COMMUNITIES

Predicting the shape of community change at a species level is possible at local and regional scales. A series of models simulate the growth and interactions of species (usually tree species) over time in a **forest gap**. The great-grandparent of these forest succession models was JABOWA (Botkin *et al.* 1972). FORENA (Forests of Eastern North America) was a development of JABOWA and FORET (Shugart and West 1977). It attempted to model the transient response of vegetation to future changes in climate (Solomon 1986) (Basics 7.2).

> **forest gap**: an opening produced by trees falling through old age or through natural disturbing agencies such as strong winds

An early application of FORENA simulated forest growth at twenty-one locations in eastern North America, as far west as a line joining Arkansas in the south to Baker Lake, North West Territories, in the north (Solomon 1986). All forests started growing on a clear plot and grew undisturbed for 400 years under a modern climate. After the year 400, the climate changed to allow for a warmer atmosphere. The model incorporated a linear change of climate between the years 400 and 500, the new climate at the year 500 corresponding to a doubling of atmospheric carbon dioxide levels. Climatic change continued to change linearly after the year 500, so that, by the year 700, the new climate corresponded to a quadrupling of atmospheric carbon dioxide levels. At the year 700, climate stabilized and the simulation went on for another 300 years. For each site, ten simulations were run and the results averaged. Validation of the results was achieved by testing with independent forest composition data, as provided by pollen deposited over the last 10,000 years and during the last glacial stage, 16,000 years ago.

Results for some of the sites are set out in Figure 7.19. At the tundra–forest border, the vegetation responds to climatic change in a relatively simple way (Figure 7.19a). With four times the carbon dioxide in the atmosphere, the climate supports a much-increased biomass and, at least at Shefferville in Quebec, some birches, balsam poplar and aspens. The response of northern boreal forest to warming is more complicated (Figure 7.19b). With a fourfold increase in atmospheric carbon dioxide levels, the climate promotes the

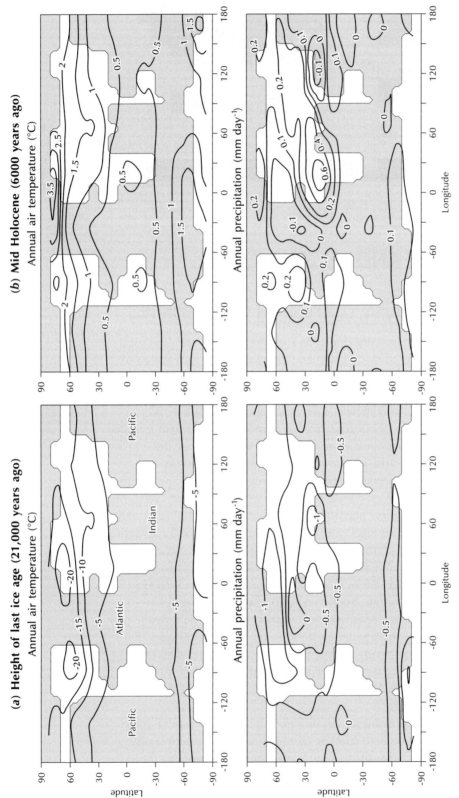

Figure 7.18 Postdicted climate during the last ice age and during the mid-Holocene compared to today. (a) Annual air temperature (°C) and annual precipitation (mm day^{-1}) 21,000 years ago. (b) Annual air temperature and annual precipitation (mm day^{-1}) 6000 years ago. *Source: After Claussen et al.* (1999)

BASICS 7.2

FORENA MODEL OF FOREST SUCCESSION: BASIC OPERATION

FORENA mimics forest succession in three stages: the establishment of new seedlings, the growth of the trees and the death of the trees. In doing so, it keeps account of trees belonging to a maximum of 72 species in a stand with an area of 0.125 ha. For each year in a simulation run, the model 'sows' a cohort of seedlings, the species composition and size of which is determined stochastically. The available species pool varies from one year to the next, depending on the plot conditions. The plot conditions include light levels, the presence of leaf litter (related to the leaf area of the plot), the presence of mineral soil (related to the death of a large tree) and the death of a tree of a species capable of sprouting from roots when the aboveground parts are lost. The maximum growth achievable in a year by each tree under optimum conditions depends upon a 'tree growth equation', with each tree species having its own size and growth rate parameters.

Multiplying the optimal growth by parameters for stand crowding, shade tolerance, temperature and drought yields the actual growth. The model sets temperature as a stochastic variable: degree-days are included as sums of monthly random values drawn from predefined means and standard deviations. A similar stochastic process defined cold winter temperatures (for January) and drought days. Lastly, for each simulated year, an age-dependent mortality function and an age-independent mortality function determine the trees that die. To make the model operational, the researchers culled parameters characterizing the limits to the establishment, growth, longevity and mortality of each of the 72 species of tree from silviculture manuals, and the values of the extrinsic variables, such as degree-days and drought days, are mapped using existing data sources.

expansion of ashes, birches, northern oaks, maples and other deciduous trees at the expense of birches, spruces, firs, balsam poplar and aspens. In all northern boreal forest sites, the change in species composition is similar, but takes place at different times. This underscores the time-transgressive nature of vegetational response to climatic change. The southern boreal forest and northern deciduous forest (Figures 7.19c and d) also have a complicated response to climatic change. In both communities, species die back twice, the first time around after just 500 years, and the second time from about 600 to 650 years. In the southern boreal forest, the change is from a forest dominated by conifers (spruces, firs and pines) to a forest dominated initially by maples and basswoods, and later by northern oaks and hickories. At some sites in the northern deciduous forest, species appear as the climate starts to change then vanish once the stable climate associated with a quadrupled level of atmospheric carbon dioxide becomes established. Species that do this include the butternut (*Juglans cinerea*), black walnut (*Juglans nigra*), eastern hemlock (*Tsuga canadensis*) and several species of northern oak. In western and eastern deciduous forests, the response of trees to climatic warming is remarkably uniform between sites (Figure 7.19e and f). Biomass declines everywhere, generally as soon as warming starts in the year 400. The drier sites in the west suffer the greatest loss of biomass, as well as a loss of species, corresponding to a takeover by prairie vegetation. The decline of biomass in the eastern deciduous forest probably results chiefly from the increased moisture stress in soils associated with the warmer climate. The increased moisture stress also gives a competitive edge to smaller and slower-growing species. These include the southern chinkipin oak (*Quercus muehlenbergii*), post oak (*Quercus stellata*), and live oak (*Quercus virginiata*), the black gum (*Nyssa sylvatica*), sugarberry (*Celtis laevigata*) and the American holly (*Ilex opaca*). Such slow-growing species come to dominate the rapid-growing species such as the American chestnut (*Castanea dentata*) and various northern oak species.

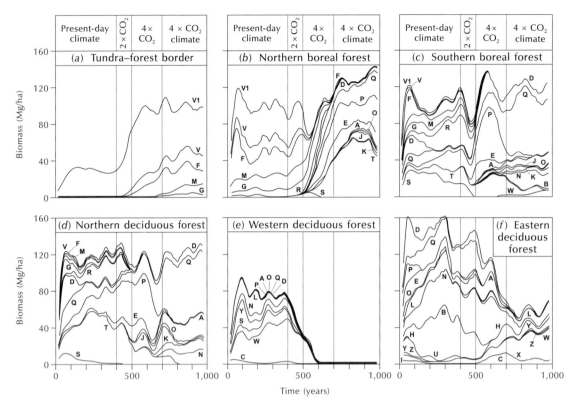

Figure 7.19 Simulations of forest biomass dynamics over one millennium in response to climatic change, induced by increasing levels of carbon dioxide in the atmosphere, at six sites in eastern North America. (a) Shefferville, Quebec (57°N, 67°W). (b) Kapuskasing, Ontario (49°N, 83°W). (c) West upper Michigan (47°N, 88°W). (d) North central Wisconsin (45°N, 90°W). (e) South central Arkansas (34°N, 93°W). (f) Central Tennessee (36°N, 85°W). The tree species are as follows: A. American beech (*Fagus grandifolia*); B. American chestnut (*Castanea dentata*); C. American holly (*Ilex opaca*); D. ashes: green ash (*Fraxinus pennsylvanica*), white ash (*Fraxinus americana*), black ash (*Fraxinus nigra*), blue ash (*Fraxinus quadrangulata*); E. basswoods: American basswood (*Tilia americana*) and white basswood (*Tilia heterophylla*); F. birches: sweet birch (*Betula lenta*), paper birch (*Betula papyrifera*), yellow birch (*Betula alleghaniensis*) and gray birch (*Betula populifolia*); G. balsam poplar (*Populus balsamifera*), bigleaf aspen (*Populus grandidentata*), trembling aspen (*Populus tremuloides*); H. black cherry (*Prunus serotina*); I. black gum (*Nyssa sylvatica*); J. butternut (*Juglans cinerea*) and black walnut (*Juglans nigra*); K. eastern hemlock (*Tsuga canadensis*); L. elms: American elm (*Ulmus americana*) and winged elm (*Ulmus alata*); M. firs: balsam fir (*Abies balsamea*) and Fraser fir (*Abies fraseri*); N. hickories: bitternut hickory (*Carya cordiformis*), mockernut hickory (*Carya tomentosa*), pignut hickory (*Carya glabra*), shagbark hickory (*Carya ovata*), shellbark hickory (*Carya laciniosa*) and black hickory (*Carya texana*); O. hornbeams: eastern hornbeam (*Ostrya virginiana*) and American hornbeam (*Carpinus caroliniana*); P. maples: sugar maple (*Acer saccharum*), red maple (*Acer rubra*), and silver maple (*Acer saccharinum*); Q. northern oaks: white oak (*Quercus alba*), scarlet oak (*Quercus coccinea*), chestnut oak (*Quercus prinus*), northern red oak (*Quercus rubra*), black oak (*Quercus velutina*), bur oak (*Quercus macrocarpa*), gray oak (*Quercus borealis*) and northern pin oak (*Quercus ellipsoidalis*); R. northern white cedar (*Thuja occidentalis*), red cedar (*Juniperus virginiana*), and tamarack (*Larix laricina*); S. pines: jack pine (*Pinus banksiana*), red pine (*Pinus resinosa*), shortleaf pine (*Pinus echinata*), loblolly pine (*Pinus taeda*), Virginia pine (*Pinus virginiana*), pitch pine (*Pinus rigida*), and, T. white pine (*Pinus strobus*); U. yellow buckeye (*Aesculus octandra*); V. spruces: black spruce (*Picea mariana*), and red spruce (*Picea rubens*); and V1. white spruce (*Picea glauca*); W. southern oaks: southern red oak (*Quercus falcata*), overcup oak (*Quercus lyrata*), blackjack oak (*Quercus marilandica*), chinkipin oak (*Quercus muehlenbergii*), Nuttall's oak (*Quercus nuttallii*), pin oak (*Quercus palustris*), Shumard's red oak (*Quercus shumardii*), post oak (*Quercus stellata*) and live oak (*Quercus virginiana*); X. sugarberry (*Celtis laevigata*); Y. sweetgum (*Liquidambar styraciflua*); Z. yellow poplar (*Liriodendron tulipifera*). *Source:* After Solomon (1986)

Models that are more recent do not all fit the JABOWA mould. FORSKA 2, for example, simulates forest landscape dynamics by focusing on a gap model. Competition between species for light and nutrients normally takes place within a 0.1 ha patch. Such climatic data as mean monthly temperature, precipitation and sunshine hours modulate the establishment and growth of the species, each species responding differently to the climatic drivers. Using an array of replicate patches simulates the dynamics of an entire landscape. Fire and logging may disturb the simulated landscape. The model usually runs for hundreds of years and predicts species diversity, biomass, productivity, leaf area and density over a set interval of time. Figure 7.20 shows a 1000-year FORSKA 2 simulation for the Boa Berg Halland region in Sweden. The model ran for 400 years under the present climate. Between 400 and 500 years, a linear warming corresponded to a doubling of atmospheric carbon dioxide levels (derived from the Hamburg ECHAM3 climate model). The remaining 500 years used the new doubled-carbon dioxide climate. Notice that Norway spruce (*Picea abies*) dominates the landscape under present conditions, with some Scots pine (*Pinus sylvestris*) and a few deciduous tree species. With climatic warming, Norway spruce disappears, Scots pine becomes dominant, and the deciduous species – silver birch (*Betula pendula*), pedunculate oak (*Quercus robur*), sessile oak (*Quercus petraea*) and beech (*Fagus sylvatica*) – become more important.

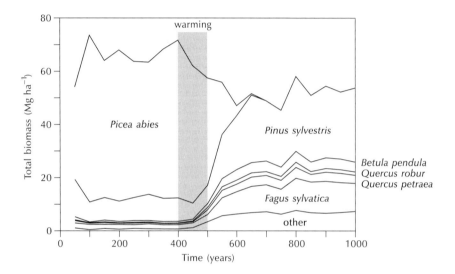

Figure 7.20 A 1000-year simulation of forest dynamics using the FORSKA 2 model at Boa Berg Halland, Sweden. The model simulated present climate for 400 years, then imposed a linear warming between 400 and 500 years, after which time it ran a climate associated with a doubling of present atmospheric carbon dioxide levels. *Source:* After Sykes and Haxeltine (2001)

SUMMARY

Scientists have found several kinds of models helpful in understanding the environment and environmental change. They have built hardware models, conceptual models and mathematical models. Hardware models may be small and simplified versions of environmental systems – such as scaled-down replicates of river estuaries or a microcosmic reconstruction of the biosphere. Analogue models are akin to scale models but with more refinements, and include maps, remotely sensed images and copies of dynamic natural systems where a different material substitutes for natural material – clay as an analogue of ice for example. Conceptual models come in a variety of guises, but all of them are verbal, pictorial or numerical representations of an environmental system that suggest how the system is put together and how it works. Of all conceptual models, mathematical models are the most sophisticated and powerful. They come in three main varieties – stochastic models, statistical models and deterministic models – although many models are hybrids and use elements of, say, a deterministic model and a stochastic model. With the advent of fast computers with huge memories, mathematical models have come into their own. Today, scientists use them to study past, present and future changes in the global ecosystems, and to focus on changes at regional and local scales. Global models can be highly simplified portrayals of the Earth system (tutorial models), richly detailed representations of the Earth system (comprehensive models), or descriptions of the Earth system lying somewhere between very simple and highly complicated (models of intermediate complexity). Comprehensive models include climate models, the most refined of which are the various general circulation models used to predict climatic change (global warming for instance), and they include dynamic global vegetation models that simulate the global patterns and dynamics of world vegetation. Earth Simulator, a huge and super-fast Japanese computer, is likely to run the most comprehensive of the models. Models of intermediate complexity try to bridge the gulf between model simplicity and model complexity. An example is the Climate and Biosphere Model (CLIMBER) used, for instance, to study world climates during the last 10,000 years. At regional and local scales, mathematical models have several applications. They predict the changing composition of forests in response to climatic changes.

Questions

1 Discuss the problems of making small-scale hardware and analogue models of environmental systems.
2 For an ecosystem of your choice, describe it in words, as a pictorial model and as a box-and-arrow diagram.
3 Explain the use of atmospheric general circulation models in predicting the likely climatic consequences of a nuclear war.
4 To what extent is 'global warming' the result of natural climatic cycles?
5 What lessons can be learnt from models simulating change in community composition?

Further Reading

Ford, A. (1999) *Modeling the Environment: An Introduction the Systems Dynamics Modeling of Environmental Systems*. Washington DC: Island Press.
A good set of ecological modelling applications, but not simple.

Huggett, R. J. (1993) *Modelling the Human Impact on Nature: Systems Analysis of Environmental Problems*. Oxford: Oxford University Press.
A straightforward introduction to the subject. Does not require much mathematical knowledge.

Odum, H. T. and Odum, E. C. (2000) *Modeling for All Scales: An Introduction to System Simulation*. San Diego: Academic Press.
A book based on Odum's energese. Includes a CD.

Wainwright, J. and Mulligan, M. (eds) (2003) *Environmental Modelling: Finding Simplicity in Complexity*. Chichester: John Wiley & Sons.
A useful collection of essays revealing the state-of-the art in physical geographical modelling applications. Not easy going for the novice, but worth a look.

Useful Websites

http://www.meto.govt.uk/research/index.html
The UK's Met Office site, with lots of information on climate models and their use.

http://www.pik-potsdam.de/cp/poem/welcome.3.html
A Dynamic Global Vegetation Model explained on the Potsdam Institute for Climate Impact Research website.

http://www.cs.utk.edu/~dongarra/esc.pdf
Useful information on Earth Simulator.

http://cpb.bio.ic.ac.uk/ecotron/ecotron.html
Details on the Ecotron from the NERC Centre for Population Biology.

http://www.face.bnl.gov/
Information on free-air carbon dioxide experiments from Brookhaven National Laboratory.

3

HUMANS AND THE PHYSICAL ENVIRONMENT

CHAPTER 8
AIR

OVERVIEW

As with water, air is an essential requirement for life on Earth and by extension, critical for human society. The atmosphere is a complex, dynamic and inherently unstable system, and one that is constantly changing in time and space. To understand the impact of human activities on the atmosphere it is useful to consider the basic principles of the system since these help to explain the relationship between the atmosphere and the water, land and life of the globe. Chapter 2 identified some of the key characteristics of the atmosphere, particularly those relevant for an understanding of human impacts. The nature of the atmosphere is changeable over time due to natural drivers, but it is also true that human activity has created change. Some of the most important changes have been in relation to our influence on the chemical composition of the atmosphere, and this chapter will explore these elements in detail.

LEARNING OUTCOMES

After reading this chapter, you should be able to build on the ideas about the atmosphere in Chapter 2 to better appreciate what the atmosphere is and how it works as well some of the ways in which humans interact with it. In particular:
- Energy and the atmosphere – how and why air moves
- Pollutants and the atmosphere – sources, characteristics and impacts
- A number of key environmental issues primarily associated with the atmosphere – acidification, air pollution and climate change.

8.1 INTRODUCTION

> The atmosphere has always been linked with life. Humans breathe and the relationship with the air is so readily apparent. The connection became so strong within some elements of 19th C science that the human soul was even thought of as a gas (Brimblecombe 2001, 35).

Air is a mixture of elements and compounds that exist as gases and surround the Earth's surface. It has long fascinated scientists, artists and poets and there is a rich tradition and long history associated with appreciating and understanding air and the atmosphere.

Increasingly, researchers see the value in trying to understand the whole system when seeking solutions to environmental problems. For example, Chameides, quoted in Ginsburg and Cowling (2003), refers to the value of using natural chemical cycles and the 'one atmosphere paradigm' integrated with meteorological and climatological paradigms to facilitate our understanding of air pollution. Chameides also notes that we are only really interested in the trace constituents of the atmosphere because of their effects on public health and the productivity and stability of agricultural and natural ecosystems. It is important to understand the scientific basis of processes as a prerequisite to understanding particular environmental problems. To do this fully it is necessary to consider the interrelationships and the different temporal and spatial scales over which phenomena and related impacts occur. All of this is the foundation upon which 'solutions' or 'sustainable management' can be based, and this is where the role of future physical geographers, environmental scientists and environmental managers will be most needed. This ongoing research work will continue the long history of scientific endeavour into understanding our atmosphere (Focus 8.1).

This chapter will build on the introduction to the nature and characteristics of the atmosphere given in Chapter 2. Where possible, the interconnections between the atmosphere and the other spheres mentioned in this book will be demonstrated, as drivers in one sphere often lead to impacts in others. As is shown throughout the examples in this book these interlinkages are not simple and it is necessary to appreciate the many feedbacks in operation to fully understand the relationship between human activity and environmental impact. This process of understanding relationships is currently incomplete and the subject of much research activity. There is also much endeavour aimed at trying to control and manage processes and the relationship between inputs and impacts, and this chapter will touch on some of the ways in which this has been done in different parts of the world.

Take the example of the current concern about climate change. This concern is not so much about the changes themselves, as the atmosphere on Earth in the far past has had much higher concentrations of carbon dioxide (CO_2) than are anticipated in the near future. Rather, it is for the rate at which these changes are taking place and the ability of natural ecosystems (and human society) to respond to them in an increasingly crowded and 'stressed' world. There are also issues associated with the introduction of entirely artificial substances which change the chemical processes within the atmosphere and which over time have impacts on the world's ecosystems. Still other drivers modify the physical characteristics of the atmosphere. The movement of air is modified through the influence of natural barriers such as mountain ranges but can also be influenced by human activity. Sometimes physical structures are built specifically to

FOCUS 8.1

A TIMELINE OF THE HUMAN ENDEAVOUR IN UNDERSTANDING THE ATMOSPHERE

The development of atmospheric science draws heavily on developments in physics and chemistry. It has a very long history and some very great challenges for the future.

340 BC Aristotle publishes *Meteorologica*, his theories form the basis of understanding of atmospheric processes for two millennia.

1643 Torricelli carries out atmospheric pressure experiments.

1686 Halley discovers that there is a difference between amounts of solar energy received at different latitudes on the globe, and that this is a driver of atmospheric circulation.

1714 Fahrenheit develops the thermometer and a measurement scale for temperature.

1735 Hadley shows the importance of Earth's rotation for global wind patterns.

1750s Black identifies carbon dioxide in the atmosphere.

1770s Rutherford identifies nitrogen in the atmosphere.

1781 Cavendish measures the amount of nitrogen and oxygen in the air.

1839 Schonbein discovers – literally smells – ozone in his laboratory and names it after the Greek for 'ill-smelling'.

1859 Tyndall suggests that water vapour and carbon dioxide have a warming effect on the atmosphere.

1878 Cornu measures the spectrum of solar radiation and suggests that the lack of ultraviolet radiation must be due to something absorbing it in the atmosphere.

1896 Arrhenius models the relationship between carbon dioxide and surface temperature.

1919–37 Bjerknes theorizes and describes mechanisms of atmospheric motion used to understand general circulation and weather.

1920 Milankovitch suggests that ice ages are related to variations in the Earth's orbit over time.

1924–28 Dobson describes patterns of ozone in the atmosphere in response to latitude and season.

1930 Chapman proposes a theory for upper atmosphere ozone production.

1938 Callendar calculates emissions of carbon dioxide and relates this to increased global temperatures.

1941 First use of radar (remote sensing technique) for tracking weather.

1949–56 Brewer and Dobson pave the way for a model of stratospheric general circulation.

1957 Revelle and Suess warn that human society is conducting a huge experiment with the Earth by altering the chemistry of the atmosphere without knowing what the impacts will be.

1959 The first satellite images of the cloud cover of the globe are produced. Suomi estimates the Earth's heat budget.

1967 Manabe and Wetherald produce radiative–convective atmospheric model of the relationships between carbon dioxide and global warming and warn that a doubling in CO_2 concentrations will lead to an average 3°C rise.

1969 Bjerknes links the southern oscillation and El Niño phenomena.

1970–74 Crutzen and Johnson and Stolarski and Cicerone identify ozone destruction mechanisms involving the oxides of nitrogen and chlorine respectively. Molina and Rowland then discover the role of chlorofluorocarbons in destroying stratospheric ozone.

1985 The British Antarctic Survey reports dramatic reductions in springtime stratospheric ozone over the previous 30 years.

1986 Global action takes place to counteract the observed problem of stratospheric ozone depletion through the Montreal Protocol on Substances that deplete the ozone layer.

1990s Researchers show that atmospheric aerosols (solid matter) have a cooling effect on the atmosphere.

1995 The Intergovernmental Panel on Climate Change (IPCC) states that the evidence suggests a clear link between human activities and global warming.

Year 2000 and beyond Researchers and governments work on trying to adapt to a changing climate as trends continue and impacts are felt.

(See Crutzen and Ramanathan (2000) and Hardy (2003) for more details.)

influence air movements, for example, through the construction of windbreaks. Issues associated with the deliberate modification of the atmosphere using artificial structures are not specifically covered in this text. Some modification that takes place is unintentional, artificial structures such as large built-up areas can create a localized effect not just on the movement of air, but also its temperature and rates of precipitation. The impacts of urban areas on some aspects of the atmosphere are considered in Chapter 14.

This chapter will give an overview of some further 'atmospheric' environmental problems (these are 'atmospheric' in the sense that they are usually considered as associated with the atmosphere, but as you will see they are also referred to in other parts of this book because it is extremely difficult – even misleading – to consider these issues in complete isolation). The particular issues or syndromes to be considered are: acidification, air pollution and climate change. This list is not by any means exhaustive, and the wealth of books and articles will provide more examples and more details. Crutzen and Ramanathan (2000) provide an interesting and accessible summary of past, present and future atmospheric issues as a general starting point.

8.2 THE MOVING ATMOSPHERE

The atmosphere is in constant motion. This motion is most pronounced in the lower part of the atmosphere, the troposphere. This is where most activity takes place, where most weather happens and where most of the contact and influence of humans is felt. As a starting point for further understanding some of the environmental issues associated with the troposphere, it is useful to investigate first how and why air moves from one place to another.

Atmospheric movement can be viewed as operating on several different scales. These scales of motion (see Figure 8.1) vary from very local (micro) scale movements such as smoke from a chimney stack, to medium (meso) scale movements of air around a city, to regional (synoptic) scale movements of weather systems to, finally, global (planetary) scale patterns (Ahrens 1998).

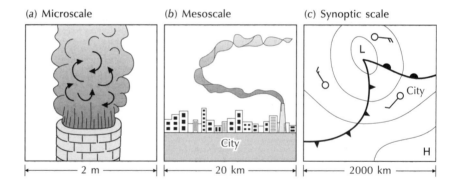

Figure 8.1 The sub-global scales of motion. *Source:* From *Essentials of Meteorology: Invitation to the Atmosphere,* 2nd edn, by Ahrens. © 1998. Reprinted with permission of Brooks/Cole, a division of Thomson Learning: www.thomsonrights.com. Fax 800 730-2215

turbulence: irregular movement of the atmosphere, perhaps as a result of a physical disturbance of the flow of air

In the discussion of the vertical structure of the atmosphere in Chapter 2, the point was made that much of the motion, and associated **turbulence**, of the atmosphere takes place near the Earth's surface. The discussion here will focus on these tropospheric air movements. Movement occurs both laterally and vertically, but it is horizontal motion that predominates. Vertical motion is in general balance with gravity but is also very important in assisting the circulation of air around the troposphere. All motion in the atmosphere occurs in order to balance forces and redistribute energy. Vertical and horizontal movements will now be considered in turn.

VERTICAL MOVEMENTS IN THE ATMOSPHERE

Vertical movement is primarily driven by the energy provided by the Sun. Solar radiation is therefore a key driver of the atmospheric system, and it is useful to consider the nature of this energy before discussing how this leads to vertical movement. This discussion will also be of value when considering the impacts of different constituents of the atmosphere later in the Chapter.

Solar energy: a key driver of atmospheric movement and processes

Solar radiation (introduced in Chapter 2) is energy that takes a range of different forms across a spectrum of **wavelengths** (Figure 8.2). This differentiation of incoming energy is very important. Different constituents of the atmosphere have different reactions to them. For example, as we have seen, ozone has the important role of absorbing ultraviolet (very short wavelength) radiation that has the happy consequence of reducing the amount of this potentially extremely harmful material that penetrates to the Earth's surface. You can see on Figure 8.2 that ozone absorbs some wavelengths of outgoing terrestrial radiation too, so it is

> **wavelengths**: a measure of the distance covered by one 'wave' of electromagnetic energy, for example crest to crest or trough to trough. Measured in micrometres (10^{-6} metres) (μm) or nanometres (10^{-9} metres) (nm)

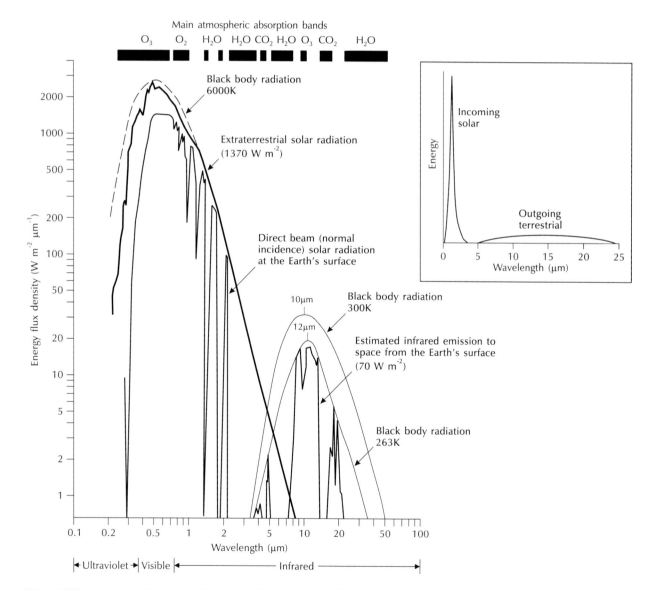

Figure 8.2 The electromagnetic spectrum. *Source:* After Barry and Chorley (2003)

a greenhouse gas, and one of its roles is to heat up the atmosphere. Figure 8.2 demonstrates that the spectrum of wavelengths associated with incoming solar radiation is quite large but it is not as large as the range of wavelengths (the spectrum) for outgoing terrestrial radiation (note the arithmetic scale used for the inset graph as opposed the logarithmic scale for the main graph). Although too much of the wrong sort of energy is disastrous, needless to say it is also critical that some of this energy does reach the Earth's surface as this is the primary energy source for plants and therefore also animals and humans.

When energy reaches the surface, it is reflected and/or absorbed to different degrees depending on what it hits on the surface. Humans are able to see a proportion of the reflected energy and we refer to these wavelengths as the visible spectrum. A simple example explains how this works. Green vegetation appears green to humans because the light in the visible spectrum that is reflected is the green component (around 500–550 nm wavelength) whilst other wavelengths of visible light are absorbed. Further (invisible) light energy is reflected or absorbed but it is not possible for humans to see this without artificial aid.

Another word for the reflectivity of an object or surface is its **albedo**. For example, thick clouds have an albedo of around 60 to 90 per cent, sand 15 to 45 per cent and forest 3 to 10 per cent. It is difficult to be more precise for general land cover types since the actual values depend on the specific characteristics of the material itself. So, for forest, its water content, health, species, time of year and so on all affect its albedo.

Remote sensing from satellites enables humans to 'see' much more information about reflected light from vegetation and other surfaces by using special sensors that detect different parts of the electromagnetic spectrum. By using satellite imagery, it is possible to look at the relative reflectance in the visible and non-visible spectrum that enables an assessment to be made of different species of plants, total amounts of biomass and the relative moisture content of vegetation. For this reason, remote sensing is an incredibly valuable surveying tool for watching the Earth as well as watching the weather (see Chapter 6).

The energy absorbed by objects and surface cover is effectively stored energy that is later released or emitted from the surface back into the atmosphere. This emitted long-wave energy is invisible to the naked eye but we can sometimes still detect it. Relative to natural surfaces, surfaces in urban areas tend to absorb and store energy over the course of a day. This is one of the reasons that urban areas can have distinctly different 'climates' compared to the surrounding countryside.

The storage of heat and transfer of associated energy from the Earth's surface to parcels of air is an important driver of atmospheric movement. Since the Earth's surface is a patchwork of different types of cover all with different properties of albedo, the surface warms unevenly. When molecules of air strike particularly warm parts of the surface they gain some of the energy through **conduction**. This then, eventually, creates an area or parcel of relatively warm air. This warm air rises and expands and cooler air falls and fills the gap that is left. This air will in turn then heat up and rise. The original parcel of warmed air will eventually cool relative to the surrounding air and then fall back towards the ground, the action of which will make the air denser and warmer (but still cool relative to the air parcels around it) until it reaches the surface once again. This process is called convective circulation and is particularly strong on bright, warm and clear days (Figure 8.3). Most of the heat/energy transfer that occurs in the atmosphere occurs through **convection** because air is a very poor conductor of

albedo: reflectivity of a material measured as the amount of energy returned compared to that received

conduction: the process of energy transfer through the contact of vibrating molecules

convection: the process of energy transfer through mass movement or mixing of a substance (gas or liquid)

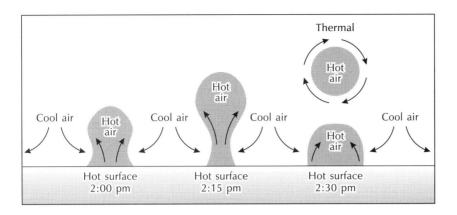

Figure 8.3 Generalized schematic of the processes of vertical movement of air parcels through convective circulation. *Source:* From *Essentials of Meteorology: Invitation to the Atmosphere*, 2nd edn, by Ahrens. © 1998. Reprinted with permission of Brooks/Cole, a division of Thomson Learning: www.thomsonrights.com. Fax 800 730-2215

heat. As well as convective circulation, air also heats and cools due to coming into contact with barriers and being forced to rise and fall as a result. One such barrier might be a large mountain. Air forced up the side of a mountain becomes increasingly cold and dense with increasing elevation (Figure 8.4).

Vertical movement takes place due to processes of heating and cooling and due to the mechanical turbulence associated with natural or artificial topographic features. The degree of instability at any point of time and space is a function of how much temperature reduces with height relative to the appropriate adiabatic lapse rate. Generally, the **adiabatic** lapse rates describe the rate of transfer of energy between different parcels of air; the rate at which temperature decreases in a rising (expanding) air parcel. If the upward movement of air does not release moisture through condensation then the energy used is equivalent to a temperature fall of 9.8°C per km of vertical distance travelled (the dry adiabatic lapse rate). If condensation is produced (which, eventually, it invariably is), latent heat is released which counteracts the rate of temperature loss (see p. 46). Given this, saturated air has a rate of temperature loss with height that is lower than dry air (the saturated adiabatic lapse rate). Since air at different temperatures has a different capacity to retain moisture, this rate varies whereas the dry rate is a constant. This can be as low as 4°C per km for warm tropical air parcels and as high as 9°C per km for very cold air parcels such as you might find in the arctic. The environmental lapse rate is the actual rate of change observed in a particular area. Actual temperature change with height can be far in excess of the idealized rate and is termed superadiabatic.

> **adiabatic**: occurring without a physical transfer of heat between an air parcel and its surroundings

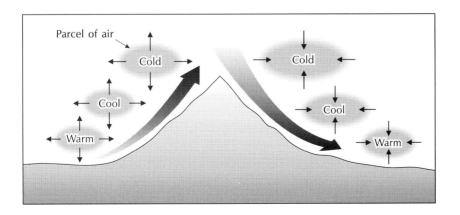

Figure 8.4 Processes of vertical movement of air parcels because of topographic features. *Source:* From *Essentials of Meteorology: Invitation to the Atmosphere*, 2nd edn, by Ahrens © 1998. Reprinted with permission of Brooks/Cole, a division of Thomson Learning: www.thomsonrights.com. Fax 800 730-2215

latent energy: energy associated with the change of state of a substance, for example liquid to solid (water to ice)

geopotential energy: energy associated with gravity

kinetic energy: energy associated with movement

advection: the lateral (horizontal) transfer of air

Given solar energy input as the ultimate driving force, this section has identified internal (heat) energy from the movement of air molecules and **latent energy** associated with the transfer of water from gas to liquid phase. Two other types of interrelated energy are also important: **geopotential energy** (associated with gravity) and **kinetic energy** associated with air movement. The receipt of heat across the globe varies enormously and there is much movement of air laterally across the globe in order to transfer energy from one part to another and keep the globe in a thermal equilibrium. This must occur otherwise we would experience the poles getting colder and colder and the tropics getting warmer and warmer. Heat from the equator is shed and distributed to the poles in order to achieve this balance, so the poles can be considered as a global sink for heat energy. On average, about two thirds of this horizontal transfer (**advection**) occurs in the atmosphere either through the physical movement of warm air masses (sensible heat) or through the condensation of water contained in air thereby releasing latent heat energy. The strength of the transfer depends on season and latitude.

Diurnal patterns of temperature and their influence on vertical motion

The maximum solar radiation received at any point on the globe occurs at noon. However, this is not usually the hottest part of the day. Instead, it is common for the maximum temperature to be reached in the afternoon. This is due to the time lag between energy being received and the air being warmed. This warming takes place because of convective vertical movement, as described above, or forced convection through mechanical mixing of air parcels on windy days. Air is not a very efficient substance through which to transmit heat – it does this best when helped by unstable air and least well when conditions are stable and calm. The degree to which this energy is translated into heat to warm the atmosphere also depends on other factors, such as cloudiness and the proximity of water bodies. During the day, clouds act to prevent energy penetrating to the surface (they reflect much of the light they receive back out to space). This will then tend to reduce the maximum temperature during the day (although this is a generalization as specific air temperatures are affected by other parameters too). During the night the opposite may happen in that clouds help 'trap' in heat that is emitted by the Earth (i.e. which has been absorbed by various surfaces during the day) making the night warmer than it would be without the clouds (where the loss of heat back out to space would be faster).

So, during the night the Earth's surface loses the energy it has absorbed over the course of a day. The air also loses the energy that it has absorbed. Since air is a less efficient conductor of energy, it loses it more slowly than the surface. By extension, the surface gets cooler quicker than the surrounding air. Some of the energy from the air is re-absorbed by the surface but then quickly re-emitted. By early morning, the surface is cold and the layers of air closest to the surface are also cold, overlain by air that is relatively warm since it has not managed to lose as much energy as that in contact with the surface. This process can be thought of as the opposite to the heating effect during the day (but with very little vertical movement of energy). This relative increase in temperature in the very lowest parts of the atmosphere is called a **radiation inversion** (or nocturnal inversion). This happens most noticeably when there is little mechanical

radiation inversion: where heat loss from the Earth causes temperatures to rise with increasing height rather than falling with height as is more usual

turbulence (wind) forcing the mixing of air. As the cold air at the surface is denser than the air above it, it will tend to stay at the surface (assuming that nothing forces it to move). During the long nights of winter, there is a longer time period for this process to take place and therefore more of the air is able to lose energy and cool. Under these conditions, and with dry air, the band of cool air can extend to several metres before relatively warm air is reached (Figure 8.5). If the air has moisture in it then when the **dew point** is reached some of the moisture will condense, which in turn (as we have seen) releases latent heat, and therefore passes energy back to the atmosphere so that it cools less quickly. The occurrence of these nocturnal inversions (and other inversions caused by meteorological conditions) can be

> **dew point**: the temperature at which saturation of the air occurs

important in terms of urban air quality since any pollutants emitted into this cold band of air cannot disperse effectively, because vertical movement from relatively cold air below to relatively warm air above is greatly inhibited. It is only when the surface heats up and allows the lowest parts of the atmosphere to warm up that the inversion begins to break up and the pollutants can disperse.

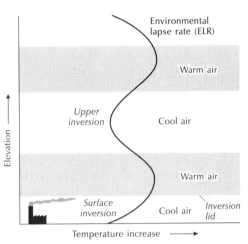

Figure 8.5 Characteristics of temperature inversions

HORIZONTAL MOVEMENTS IN THE ATMOSPHERE

At the beginning of this section, it was noted that, although vertical movement is very important, horizontal movement leads to most of the transfer of air around the globe and the associated transfers of energy. Barry and Chorley (2003) identify four controls on the lateral or horizontal movement of air: the pressure gradient force, the Coriolis force, centripetal acceleration forces and frictional forces. The laws of horizontal motion associated with these are described below (Barry and Chorley 2003; Ahrens 1998):

1. *Pressure gradient force* – Differences in pressure due to thermal or mechanical mixing of air masses. Movement occurs from high pressure to low pressure and would take place along lines of equal pressure were it not for the other forces discussed below. Where there is a steep change in pressure this movement is fast thereby giving rise to high winds. Where there is less difference then the process is slower and results in more gentle winds. It is differences in pressure that weather forecasters are referring to when they talk about tightly packed **isobars**, meaning strong or gale force winds for an area.

> **isobars**: graphical depictions (lines) denoting zones of equal pressure

2. *Coriolis force* – Air masses travelling on a spinning Earth appear to be deflected from a straight line because the area over which they travel is moving with the spinning Earth. If you were near the middle of a large spinning roundabout and tried to throw a ball for another person standing on the same roundabout but further towards the edge, you

would find that your normal judgement would cause you to miss the person and the ball would appear to be deflected from your intended path. You would miss for two reasons. Firstly, you were actually moving when you threw the ball, affecting the direction of your throw and secondly because the person you were aiming for has also since been moved relative to their original position. This acts at right angles to the direction of travel so that air masses are deflected to the right in the northern hemisphere and the left in the southern hemisphere. There is no deflection at the equator.

3. *Centripetal acceleration* – Anything travelling in a curved path has an inward acceleration that is directed towards the centre of rotation. In the northern hemisphere, a low-pressure system (known as a cyclone) bends to the left as it moves anticlockwise around the centre of the low. The reason for this is a combination of the effect of the pressure gradient causing air to move towards the area of lowest pressure; the Coriolis effect that deflects the air to the right (outwards) and finally the centripetal force that then exerts an inward pressure as the air moves around the low pressure centre. Motion is inwards as the strength of the two inward forces is greater than the strength of the outward force. In terms of high-pressure systems, the pressure gradient is an outward force, as air wants to move from high pressure to low pressure conditions. As the air moves again, it is deflected to the right and the inward centripetal force exerts an inward force to keep the air circulating. Inward forces must still predominate because the winds still circulate. In reality there is a final force to consider which gives the last piece of the jigsaw to explaining the way wind patterns behave in relation to low and high pressure systems.

4. *Friction* – In the real world some of the air circulating around an area of low or high pressure is in contact with the Earth's surface. This contact exerts friction on the travelling air that acts to slow it down. The zone above the surface that this layer influences is called the planetary **boundary layer**. The height of the boundary layer is variable in time and space from a few hundred metres under very **stable** night-time conditions with little vertical movement in winter, to 1–2 km during afternoon **unstable** convective conditions with strong vertical movements. Occasionally, the boundary layer is neither predominantly stable nor unstable and under these conditions, the effect of frictional slowing of the air travelling around an area of low pressure causes a lesser deflection due to the force of the Earth spinning. With a weakening outward force, air is carried into the centre of the low pressure and, when it reaches the centre, it is eventually forced upwards in a fountain-like motion (divergence). For high-pressure systems, it is a similar story. Friction slows the speed of the air closest to ground level and this bends the path of air trying to move from the high to low pressure less. Outward forces predominate so that air spirals outwards. As this occurs, so air from higher parts of the system is sucked down to the centre (convergence). Different types of surfaces cause different amounts of friction (or drag). The degree of drag caused by a particular surface can be quantified through a measure

boundary layer: the part of the atmosphere within which the effects of friction or drag of the surface can be felt

stable: resistant of vertical motion so if a parcel of air is lifted it remains relatively cool compared to its surroundings and tends to fall back to its original elevation

unstable: not resistant to vertical motion so if a parcel of air is lifted it remains relatively warm compared to its surroundings and tends to continue to rise itself as a result of its own buoyancy

called **surface roughness**. Surface roughness length defines the height at which 'the neutral wind profile extrapolates to zero wind speed' (Oke, cited in Barry and Chorley 1998, 108) in response to different surface types.

> **surface roughness**: quantification of the amount of friction or drag exerted by a particular type of land cover

ATMOSPHERIC CIRCULATION PATTERNS

Given the previous discussions on the laws of temperature, pressure and height, and what has been said about how vertical motion can occur in the air, on a non-rotating Earth you might expect (as did George Hadley initially) that unequal amounts of solar energy reaching the equator and the poles would lead to a redistribution of heat and energy through an enormous convection process. Warm, high pressure air would rise from the equator and travel (whilst cooling) poleward before sinking at the poles. The coldest air at the poles would be forced back towards the equator where it once again warms and the cycle continues (Figure 8.6). In fact, though, this is not quite the case. Instead, three general zones (or cells) can be generally observed. Between the equator and the approximately 30° latitude, this cycle does seem to occur (it is termed a Hadley cell). Air subsides or falls back to the surface somewhere between 20 and 35 degrees latitude due to rapid cooling and the influence of the Coriolis effect. This subsided air splits when it reaches the surface, some pushing poleward and some towards the equator (Figure 8.7). The prevailing wind patterns in the particular bands are due to the balance of the initial direction of the movement towards the equator or pole and the Coriolis effect. In the Northern hemisphere there is a

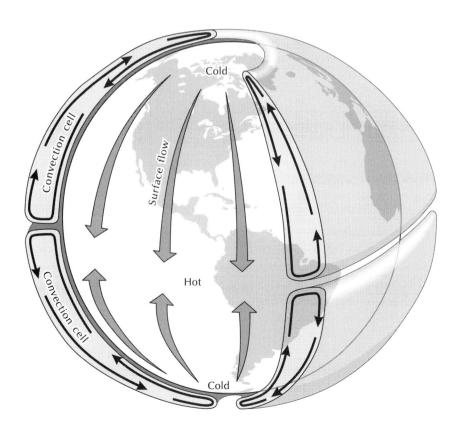

Figure 8.6 Global circulation on a non-rotating Earth. *Source:* After Lutgens and Tarbuck (1998)

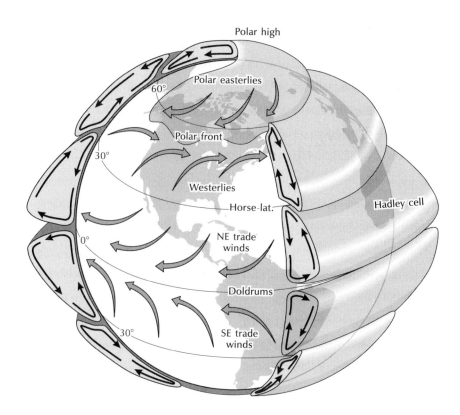

Figure 8.7 Idealized global circulation. *Source:* After Lutgens and Tarbuck (1998)

deflection to the right, and so air moving towards the poles becomes a south-westerly wind (the 'westerlies') and air moving towards the equator becomes a north-easterly wind (NE 'trade' winds). In convergence or divergence zones, where air tends to move upwards or downwards, there can be long periods with little wind e.g. the horse latitudes and the doldrums (equatorial zone). This now gives more meaning to the latitudinal patterns discussed in Chapter 2 (see p. 49).

OCEANS AND THE ATMOSPHERE

There are similar sorts of processes and forces taking place in the world's oceans, with transfers of energy taking place from one part of the globe to another. In this case the presence of land has a much stronger influence over warm and cold circulation patterns (Figure 8.8). Just as there are horizontal movements, there are also vertical cyclical processes in operation. The influence of ocean processes on weather patterns can be illustrated through the example of the Southern oscillation and the phenomenon known as El Niño (Focus 8.2). A similar process occurs in the Northern Hemisphere – the North Atlantic Oscillation. The implications of global warming on these processes are still a subject of considerable research.

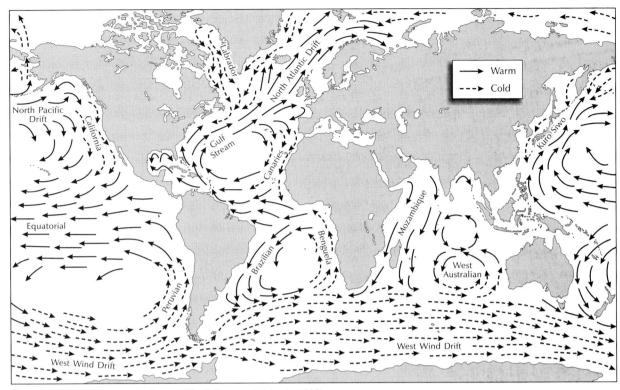

Figure 8.8 Major ocean currents. *Source:* After Lutgens and Tarbuck (1998)

FOCUS 8.2

WHAT IS THE ENSO?

ENSO refers to a regional ocean–atmosphere relationship occurring over the Equatorial Pacific Ocean (Wells 1997). El Niño refers to the oceanic component and the Southern Oscillation refers to the atmospheric component. Whereas much atmosphere–ocean interaction takes place over relatively short timescales (for example over the course of months and seasons), this phenomenon takes place on average every four years. Having said this, it is an irregular occurrence that, in reality, takes place anytime between two and seven years.

The 'usual' El Niño involves the oceans near to Ecuador and Peru undergoing a warming around December or January. Most years there are cool waters in the area of the ocean just off the coast due to the influence of upwelling currents. This cool water is rich in nutrients and on warming at the surface has ideal conditions for the growth of phytoplankton that supports the fisheries in the area. However, during ENSO events, the upwelling of water from cool currents does not occur; instead, warmer, less nutrient rich currents influence the water in the area. Consequently, the fisheries experience significantly reduced catches and there is a considerable knock-on effect on the wider economies of the affected nations.

These events have an influence on the atmosphere in that there must be a response to the different oceanic warming effects in ENSO years through changes to the surface pressure gradient in the region (and hence the south-east trade winds). During normal years, the surface pressure over Australia and Indonesia is low and there is high rainfall, but over the south-east Pacific the surface pressure is high and there is low rainfall. The result is strong south-east trade winds. During El Niño years, the tradewinds are weaker. This is because the surface pressure is higher than usual over North Australia and Indonesia (with relatively low rainfall) and the surface pressure is lower than usual (with relatively high rainfall) in the south-east Pacific area.

The overall impact of ENSO, although most strongly felt in the Pacific region, affects circulation systems more widely and even the global atmospheric system as a whole. For example, during El Niño years, winter temperatures in the west of Northern America tend to be relatively warm. The phenomenon, together with similar ocean–atmosphere relationships in the northern hemisphere, is still the subject of much research activity.

8.3 THE COMPOSITION OF THE ATMOSPHERE

Chapter 2 gave an overview of the composition of the atmosphere. This section will look at some of the specific components in a little more detail and identify the environmental issues with which they are each associated.

GREENHOUSE GASES

Although not commonly discussed within the general climate change literature, the most abundant greenhouse gas is in fact water vapour (Barry and Chorley 2003). Clearly, water vapour is an essential component of the atmosphere and it is important not only because it is a greenhouse gas but also because of its role in the hydrological cycle. The amount of water vapour present in the atmosphere varies with location within the range of 1 per cent and 4 per cent and it is concentrated in the part of the atmosphere that is in direct contact with the Earth.

However, there are other gases which are both natural (e.g. carbon dioxide (CO_2), methane (CH_4) and nitrous oxide (N_2O)) and human-made (e.g. HFCs, SF_6 and PFCs) that also act as greenhouse gases. The natural greenhouse gases, and carbon dioxide in particular, are essential in this role, but the artificial greenhouse gases are not. It is also important to note that each greenhouse gas has a different capacity for absorbing heat and this is known as its **global warming potential** (expressed relative to CO_2 over a 100-year time horizon). This means that a relatively small quantity of emissions of one type of pollutant can have a much larger effect compared to the same amount of mass emission of another species. Global warming has a global scale, long-term impact but its management requires action at local and regional scales (that is, the sources and sinks of greenhouse gases). It has been (and continues to be) difficult to get international consensus on measures to reduce anthropogenic emissions of greenhouse gases simply because of this fact. It has been one of the successes of the sustainability movement that real efforts are now being made to try to tackle this problem, epitomized by the phrase 'think global, act local' associated with Agenda 21. To explain further the complexities of this issue of managing the global problem of the accelerated greenhouse effect and global warming, the characteristics and sources of the main greenhouse gases are now examined in turn. Climate change itself is discussed in more detail later in the chapter, where some of the essential impacts will be examined.

> **global warming potential**: the relative capacity of a substance to absorb heat energy and thereby warm the atmosphere, usually expressed relative to the warming effect of carbon dioxide over a century

Carbon dioxide

In its pure form and at environmental temperatures, carbon dioxide is a colourless, odourless, non-flammable gas that is slightly denser than air. Since it is slightly soluble, it reacts with water to form carbonic acid (this is why rainwater is weak carbonic acid). Pure CO_2 is used as a coolant, a preserver, in fire extinguishers and in carbonated drinks. In a cooled solid form it is used as 'dry ice' (cardice) (Environment Agency 2003a).

Average global concentrations of CO_2 are currently around 0.037 per cent by volume and have increased by around 0.009 per cent by volume from pre-industrial times. Although at high concentrations (e.g. in the occupational environment), it can have a detrimental impact on health, the main rationale for the control of CO_2 is in view of its importance as a greenhouse gas. It is included under the UN Framework Convention on Climate Change (Kyoto Protocol).

The most important sources of CO_2 are natural. It is formed from the respiration of plants and animals and consumed by photosynthesis by plants (a process largely in balance). Other natural sources include volcanoes and forest fires. Globally, human-made sources contribute around 4 per cent of total CO_2 emissions, predominantly from fossil fuel combustion. In the UK, carbon dioxide represents almost 80 per cent of the total UK contribution to global warming emissions (Environment Agency 2003a).

Within the context of controlling global warming, natural ecosystems are considered as important carbon sinks (see p. 224 and Focus 8.3). The carbon cycle is a very useful starting point for beginning to appreciate the sources, pathways and sinks of carbon. It has also been recognized that ecosystems will respond to elevated CO_2 by increasing rates of development and overall biomass through generations (Kohut 2003). A study of the impact of elevated CO_2 concentrations (200 and 700 ppm) on *A. thaliana* over five generations of exposure, showed increasing trends in terms of biomass, rates of development and quantity of seed produced (Ward *et al.* 2000 cited in Kohut 2003). However, as Kohut (2003) notes, the long-term effects of gradually increasing concentrations on plant and animal ecosystems and their genetic diversity are currently unknown.

FOCUS 8.3

CARBON DIOXIDE ISSUES AND TRENDS

Sources
Around 75 per cent of emissions of CO_2 from anthropogenic sources over the last 20 years are from the combustion of fossil fuels. The remainder is the result of land use changes and in particular deforestation. These emissions occur due to the release of carbon that has been stored up in the fuel, land and vegetation – in the case of fossil fuels this carbon had, over geological time, been effectively removed from the system, but due to human activity has now been reintroduced. The top ten emitting countries account for two thirds of all emissions (Hardy 2003). The main contributor in terms of absolute emissions and per capita (per person) emissions is the United States. These countries use their own, and other countries', carbon stores for energy for light, transport and heating/cooling systems (see Figure 8.9).

Sinks
The land and oceans take up some 50 per cent of these emissions. The deep water of the ocean contains some 38,100 Gigatonnes of carbon (GtC). Terrestrial vegetation and soils are the next most important

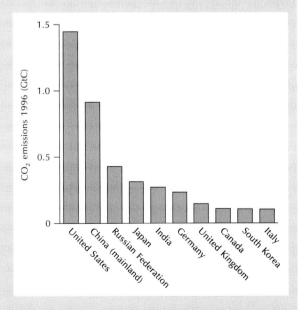

Figure 8.9 The top ten emitters of carbon dioxide. *Source:* After Arland *et al.* (2002)

(2190 GtC). The atmosphere holds 750 GtC. (See Figure 8.10 for details of quantities of carbon associated with different processes within the carbon cycle.)

Concentrations

Since the middle of the eighteenth century, CO_2 concentrations have increased by almost one third (Figure 8.11). This is the highest that it has been for 420,000 years. Rates of change are quicker than at any point in at least the last 20,000 years. Average rates of increase are about 1.5 ppm per year over the last 20 years, but with variations year on year due to climatic processes and their influence on the efficiency of sinks. The Scripps Institution of Oceanography has been monitoring concentrations of carbon dioxide at the Mauna Loa observatory on Hawaii since 1958. It is the longest continuous record that exists. The data reveal

an indisputable rise in global background CO_2 over time (Keeling 2003).

Implications

The trend in CO_2 concentrations is mirrored by those exhibited by other key greenhouse gases. Similar trends are observed in global temperatures and exactly what the impacts will be are unknown, although we now have some better ideas of what they may be. In the view of the Inter Governmental Panel on Climate Change (IPCC) – comprising some thousand scientists around the globe – the evidence strongly suggests that these trends are not natural and that the activities of human society are the root cause.

(See IPCC (2001), Keeling (2003) and Hardy (2003) for more details.)

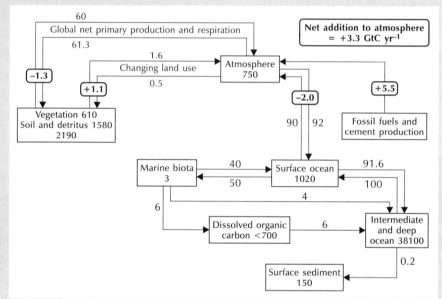

Figure 8.10 The importance of different parts of the carbon cycle for the exchange of carbon. *Source:* After Schimel *et al.* (1996)

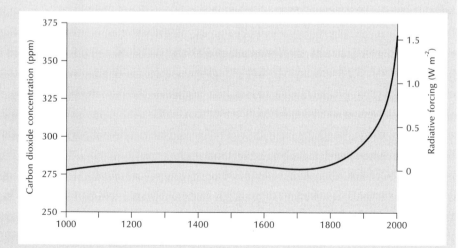

Figure 8.11 Global trends in concentrations of carbon dioxide. *Source:* After IPCC (2001)

Methane

Methane is a colourless gas and is odourless at low concentrations. It is highly combustible and can be used as a source of energy. It is produced by the anaerobic decomposition of organic matter (i.e. without the presence of oxygen). Accordingly, it builds up in places with limited oxygen and a lot of organic waste. Two good examples are landfill sites and the insides of animals, and these represent two key sources of methane gas to the atmosphere. Indeed, if landfill sites are not properly ventilated then concentrations of methane can build up to explosive proportions. In the UK, agriculture and landfill sites were two of the most important sources of methane, the former contributing almost half of the total emissions (Environment Agency, 2003b). From this it can be seen that sources are both natural and human-made and this is reflected in the global source split. The most important sources are livestock farming (around 30 per cent), oil and gas production (around 16 per cent), coal mining (around 13 per cent), wet rice cultivation (around 25 per cent) and solid waste (around 16 per cent) (based on 1996–97 figures from the World Resources Institute cited in Hardy 2003).

Over the last 200 years, concentrations of methane in the atmosphere have doubled (Figure 8.12). This is of particular concern as the warming effect associated with methane is over 20 times that of carbon dioxide over the same period. In view of this, emissions of the gas are now the subject of international agreements (Kyoto Protocol).

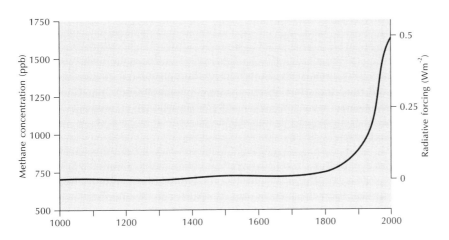

Figure 8.12 Trends in methane concentrations. *Source:* After IPCC (2001)

Nitrous oxide

In its pure form and at environmental temperatures, N_2O is a colourless, non-flammable gas with a slightly sweet taste and odour and slightly soluble in water. Owing to its anaesthetic properties, it was historically used for pain relief in medicine and dentistry and also for recreational purposes ('laughing gas'). It is also used in the dairy industry as a mixing and foaming agent where it inhibits the growth of bacteria in products (Environment Agency 2003c).

In view of its uses as an anaesthetic, N_2O can have detrimental impacts on health at high concentrations, e.g. in the occupational environment. However, the primary reason for control of N_2O is due to its role as a greenhouse gas. Concentrations of nitrous oxide have shown a similar upward trend to the greenhouse gases already discussed (Figure 8.13). Although the mass emissions of

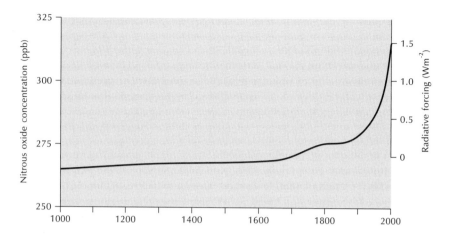

Figure 8.13 Trends in nitrous oxide concentrations. *Source: After IPCC (2001)*

> *emissions inventory*: a schedule of sources and associated mass emissions of pollutants for a particular area over a particular period of time

N_2O are relatively small, the impact it has on global warming is much larger, 310 times that of CO_2 over 100 years. It is the third most important greenhouse gas in the UK. It is also worth pointing out that nitrous oxide also depletes stratospheric ozone if it reaches the upper atmosphere. Given these characteristics, it is controlled through the UN Framework Convention on Climate Change (Kyoto Protocol) and United Nations Economic Commission for Europe (UNECE) Convention on Long Range Transboundary Air Pollutants (LRTAP).

As with the other major greenhouse gases, it is produced by natural and human-made processes. The soil and water microbial processes of nitrification and denitrification produce it naturally. These processes are enhanced by the application of fertilizer to agricultural soils, making this a significant human-made source. Other human-made sources include industrial processes, power generation and road transport (Environment Agency 2003c). The Global **Emissions Inventory** Activity project (GEIA) identify nine key sources: natural and agricultural soils; animal excreta; deforestation; oceans; industry; fossil fuel combustion; biofuel combustion; agricultural waste combustion and biomass wasting (Kroeze 2003). The GEIA provides maps of the spatial distribution of many of the pollutants of interest on a 1° by 1° latitude and longitude basis. Further information on the project and its results can be found at www.geiacenter.org.

Nitrous oxide is a compound of nitrogen and as such is a component of the nitrogen cycle. As with carbon it is therefore useful to map out the processes and pathways associated with nitrogen in the atmosphere as a means of getting a full picture of the interactions that take place (see pp 31–32). Note that there are many forms of nitrogen in the atmosphere and these can have an influence on different aspects of the atmosphere. Nitrous oxide is a greenhouse gas but there are other reactive species that have an acidifying effect on the atmosphere.

Hydrofluorocarbons

HFCs are a large group of gaseous compounds that all contain carbon, fluorine and hydrogen and have a similar composition. These are human-made products which do not occur naturally in the environment. They have become important, and are now emitted to the atmosphere, because they are used as a replacement for HCFCs that are linked to the depletion of stratospheric ozone (see Chapter

2). Although the specific characteristics of each chemical do vary, as a group they are generally relatively unreactive, colourless and odourless gases. In view of their replacement to CFCs and HCFCs, they are used as coolants in refrigerators and in air conditioning equipment. They are also used as propellants in industrial aerosols and in Metered Dose Inhalers (in the treatment of respiratory problems such as asthma). They can also be used on a more minor scale for the production of packaging foam, solvent cleaning and fire extinguishing equipment (Environment Agency 2003d).

Some HFCs can cause severe detrimental effects to human health in high concentrations (some species are categorized as hazardous Volatile Organic Compounds (VOCs) by the Environment Agency). The whole group of substances are also controlled in view of their impact on global warming, estimated to contribute some 2.2 per cent of the UK's global warming total. Actual global warming potentials vary widely but are at least 100 times more potent than CO_2 over 100 years (GWPs range from 140–11,700) (Environment Agency 2003d; Salway *et al.* 2003). As relatively stable compounds, they have a long lifespan in the atmosphere. They are controlled under the UN Framework Convention on Climate Change (Kyoto Protocol). In view of the range of compounds in this category and the different molecular weights associated with each, mass emission statistics are expressed in terms of CO_2 equivalent. This is an important point to note when considering any quantification of the mass emissions of chemical compounds. It reflects the importance of units and making sure that when making comparisons these comparisons are actually valid.

Perfluorocarbons

As with HFCs, PFCs are entirely human-made and do not occur naturally in the environment. PFCs are a group of generally odourless, colourless and non-flammable gaseous compounds containing just carbon and fluorine. They are used with HFCs in refrigerants and have increased in usage with the decline in use of chlorinated species. Other uses include aluminium production and, to a lesser extent, fire extinguishing systems and medical and cosmetic applications. The electronics industry is an important user of PFCs where compounds are used in the manufacture of semiconductors, and is actively seeking alternatives as well as following stricter recycling and recovery processes (Environment Agency 2003e). In addition to the uses of PFCs indicated above there are also emissions sources associated with the manufacture of compounds.

The environmental impacts of PFCs are generally more severe than HFCs. The former have global warming potentials that range from 6500 to 9200 relative to CO_2 over 100 years (Salway *et al.* 2003). Fluorinated organic compounds have also been linked to the occurrence of trifluoroacetate (TFA) in rainwater that is harmful to plant life (Environment Agency 2003e). PFCs are currently controlled under the UN Framework Convention on Climate Change (Kyoto Protocol). As with HFCs, in view of the range of compounds in this category and the different molecular weights associated with each, mass emission statistics are expressed in terms of CO_2 equivalent.

Sulphur hexafluoride

Sulphur hexafluoride is entirely human-made and not found naturally occurring in the environment. In its pure form and at environmental temperatures, SF_6 is a colourless, odourless, non-toxic and very dense gas. It is also very stable chemically and highly inert. These properties have led to its use as an electrical insulator in electrical switchgear and within the magnesium smelting processes. It is also used in trainer manufacture for cushioning, although since its negative environmental consequences were discovered its use is being phased out. The main sources of SF_6 are therefore associated with these processes and activities and, despite a high level of recycling, in its manufacture (Environment Agency 2003f).

At high concentrations, SF_6 can have detrimental impacts on human health, but it is principally controlled in view of its role as a greenhouse gas. Indeed it has the highest known global warming potential, it is some 23,000 times more potent than CO_2 over 100 years, and has a very long atmospheric lifetime due to its very stable and inert properties. Fortunately, its emissions are much lower by mass than the other six main greenhouse gases in the UK; for example, it contributes only around 0.2 per cent of the total UK contribution to global warming. Reporting for this pollutant is required under the UN Framework Convention on Climate Change (Kyoto Protocol).

REACTIVE GAS SPECIES

Greenhouse gases have a particular role to play in the atmospheric system and a particular set of impacts. Fundamentally, as we have seen, this relates to their capacity to absorb outgoing energy and therefore warm the atmosphere. This warming then has other impacts on weather and climate (and other Earth processes) that human beings will have to adapt to over the coming years. This is not the only group of components of the atmosphere that have affected it. There are also reactive gas species that have an impact through chemical reactions with the natural and other components of the atmosphere, sometimes producing compounds that have a detrimental effect on humans, animals and plants.

While there are trace gases in the atmosphere that are inert and which do not change over time (such as argon, neon, krypton and xenon), most of them will react with the other constituents of the atmosphere to form new species or to catalyse 'natural' reactions. Reactive gas species can affect the environment at all spatial scales, from global and continental through to regional and local. One notable process associated with reactive gas species is acidification. Acidification has impacts not only on the atmosphere itself but also on the other spheres covered in this book.

Since anthropogenic emissions have an influential input into processes associated with reactive gas species, they are subject to a range of source controls and some of these are outlined later in the section. However, it should be realized that there are also significant natural sources of some pollutant species and there may be few, if any, opportunities to influence them. Volcanic eruptions, for example, give rise to enormous quantities of sulphur dioxide being emitted into the atmosphere, yet clearly there is not much that can be done about it. See Focus 8.4 for a focused discussion on selected natural and human-made sources.

It is not possible in a chapter of this length and breadth to consider all the multitude of reactive gas species present in the atmosphere. Fenger (1999) notes that to date there have been almost 3000 anthropogenic pollutants identified, 500 of which are associated with motor vehicles alone. Fenger (1999) also notes

FOCUS 8.4

FOCUS ON EMISSIONS SOURCES

This box provides an overview of two contrasting sources of atmospheric pollutants, one natural and one anthropogenic. Some emissions are more easily controlled than others but an appreciation of all types of sources is required in order to understand why the composition of the Earth's atmosphere is as it is. It is important to recognize the role of large natural sources in contributing to environmental problems as well as the contribution of human society. The two sources are volcanoes and transport.

Volcanoes

Volcanoes are hugely important sources of atmospheric pollution. The main emissions from volcanoes contain sulphur, in particular sulphur dioxide and hydrogen sulphide (Andres and Kasgnoc 1998), although they are also important sources of chlorine compounds (RCEI 2003). Different volcanoes exhibit different types of emission behaviour depending on their activity patterns. Volcanic sulphur (S) emissions typically have a tropospheric lifetime in the order of a few weeks. Although these emissions can ultimately help through adding nutrients to the soil near the volcano, in too large quantities they may also have an acidifying effect. They can also have other regional impacts, sometimes affecting temperature and precipitation patterns (Andres and Kasgnoc 1998). Researchers in Taiwan have correlated S-rich eruptions over the last century with the incidence of drought in Taiwan (Li et al. 1997). Others have made links between volcanic activity and diurnal temperature ranges in the south-eastern United States (reduced maximums and increased minimums following emission events) (Saxena et al. 1997). The degree of impact that volcanoes have depends on three main parameters: the quantity of SO_2 emissions, volcano latitude and volcano altitude. Since emissions are associated with high temperatures and pressures they are more likely to be able to penetrate into the stratosphere. Here SO_2 is converted to sulphate aerosol (solid particles) that has a number of global impacts, for example on albedo, atmospheric warming, other chemical reactions and cloud properties (Andres and Kasgnoc 1998). Table 8.1 shows the SO_2 flux (mass emission of SO_2 per unit time) associated with the world's top five continuously emitting volcanoes (taken over a 25 year period).

Transport

All forms of transport that use fossil fuels (either directly, e.g. through the internal combustion engine, or indirectly through the use of electricity) give rise to emissions of atmospheric pollutants and have influenced acidification, ozone depletion and climate

Table 8.1 The five largest continuously emitting volcanoes in terms of sulphur dioxide (tonnes per day)

Volcano	SO_2 (Mg per day)
Etna	4000
Bagana	3300
Lascar	2400
Ruiz	1900
Sakura-jima	1900

Source: Data compiled from Andres and Kasgnoc (1998)

change to some degree (Colville et al. 2001). Although the specific nature of emissions and their relative quantities are a function of fuel type, engine type, operating conditions and controls, it is possible to make some generalizations. By mass, the most significant emissions are carbon dioxide and water vapour – products of complete combustion of fuel. However, most engines do not ensure that there is enough air in the system to result in complete combustion of all of the available fuel, and so a small fraction remains only partly oxidized so that carbon monoxide, volatile hydrocarbons and carbonaceous particles are also emitted. Furthermore, since all common fuels have trace impurities, these are also released, so any sulphur present may be oxidized to sulphur dioxide or converted to sulphate. At the high combustion temperatures associated with engines, atmospheric nitrogen present in air is readily oxidized as nitric oxide (NO) and some nitrogen dioxide (NO_2). Once the exhaust gases leave the engine, the nitric oxide undergoes oxidation to nitrogen dioxide too. Nitrous oxide is emitted in small quantities but has been found to be more prevalent in the exhausts of vehicles fitted with catalytic converters (Colville et al. 2001). The impacts of transportation emission sources are felt at a number of spatial scales. Since they are generally at ground or near ground sources, road transport can have very significant impacts on air quality in the immediate area of the road. Specific meteorological conditions such as inversions and/or very stable air, as well as aspects of the built environment (such as street canyons, see Vardoulakis et al. (2003) for details and a review of means to estimate their impacts) can inhibit the dispersal of pollutants and lead to concentrations that affect human health. Over time, dispersion occurs and the pollutants then have an impact at the regional scale, and eventually, for some pollutants, there will be global impacts as suggested at the beginning of this section. Aircraft are a slightly different case to surface transport since, although the products of combustion of aviation fuel (kerosene) are similar, the impacts of emitting these in the very top areas of the atmosphere are still being investigated. What is certain is that these will be having an impact and this will have consequence for the global atmosphere.

that the impacts of a mere 200 of these have actually been investigated and we know even less about their ambient concentrations. Further information on a wider selection of pollutants can be found in the first instance in the suggested further reading at the end of the chapter. Here, it is sufficient to discuss general processes in relation to the human-induced environmental issues that are the chief focus of the remainder of this section of the chapter. The reactive species groups considered in particular detail are sulphur, nitrogen and chlorine species. These relate to the sulphur, nitrogen and chlorine cycles respectively. There are also important influences associated with atmospheric particles (solid matter held in the air), but these are not considered separately in this text.

Sulphur species

The primary concern about sulphur dioxide concentrations has traditionally been in view of human health impacts and its role in the process of acidification. The sulphur cycle (see Chapter 2) illustrates sources, reactions and deposition pathways in the environment. The main pathway associated with acidification is the wet or dry deposition of sulphuric acid that is predominantly formed through oxidation of emitted SO_2.

Health effects of SO_2 are associated with the respiratory system including, for example, asthma. Levels of sensitivity vary, with some individuals experiencing effects at concentrations of 100 ppb (parts per billion) while others appear tolerant up to as much as 1000 ppb (Border 1994). As with all air pollutants it is principally the very young and very old who are the most sensitive to elevated concentrations. In addition to these impacts on humans, plants and animals, sulphur dioxide and its pollutant products also have an impact on the built environment. The UK National Materials Exposure Programme (NMEP) assessed the extent of impacts of a range of pollutants on specific materials such as stone, clothing and rubber (e.g. Butlin *et al.* 1992a and 1992b). They found that sulphur dioxide had an important role in corroding stone and metal. Together with other pollutants, historic as well as contemporary, urban pollution has led to the degradation of historical and cultural artefacts in many towns and cities across the globe.

Fossil fuel combustion is the largest anthropogenic source of SO_2. Accordingly, power stations and domestic fuel combustion remain important sources. Furthermore, shipping can be important in port areas due to the combustion of sulphur rich fuels by large ships (known as bunker). It has been estimated that as much as 5 per cent of global sulphur emissions come from shipping sources (Corbett and Fischbeck 1997). Emissions controls in both North America and Europe have seen a dramatic reduction in sulphur dioxide emissions (particularly in Europe). Cape *et al.* (2003) note reductions of 43 per cent (EMEP countries), 52 per cent (EU countries) and 17 per cent (North America) in emissions of SO_2 between 1990 and 1998. Lefohn *et al.* (1999) trace the historical trends in SO_2 emissions back even further from 1850, and note the rise and fall in emissions from individual countries across the globe.

Nitrogen species

Chapter 2 also discussed the nitrogen cycle and the various forms and pathways of nitrogen in the environment (of which there are too many to fully discuss in detail here). One of the oxidized forms of nitrogen, nitrous oxide, has been covered in a previous section and has been shown to be important as a greenhouse gas due to its role in depleting stratospheric ozone.

Nitric oxide is readily oxidized to nitrogen dioxide in the ambient air. It is therefore a secondary pollutant rather than one that is emitted directly from a particular source (although a small proportion usually is). Nitrogen dioxide (NO_2) is a reddish-brown gas and can sometimes be seen in the atmosphere, either through releases from stacks associated with large combustion sources, or as smog in air around large urban areas. The oxides of nitrogen have far reaching effects and are influential in all the key environmental issues considered in this chapter. They also have impacts at all spatial scales from the very local, such as the build up of nitrogen dioxide in the vicinity of a busy road, to the regional, such as the deposition of nitric acid, through to global effects such as the role of NO_x in stratospheric ozone depletion and N_2O as a greenhouse gas. The major anthropogenic sources of the oxides of nitrogen are fuel combustion, biomass burning and some non-combustion related industrial processes. Natural sources include lightening (very small) and the actions of soil microorganisms.

Whereas many urban pollutants have been effectively controlled in western European countries, nitrogen dioxide is one pollutant that remains a considerable issue (AQEG 2004). A number of health related air quality standards have been developed by various organizations, such as the World Health Organization and the European Union, in order to help protect vulnerable members of society. An individual's exposure to nitrogen dioxide varies significantly during the course of a day. In a study of daily exposure patterns carried out by Sexton and Ryan (1988) peak exposure to nitrogen dioxide actually occurred as a result of meal preparation, and was therefore associated with indoor air pollution sources rather than outdoor sources.

There are targets to reduce both emissions and concentrations in the atmosphere as a means of trying to manage the problem and find workable solutions. Some of the main emissions reduction targets in the EU are shown in Table 8.2. Most European countries are successfully reducing emissions of NO_x. However, the trends in concentrations of NO_2 are proving more difficult to reduce, especially in urban areas (AQEG 2004). The full reasons for the lack of a significant response of nitrogen dioxide concentrations to reductions in estimated emissions are not yet fully understood.

Chlorine species

The environmental significance of reactive chlorine has already been touched upon elsewhere (see Chapter 2). In terms of global totals, industrial sources in 1990 accounted for only about one eighth of the amount produced by combustion processes (largely biomass burning, much of which is associated with human activities, but also including waste incineration). Natural sources are important to total global emissions and are derived from oceans and terrestrial sources. Terrestrial sources include fungi, soils and volcanoes (Graedel and Keene 1999).

Table 8.2 Air emission reduction targets in the EU

Policy/pollutant	Base Year	Target Year	Reduction (%)
UNECE/CLRTAP			
Sulphur dioxide[a]	1980	2000	62
Sulphur dioxide[b]	1980	2000	62
Nitrogen oxides[c]	1987	1997	Stabilization
Nitrogen oxides[b]	1987	1997	Stabilization
Non-methane volatile organic compounds[d]	1990	2110	30
Non-methane volatile organic compounds[b]	1990	2110	58
Ammonia[b]	1990	2010	12
Fifth Environmental Action Plan			
Sulphur dioxide	1985	2000	35
Nitrogen oxides	1990	2000	30
Non-methane volatile organic compounds	1990	2110	58
Directive on National Emissions Ceilings (NECD) (proposed)[e]			
Sulphur dioxide	1990	2010	78
Nitrogen oxides	1990	2010	55
Non-methane volatile organic compounds	1990	2110	62
Ammonia	1990	2010	62

Notes

[a] from 1994 Sulphur Protocol overall emission reduction (varies between individual members

[b] from multi-pollutant 1999 Gothenburg Protocol (same across all member states)

[c] first NO_x protocol

[d] from NMVOCs protocol

[e] from National Emissions Ceiling Directive, different ceilings for individual member states

Source: After Erisman *et al.* (2003)

8.4 HUMANS AND THE ATMOSPHERE

This chapter has already highlighted and explained how humans influence the atmosphere, particularly in relation to influences on the composition of the atmosphere. This final section brings together this information with the other aspects of the atmosphere discussed, in order to focus on three key environmental issues; air quality, acidification and climate change.

AIR QUALITY: A SYNTHESIS

Much of this chapter has made reference to processes and ideas that are important in understanding what air quality is, what influences air quality and what air quality issues face us in the future. The sections on individual pollutants have helped to show the types of issues associated with each. However, in reality, all of these impacts and processes occur together. The cycles of sulphur and nitrogen are interlinked and the presence of one pollutant may either lead to the creation of another or help catalyse reactions between other pollutants. Furthermore, it is also true that sometimes the presence of one pollutant may actually increase the impact of another so that the combined effect is greater.

Urban air is generally more polluted than the air in the less built-up zones that surround it. Cities are major sources of many air pollutants and these can be emitted, dispersed and deposited without ever leaving the city boundary. There are particular meteorological conditions that occur in and around cities that also play a role. This, and the fact that cities have many complex structures associated with them, can make understanding and modelling air quality very difficult. Nevertheless, since most of the globe's population live and work in the city, these issues need to be tackled. There are many books and papers dedicated to urban air pollution and urban air quality issues (see, for example, Elsom (1996) or Fenger (1999) as a starting point).

The point has already been made that ozone in the troposphere has negative effects on human health and vegetation. It is formed through reactions between the oxides of nitrogen and various volatile organic compound species that are released from industry and transportation sources in the presence of sunlight, and so is termed photochemical smog. It is therefore a secondary pollutant and one that has an impact over large areas, not necessarily in the area in which the precursors are emitted. Given this, cities can pollute air that eventually leads to high ozone concentrations on the outskirts of that city and beyond into the countryside.

Sources and impacts can be very different in different parts of the world. For example, in Australia and New Zealand emissions from barbecues, lawnmowers and pleasure boats can create localized air quality problems. Here nearly 90 per cent of the population is concentrated into 5 per cent of the landmass, mostly along coastal regions (Crabbe 2003).

Baldesano and Jimenez (2003) present an interesting analysis and cross-comparison of air quality between 1990 and 2000 in some of the world's megacities (São Paulo, New York, Mexico City, London, Paris and Los Angeles). They also relate their findings more widely to the current air quality problems experienced by other large cities across the globe. Their results can be summarized in terms of specific pollutants as follows:

Ozone

All the examined megacities, except London, in 2000 exceeded the World Health Organization (WHO) hourly average Air Quality Guideline value (based on health) for ozone. The analysis revealed that of the six cities examined, Mexico City had the highest average hourly concentrations, some 2.5 times the WHO guideline. São Paulo and New York were seen to have increasing trends, but most other cities showed a downward trend over the same period. The study also pointed out that this problem seems to be equally important across the globe, regardless of general levels of economic development.

PM_{10} (Particulate matter with a diameter of less than 10 micrometers)

All examined megacities exhibited a downward trend, although some (Mexico City and São Paulo) still had annual average concentrations that were higher than the WHO annual average guideline. There were large exceedences of the WHO guideline in most Asian cities.

Sulphur dioxide

By the year 2000, all the megacities considered had concentration values lower than the WHO mean annual concentration guideline for sulphur dioxide. The fall in annual average concentrations was particularly marked in Mexico City between 1990 and 1995. It was noted by the authors, however, that other cities in South-East Asia and South America did still have high concentrations and concentrations that were greater than the WHO guideline.

Nitrogen dioxide

Annual average concentrations for all the surveyed cities exhibited a downward trend, although all also remained higher than the recommended WHO annual average nitrogen dioxide guideline in the year 2000. Some other cities actually had values that were three times higher than the recommended maximum annual concentration (for example, Kiev and Beijing).

ACIDIFICATION

The concern about acidifying species, such as SO_2 and NO_2, and their reaction products was a response to observed impacts on plants, animals and humans linked to the combustion of fossil fuels. Concern initially arose during the industrial revolution in western Europe and, following economic development, has since become important in many other parts of the globe. Two early studies cited in Mansfield and Lucas (1996) made the link between increasing urban pollution (largely SO_2 and black smoke from domestic and industrial coal combustion) and reduced growth and variety of vegetation in the cities of Leeds and Manchester in England at the turn of the nineteenth century (Cohen and Ruston 1912; Pettigrew 1928). These localized effects were 'controlled' through legislation that encouraged a 'dilute and disperse' solution. Urban areas increasingly became smoke (and sulphur) free, eventually with power generation being concentrated in an increasingly small number of very large point sources. However, by the 1960s it became clear that the solution of increasing chimney heights to disperse pollutants over wider areas merely shifted the problem to a continental scale and the northern European countries increasingly saw the impact of their 'dirty' neighbours. There was much scientific work and political debate to establish a link between sources and impacts. Two particular developments were very influential during this period. The first was the use of back-trajectory models to trace parcels of air that arrived in one area back in time and space to where they originated, thereby enabling cause and effect to be linked. The second was the development of the concept of critical loads – the amount of deposition of acidic (or in fact any) compound that could be handled by the environment without any observed detrimental effect. Critical loads could be estimated for soils, water, vegetation and their related ecosystems and enabled a quantification of the amount of reduction in emissions needed to protect the most vulnerable areas.

Eventually, in the 1980s, there was broad consensus about the problem and its causes followed by a concerted effort by the industrialized nations to try and counteract observed acidification responses in lakes, forests and associated ecosystems. The European Union (EU) imposed stricter controls over large combustion sources, such as power stations, to try to reduce the overall emissions of sulphur dioxide from member states. Linked to this, but covering a larger area (the 15 member states of the EU, Eastern Europe and the Scandinavian countries), the European Monitoring and Evaluation Programme (EMEP) helped through bringing in sophisticated modelling techniques to help establish the extent of the problem and the best means of attaining solutions. The EMEP programme was also influential in helping develop much contemporary knowledge about emissions sources and releases, for example, through the development of the European Emissions Inventory Guidebook, which brings together relevant information for the whole of Europe and which promotes consistency of approach through the nation-states. At the same time, similar pollution issues were being handled in the United States through the Environmental Protection Agency.

It has been suggested that there is an evolution of the importance of different environmental concerns with stages of development. Given this, it is not surprising to find out that there are now major acidification concerns in the developing world (Emberson *et al.* 2001; Kuylenstierna *et al.* 2001) even as the developed world is beginning to see a recovery in some of the ecosystems affected by previous acidification events. This process is not automatic and recovery, if it occurs at all in some areas, is likely to be a slow process. There are monitoring programmes in place to help assess the degree to which a recovery is underway and what the main issues that might prevent a recovery are (Gunn and Sandoy 2001).

CLIMATE CHANGE

Climate change is a recurring theme in many chapters of this book since it is so clearly influential on many of the Earth's spheres. Since this is primarily an atmospheric issue, this section will provide an overview of the changes to the climate that are expected and give a very brief review of the key impacts that are anticipated as a result. These impacts are covered in more detail elsewhere in the book and the drivers associated with human society are discussed further in Chapter 15.

The issue

The upward trends in the emissions of greenhouse gases, the significance of these gases and their rising concentrations have already been covered in some depth. What then has been the Earth's response to date and what do we expect the future to hold?

Observations of a warming world

The IPCC reports that the twentieth century has seen a rise in global average surface concentrations by about 0.6°C (±0.2). This figure masks a high degree of variability with time but the trend is indisputable. Surface concentrations have increased most markedly but measurements also suggest that warming has occurred up to around 8 km (IPCC 2001).

Other observations suggest that there have also been changes in precipitation rates. IPCC (2001) work suggests that mid and high latitudes of the northern hemisphere have experienced a 0.5 to 1 per cent increase per decade (together with increased frequency of heavy precipitation events) and tropical areas (between 10° latitude north and south) in the region of 0.2–0.3 per cent during the twentieth century. Between 10° and 30° north there have been some reductions in precipitation rates. No discernable trends were found in the southern hemisphere. The mid to high latitudes have seen an increase of cloud cover by about 2 per cent and a reduction in diurnal temperature ranges. There are fewer occurrences of extreme low temperatures and slightly more occurrences of extreme high temperatures (IPCC 2001). Warm episodes of the El Niño–Southern Oscillation have been more frequent and more intense over recent years and there has been some evidence to suggest that the frequency of drought conditions are increasing in parts of Asia and Africa.

Modelling the future

Complex modelling research suggests that there will be a further increase in the concentrations of greenhouse gases; carbon dioxide for example has been estimated to rise by 540 to 970 ppm depending on the emissions and development scenario that is taken (IPCC 2001). As a result, projected temperature changes are expected to be in the order of 1.4 to 5.8°C for the years 1990 to 2100 (see p. 195). These are much higher than those that the twentieth century have brought and are unprecedented, in terms of the rate of change, over the last 10,000 years (based on palaeoclimate data). These average temperatures will have distinct geographical variability and it is likely that there will be changes in precipitation and ocean processes. Many countries have produced their own accounts of the likely implications. The actual changes that will take place are uncertain. This is due to impacts from other processes and the presence of pollutants in the atmosphere (e.g. aerosols that have a counteracting cooling effect). However, the following extreme events have been highlighted as likely or very likely to occur during the twenty-first century (IPCC 2001).

- Higher maximum temperatures and more hot days nearly everywhere on land

- Higher minimum temperatures, fewer cold days and frost days nearly everywhere on land

- A smaller temperature range during the day

- More intense precipitation

- Increased drying and risk of drought.

The implications of these trends are dealt with in each of the specific chapters devoted to that media. Some of the changes may seem very small, but small differences can lead to quite radical changes in the atmosphere and oceans as the Earth attempts to redistribute the energy and find a new equilibrium or balance. The only certain way of finding out what will happen is to wait and see.

SUMMARY

This chapter continued from the introduction to the atmosphere given in Chapter 2 and considered the processes of movement in the atmosphere. Since the troposphere is where most of the impacts of human activity are felt, at least initially, this was the focus of the discussion. The chapter went on to consider trends in concentrations associated with different constituents of the lower atmosphere. The sources, trends and characteristics of greenhouses gases in terms of climate change, and reactive species in terms of urban pollution and acidification, were presented in some detail. The chapter ended with a broader consideration of these environmental issues, looking at air quality across the globe's megacities and the likely impacts of climate change.

Questions

1 Describe the structure of the atmosphere. How does this help with the notion that pollution is just matter out of place?
2 What are the key drivers and modifiers of atmospheric movement in the troposphere?
3 Compare and contrast air pollution problems in two locations around the globe in contemporary times. How do these differ to conditions 50 years ago?
4 Using the data from the internet, plot a graph of carbon dioxide concentrations over time. The suggested website for this work is http://ingrid.ldgo.coluimbroia.edu/sources/.keeling/.mauna_loa.cdf

5 Using the Global Sulfur Emissions Database (http://www.asl-associates.com/sulfur1.html), plot a chart demonstrating the reduction of emissions of sulphur dioxide for one location over time. What improvements for the environment and human health do you think will have happened as a result of the changes you observe?
6 Why were dilute and disperse policies not a successful means of overcoming problems associated with acidification?
7 With reference to the suggested texts and references in this section, outline the implications of climate change for your continent, your country and your nearest city.

Further Reading

The text in this chapter has drawn heavily on a number of introductory texts describing different aspects of the atmosphere. These are recommended as starting point for further study and to provide greater depth to the ideas that have been introduced in this section. Although all are introductory texts, they differ in terms of overall focus and level of detail. They are listed in order of the author's assessment of the general level, accessibility and detail of material they contain:

Crutzen, P. J. and Ramanathan, V. (2000) 'The ascent of the atmospheric sciences'. Science 290, 299–304.
A paper that summarizes the key research findings that have developed humankind's understanding of the atmosphere.

Ahrens, C. D. (1998) Essentials of Meteorology: An Invitation to the Atmosphere. Belmont, USA: Wadsworth Publishing.
A readable, richly illustrated introductory text covering the principles of the atmosphere and their relationship to meteorology.

Lutgens, F. K. and Tarbuck, E. J. (1998) The Atmosphere: An Introduction to Meteorology. Upper Saddle River, New Jersey: Prentice Hall International Inc.
Another nicely illustrated introductory text covering a wide range of issues and concepts associated with the atmosphere.

Useful Websites

http://www.Geiacenter.org
The GEIA Inventory website contains reviews of a wide range of pollutants/pollutant groups covering ozone depleters, greenhouse gases and health related pollutants. It also provides a summary of the major sources and their distributions.

http://www.uea.ac.uk/~e044/apex/global.html
Peter Brimblecombe at the University of East Anglia in the United Kingdom has an excellent collection of links to global databases.

Here are some other suggested sites from that resource:

http://www.ncl.ac.uk/airweb/
Air Pollution Effects on Plants. Photographs demonstrating the range and extent of impacts of air pollutants (covering sulphur dioxide, ozone and fluoride) on different plant species.

http://www.grid.unep.ch/fires
Global Forest Fire Data (GRID:UNEP). A record, updated every fortnight, of the principal forest fires occurring around the globe.

http://www-cger.nies.go.jp/grid-e/gridtxt/grid7.html
Matthews Global Seasonally Integrated Albedo. Data files describing estimated seasonal percentages of incoming solar radiation reflected back into space for different seasons.

http://www-cger.nies.go.jp/grid-e/gridtxt/grid5.html
Matthews Global Vegetation Data. A dataset of global vegetation types, based on dominant class, from 32 different classifications, on a $1 \times 1°$ latitude spatial resolution. This was one of the datasets used to estimate

Thompson, R. D. (1998) *Atmospheric Processes and Systems* (Introductions to Environment). London: Routledge.
A good general foundation text covering atmospheric concepts and processes in some detail.

Barry, R. G. and Chorley, R. J. (2003) *Atmosphere, Weather and Climate*, 8th edn. London: Routledge.
One of the mainstay texts on the atmosphere that takes the reader from general principles to a more in-depth understanding.

On specific environmental themes, the following are recommended:

Elsom, D. (1996) *Smog Alert: Managing Urban Air Quality*. London: Earthscan.

Fenger, J. (1999) 'Urban air quality'. *Atmospheric Environment* 33, 4877–900.

Hardy, J. T. (2003) *Climate Change: Causes, Effects and Controls*. Chichester: John Wiley & Sons.

Gribbin, J. (1998) *The Hole in the Sky*. London: Corgi Books.

global albedo in the Matthews Global Seasonally Integrated Albedo dataset mentioned above.

http://www.grid.unep.chh/datasets/gnv-data.html
TOMS Ozone Image Data 1978–91. GRID:UNEP Satellite based global ozone measurements.

http://www.wdc.rl.ac.uk/wdcmain/wdca/wdca_meteor.html
World Data Center for Meteorology. Datasets and other information from a range of research programmes. There are other world data centre sites that can be accessed from the root directory, including information on sunspots, atmospheric trace gases and aerosols.

CHAPTER 9
WATER

OVERVIEW	LEARNING OUTCOMES
Of all the natural resources, water perhaps demonstrates most fully the interdependence of the physical and human spheres. Human societies have a great impact on the hydrological cycle, diverting rivers, damming flow and abstracting water from surface and groundwater stores. Human societies are also greatly susceptible to changes in water resources such as flooding, droughts and the availability of potable water. This chapter discusses basic hydrological processes and examines wetlands as aquatic habitats of international importance. Extreme flows of floods and droughts are considered, and also how humans affect water quantity and quality.	After reading this chapter, you should be able to understand: • Basic hydrological processes • The nature of extreme flow: floods and droughts • Wetlands as fragile and diverse habitats • The impact of human activities on the hydrological cycle from the creation of urban surfaces, building of dams, groundwater abstraction and the deterioration of water quality • International concerns facing water supply and management.

9.1 HYDROLOGICAL PROCESSES

Precipitation, evaporation, infiltration, soil moisture, groundwater and runoff are the key hydrological processes (Figure 9.1).

Figure 9.1 Runoff: the mighty Zambesi plunging over the Victoria Falls, Zimbabwe. Photograph by Mike Acreman

PRECIPITATION

Precipitation is one of the most important factors controlling hydrology and water processes, as it is the main input of water to the Earth's surface. The amount of precipitation received at the Earth's surface varies in both time and space, and knowledge of its distribution is essential to the understanding and sustainable management of the fluxes of soil moisture, groundwater and river flow. As precipitation is a visible and easily measured occurrence, data are more readily available than for other components of the water cycle.

Precipitation can fall in liquid or solid forms. Liquid forms (of different drop size and intensity, such as rainfall, drizzle and mist) play an immediate part in the hydrological cycle once the water reaches the surface. Solid precipitation (snow) may remain outside the hydrological cycle for long periods. For example, precipitation falling as snow in areas where there is an overall accumulation (when more snow falls in winter than is melted in summer) and glacier formation can remain out of the hydrological cycle for many years.

Precipitation occurs because of **synoptic weather conditions** and the properties of air masses. Over short time periods, precipitation totals can vary considerably; however, long-term data show broad climatic trends. Figure 9.2 shows that the equatorial regions receive the highest annual precipitation totals, over 2000 mm, due to the continuous uplift of air resulting from the convergence of the trade winds and strong convectional currents (p. 220). Subtropical and polar regions have values below 250 mm. In the subtropics, this is due to the presence of high-pressure zones of subsiding air, and the absence of uplift bringing air to

synoptic weather conditions: weather conditions over a large area of the atmosphere.

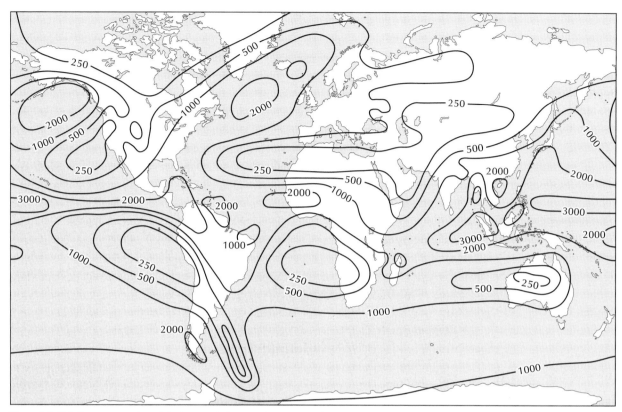

Figure 9.2 Global mean annual precipitation (mm). *Source:* After Riehl (1965)

saturation levels. In the polar zones, it is due to the lack of uplift mechanisms as well as a lack of moisture in the air. The mid-latitudes or temperate latitudes receive approximately 1000 mm per year due to cyclonic and **orographic rainfall**.

> **orographic rainfall**: precipitation created by the forced ascent of moist air over topographic obstacles such as mountains

EVAPORATION

It is difficult to measure directly the loss of water from open water, soil and vegetation. This is because the process of evaporation depends upon the relationship between many factors, such as the latent energy required for water to vaporize, the atmospheric demand for water, the availability of water for evaporation and the resistance to water loss by the evaporating surface such as plant stomata. The terms potential and actual evaporation distinguish between the amount of water that could evaporate and that which actually does evaporate. **Potential evaporation** means the amount of water that would evaporate due to atmospheric demand if water supply were unlimited, as from a freshwater surface. **Actual evaporation** is lower than potential evaporation, as the evaporation flux lost is constrained by low soil moisture, or by plant regulation such as the closing of stomata. In addition, the term evaporation usually refers to an open water surface; transpiration refers to water loss from vegetation. These terms combine to form **evapotranspiration** to describe water loss from the Earth's surface. Figure 9.3 shows the global pattern of evaporative loss.

> **potential evaporation**: the amount of water that could be evaporated from water or soil according to the atmospheric demand, if water supply were unlimited
>
> **actual evaporation**: the water that is actually lost given environmental constraints
>
> **evapotranspiration**: the loss of water from soil, water and plant surfaces (transpiration)

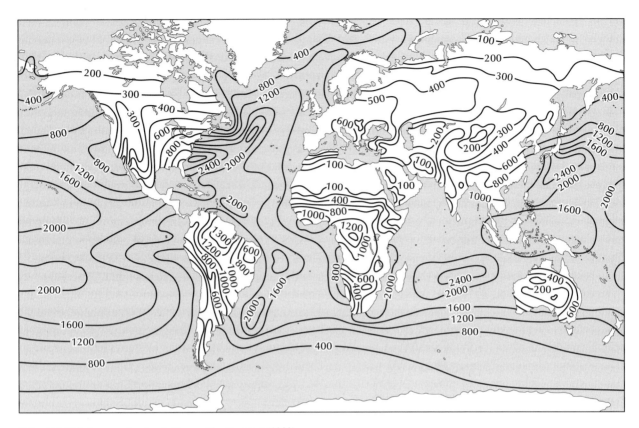

Figure 9.3 Global evaporation (mm). *Source:* After Brutsaert (1982)

INFILTRATION

antecedent conditions:
the conditions before an event that govern its effect or impact. For example, antecedent conditions of a lengthy wet period before a thunderstorm result in little infiltration and large amounts of overland flow and runoff as the saturated ground cannot accept any more water, and flooding may occur. If the antecedent conditions were dry, infiltration could occur over the unsaturated ground with less overland flow and runoff, minimizing flooding downstream

infiltration rate: the quantity of water that infiltrates the soil over a certain time period

Following a rainstorm, the amount of overland flow or runoff depends upon the nature of the surface in terms of its permeability to water infiltration (e.g. bare rock, dense vegetation, permeable soil) and the **antecedent conditions**. The terms **infiltration rate** and **infiltration capacity** are used in the literature, and it is necessary to distinguish between them. They are not interchangeable, as the infiltration rate is a measure of the velocity of water infiltrated over time, while the infiltration capacity is the maximum infiltration rate as a function of the soil moisture content. Infiltration rates may be less than maximal, due to the inadequate supply of water, and the infiltration capacity can change over time due to the ability of the soil to absorb water, which is partly determined by the duration of precipitation and antecedent conditions.

The infiltration capacity of the surface depends on a number of factors, varies in space and time and decreases during storm events. The **permeability** of the surface to infiltration is of primary importance. This relates to not only the makeup and composition of the surface (for example, permeability and porosity of rock, the depth of soil and the soil texture, urban or compacted surface), but also to the antecedent conditions before the rain event. Wet antecedent conditions reduce the infiltration capacity, allowing only a small amount of infiltration to occur before reaching saturation. Conversely, if the antecedent conditions are dry, the infiltration capacity of the surface would be large allowing more water to infiltrate. Temperature also has an effect as frozen soils allow little infiltration to occur.

The amount of precipitation that reaches the ground surface is dependent upon the nature and type of **interception** from vegetation. Vegetation intercepts rainfall and stores it on plant surfaces. The water then either travels slowly to the ground surface or evaporates back to the atmosphere. This can significantly reduce the amount of water that reaches the ground. If there is only a small amount or no interception, the **rainfall intensity** is also influential upon infiltration: if the rainfall rate exceeds the infiltration rate of the surface (i.e. the speed at which the soil can absorb water), water will not be able to infiltrate the surface. Finally, the slope of the ground is important: water will pond on a flat surface and infiltrate later, while precipitation falling on a sloping surface will give rise to **overland flow** (also termed quickflow).

SOIL MOISTURE

The uppermost layer of the Earth consists of a system of solids (mineral and organic matter), liquid (water with dissolved solutes) and gases (atmospheric gases plus water vapour, amongst others). This layer of the Earth is termed the **unsaturated zone**, and the amount of water stored in this layer is very small compared to the other water stores of the planet. The soil-moisture water store, however, is of crucial importance, as it supplies vegetation and agricultural crops, and influences overland flow and runoff. Water enters the soil through infiltration and is held there through **sorption** (both adsorption and absorption), osmotic forces and capillary forces. The movement of soil water is mainly downwards to the **water table**, at a rate dependent upon several factors, including the soil characteristics of permeability, texture and saturation. Chapter 10 explores these soil characteristics in more detail.

GROUNDWATER

Water that infiltrates into the ground through the soil will percolate downwards to the groundwater. The water table is the point that demarcates the unsaturated, or soil-moisture zone, from the saturated rock and soil, termed groundwater or saturated zone. Water in the surface drainage system will eventually flow into the sea where the cycle of evaporation, precipitation and movement over and through the land begins again. Water in groundwater also flows towards the sea, although on a different timescale, and can emerge in springs, streams or wetlands (see Concepts 9.1) where the land surface is at or below the position of the water table.

Groundwater movement is slower than surface water movement, as water must move through interconnected pore spaces within the rock. The rate of groundwater movement depends upon the porosity and permeability of the rock. **Porosity** determines how much water the rock can hold and varies between rock types, with typical values of 25–50 per cent for well-sorted sand or gravel and 33–60 per cent for clay. The permeability of a rock or sediment is the degree to which water can flow through it, and is related to the size and interconnectedness of pore spaces. Well-sorted sand has large interconnected pores and as a result is more permeable, with water moving quickly through it. In comparison, a clay rock or sediment may contain a high number of pores but they are small and poorly interconnected, slowing water movement.

Fundamental to saturated and unsaturated flow is the **hydraulic conductivity**,

---

infiltration capacity: the maximum infiltration rate at a given point under certain conditions

permeability: the ability of liquids or gases to move through soil and rocks, which depends upon the interconnectivity of rock spaces, and the shape and packing of the grains

interception: the interruption of the pathway of precipitation to the ground surface by vegetation. Water is either held on plant surfaces, from where it will evaporate, or will move at a slower rate towards the ground

rainfall intensity: the rate at which rain falls, expressed as total rainfall over a unit of time

overland flow: the movement of water as sheet flow over land due to ground saturation or impermeability

---

unsaturated zone: the area between the ground surface and the water table

sorption: the uptake of an ion or molecule by adsorption and absorption. It is often used when the exact nature of the removal is unknown, or as a collective term for both processes

water table: the upper limit of groundwater saturation in permeable rock; the level at which groundwater pressure is equal to atmospheric pressure

---

porosity: the volume of open pores compared to the volume of soil or rock

hydraulic conductivity: the rate of water movement through soil or rock

which is the rate of water movement through the rock or soil. The hydraulic conductivity depends upon texture and structure, degree of saturation, tortuosity of the pore system and the size of the water-filled pores. The hydraulic conductivity also depends upon moisture content, being fastest in saturated conditions and at a much reduced velocity in unsaturated conditions. The reduction in hydraulic conductivity with moisture content is due to flow taking place only through water films: at saturation, all pores have a large water surface area allowing maximum velocity. At saturation, soils with coarse textures and low retention capabilities, such as sand, will exhibit higher values of hydraulic conductivity than fine-grained highly retentive soils such as clay. However, owing to the higher hydraulic conductivity and lower retention, the sandy soil will dry out faster than the clay soil, and a crossover point will be reached with continued drying, when the clay soil has a higher hydraulic conductivity compared to the sand due to its greater proportion of water filled pores.

A large proportion of water supplies for domestic, agricultural and industrial use comes from aquifer systems (Figure 9.4). An **aquifer** is a saturated permeable rock. Sand and gravel make good aquifers due to their high permeability. Different types of aquifer exist according to their geological setting. An **unconfined aquifer**, or water-table aquifer, is one with no confining layers between the aquifer and the surface. A **confined aquifer** is one overlain by a layer of lower hydraulic conductivity, that is, an **aquitard** or an impermeable layer. **Recharge** occurs either in areas where the confining layer is absent or by the slow infiltration through the confining layer. Confined aquifers can show **artesian** properties if the pressure due to the weight of water in the overlying layers is high enough to drive water upwards, and may form artesian springs if the pressure forces water to the surface.

> *aquifer*: a saturated permeable rock
>
> *unconfined aquifer*: an aquifer with no confining layers between the aquifer and the surface
>
> *confined aquifer*: an aquifer overlain by a layer of lower hydraulic conductivity
>
> *aquitard*: a layer of low permeability that can store water; water can move through the layer slowly
>
> *recharge*: the addition of water to an aquifer. Can either occur naturally through precipitation infiltrating the soil and percolating down to the underlying aquifer, or it can be done artificially by pumping water into the rock
>
> *artesian*: where the water pressure is high enough to force the water upwards

RUNOFF

Runoff can also be termed river discharge or streamflow. Water enters channels by direct input (precipitation falling into the stream), overland flow, or throughflow through the subsurface soil layers. Water also follows slower

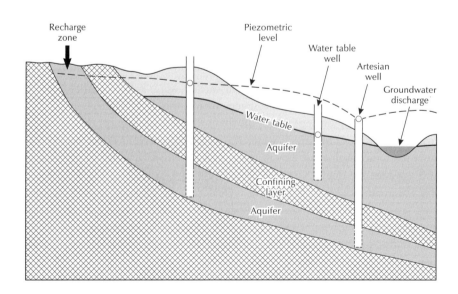

Figure 9.4 Groundwater recharge and different aquifer systems. *Source:* After Newson (1994) by permission of Oxford University Press

pathways through the soil or underlying geology to join the channel, which is termed **baseflow**. The main controlling factors governing the production and rate of runoff include rainfall intensity, snowmelt, slope, vegetation cover, land use, drainage basin shape, channel network, soil infiltration rate and antecedent conditions.

Once water has joined a channel, there are a number of possibilities as to its destination. Where channels feed surface water features, the water is stored within the lake or wetland (see Concepts 9.1), for example. The water then either evaporates and rejoins the atmosphere, or percolates downwards through the soil or sediment layers to recharge the underlying aquifer. Alternatively, water moves through these aquatic habitats and flows towards the sea, joining on route, to form progressively larger rivers. There, water evaporates from the sea surface and the hydrological cycle starts again.

A plot showing discharge against time at a particular location on the river is termed a **hydrograph** (Figure 9.5). The plot shows the total runoff or streamflow and can be broken down into the components of overland flow, throughflow, baseflow and direct input. Storm hydrographs show the response of the river to particular rainfall events. Hydrologists use the **unit hydrograph** to determine the rainfall–runoff relationship, calculating how much precipitation flows over the land as streamflow. The unit hydrograph can be used to predict the time and volume of peak discharge, which is of immense value for flood forecasting.

> **baseflow**: the contribution to stream flow from upwelling groundwater into the channel
>
> **hydrograph**: plot of discharge against time at a particular location on the river
>
> **unit hydrograph**: the conceptual process where precipitation is converted into runoff by assuming rain events of equal duration produce the same hydrograph. The unit hydrograph is produced from rainfall falling in a unit of time uniformly over the catchment to a certain depth. The area under the curve of the hydrograph gives the volume of surface runoff and is equivalent to the effective depth of rainfall over the catchment area

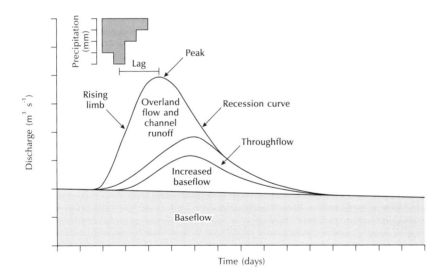

Figure 9.5 A hydrograph and component contributions to it

CONCEPTS 9.1

WETLANDS AS HABITATS OF INTERNATIONAL IMPORTANCE

Traditionally perceived as wastelands, wetland environments are becoming increasingly valued due to the increased awareness of their functions and benefits both to humankind and to the wider environment (Mitsch and Gosselink 2000). Wetlands cover an estimated 6 per cent of the world's surface (Maltby 1986), and the range of habitats classified as wetlands are extremely diverse, as exemplified by the definition given by the **Ramsar Convention** on Wetlands of International Importance especially as Waterfowl Habitat.

What are wetlands?

The Ramsar Convention defines wetlands as:

'Areas of marsh, fen, peatland or water, whether natural or artificial, permanent or temporary, with water that is static or flowing, fresh, brackish or salt, including areas of marine water the depth of which at low tide does not exceed six metres. Wetlands may incorporate riparian and coastal zones adjacent to the wetlands, and islands or bodies of marine water deeper than six metres at low tide lying within the wetland.'

Under this definition, the term 'wetland' includes such diverse habitats as mangrove swamps, tidal saltmarshes, upland peat bogs and wet grasslands. Wetlands that fulfil certain criteria can be nominated for inclusion in the Ramsar List of Internationally Important Wetlands.

The above criteria focus upon the ecological role of wetlands as habitats for populations of waterfowl, plants or animals. While the Ramsar definition and criteria of designation convey much of the essential character of wetlands, they can be criticized for lacking scientific exactness. However, the definition and classification of wetlands is fraught with problems due to their diversity and dynamic nature. Other definitions and classifications focus upon the physical characteristics of wetland environments, such as the presence of water during the growing season, **hydric soils** and **hydrophytic vegetation.**

Hydrology is probably the single most important determinant for the establishment and maintenance of specific types of wetland and wetland processes. A change in the hydrology of the wetland can impinge upon, and lead to changes in, its ecological and physical functioning. Hydrology affects primary productivity, decomposition and export of matter, and hence organic matter accumulation and nutrient cycling and availability. Hydric soils are typically anaerobic in nature and form under saturated conditions. Wetland soils are saturated for at least part of the year. Gas transfer between air and soil is drastically reduced in waterlogged and submerged sediments and so oxygen diffusion is significantly reduced. They often have a thick organic layer and can be groundwater or **surface water gleys**. Hydrophytic vegetation are aquatic plants that have morphological adaptations to grow in waterlogged soils, such as the presence of internal air spaces that can transport oxygen to the roots through diffusion and convective air flow, and an extensive root and rhizome system.

Ramsar sites

Currently there are 136 Contracting Parties to the Ramsar Convention and 1284 sites listed as Wetlands of International Importance, covering a total of 108.9 million hectares. Ramsar sites represent some of the best examples of the world's wetland ecosystems in terms of biodiversity, ecological and hydrological functions, and socio-economic functions. They vary widely in size and are distributed unequally over the world (Figure 9.6). Some areas are more fully represented than others. For example, of the seven world regions of Africa, Asia, Eastern Europe, Western Europe, the Neotropics, North America and Oceania, Western Europe has the greatest number of listed sites, while the greatest area of sites is found in North America.

Ramsar sites:

- Are a particularly good representative example of a natural or near-natural wetland, characteristic of the appropriate biogeographical region

- Support an appreciable assemblage of rare, vulnerable or endangered species or subspecies of plants or animals, or an appreciable number of individuals of any one or more of these species

---

Ramsar Convention: established in the city of Ramsar, Iran 1971, it initiated the international cooperation in the designation, conservation and wise use of wetlands

hydric soils: soils that have formed under saturated conditions and which experience saturation during the growing season for long enough to develop anaerobic conditions

hydrophytic vegetation: vegetation adapted to grow in saturated soil conditions

surface water gley: a gley soil is one which exhibits prolonged soil saturation resulting in the presence of blue/green colour caused by the reduction of iron in the soil to its ferrous state in the anaerobic conditions. Surface water gleys are soils in which poor drainage or the presence of lenses or pans causes restricted water movement down the profile causing water logging in the upper horizons

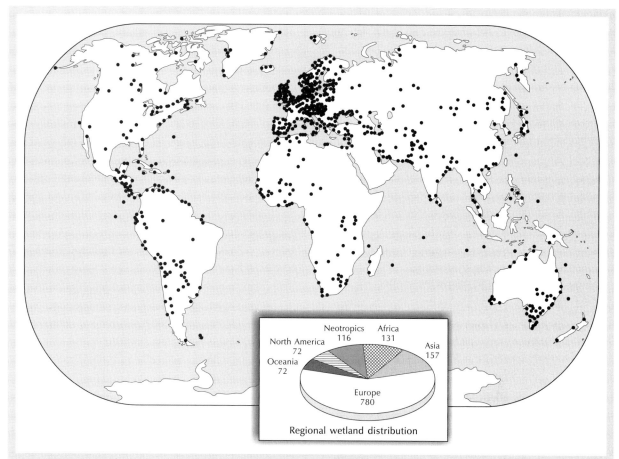

Figure 9.6 The global distribution of Ramsar sites. *Source:* After Wetlands International

- Regularly support 20,000 waterfowl; or where data on populations are available, regularly support 1 per cent of the individuals in a population of one species or subspecies of waterfowl.

Wetland goods

In addition to ecological benefits, many wetland processes and functions benefit society and provide a livelihood and materials for millions of people. The concept of **wise use** arises from this link between wetlands and humans and can be defined as the sustainable use of wetlands for the benefit of present generations while maintaining the potential to meet the needs of future generations. The Ramsar Convection considers activities within wetlands that are harmonious with, or do not adversely affect, the ecological character of the wetland, and are undoubtedly compatible with conservation aims.

There is growing awareness, that while wetlands exert a strong influence on the hydrological cycle, the nature of this influence can change depending upon the environmental conditions or time of year, for example. This is examined by Bullock and Acreman (2003), who reviewed 169 worldwide studies of wetland water-quantity functions to establish a benchmark of aggregated knowledge. The evidence shows both a strong concurrence for some hydrological measures and a diversity of functions for apparently similar wetlands.

Wetlands also influence water quality. They can lock nutrients out of incoming water and remove them from the system, but can also release the nutrients at certain times or conditions, such as over winter from decaying vegetation or under high-flow conditions that disturb sediment and dislodge nutrients locked within. Greater understanding is required of wetland hydrology, sediments and chemical processes to fully understand how wetlands work under different hydrological or seasonal regimes. This is not only to take advantage of wetland 'goods' under the concept of the 'wise use' of wetlands, but also to enable the sustainable management and restoration of wetlands that have become degraded by human activities.

An important wetland function is the support of rural livelihoods in many countries, particularly in developing countries (Plates 7 and 8). The support of fisheries through the provision of habitat, nursery or spawning

---

wise use: the sustainable use of wetlands for the benefit of humankind in a way compatible with the maintenance of the natural properties of the ecosystem

grounds is particularly important in many different wetlands from the coastal mangrove areas of Thailand to the freshwater fish of lagoonal or riverine wetlands along the Niger River in Mali. Wetland floodplains provide grazing lands for herded animals and wild herbivores, and grasses and seeds are collected as fodder for sale or dry season feed. Many products are sourced from wetland forests including timber, fuel wood, resins and medicines, and rice crops can be cultivated in swamp areas. In addition, wetlands are used the world over for recreation, nature watching and hunting.

Flood storage and **flood desynchronization** are considered by many as among the most economically important wetland functions. Wetlands are like 'sponges' in the landscape – they store rainfall and runoff and release water slowly, reducing the magnitude of the flood peak. Economic assessments have shown that preserving wetlands in flood-prone areas, rather than draining or converting them to other uses, is a much more economical means of flood defence than artificial barriers. This is also true for coastal zones, where systems of **managed realignment** are adopted in areas where maintenance of sea defences is becoming increasingly expensive, particularly in the face of climate change and sea level rise. In these areas, sea defences are no longer maintained or are removed, allowing the sea to encroach onto the land. Natural processes encourage the formation of estuarine habitat such as saltmarsh, which gives a certain level of flood protection. In many cases, stakeholder participation is imperative for managed realignment as the change in flood protection and the concept of losing land to the sea meet with strong public opposition (Myatt et al. 2003).

Wetlands can also play a role in **groundwater recharge** whereby the underlying aquifer is replenished by water moving slowly through the wetland into the underlying rock layers. Wetland also encourages sedimentation: the slow velocity of water, due to a low gradient and vegetation acting as baffles, encourages particles to drop out of suspension. Wetlands can filter the water and remove any nutrients or contaminants: they have the ability to transform these in incoming

water through processes of sediment deposition and chemical oxidation and reduction. Wetlands can act as a pollutant sink either by **sequestration** of pollutants in sediment, **bioaccumulation** in plant biomass, or through transformation into less harmful compounds. This ability of wetland systems to purify water has long been recognized and exploited for the treatment of domestic sewage, and wetlands are now increasingly being used to treat wastewaters from a variety of sources, including urban runoff and highways discharge, as well as for sediment trapping. Recently, concerns have focused on the use of natural wetlands as water-treatment systems, as long-term discharge of wastewater can result in a degraded conservation value of the habitat. In addition, natural wetlands do not always function efficiently for the purposes of permanent storage of controlled discharges of nutrients and pollutants. These concerns have led to the development of constructed wetlands to simulate and enhance the optimal properties of natural systems for wastewater treatment (Figure 9.7).

Shallow-flooded or saturated soil conditions and water tolerant rooted plants used to stabilize and oxygenate the sediments characterize **constructed wetland systems**. Quiescent conditions allow sedimentation and intense biological processes to occur that transform pollutants into less harmful compounds. Some wetland plants 'leak' oxygen from their root systems into the adjacent sediment, creating aerobic conditions in the sediment and so allowing oxidation processes to take place (e.g. organic matter decomposition and the growth of nitrifying bacteria). The rhizosphere also allows diffusion of carbon dioxide, hydrogen sulphide and even methane from the sediment back to the atmosphere. Studies have shown common reed (*Phragmites australis*) and reed mace (*Typha latifolia*) allow this diffusion to occur, and they are frequently cited as the optimal plants (e.g. Shutes 2001; Ellis et al. 1994; Cooper and Findlater 1990) in Europe for pollution treatment in wetlands. Other plants include water hyacinth (*Eichornia crassipes*) used in tropical climates, but it can become a nuisance as it spreads rapidly,

Figure 9.7 A small, constructed wetland to treat water quality in Canada. Photograph by Mike Acreman

Figure 9.8 Many wetlands have been lost through the exploitation of peat reserves. Here, peat is cut from Shapwick, on the Somerset Levels and Moors, UK. Photograph by Mike Acreman

restricting navigation and altering the natural ecosystem, as has happened in Lake Victoria, Africa (Kivaisi 2001).

Wetland loss

The increased value placed upon wetland areas means that rates of destruction and drainage have slowed but not ceased. In the future, wetland loss will result from continued economic pressures upon farmers, the demand for land for housing, agriculture and industry, and sea level rise in coastal areas. Climate change will also have a significant impact. The loss of wetlands in Europe has been attributed to many factors, including changes in agricultural practices, improvements in land drainage and sea defence practices, water abstraction, industrial development and fragmentation and isolation of sites (Figure 9.8). The dominant forces are attributed to national and EC agricultural policies and the Common Agricultural Policy (CAP) that have encouraged the conversion of grassland areas to arable crop production and offered grants to install field-drainage systems. In other areas of the world, the construction of dams to generate hydroelectric power, regulate or divert flow regimes, and store water for irrigation purposes have also resulted in significant effects on downstream wetland environments. Downstream of dams, the natural variability of flow (between seasons and over years) is subdued or transformed with immediate effects on water quantity and quality, which in turn impinges upon the ecological system and human communities depending upon the water regime.

9.2 EXTREME FLOWS: FLOODS AND DROUGHTS

The pattern of flooding and drought is one of natural variability driven by global atmospheric circulation systems and periodic events such as El Niño. Recently, many people have begun to believe that the impacts of climate change are also important. The Intergovernmental Panel on Climate Change (IPCC) expects significant changes in the magnitude and frequency of both floods and droughts due to climate change affecting the distribution, intensity and magnitude of precipitation events (IPCC 2001) (p. 196).

FLOODS

A waterway is unable to contain the volume of water passing through for all flows, and so rivers can be conceptualized as having two channels: the summer channel, which is the familiar riverbed and sides; and the winter channel, which includes the floodplain. The use of the floodplain by the river is part of its natural system. A flood is an excess volume of water, but the threshold defining what is a flood and its severity is largely determined by the impact of the flood on human societies (Figures 9.9 and 9.10).

The World Meteorological Organization estimates that floods affected 1.5 billion people from 1991 to 2000. The ability to forecast the occurrence, severity and impact of flood events is a major challenge to human societies, and one that is becoming increasingly important with the uncertainties of climate change. Flooding is part of the natural hydrological cycle and only a hazard when it affects human societies through the inundation of settlements or agricultural areas.

Floods are the result of an array of interlinked environmental factors, but the increased risk and hazard to humans is due to unpredictable climate behaviour and increased urban encroachment upon the floodplain. In some countries, the floodplain is the location of squatter or shanty settlements, as there is little land

Figure 9.9 Floods can have major economic, as well as social, impacts. Here a shop sells flood-damaged stock. Photograph by Mike Acreman

Figure 9.10 Images of the River Dulain at Carrbridge under (a) normal flow in the early 1980s and (b) flood flow in August 1970. Photographs by Mike Acreman

elsewhere for the poorest members of society to live. The vulnerability of buildings in these areas, generally not covered by any planning or flood protection works, render these areas extremely sensitive to flood events, with the potential for immense loss of life. Elsewhere, particularly in developed countries, development upon the floodplain results from a negative feedback cycle wherein existing flood defences encourage greater development due to better hazard forecasting and drainage infrastructure (Parker 1995). As development continues and the natural catchment drainage becomes increasingly modified, the nature of the flooding will change: floods may occur at new locations with greater magnitude or frequency, at increasingly smaller storm events. This change may be beyond the ability of planners to cope or predict, and have an adverse impact upon people affected by flooding in their home or business.

Causes of floods

Flooding follows heavy rainfall or snowmelt and can either be fluvial or groundwater in origin. Flooding can also occur as a result of sea surges in coastal zones, of hurricanes and of changes in the catchment area (such as deforestation or the drainage of wetlands).

Fluvial flooding occurs when the channel cannot accommodate the volume of runoff. Groundwater flooding can occur when water infiltrating the ground causes the water table to rise above the ground surface or to activate spring lines. The size and duration of a flood depends upon the nature of the storm, the catchment characteristics and the antecedent conditions. The severity and duration of the storm is important: a light drizzle over many days will not cause the same flood reaction as an intense storm lasting a few hours that exceeds the infiltration rate of the soil. The amount of interception and infiltration of incoming precipitation greatly affects the volume of runoff and hence propensity to flood. The catchment characteristics – shape, slope, vegetation, size, antecedent conditions and so on – are all influential factors in flood generation. Flooding can also occur from snowmelt, landslides and seismic activity. Deforestation and overgrazing can severely reduce the interception and infiltration processes and thus increase the severity and magnitude of flooding.

Movement of water out of the channel and over the floodplain reduces the velocity of the flood due to the reduced gradient and greater surface area over which flow occurs. Material transported in the water drops out of suspension as there is now insufficient energy to transport the load. This periodic deposition of rich fertile sediment is of immense benefit in some regions, boosting biodiversity and agricultural productivity. The Nile valley is the classic example of this process.

Coping with floods

In the twentieth century, engineering solutions to prevent or control flood hazards have included embankments or dams to prevent water entering on the floodplain. Channels have been 'improved' by straightening reaches, sometimes replacing the bed and sides with human-made material such as concrete. This approach has been generally successful in protecting land, buildings and infrastructure in flood prone areas, but there are a number of disadvantages to this approach. The creation of walls and embankments in one stretch of the river prevents any flood storage attenuation and the water continues to flow downstream. This protects the land behind the floodwall but may exacerbate flood risk downstream. The success of the flood defences may lead to a certain false sense of security of minimized flood risk allowing intensive development of the floodplain. Overtopping of embankments during a great flood could be catastrophic and result in possible loss of life. The defences also prevent the movement of water back into the channel, causing water to remain on the land long after the river has subsided. The annual cost of maintaining and upgrading flood embankments is considerable, and the modified channel supports a reduced biodiversity.

Alternative approaches to minimize and cope with flooding are being adopted. The catchment is now the fundamental unit, with rivers conceptualized as an interlinked system of upstream and downstream reaches connected to the floodplain.

Flood estimation

The ability to estimate floods using rainfall and river-flow data means it is possible to assess the frequency of certain flood magnitudes. This is important in managing risk, determining areas that require flood control or restrictions on development, and checking the capacity of infrastructure such as bridges and drainage systems to cope with storm events. The **flood return period** is the average period of time within which the event occurs once. The return period does not mean that the event cannot occur more than once, or, conversely, not at all, which can make it initially difficult to comprehend. While the use of the return period is useful and necessary in order to design engineering works without excessive cost and extent, it is only a statistical measure based on past records and assumes that that there are no problems using past records to predict the future.

> *flood return period*: the statistical risk of flooding from events of a particular magnitude described in terms of a return period. For example, a large flood estimated to be a 1 in 100 year flood (1:100) has a one per cent chance of occurring in any one year

A short dataset incurs problems, however: as the data period extends so the return period may change. The accuracy of flood-frequency estimation increases with longer datasets. For example, in a 10-year dataset, the highest flood is representative of a 1 in 10 year flood (1:10): as the dataset lengthens, it may be apparent that the same flow is more representative of a 50 year flood (1:50). The standard design periods for fluvial flood defences in the UK are in the order of 1:50 or 1:100 if there is a threat to life; these double for tidal storm events. Standards are higher for dams, which are designed for a 1:10 000 year flood or greater due to the catastrophic damage that could occur if it were breached.

In addition, the assumption that past records are suitable to predict the future is only valid under stable climatic conditions. Historic data may not give a true measure of contemporary climate, particularly as the effects of climate change are still under discussion. For example, the Thames Barrier (operational in 1982) (Figure 9.11) was designed to protect against a tidal flooding return period of up to 1 in 1000 years until 2030. The very high design return period is indicative of the severe impact upon property and life that a flood event would cause in London. However, it is now thought that the Barrier would only be half as effective by 2030 due to climate change and potential sea level rise.

Figure 9.11 The Thames Barrier, London. Photograph by Nicholas Bailey/Rex Features

The current approach for flood estimation in the UK uses the *Flood Estimation Handbook* (FEH) and software, published by the Centre for Hydrology and Ecology in 1999 (Institute of Hydrology 1999). It is often said that hydrology is an inexact science, and this is reflected in that floods are *estimated* using FEH. It is important to note that it is not possible to calculate *exactly* the size of a flood due to the limitations of current understanding, the extent of data available on which to base our predictions, and the dynamic nature of the climatological and hydrogeomorphological processes governing the impact of the flood. The FEH uses flow and rainfall data and a host of catchment descriptors to estimate flood flows. The accuracy of river-flow data has a direct impact on the accuracy of flood estimation. For high volume flows, however, the rating curve (p. 123) of the channel might be 'drowned' out, and additional readings are required to extend the relationship. It can be dangerous to take manual readings in rivers in high flows. The catchment descriptors important in estimating flood risk and extent include catchment drainage area, distance and slope of drainage paths, the amount of flood attenuation due to water storage in reservoirs and lakes into the catchment, extent of urban and suburban land-cover, and average rainfall. Soil hydrological characteristics – moisture content and permeability for instance – are also important.

Flood events

Recently, disastrous floods affected many parts of the world including Europe, Asia and South America. Recent flood events in the UK in 1998, 2000 and 2002 have had widespread social and economic implications. These include the withdrawal of insurance policies on properties that have flooded, and the rethinking of flood defence policy in structural (e.g. prioritize the repair of flood defences, identify areas for flood storage within the catchment) and non-structural terms (e.g. increase funding to improve the accuracy of flood forecasting). New planning rules require local authorities to manage development in a way that minimizes the risk of flooding. The Environment Agency of England and Wales has drawn flood risk maps to identify susceptible areas, which must be subject to a detailed flood risk assessment if any development is proposed. The definition of 'flood risk' covers land lying within the area enclosed by a flood with a 1 per cent annual probability of occurrence for river, and a 0.5 per cent annual probability of occurrence for coastal areas.

DROUGHT

Drought results from a combination of meteorological, physical and human factors. The primary cause is the lack of precipitation, in terms of timing, distribution and quantity in relation to human needs, as it directly influences soil moisture, groundwater and runoff. Temperature is also an important factor, affecting the evaporation of water. Humans can create and exacerbate droughts through water mismanagement, as in the exploitation of groundwater.

Types of drought

Drought is a subjective concept, as different activities require different amounts of water and so the thresholds when water becomes scarce occur at different times. Thus, there exist a number of different types of drought. Figure 9.12

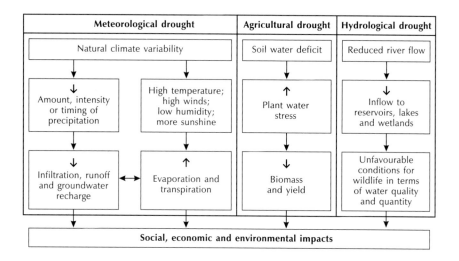

Figure 9.12 The main types of drought and their impacts. The timescale of drought severity and persistence increases from left to right. The arrows in the boxes refer to the increase (↑) or decrease (↓) of the components

details meteorological, agricultural and hydrological droughts. In addition to these types there are also supply droughts, when water for public consumption is scarce and is rationed, and ecological drought, indicative of drought conditions affecting wildlife, in particular those dependent upon aquatic ecosystems.

There is no agreed definition of drought, but definitions generally include rainfall indicators, such as a reduction of water availability over a certain period over a certain area, or a lengthy period in which no significant precipitation is recorded. A period of rainfall deficit is one where the receipt of precipitation is 75 per cent or less than the long term mean (EEA 2001). Thus, any region can experience drought: it is not always the areas with the lowest effective rainfall, as socio-economic factors mean that even the wettest areas are susceptible to drought due to the high demand from large populations and irrigated agriculture.

Impacts of droughts

Droughts have severe impacts on water resources, impinging significantly on health and agricultural productivity, but also on industrial output and hydropower generation.

The World Meteorological Organization estimates that drought caused 280,000 deaths between 1991 and 2000, and that the estimated population living in water-scarce countries will rise to between 1.0 billion and 2.4 billion by 2025. Africa and parts of western Asia are particularly vulnerable to water shortages. In addition to severity, it is the extent and timing of the drought that greatly affects human ability to cope. Droughts in summer can result in serious loss of yield in agricultural systems, while drought in winter or spring can mean unreplenished reservoirs and aquifers causing supply problems later in the year.

The main impacts of droughts include water-supply problems, decline in agricultural productivity, the potential loss of crops and cattle and increased pollution of freshwater ecosystems. Planning for drought depends upon sustainable water management, considering both water quantity and quality factors. The need to abstract water (for public supply, irrigation and industry) must balance with the need to maintain flows for ecosystem health and to minimize the impact of effluent and diffuse pollution sources.

Droughts are more difficult to predict than flood events. New tools, such as remote sensing, help to identify and monitor areas affected by drought (McVicar and Jupp 1998). The advantage of remote sensing is the ability to cover large areas over regular time intervals, and estimate parameters such as vegetation cover and condition, soil moisture and the spatial limits to drought affected areas.

Preventative measures can be implemented, including the building of storage reservoirs to offset uneven receipt of water over time, storing water in wet periods and releasing flows during dry seasons. Improvements in the efficiency of water distribution and reuse are also important measures to conserve resources. Groundwater supplies become important during drought conditions if surface water stores become dry.

9.3 HUMAN IMPACTS ON THE HYDROLOGICAL CYCLE: WATER QUANTITY

Human activities have significant impacts on water quantity, quality and aquatic ecology. These impacts are either direct, such as the damming of a river and the pumping of groundwater, or indirect, such as a change in land use that impacts upon the relationship between rainfall runoff and flooding. Table 9.1 shows a tripartite categorization of the impact of human activities: river, lake and estuary regulation; water abstraction and release; and activities in the catchment. Some of these impacts and their effects upon the water cycles are discussed below.

The level of human intervention in the hydrological cycle operates on all scales: local, regional and global. On local scales, human actions exacerbate floods, alter the water cycle through irrigation schemes, cause environmental changes by building dams and channelling rivers, and pollute rivers, lakes and groundwater by using them as a waste dump. At regional scales, humans have created shortfalls in supply, lowered water tables and planned interbasin transfers of water. On a global scale, there are real and accelerating problems of water supply and water quality. Climatic changes are set in motion that can lead to disruption and alterations in the global water cycle. These in turn create a number of connected problems from the increased frequency and intensity of thunderstorms and hurricanes, to extreme differences between floods and droughts.

WATER USE AND DEMAND

Figure 9.13 contrasts the magnitude of different uses of water in Europe and Africa. In Europe, the dominant use is of water power (38 per cent) and

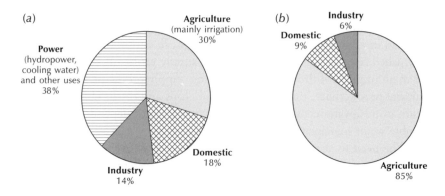

Figure 9.13 The main uses of water in (a) Europe and in (b) Africa. *Source:* After Nixon *et al.* 2000 (Europe) and FAO 1995 (Africa)

Table 9.1 Human interventions in the hydrological cycle

Intervention	Reason	Notes
River, lake and estuary regulation		
Damming, building and management of river reservoirs	Public, industrial and irrigation water supply	Owing to a lack of freshwater or water of sufficient quality
	Cooling water for power plants	
	Hydropower	
	Navigation/Transport	
	Flood control	
	Low flow enhancement to provide a reliable downstream water supply	To protect the downstream river habitat, maintain alluvial groundwater tables and enhance flows in the event of a pollution event
	Fish farming and fishing	
	Recreation	
	River channelization	For flood control, flooding for irrigation, drainage of surrounding land and navigation
	Building of weirs	To regulate flow and water level, improve fish habitats, production and fishing possibilities
	Dredging of river channels	For navigation, drainage of the surrounding land and mining of river bed gravel
Lake regulation	Lake shore modification	To prevent erosion, and encourage tourism and recreation
	Fishing	
	Water storage	For public supply, hydroenergy and flood defence
Estuary regulation	Estuary barrages	For hydropower generation, flood defence, land reclamation, navigation, industrial development, water storage and tourism
	Upstream modifications of tidally influenced river reaches	To control tidal effects in river and flood effects in estuary areas
Water abstraction or release		
Surface water abstraction/release	Public, industrial and irrigation water supply	Owing to a lack of freshwater or water of sufficient quality, or to meet public taste preferences
	Recharge groundwater	
	Fish farms	
	Cooling water for power plant	
	Generation of hydroelectric power	
	Store water in reservoirs	
	Interbasin transfer	
Groundwater abstraction/release	Public, industrial and irrigation water supply	Owing to a lack of freshwater or water of sufficient quality
	Cooling water for power plants	
	Fish farming	
Activities in the catchments		
Change in land use	Intensification of agriculture and water regulation	Cultivation of crops with high water requirements (e.g. sugar beet, potatoes) or a dependence upon irrigation (e.g. vegetables in arid areas). Conversion of grassland to arable land, drainage of wetlands crop rotation, set-aside and soil compaction
	Land drainage	For cultivation, flood control, urbanization and infrastructure
	Deforestation	Owing to land cultivation, urbanization and tourism
	Afforestation	For the production of raw material (pulp, paper, energy) and prevention of erosion

Source: Adapted from Scheidleder *et al.* (1996)

agriculture (30 per cent). In Africa, 85 per cent of all water used is for agriculture. As agriculture depends heavily on water, should supplies become scarce, agricultural productivity will decline causing problems of food shortage and famine conditions as experienced so often in some parts of the continent.

Figures 9.14 and 9.15 show the amount of water withdrawn from surface and groundwater and its use for different geographical and climatic regions of Africa. Overall, the majority of water withdrawn is for agricultural use, but this varies according to the climate of each region. The arid lands of the northern region and Sudano-Sahelian are the areas of most water withdrawn, mainly for agriculture, compared to the humid central regions of the continent where the greater water availability results in less water abstraction and a lower share of that water for agricultural use. The quantity of water as a percentage of internal renewable water resources can indicate the potential of water exploitation and the use of non-renewable sources and the importance of international water transfers or river systems for some countries.

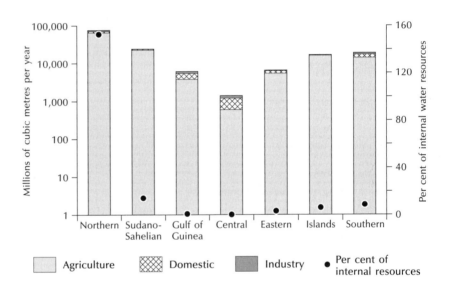

Figure 9.14 The volume of water annually abstracted from surface and groundwater for different geographical and climatic zones in Africa. *Source:* After FAO (1995)

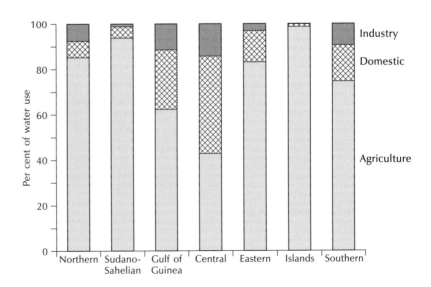

Figure 9.15 The different uses of water abstracted from surface and groundwater for different geographical and climatic zones in Africa. *Source:* After FAO (1995)

IRRIGATION

The availability of water is a major constraint to agriculture. Irrigation, which boosts plant productivity, is a major water user in agricultural systems. Irrigated agriculture accounts for more than 35 per cent of global agricultural output, a figure likely to increase with potential climate change. The benefits of irrigation are that cultivation is allowed in areas where the natural input from precipitation is insufficient for crop growth, or too variable to obtain the optimum yield. Irrigation also allows the opportunity to cultivate two crops a year. The reasons behind applying irrigation water in western countries are mainly profit related, whereas less economically developed countries, broadly speaking, irrigation is used to aid food production for the local population (Figure 9.16).

Figure 9.16 Irrigation allows the agricultural development of dry regions such as the Messara valley, Crete, Greece. Photograph by Mike Acreman

The area of land under irrigation has risen considerably since the 1900s (Figure 9.17), but is spread unequally over the world's surface. Rosegrant *et al.* (2002) combined two models to estimate water consumption and irrigation in 2025. The models were the International Model for Policy Analysis of Agricultural Commodities and Trade, and a new Water Simulation Model that uses up-to-date water databases, and these estimates are presented in Table 9.2 for developed and developing countries. Three scenarios are presented. Business-as-usual (BAU) is an indication of likely water and food outcomes based on the maintenance of current trends. The water crisis scenario (CRI) projects deterioration of current water trends and government policies, and the sustainable water use scenario (SUS) forecasts the effect of improved water sector policies and trends. Table 9.2 shows that the greatest impact is upon the developing as opposed to the developed world, and that the scenarios with the greatest impact on projected estimates are CRI and SUS. CRI incurs an increase in water use, while SUS allows water savings. By 2025, the volume of water used for irrigation is increased because of reduced efficiency, which in turn requires higher water withdrawals to compensate. Under SUS, irrigation consumption declines through the impact of higher water prices and water efficiencies.

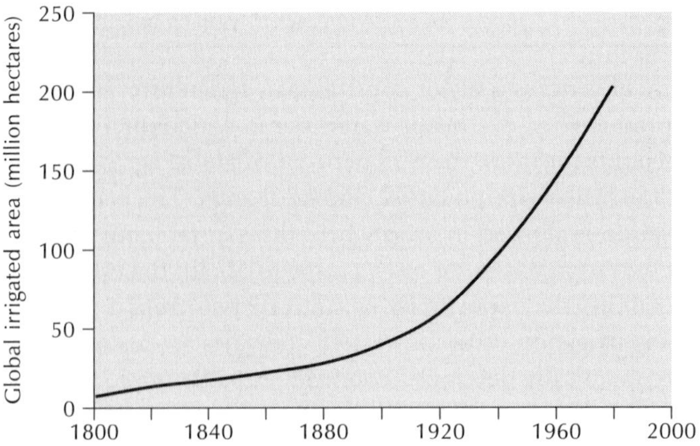

Figure 9.17 Increase in global irrigated area throughout the nineteenth and twentieth centuries. *Source:* Adapted from Grigg (1995)

Table 9.2 Irrigated area in 1995 and projected estimates for 2025

Area	1995 baseline estimates (km³)	2025 estimates (km³)		
		BAU	CRI	SUS
Developed Countries	272	277	304	258
Developing Countries	272	1216	1440	939

Source: After Rosegrant *et al.* (2002)

Types of irrigation

There are many different types of irrigation, developed to fit with the geographical and environmental setting of the application area. The main types are **gravity-flow systems**, **drip irrigation** and **spray systems**. Gravity-flow systems involve water from a reservoir or river moved by gravity to and through the cultivated area. This is a cheap but inefficient way of transport, as a lot of the water fails to reach the crop due to evaporation. Drip or trickle irrigation consists of a network of pipes, either above or below the ground, pierced with holes to let water escape. The holes are positioned to feed water directly to the plant roots, minimizing water wastage through leakage or evaporation. The most expensive irrigation methods involve spray systems that either move mobile booms over the field, or are stationary and spray a circular area. Spray systems in very large-scale agricultural fields can be entirely mechanized and automated with computers controlling all aspects, from the area covered to the rate of application per nozzle. These systems can also deliver water laden with fertilizers, pesticides or other substances to encourage plant growth, a practice termed 'fertigation'.

Problems with irrigation

Problems experienced with irrigation schemes include the exploitation of groundwater (discussed below). In arid areas, care must be taken under irrigation to prevent the concentration of soluble salt near the soil surface. Water applied to the surface can leach the salts downwards, leaving the surface soil relatively salt free. However, as the soil dries out, capillary movement of salt-laden water upwards can concentrate salts in the root zone. Irrigation can alleviate or

gravity-flow systems: an irrigation term relating to the distribution of water in an irrigated field using a network of ditches and small channels through which the water moves by gravity

drip irrigation: an irrigation term referring to a form of irrigation where the water is delivered using pipes with holes which allow water to escape at the base of the plant

spray systems: an irrigation term referring to a non-specific mass delivery of water to a crop which mimics the effect of rainfall

compound the problem, depending upon the amount of water used and the timing and method of application. Irrigation schemes are criticized for increasing malaria within the region by providing breeding grounds, and for increasing schistosomiasis (a disease transmitted by water contaminated with the Schistome worm) by harbouring large populations of snails. Irrigation can also increase the problem of liver fluke.

Irrigation is a major consumer of water, and thus another problem of its increased use is how to manage water resources in a sustainable way and improve water efficiency. The demand for water for agricultural production in water-scarce regions may mean that, increasingly, wastewater and water of low quality is used (Pereira *et al.* 2002) with impacts on health and the environment (see Focus 9.1). An equitable allocation of water between different uses and users where water is affordable is the optimal goal, but one that faces social, institutional and political problems (Johansson *et al.* 2002).

The Aral Sea Syndrome is an example of environmental damage resulting from a large-scale irrigation project to support cotton monoculture. At 68,000 square kilometres, the Aral Sea was until recently the world's fourth-largest lake. From the 1960s, the tributaries Amu Darya and Syr have been diverted for cotton irrigation. Water flow into the lake is insufficient to match evaporation losses and the lake has shrunk to less than half its size, with the shoreline receding by up to 60 km in places. The use of pesticides on the cotton crop has greatly affected the ecosystem: all twenty-four fish species native to the lake are considered to be extinct. The environmental impact has also had a significant effect on the health of local people (Crighton *et al.* 2003; Jensen *et al.* 1997). The reduction of the lake surface area has also affected the climate: less water evaporates into the atmosphere to fall as rain, which has resulted in the development of dust storms in the area. Modelling has shown that the maintenance of current irrigation practices will lead to the worsening of environmental conditions, and infrastructure improvements and changes in crop production are vital to sustainably manage the current irrigation schemes (Caia *et al.* 2003).

EXPLOITATION OF GROUNDWATER

Exploitation of groundwater is a major problem facing the world today. Pumping water out of aquifer systems lowers the groundwater level, creating a **cone of depression** around the pumping well. The size of the cone of depression is dependent upon such aquifer characteristics as the permeability and porosity of the rock, the thickness of the aquifer and the hydraulic gradient, as well as upon the rate of pumping and the rate of aquifer recharge. The cone of depression grows in extent when water is abstracted at a rate that exceeds recharge, and the level of the water table is progressively lowered. If the cone of depression grows sufficiently and intercepts areas of groundwater discharge, such as rivers, lakes and wetlands, the area will become a recharge zone. This can have significant effects on the health of aquatic environments dependent upon groundwater inputs in summer to maintain water levels and flow, as the water quantity and quality can significantly decline.

The impact of human-induced **low flows** on the flora and fauna of river and wetland systems is a new area of concern (Smakhtin 2001). The increased awareness of the effects on aquatic ecosystems from abstractions, changes in land use that use or intercept more water (e.g. afforestation) and drought conditions

cone of depression: the localized depression around a well, caused by the rate of water abstraction exceeding that of recharge

low flows: the river flow during the dry season; anthropogenically induced low flow is the reduction in discharge and other aspects of the flow regime caused by human activities such as abstraction, flow diversion and land use change e.g. afforestation, cropping and urbanization that affect interception, storage and evaporation processes

ecological flow: the minimum flow required to maintain ecological quality and processes, and ecosystem sustainability

saltwater intrusion: the movement of saline water into an aquifer formerly holding freshwater

(when the water needs of the environment are a low priority) has resulted in the development of **ecological flow** requirements. These are the conditions required to support aquatic biota, such as fish, and prevent any deterioration or reduction in the habitat required for their life cycle. If monitoring data show flow conditions are too low to support desired species (e.g. Dunn *et al.* 2003) steps can be taken to issue or reissue abstraction licenses, with restrictions that water can only be taken if the ecological flow requirements are not affected.

Aquifer overexploitation in Europe is particularly acute in Mediterranean countries, where excessive abstraction for irrigation is commonplace. Owing to groundwater overexploitation, about 50 per cent of major wetlands in Europe are affected and saltwater intrusion is severe along the coasts of the Mediterranean, Baltic and Black Seas.

One of the main problems of overexploitation of groundwater is a reduction in water quality and **saltwater intrusion** in coastal areas, for example in Denmark and Portugal. Pollution of groundwater is more difficult to treat than that of surface waters and persists over longer timescales. This is because groundwater is filtered slowly and has reduced levels of biological and chemical reactions (caused by lower temperature and oxygen levels) compared to surface water. Subsidence can also be a problem where water is abstracted from poorly consolidated rock, and is caused by the inter-granular water pressure reducing and the overlying weight of sediments causing the sediment grains to pack more tightly, leading to a fall in the vertical height of the ground. The ground shrinks in height but can also shrink laterally too, causing crevasses to appear at the surface. This results in major devastation, and can destroy buildings, roads and rail lines, as well as underground pipe networks.

DAMS

Damming rivers is one of the most dramatic human impacts on the hydrological cycle and natural environment, and is one of the most controversial aspects of the global water debate. Globally, there are more than 45,000 large dams (more than 15 m high) with a combined storage capacity of 6000 km^3 of water.

Dams and reservoirs provide water storage so that the water received in the wet season can be stored for use in the dry season. In this way, water supplies are secured for public and industrial supply during dry periods, as well as providing reserves for hydropower generation, flood control and irrigation. Dams have been used to demonstrate the importance and technical ability of a country, but, at the same time, millions of people displaced from dam sites have been forcibly resettled without adequate compensation (land and/or money), aggravating poverty. Dams also drastically change environmental conditions for all riparian and aquatic organisms and processes, such as nutrient cycling, in both the standing water body and downstream, severely affecting rural livelihoods and local economies dependent upon the natural flows of the river. A major criticism of large dams focuses on their impact upon local populations. While dams bring benefits in terms of water, and therefore food security, local populations suffer from increases in vector-borne diseases, such as schistosomiasis, and social disruption caused by construction and involuntary resettlement. In addition, local populations seldom benefit from any revenue generated by the dams from water transfer and electricity generation (Lerer and Scudderb 1999).

In recent years, the value of dams has been the subject of much examination, with environmental and social costs weighted against the socio-economic benefits. As a result, the World Commission on Dams was established in 1998 to review the impact upon development and social and environmental conditions. It also created guidelines for the management of dams, from their planning and construction to the operation, monitoring and decommissioning of dam structures to make the best use of the positive aspects of dams. The Commission produced its report in 2000, concluding that, while dams have made an important contribution to human development, the social and environmental costs have been unacceptable in many cases (WCD 2000). The recommendations of the report centre on the development of water resources while incorporating properly social and environmental considerations. An inclusive participatory decision-making process is required, based on efficiency, equity, accountability and sustainability. Seven strategic priorities are identified, with good practice guidelines set out for each (WCD 2000):

1. *Gaining public acceptance*. The good practices here are stakeholder analysis, negotiated decision-making and free prior and informed consent.

2. *Comprehensive options assessment*. Good practices are strategic and project-level impact assessment on environmental, social, health and cultural issues; undertaking multi-criteria analysis; life-cycle assessment; considering greenhouse gas emissions; the distributed analysis of projects; the valuation of social and environmental impacts and improving economic risk assessment.

3. *Addressing existing dams*. The good practice associated with this strategic priority is ensuring that operating rules reflect social and environmental concerns.

4. *Sustaining rivers and livelihoods*. Good practices here are making baseline ecosystem surveys; assessing environmental flows and maintaining productive fisheries.

5. *Recognizing entitlement and sharing benefits*. Good practice here involves establishing baseline social conditions; undertaking impoverishment risk analysis; implementing mitigation; drawing up a resettlement and development action plan and introducing a project benefit-sharing mechanism.

6. *Ensuring compliance*. Good practices in this strategy are making compliance plans; establishing independent review panels to deal with social and environmental matters; issuing performance bonds to reward compliance; setting up trust funds and instigating integrity pacts to avoid corruption.

7. *Sharing rivers for peace, development and security*. Good practices here are establishing national legislative provision where dams are built on transitional rivers; not building dams where neighbouring states raise objections that are upheld by independent bodies and the even-handed allocation of the water and the benefits derived from it.

The international response to the WCD principles and guidelines was mixed: some countries accepted the basic principles, while others held serious reservations. Some have argued that it would be difficult to implement the detailed guidelines as stakeholders would be provided with a veto over any dam development plan, and would impose massive costs and delays on developing countries. The United Nations Environment Programme (UNEP) Dams and Development Project is supporting further international dialogue of the Commission's recommendations, with the aim to reconcile opposing views and implement the recommendations.

URBANIZATION

The biggest impact of urbanization upon hydrology is the increase in impermeable surfaces. The impermeable surfaces prevent precipitation from infiltrating the soil and percolating to groundwater, instead channelling it away from the surface by sewers and storm drains. Storm drains are designed to carry water away quickly to rivers or other watercourses. There are two major impacts from this in terms of water quantity and quality. Firstly, the short-circuiting of the storm water by drains greatly reduces the time taken for the storm water to enter the watercourse. This causes the receiving watercourse to respond rapidly to the storm (causing a flashy hydrograph), with peak discharge levels occurring more quickly than if the water had flowed over or through 'natural' surfaces. These effects are compounded by any development on the floodplain that constricts the natural flow and capacity, causing floodwater to backup and overtop the banks. Secondly, storm water flowing over urban surfaces effectively 'washes' them of pollutants (including suspended solids, particulate matter and dissolved chemicals from vehicle exhausts) and transports them to the receiving watercourse. A flush of pollutants occurs at the beginning of the storm event, which may reach toxic levels.

9.4 HUMAN IMPACTS ON THE HYDROLOGICAL CYCLE: WATER QUALITY

As the hydrological cycle progresses, pure (distilled) water from evaporation becomes concentrated with insoluble and dissolved substances as it moves through the atmosphere, vegetation, soil and rock by the processes of precipitation, groundwater storage and river flow. In effect, the water picks up the chemical and physical properties of the atmosphere, rock or sediment through which it moves.

Human societies have had an increasing impact upon water quality, directly through the introduction of chemicals into the atmosphere, land or watercourses, and indirectly by land use changes. Owing to the visible environmental and health effects of poor water quality, concerns and standards set for water quality variables focus on surface waters such as rivers, lakes and groundwater used for drinking water.

DEFINITION OF WATER QUALITY

There is no simple measure of the purity of water; the term 'quality' only has meaning when it relates to a specific use of water. Chemical loading in water can be considered as either a contaminant or pollutant. Newson (1997) distinguishes between **contamination**, referring to the introduction of new material and compounds into the water system, and **pollution** as the introduction of damaging loads or concentration. Thus, whether a stream is contaminated or polluted depends upon the intended use of the water. For example, the quality of water desired for industrial use is not the same as the quality required for human consumption. Water suitable for industrial purposes would be considered polluted and unsuitable for drinking. Water quality thus impinges upon water resources: the higher the water quality standard, the lower the volume available, and so the amount available for drinking water may be significantly less than that available for industrial use. In ecological terms, the sensitivity of the receiving water determines the level at which a contaminant will become polluting. Sustainable water resource assessment must not only account for the availability of the resource, but also the quality, as poor water quality will reduce the availability of water for different uses.

> *contamination*: the presence or introduction of unwanted materials and compounds into the water system
>
> *pollution*: the presence or introduction of damaging loads or concentrations of unwanted materials and compounds into the water system

POLLUTION PATHWAYS

Pollutants fall into two categories – insoluble and soluble. The two main pathways for these chemicals to enter the water system are via transport by suspended particles in the case of insoluble chemicals, and the dissolution of minerals and chemicals in water in the case of soluble chemicals.

The introduction of new materials and compounds to surface water systems can be classified into two sources: **point** and **diffuse**. Point sources refer to a point in space such as an effluent or outflow pipe. Diffuse sources refer to non-point introductions, such as water flowing over field or forest surfaces entering streams along their margins over some distance. Consequently, diffuse pollution sources are much more difficult to isolate, identify and treat.

> *point pollution*: the discharge of pollutants into the water system at a specific point in space
>
> *diffuse pollution*: the discharge of pollutants into the water system from a large area where the source is not readily identifiable

ATTRIBUTES OF WATER QUALITY

The term 'water quality' covers both physical factors and chemical characteristics.

Physical factors

Physical factors of water quality include colour, temperature and olfactory aspects such as taste, smell and odour. The temperature of water is important as it influences the levels of dissolved oxygen, the level of chemical activity and habitat availability. Solids in water affect the physical nature of water quality and can range in size from boulders to fine particulate matter. Small sediment particles also have an influence on the chemical quality due to the presence of chemicals adhered onto the sediment particles, and they also affect the turbidity of the water column along with the suspended matter, such as microorganisms.

Chemical characteristics

Chemical characteristics include levels of acidity, oxygen and the concentrations of solutes, such as nitrate, and levels of pesticides and metals.

One of the most important aspects of water quality is the acid level, measured in pH units. Water with a pH of 7 is said to be neutral; a pH greater than 7 is alkaline and below pH 7 is acid. Water should generally have a pH between 6 and 8.5. The importance of pH is that the water acidity affects the solubility and bioavailability of certain pollutants, for example, metals such as aluminium become soluble at low (<3) pH levels. The anthropogenic emission of nitrogen oxides and sulphur dioxide has caused the acidification of soil lakes and rivers through the creation of **acid rain** with massive detrimental impacts upon flora and fauna (p. 375).

The level of dissolved oxygen is important for sustaining aquatic life: one of the major impacts of organic pollution events is the decrease of oxygen in the water column to levels fatal for, for example, some fish species. The impact of organic matter pollution can be measured in terms of Chemical Oxygen Demand (COD) and Biochemical Oxygen Demand (BOD): the higher the COD and BOD the greater the amount of biodegradable organic matter and thus the higher the pollution level. COD is a measure of the total quantity of oxygen required to chemically oxidize all organic material into carbon dioxide and water. BOD is the amount of oxygen that would be consumed if bacteria and protozoa oxidized all the organic matter in one litre of wastewater.

The number of chemicals that can exist in solution at normal temperatures is high. For any particular water body, it depends upon the pathway followed by the water and the time spent in contact with the soil and rock. Of increasing concern is the impact of human actions upon chemical loading of water bodies, and there are now international guidelines for water quality set by the World Health Organization to try to prevent damage to the environment and human health.

Broadly, the compounds of most concern have been the loss of fertilizers or nitrogen and phosphorus from agricultural areas. Recently, attention has focused on the long-term problems of compounds that remain within the environment for long periods, such as pesticides and organic pollutants. These compounds are stable and thus persist in the environment for long periods. They reside in sediments and are either transported into groundwater and aquifer systems or are ingested by benthic feeders and proceed to accumulate in the food chain. Examples are the pesticides atrazine and simazine, levels of which are increasing in European groundwater (Nixon *et al.* 2000). The long-term deleterious effects of these pollutants are unknown and they are difficult to monitor and remove.

acid rain: precipitation where the raindrops have been contaminated by sulphur dioxide or nitrogen oxide in the atmosphere, causing the pH level of the rain to fall below 'normal' levels

NITROGEN AND PHOSPHORUS

Nitrogen and phosphorus provide good examples of the problem of human enhancement of chemicals in the environment. These nutrients can exist in different forms (or 'species'); for example, nitrogen exists in the form of nitrate, nitrous oxides, ammonia and nitrogen gas (p. 32). Sources of these nutrients include sewage effluent, detergents and industrial effluent (particular food processing and agricultural sources, such as the application of fertilizers and animal farming). They are lost from agricultural areas as leachate from water percolating through the soil into the drainage systems, erosion of soil particles containing adsorbed phosphorus, and the inefficient return of slurry over the land.

High levels of nitrogen and phosphorus in water bodies can cause **eutrophication** to occur (p. 380). In inland waters, phosphorus is the limiting factor for growth, as the amount biologically available is small in relation to

eutrophication: the enrichment of water bodies with mineral and organic nutrients. This initially supports a proliferation of plant life but can lead to reduced dissolved oxygen levels and changes in the dominance of plant groups

demand, while nitrogen is usually deficient in marine waters. Naturally eutrophic water bodies occupy areas where the underlying geology contains phosphorus (apatite), but the majority of eutrophic waters result from nutrients entering the system. The initial effect of nutrient loading on a water body is the support of a high diversity of flora and fauna. Oxygen levels within the water decline rapidly, however, leading to a reduced biodiversity with a change from **macrophytes** to **phytoplankton** and **cyanobacteria**. Turbidity increases, as does the rate of sedimentation, and anoxic conditions may develop. The water may become injurious to health of both humans and animals, decreasing the amenity value and suitability for potable water supply and fisheries.

High levels of nitrate and phosphorus are a common feature of many European water bodies (Figure 9.18) and monitoring programmes have been initiated to assess the level of pollutants and contaminants against regulatory targets. The concern is that, due to the time taken for the nutrients to reach aquifer systems, groundwater concentration will continue to rise for many years, causing its value as potable water supply to reduce. Owing to the impact on health, the World Health Organization and the EC Drinking Water Directive sets a limit of 25 mg NO_3 l^{-1} for surface and groundwater. The natural level of nitrate in groundwater is generally below 10 mg NO_3 l^{-1}, with elevated values caused by human activities.

Monitoring records show that phosphorus levels fell between the late 1980s and mid-1990s. Nitrate concentrations increased rapidly between 1970 and 1985, remaining relatively stable since (Nixon *et al.* 2000). Nutrient levels in coastal waters show little improvement, however. Decreases in nutrient levels are associated with improvements in wastewater treatment, particularly sewage, and reduced use of phosphorus in detergents.

> **macrophyte**: a large plant; a member of the macroscopic plant community
>
> **phytoplankton**: the plant community in marine and freshwater systems which floats free in the water and contains many species of algae and diatoms
>
> **cyanobacteria**: unicellular organisms formerly included in the plant kingdom as blue-green algae, but now classed as a phylum in their own right under the Monera Kingdom. They feed by photosynthesis with photosynthetic pigments of chlorophyll, phycoerythrin (red) and phycocyanin (blue)

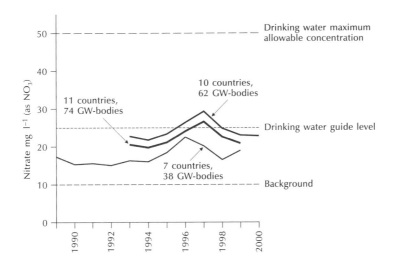

Figure 9.18 The trend of nitrate contamination of European aquifers over time. *Source:* After Scheidleder (2003)

WATER AND HEALTH

Water is essential to life, and its quality is of utmost importance. It is estimated that, worldwide, approximately one billion people lack access to improved water supply and 2.4 billion people lack access to adequate sanitation (UN/WWAP 2003; Gleick 1998). In some countries, only 20 per cent of the rural population have access to water of satisfactory quality. The World Health Organization

(WHO) estimates that contaminated water in developing countries transmits 80 per cent of illnesses, with many people having no access to water of acceptable quality, and accounts for a third of deaths.

Increased health risks are strongly associated with the lack of access to clean water and sanitation. In addition to health risks posed from drinking water polluted with excessive chemical loads, the dangers from drinking water contaminated with biological agents, such as viruses, are severe. Webb and Iskandarani (1998) identify five main categories of diseases related to water:

1. Waterborne diseases (typhoid, cholera, dysentery, gastro-enteritis and infectious hepatitis).

2. Water-washed infections of the skin and eyes (trachoma, scabies, leprosy, conjunctivitis and ulcers).

3. Water-based diseases (schistosomiasis and guinea-worm).

4. Diseases from water-related insect vectors such as mosquitoes.

5. Infections caused by defective sanitation.

Epidemics of waterborne diseases occur frequently in areas with an inconsistent water supply and poor infrastructure. Many illnesses such as cholera, typhoid or dysentery, for example, relate to drinking water supplies contaminated with pathogenic organisms. In addition, increasing pollution and contamination of water, particularly of aquifers used for drinking water, results in a reduction in the amount available for consumption and other uses in many areas. An example of this is in Bangladesh and India where an estimated two million people drink water from groundwater sources that are contaminated with arsenic (Gleick 2001) (Focus 9.1).

FOCUS 9.1

ARSENIC IN DRINKING WATER

Since the 1970s, between 8 and 12 million groundwater wells have been drilled in Bangladesh to provide a water source free of pathogenic bacteria and waterborne diseases. As a result, high mortality rates caused by diarrhoeal disease have sharply fallen. However, cases of arsenic poisoning started to appear in the 1980s and it was discovered that shallow groundwater in parts of the Bengal Delta is contaminated with naturally occurring arsenic. The arsenic is present at levels many times higher than the WHO recommendation of 0.01 mg l^{-1}, and half the population of Bangladesh is potentially at risk from exposure to unsafe levels.

Area at risk

The area at risk was thought to be limited to the Ganges Delta (the Lower Ganga Plain). The results of a number of studies indicate that a much wider area may be affected than originally thought, including the upper and middle plains of the Ganges, parts of northern India, and the lower plains area of Nepal. Adverse health effects from arsenic have been documented in Bangladesh, China, India (West Bengal) and the USA. In addition to these countries, the WHO reports that levels of arsenic in drinking water exceed 0.01 mg l^{-1} in many others, including Argentina, Australia, Chile, Hungary, India, Mexico, Peru and Thailand. For example, research by Chakraborti et al. (2003) in the Bhojpur district, Bihar state, India, indicates that 57 per cent of wells have levels of arsenic of 0.05 mg l^{-1} and 20 per cent have levels above 0.3 mg l^{-1}. This equates to 5 and 30 times the WHO acceptable level. Skin lesions typical of arsenic poisoning affected 13 per cent of adults and 6 per cent of children. Nearly every new study identifies more villages at risk.

The delayed effects of arsenic poisoning, and the lack

of awareness of the problem, are factors in determining the extent and nature of the arsenic pollution in water. For example, it had been thought that water identified as containing arsenic could be used for purposes other than drinking water. So when identified, contaminated wells were disused for drinking water and used instead as a source of irrigation water. This was considered a safe use of the water but recent studies show that well water for irrigation has led to soil and food contamination. Arsenic is being consumed, not only directly from drinking water, but also from food (mainly) irrigated with, and cooked in, contaminated water (Baea et al. 2002).

The extent of the arsenic problem and the symptoms of arsenic poisoning are not widely known. Chronic arsenic poisoning manifests itself as skin lesions, neuropathy, increased miscarriage and premature delivery and cancers. Concomitant with these impacts on health are social and economic hardships such as health care costs, lack of food or income through disability and potential social exclusion.

It was originally thought that 60 million people regularly ingest arsenic through drinking contaminated groundwater in Bangladesh alone, and that up to 150 million people were at risk. Some recent estimates have placed up to 330 million people at potential risk. Difficulties in testing for arsenic at levels that cause human health issues have hampered efforts to determine the extent of the problem. Laboratory analysis is the most accurate method, but it is costly and not universally available. Field-testing kits are unreliable at lower concentrations of concern for human health.

Source of arsenic

The arsenic arises mainly from the dissolution of minerals in rocks. Industrial effluents and the combustion of fossil fuels are also contributing sources. Locally elevated concentrations are found within Bangladesh and West Bengal, where sediments containing arsenic, eroded from the Himalayas, have been deposited in the aquifers of the Ganges Basin. Studies show that high arsenic levels exist predominantly in shallow aquifers; deeper aquifers, and those separated from shallower aquifers by an aquiclude, are largely free from arsenic.

Solutions

Chakraborti et al. (2002) are critical of the fact that arsenic was identified over 20 years ago, and yet little has been achieved in finding a solution, promoting awareness of the problems and educating people about the arsenic in the water and the consequences of ingesting it. They suggest the only mechanism to combat arsenic groundwater is to adopt effective watershed management to harness the extensive surface water and rainwater resources of the region and minimize the dependence upon groundwater sources. Surface-water storage areas can be created to store and treat low-arsenic sources such as precipitation. However the dangers of bacteriological infection are high, as is the possibility of pollution.

As a short-term solution, arsenic can be removed from contaminated water at the household level using treatment systems such as co-precipitation, ion exchange and activated alumina filtration. These technologies are still being developed, however, and there are few or no proven technologies for the removal of arsenic at water collection points such as wells, hand-pumps and springs.

The long-term solution is to find a source of water uncontaminated with arsenic or to implement piped supplies of treated water. There are up to 8–12 million shallow tube-wells (c. 35 m) in Bangladesh from which about 90 per cent of the population obtain water supplies. Replacing these with a series of deeper wells (>200 m) or piping water from safe or treated sources would be costly, time consuming and require the installation of major infrastructure. The nature of water supply must change from a situation of water independence with personal wells to a community wide management. In addition, as the recharge rate of deep aquifers is much longer than that of shallow ones, the dangers of overexploitation and the potential contamination by other naturally occurring substances must be managed.

A group of Bangladeshi villagers have undertaken legal action against the organizations that drilled the wells because the presence of arsenic in the wells was not tested. The outcome of this action may set a precedent deterring other organizations from making recommendations, drilling more wells or finding solutions to the crisis.

9.5 WATER IN AN INTERNATIONAL SETTING

The year 2003 was designated the International Year of Freshwater by the United Nations in order raise the profile of the pressing global water issues of quality and quantity.

The lack of access to water and sanitation is a great source of tension and conflict both within and between countries. Water availability is already becoming a major cause of conflict, for example, in the Middle East, and it is anticipated that wars may be fought over the share of international river systems and the diversion of waters for irrigation or dam schemes. This is particularly true for shared river basins and surface-water and groundwater that are shared between a number of political entities or groups.

To reduce conflict and insecurity, water must be managed in a way that promotes equitable sharing of transnational waters (including river basins, lakes and aquifers). Equitable sharing includes not only the water for drinking water, agriculture, and industrial use but also the ancillary benefits such as power generation and infrastructure.

THE EUROPEAN WATER FRAMEWORK DIRECTIVE

It is useful here to look at water management in the developed world, using Europe as an example. The majority of EC water policy legislation developed since 1975 considers either specific issues, such as individual substances (e.g. the Nitrate Directive), or else processes (e.g. the Waste Licensing Regulations). The Water Framework Directive (WFD) can be considered the most significant piece of European water legislation for over twenty years, and heralds a new era in environmental policy, away from traditional piecemeal approaches to solving environmental problems with isolated 'end-of-pipe' solutions, towards a more holistic or integrated problem solving approach (Basics 9.1). The Directive will encompass and update all existing EC Water legislation (Figure 9.19) by setting common objectives for water for the whole of the EU.

The Water Framework Directive (WFD) has the objective of: '*establishing a framework for Community action in the field of water policy*'. The aim of the Directive is to integrate water management within the EU and to establish systems of analysis and planning using the catchment approach. Improvement in water status is to be achieved through River Basin Planning, an approach that accords closely with existing hydrological practice in the UK, which uses the catchment as the fundamental unit of planning.

Objective of the Directive

The intention of the WFD is to group inland surface and groundwater bodies, plus transitional and coastal water, into River Basin units. For each unit, the aim is to attain an ecological quality that is close to unspoiled or pristine condition

Surface water directive 1975		Dangerous substances directive 1976		Bathing water directive 1976
Freshwater fish directive 1978	Birds directive 1979	Shellfish waters directive 1979		Groundwater directive 1980
Sewage sludge directive 1986		Urban waste water treatment directive 1991		Nitrate directive 1991
Habitats directive 1992		Integrated pollution prevention and control directive 1996		Landfill directive 1999

Figure 9.19 The Water Framework Directive will absorb existing EC Water Legislation

BASICS 9.1

WATER REGULATION

Until the publication of the Water Framework Directive, the history of legislation covering water quality had been of a piecemeal approach, focusing on individual substances or processes. The objective of monitoring and legislation is to turn the water quality and ecological status of the aquatic system back to near pristine conditions. Figure 9.20 highlights the complexity of different human activities operating within the catchment that can negatively affect water quality. It is an immensely difficult task to regulate the many human impacts, water abstraction and discharge; and a technological challenge to ensure that the aquatic system remains in good ecological condition, without incurring excessive cost and social restrictions.

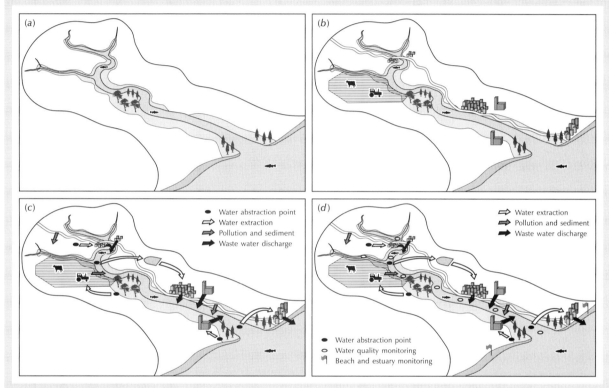

Figure 9.20 The complexity of human uses within a catchment and water quality legislation. (a) A catchment untouched by human activities. The river and its floodplain can be seen and the water quality is good enough to support fish. There are riparian and estuarine wetland systems. (b) Agricultural activities now take place in the catchment; there is a road network, two small settlements, and a large urban area built upon the floodplain. Industry is also present near the urban area and by the coast. (c) The impact of these human activities is now felt on management of the catchment. There is water abstraction, which includes surface-water from the river and a newly built reservoir. Water abstraction and the storage of water in a reservoir can have the effect of reducing water flow in rivers to such low levels that the stream biota cannot survive. The arrows indicate the pathways of the abstracted water: piped into domestic, agricultural and industrial networks, or used and discharged back into the river. The discharge of water back into the river system can affect water quality in terms of chemical loading, oxygen content or temperature at specific points along its length. Sediment and particles from agricultural and urban areas carry sorped pollutants into the river network. (d) Water quality legislation has been superimposed on these human activities. Surface-water quality is monitored downstream of abstraction and wastewater discharge points. These monitored points are located where the impact may be greatest according to the surrounding land use, to assess the maximum impact of wastewater discharge into the river. The estuary and beach water is also monitored to assess the level of pollutants carried to the sea by rivers and by discharges into near shore waters. *Source:* Kindly provided by Phil Sutherland

by the year 2015. Further, the Directive establishes a framework for the protection of all waters that:

- Protects, enhances and prevents further deterioration of water resources

- Promotes sustainable water use

- Protects aquatic environments through the reduction of discharges, emissions and losses of hazardous substances

- Reduces current pollution of groundwater

- Contributes towards mitigating the effects of floods and droughts.

The key objective of the Directive is to achieve good water status for surface water and groundwater by 2015. Good status for water bodies is the sum of 'good ecological status' and 'good chemical status'. Here the Directive advances on previous water legislation by using ecological quality as well as the more traditional assessment of chemical quality. Ecological status is determined by biological quality elements, supported by hydromorphological and physicochemical quality elements. It may not be possible to attain this goal for all waters, particularly for artificial or heavily modified water bodies (e.g. reservoirs and canals) by 2015, and so two further six-year cycles of planning and implementation are allowed to enable all water bodies to meet the standards.

GLOBAL DEMAND

Global water demand tripled between 1950 and 1990 with expectations of demand outstripping supply by 56 per cent in 2025. The World Water Council report that residential water use in Europe is approximately 200 litres per capita per day. This contrasts with the average 10–20 litres per capita per day in many sub-Saharan countries, described as 'undesirably low'. In 1999, the World Commission on Water issued a statement on World Water Day (22 March) warning that humans are currently facing a global water crisis that can only accelerate. Unless the global water gap is bridged, water scarce countries will face higher food prices and expensive freshwater supplies. As many water scarce countries are also the poorest, millions of people will face starvation and famine. The Commission estimate that nearly 450 million people in 29 countries face water shortage problems, a figure projected to increase to nearly 2.5 billion people by 2050.

WATER AS A HUMAN RIGHT

On 26 November 2002, the UN Committee of Economic Social and Cultural Rights declared water a human right. This declaration puts an obligation on governments to provide access to sufficient, affordable, accessible and safe water supplies and safe sanitation services to all citizens without discrimination. The declaration of water as a human right entitles people in countries that have ratified the declaration to demand the right to water and force their governments to adopt the necessary policies to make it a reality. Proposals to include water services in the General Agreement on Tariffs and Trades (GATTS) under the World Trade Organization (WTO), and the controversial issue of the privatization of water industry in developing countries, could cause conflict with the concept of water as a human right.

Water availability also affects community structure and the development of communities. If water is unavailable or unaffordable, it must be fetched from alternative sources, sometimes at great distances. In developing countries women and children, normally girls, usually carry the increased burden, risks and time in fetching water over long distances. This gender based role results in fewer girls attending school than boys and a lower status of women in the local community.

WATER SECTOR REFORM IN DEVELOPING COUNTRIES

Many developing countries need reforms in their water sectors. Where national water utilities have been unable to provide safe and affordable water to all, expansion of the water infrastructure is needed, as large proportions of the population do not have access to piped or clean water systems. Such communities rely upon sources such as water tankers, where the quality of the water supplied can be uncertain and more expensive than piped supplies. Alternatively, untreated water is taken from open watercourses or shallow dug wells.

Public and private control of water services

The World Bank argues that public control and subsidy of water and sanitation services is too costly and inefficient for developing countries faced with large debts. The traditional approach of treating water as a public service should be changed to favour a more economical approach.

As it is often the poorer communities which are not connected to the piped water network, the publicly subsidized supply benefits the wealthier communities and industry rather than those most in need. Sustainability can be achieved by cost recovery, treating water as an economic good in order to achieve effective water management and expand water and sanitation networks sufficiently to reach the most needy communities. Higher payments from consumers will provide the revenue and incentives to extend the water infrastructure to a greater proportion of the population. In addition, water rates can also help to promote conservation of the resource and reduction of any waste or leakage. Foreign investment in developing countries' water sectors could also be beneficial in terms of technology and skills transfer.

Conditions attached to World Bank structural adjustment loans, and water and sanitation loans, routinely include the requirement of increased cost recovery or economic pricing for water services as well as the participation of private companies or investors.

Opponents of the World Bank policy state that this theory is flawed and is concerned more with foreign debt repayment than the provision of safe, affordable water. Critics point to the little evidence of expansion of water services by multinational water companies to poor communities, where the ability to pay increased fees is limited. In addition, autonomous or community funded local water projects can be taken out of the community control and placed under the control of the private company.

Increased private company involvement in water services in developing countries is a controversial issue. The UN Committee on Economic, Social and Cultural Rights (UNESCO) recommends that where water supply is managed by profit making private companies, the State is obliged to ensure that the poor receive a minimum supply of drinking water and sanitation, and to monitor, amongst other things, the quality and quantity of water, pricing, sanitation and participation by users. The main arguments against private sector involvement centre on the economical and health consequences of raising water prices. Increased consumer fees for water can make safe water unaffordable for poor and vulnerable populations: higher water prices would result in the poor using less or using unclean sources. Those communities that obtain water from tankers would be affected, as the increased costs paid by the tanker operators to public water utility would be passed on. Action taken to reduce leakage through the pipe network would also cut off those impoverished communities that have tapped into the networks unofficially or illegally.

FOCUS 9.2

WATER PRIVATIZATION IN COCHABAMBA, BOLIVIA

Bolivia accepted comprehensive water privatization as a condition of receiving new loans from the World Bank (Kohl 2002; Komives 2001). In response to these conditions, the Government privatized the water resources of the city of Cochabamba, the third largest city in Bolivia, and one suffering from water shortages. A monopoly was awarded to the consortium Aguas del Tunari, with UK and US links, to repair and expand the water system. The consortium had control over all water in the district including not only the municipal water network but also wells financed by local communities with no public support.

The introduction of large increases in water rates to nearly 25 per cent of the minimum monthly wage sparked widespread street protests at the beginning of 2000. This contrasts with the 5 per cent of household income considered by the World Bank to be affordable for water services, and the cost of approximately 1 per cent experienced in EU Member States. The public protests spread nationally, calling for the removal of the consortium. Protestors created the 'Coalition in Defence of Water and Life' known as La Coordinadora and attracted international interest particularly among international water companies and organizations.

After months of protest, management of the water supply returned to public control. A new management committee was formed and included some members of La Coordinadora, who vowed to treat water supply as a public good. However, improvements in the delivery and expansion of water have not yet occurred and much of the population still have unstable, or no access to piped supplies. The large sums of money required for improvement cannot be generated from price increases, and investment from foreign investors has ceased due to the political instability.

SUMMARY

Water is an essential and precious resource governing human health and activities from agriculture to industrial development. The hydrological cycle has the ability to greatly affect human society through floods and droughts; and it is generally the most vulnerable groups, such as shanty dwellers on floodplains and those dependent upon rain-fed agriculture, that suffer the most. Climate change will influence global precipitation patterns and distribution, which in turn will affect the magnitude, frequency and severity of floods and droughts in the future. The effects upon human societies are likely to be immense.

In turn, human activities – the draining of wetlands, overexploitation of groundwater, pollution of watercourses and so forth – affect all aspects of the hydrological cycle and all aquatic habitats. Human activities have severely affected water quantity and quality, with concomitant negative impacts on aquatic habitats and human health. The growing understanding of water flows and processes, and the awareness of human impacts, have initiated attempts to regulate and control activities within the catchment. An example is the European Water Framework Directive, which takes a holistic approach to stop and prevent further deterioration of water resources and promote the sustainable use of water. However, the long-term effects of some impacts and pollutants mean that the problems will not be easy to solve in the short-term future. Looking ahead, global attention needs to focus on the major problems of insufficient water resources, and the provision of affordable safe drinking water, particularly in the developing world, where unsafe drinking water and waterborne diseases claim too many lives.

Questions

1 What might be the impact of climate change on precipitation and evaporation patterns and how might this affect runoff, soil moisture and groundwater levels?
2 Why have wetland environments been the focus of drainage and destruction? What benefits and goods do wetlands provide for society?
3 How can the risk of floods and droughts be managed, and what can be done to minimize their impacts?
4 What are the short- and long-term effects of the increase in irrigation in terms of water demand, supply and security?
5 What are the advantages and disadvantages of sourcing water from a large dam and from pumping groundwater?
6 What are the concerns about deteriorating water quality in surface and groundwater stores?
7 How can a secure and clean water supply be provided in an equitable and affordable way in the face of water shortages?
8 How does the position of water as a human right conflict with privatization of water supply and profit making?

Further Reading

Good texts on the topics introduced in this chapter include:

Arnell, N. (2002) *Hydrology and Global Environmental Change*. Harlow: Prentice Hall.
Full of useful information and ideas.

Jones, J. A. A. (1997) *Global Hydrology: Processes, Resources and Environmental Management*. Harlow: Longman.
An excellent text on water in a global setting.

Newson, M. (1994) *Hydrology and the River Environment*. Oxford: Clarendon Press.
A good text linking hydrology with aquatic ecology.

Ward, R. C. and Robinson, M. (1999) *Principles of Hydrology*, 4th edn. London: McGraw-Hill.
Classic hydrology text covering the subject in a detailed non-mathematical way.

Fetter, C. W. (1994) *Applied Hydrogeology*, 3rd edn. Upper Saddle River, New Jersey: Prentice Hall.
An excellent introduction to hydrogeology.

Useful Websites

www.wateraid.org.uk
A good web page detailing the work and research undertaken to improve water and associated health conditions and education internationally by WaterAid, a UK charity.

http://www.ramsar.org

http://www.wetlands.org
The world of wetlands is described in the web pages of the Ramsar Convention, and Wetlands International.

http://www.who.int/health_topics/water/en/
The World Health Organization and its pages devoted to water, health and sanitation.

http://www.unesco.org/water
The focus on water issues and development by the United Nations Environment and Social Organization are detailed in its web page.

http://www.dams.org
World Commission of Dams web page, details the priorities and good practice guidelines. The UNEP Dams and Development web page carries on the Commission's output: http://www.unep-dams.org/

Mitsch, W. J. and Gosselink, J. G. (2000) *Wetlands*, 3rd edn. New York: John Wiley & Sons.
One of the most highly regarded books detailing the range of wetlands and their natural functions.

Kadlec, R. H. and Knight, R. L. (1996). *Treatment Wetlands*. Chelsea, Michigan: Lewis Publishers.
A good text on the use of wetlands to treat and purify water.

Newson, M. (1997) *Land, Water and Development: Sustainable Management of River Basin Systems*, 2nd edn. London: Routledge.
A good text linking water, development and sustainable management.

http://europa.eu.int/comm/environment/water/water-framework/index_en.html
Good pages introducing the Water Framework Directive and its implication for European water management. An introduction to the Directive can be found at the above address.

http://www.environment-agency.gov.uk/subjects/wtd
The WFD pages of the Environment Agency, responsible for implementing the Directive in England and Wales, detail its implications.

CHAPTER 10
SOIL

OVERVIEW

The pedosphere is a highly organized physical, chemical and biological system with characteristic properties. Soils vary from place to place but several main groups are recognized in soil classification schemes. Owing to its role in sustaining life, soil should be one of humanity's most prized assets. In using soil, however, humans cause local and regional soil degradation. Soil degradation takes several forms. Erosion by water and wind are common. Biological degradation involves changes in humus content and the loss of animal and plant life. Physical degradation is the loss of structure and change in permeability, with the related changes in soil drainage. Chemical degradation includes acidification and a loss of soil fertility, alkalization, and changes in acidity, salinity and chemical toxicity. The non-sustainable use of soils leads to the Dust Bowl Syndrome and a number of other syndromes, including land contamination through industrial pollution, that are far-reaching in time and space. Soil erosion, by both wind and water, may be reduced by the application of suitable conservation strategies. Globally, soils are a crucial link in the carbon cycle. Their more climatically sensitive properties, such as nitrogen and carbon levels, can be expected to respond to a rise in global temperatures. Many soils would probably emit more carbon dioxide, so accelerating the warming effect.

LEARNING OUTCOMES

This chapter will help you to understand:
- How to describe soil properties
- Systems of soil classification
- Degradation of soil associated with mismanagement
- Environmental issues concerning soils, including soil erosion, soil salinity and soil contamination.

Soil is a precious resource for humanity as it underpins world food supply. However, soils are often overlooked in discussions of the impact of human actions upon the environment, with the result that soil mismanagement creates problems of soil degradation, with erosion being a particular difficulty. Problems of soil contamination affect the purposes for which land and soils can be used, the quality of food and associated human health. This chapter examines the nature of soil formation and important soil properties and processes. It also probes some of the environmental issues affecting soils. The mismanagement of soils, including chemical contamination and erosion, is creating problems that will linger well into the future. Soils have an important role in climate change, due to their crucial setting in carbon and nitrogen cycles. The effects of climate change, particularly temperature and precipitation, will impact upon land suitability for agriculture, and, in areas on the verge of overproduction, upon world food supply.

10.1 SOIL PROPERTIES

Soil can be seen as a three-phase system – gas, liquid and solid (Hillel 1998). This means that a column of soil is composed of gas-filled pores, soil water and soil solids. The characteristics of each of these components will differ for any given point in time, depending on a wide range of influencing variables. The proportion of each variable will also differ for different soils. Some of the main characteristics and properties of soils are discussed below.

THE SOIL PROFILE

soil horizon: a layer of soil, lying roughly parallel to the land surface, which has characteristics created by soil-forming processes

soil profile: all the soil horizons from the ground surface to the unaltered bedrock or parent material; usually divided into A ('topsoil'), B ('subsoil'), and C (weathered parent material) horizons, the A and B horizons together forming the solum

parent rock: the bedrock in which a soil profile is formed

parent material: the material in which soil profile is formed

podzol: a soil with a spodic B horizon (a horizon showing illuvial accumulation of sesquioxides, with or without organic matter), beneath a light-coloured (albic) A horizon

mollisol: a soil with a thick, dark surface horizon rich in calcium and magnesium cations (a mollic horizon)

If you take a spade and dig soil pits around the area in which you live, you will notice that the soil properties vary with depth. These different layers are termed **soil horizons**; each horizon has specific characteristics, the combination of which produces the **soil profile** specific to a given site. Soil profiles can also be seen where land has been cut through for road or railways (Plates 10 and 11). Soil profiles are spatially variable, as well as being variable with depth. Despite the many differences that can occur between different soil profiles in the natural environment, conceptually these can be reduced to a formalized description of each horizon.

Figure 10.1 shows a 'classic' soil profile comprising five horizons. The organic surface layer, O (or A_o), contains leaf litter and humus, which is divided into fresh organic litter, L, partly decomposed litter, F, and black humus, H, layers in the example. The underlying mineral horizons are A, E, B and C. The A and E horizons form the 'topsoil', the B horizons the 'subsoil'; together they constitute the solum. The C horizon is the weathered region overlying the **parent rock** or **parent material** that forms the mineral mass of the soil.

Each layer of the profile can be subdivided. In some soils, the surface organic layer is divisible into subhorizons, as in the iron-humus **podzol** (Figure 10.1). Divisions within organic layers are generally very narrow bands and difficult to define except in organically rich soils, such as peat, or natural grassland soils such as **mollisols** in the American prairies or Russian steppes. The A horizon can be divided into three parts. At the top is a zone where mineral soil is mixed with humus accumulation – the A1 horizon. This overlies a leached zone called the E or A2 horizon. The E horizon is not present in all soils and represents the zone where the products from the A horizon are leached through and down the profile.

The B horizon contains the matter washed down from the surface – organic

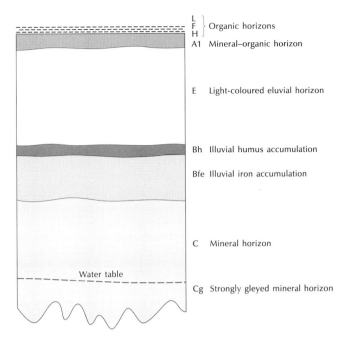

Figure 10.1 A profile of an iron-humus podzol

matter, minerals, clay particles and solutes – and then deposited. Soil pans and impermeable lenses can occur in B horizons due to the infilling of pores by clay particles washed down from the surface. The process of dissolved and suspended material being washed down the profile is termed **eluviation**. The movement of minerals in solution through the entire soil profile is termed **leaching**. Clay particles and **colloid** suspensions can move in suspension from eluvial horizons, e.g. the A horizon, to be deposited in lower, illuvial B horizons. Table 10.1 lists the names describing soil horizons in more detail.

Soil enrichment occurs with the deposition of material by wind or water onto the soil, or by human additions such as calcium or manure spread on the surface or injected into the soil. Removal of soil materials occurs by erosion, leaching of dissolved ions through the profile to groundwater, or the depletion of soil nutrients by plants or crops. Translocation processes are involved in the movement of soil materials from one horizon to another, either upwards or downwards. These processes include, for example, the washing of clay humus complexes and colloids into B horizons, or the upward movement of salts in arid soils in response to evaporative demand.

Soils that have been ploughed for agriculture lose their natural horizons and possess two basic layers: the **plough layer** (A_p) and the subsoil. The action of ploughing destroys the natural soil structure and disrupts the biological system from one of balance to one where the emphasis is on production, where soil nutrients, locked into plant biomass, are removed from the system. Modern (or western) agricultural practices of monoculture over large areas of land need to add chemical fertilizers into the system to replace the nutrients lost with every crop, and to control invasive species. The use of heavy machinery on soil with poor soil structure can lead to the creation of **plough pans**, impermeable lenses within the soil, usually at the base of the plough layer. The lenses comprise compacted soil that can impede drainage, leading to waterlogging and areas without oxygen. This can decrease the ability of plant roots to penetrate the soil, hampering growth, and in turn reducing productivity and yield.

eluviation: the washing out of soluble materials from a soil horizon

leaching: the washing out of soluble material for the entire soil profile (or sometimes a specific horizon)

colloids: tiny (1–100 nm) particles of organic and inorganic soil materials normally in suspension and highly dispersed

plough layer: a ploughed and cultivated organo-mineral surface soil horizon

plough pans: compacted layers of soil that build up below the ground surface owing to ploughing or other cultivation activities

Table 10.1 Nomenclature of soil horizons

Horizons type and symbol	Description
Organic horizons on mineral soil surfaces:	
O	Horizon formed from organic litter derived from animals and plants (mainly mosses, rushes and woody materials)
L	Largely undecomposed litter (mainly, leaves, spent fruits, twigs and woody materials)
F	Partly decomposed litter (mainly, leaves, spent fruits, twigs and woody materials)
H	Well-decomposed humus layer, low in mineral matter. Organic structures indiscernible
Mineral topsoil horizons:	
A	Mineral horizons that are rich in organic matter, or that have lost clay, iron and aluminium, or all
A1	Mineral/organic horizon formed at, or immediately below, the surface
A2 or E	Light-coloured eluvial horizon (clay or sesquioxides of iron and aluminium, or both, removed)
A3	Horizon transitional between A and B horizons. Dominated by properties of overlying A1 or E horizons, but with some B horizon properties
AB or A/B	Horizon of transitional character between A and B, with an upper part dominated by A horizon properties and a lower part dominated by B horizon properties, although the two parts cannot be separated into A3 and B1 horizons
AC or A/C	Horizon transitional between A and C and having subordinate properties of both
Mineral subsurface horizons:	
B	Altered horizons distinguished from overlying A or E horizons and underlying C horizons by colour, structure and illuvial concentrations of silicate clay, iron, aluminium or humus
B1	Transitional horizon between B and A1 or E. Properties of the underlying B2 horizon dominate, while properties of overlying A1 or E horizon are subordinate
B2	The part of the subsoil where properties are clearly expressed without subordinate properties suggestive of a transitional layer
B3 or B/C	Horizon transitional between B and C horizons. Dominated by properties of overlying B2 horizons, but with some C or R horizon properties. Only defined if a B2 horizon exists
Mineral substrata horizons:	
C	Lowest mineral horizon, excluding bedrock. Little altered save by **epimorphic processes**, irreversible and **reversible cementation**, **gleying**, and the accumulation of calcium and magnesium carbonates or more soluble salts. May or may not be the same as the material in overlying horizons
R	Underlying consolidated bedrock
Subhorizon symbols:	
b	Buried soil horizon
c	Irreversibly cemented horizon
ca	Accumulations of secondary carbonates, commonly calcium carbonate
cs	Accumulations of calcium sulphate
cn	Accumulations of concretions or hard, non-concretionary nodules enriched in iron and aluminium sesquioxides, with or without phosphorus
e	Eluvial horizon (eluviation of clay, iron, aluminium and organic matter, alone or in combination)
f	Permanently frozen horizon
g	Strongly gleyed (mottled) horizon formed under reducing conditions
h	Illuvial humus accumulation
fe (or ir)	Illuvial iron accumulation
m	Strong cementation and **induration**
p	Disturbance by ploughing, cultivation or other human activities
sa	Accumulations of secondary salts more soluble than calcium and magnesium carbonates
si	Nodular or continuous cementation by siliceous material, soluble in alkali
t	Illuvial silicate clay accumulation
u	Marked disruption by **pedoturbation** (other than **cryoturbation**)
x	Fragipan character (firm, brittle, high density, low organic matter)

epimorphic processes: processes that occur as rock is exposed to near-surface conditions. They include weathering, leaching and the formation of new minerals

reversible cementation: the formation of hard layers in soils that can subsequently be removed

gleying: a soil process in which iron is reduced under waterlogged (and therefore anaerobic) conditions to a soluble form and washed out of a horizon, leaving a bluish or greyish matrix with or without yellowish brown or brown spots (mottles)

induration: the hardening of a soil horizon owing to the cementation of soil grains

pedoturbation: a term covering all processes that tend to mix the soil by churning it

cryoturbation: a term denoting soil mixing through the action of ice

COLOUR

Soils have many different colours that reflect the nature of the parent material, the presence of minerals, the development of the soil over time or the current environmental processes at work.

Describing colour can be a subjective exercise. However, soil scientists use a standardized system called the **Munsell Soil Colour Chart System**. This colour system is based on any colour being defined according to hue (H, colour similarity), lightness (V, brightness) and chroma (C, colourfulness, intensity). The Munsell colour book displays variations of colour based on the hues of red, yellow, green and blue and variation on chroma and lightness. Each colour is coded, which allows the soil to be identified to the colours of other soils in a systematic manner. Colours are described using the unique Munsell colour notation written in the form HV/C. For example, a particularly vivid red soil would be classified as 5R 6/14.

Generally, the differences in colour between different horizons give clues that allow the interpretation of the soil history and current conditions. The uppermost horizons tend to be dark brown or black due to the presence of organic matter. Peat soils are black throughout the profile. The next horizon tends to be lighter coloured due to the result of leaching processes and translocation of particles. Red or yellow soil indicates the presence of iron oxides, while a blue-grey colour indicates anaerobic conditions from waterlogging, as in gley soils. Red mottling is evidence of a fluctuating water table, while white indicates salts brought to the surface by capillary action (e.g. saline soils).

> **Munsell Soil Colour Chart System**: a standard system for describing soil colour in terms of hue, lightness and chroma

TEXTURE

Soil texture is defined by the size of the particles in the mineral matter. The texture of the soil is of utmost importance as it influences, and is intrinsically linked to, most soil properties, from drainage to nutrient retention. Soil particles range in size from gravel, through sand, silt and clay, to colloids, which are microscopic organic and mineral particles of immense importance due to their nutrient holding abilities (Table 10.2).

When describing soil texture, only the most important of these particles are used: sand, silt and clay (which includes colloids). Soils are classified based on the proportion of these principal components: for example, a sandy soil is dominated by the sand particles; a silty clay is dominated equally by the silt and clay fraction, with only a small amount of sand; and a loam is a soil with approximately equal mixture of all particle sizes.

Figure 10.2 shows how the texture of a soil can be generally determined by the touch and malleability of the soil. This involves taking a piece of soil, moistening it and trying to roll it into shapes. A more precise method involves sieving a soil sample

Table 10.2 Soil particle-size classes

Grain	Particle size (mm)
Stones and gravel	> 2
Sand	0.0625–2
Silt	0.002–0.0625
Clay	<0.002

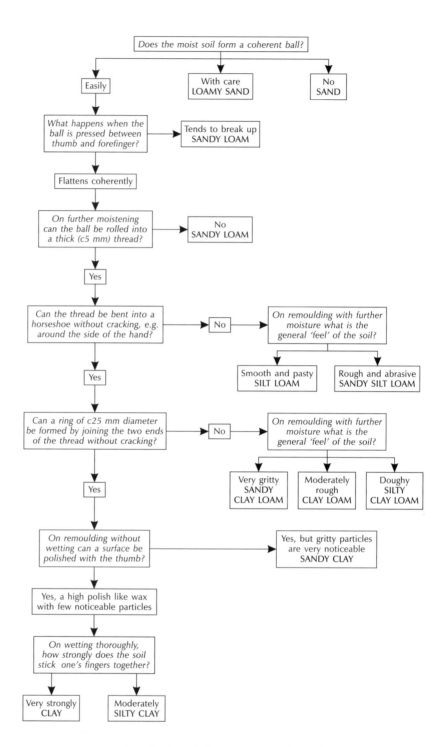

Figure 10.2 Determining soil texture by touch and malleability. *Source:* Adapted from Burnham (1980)

to get the proportions of sand, silt and clay, and using the soil texture triangle (Figure 10.3). For example, a soil with 20 per cent clay, 40 per cent sand and 40 per cent silt would be a loam. If a soil is stony, the size, shape, abundance and lithology (rock type) of the inclusions should be noted for those above 2 mm in size.

Sandy soils have large pore spaces and thus have good drainage and aeration properties. They allow air diffusion and root penetration. Rapid drainage, however, means they are susceptible to drought conditions and the lack of humus and clay colloids means that sandy soils can be nutrient poor, and need fertilizers when used for agriculture.

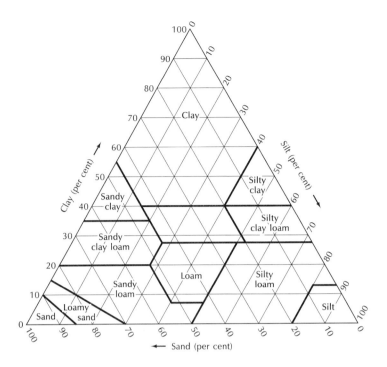

Figure 10.3 A soil texture triangle

Clay soils have small pore spaces and poor drainage and aeration properties. The large amount of colloids and organic matter means that they adsorb many nutrients and minerals, and have a large moisture content. The poor drainage of such soils, however, means that the soils become quickly waterlogged. Anaerobic conditions develop which can turn soil towards acidic conditions. The soil is heavy and difficult for roots to penetrate, or plough, without causing structural problems and smearing.

Loam soils are optimal for agriculture as they consist of a roughly equal proportion of sand, silt and clay that makes them well drained and aerated, the clay particles provide nutrients and minerals and the silt fraction binds the soil and renders it less susceptible to erosion.

SOIL STRUCTURE AND POROSITY

Another important soil property is the soil structure, which is characterized by the way soil particles group together into **peds** or **aggregates**. Biological processes bind particles together through the action of plant roots and burrowing by soil fauna, other physical processes such as shrinking and drying and the adhesive nature of clays and colloids all help create soil structure. A soil can have a number of different ped types down the profile or even in each horizon.

Soil peds can be classified into a number of different shapes such as blocky, prismatic or crumb, amongst others. Figure 10.4 shows a number of different

> ***peds***: individual soil aggregates
>
> ***aggregates***: units of many peds

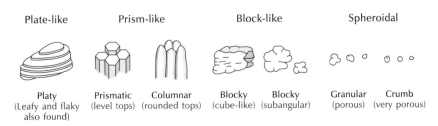

Figure 10.4 Soil peds

Plate-like — Platy (Leafy and flaky also found)

Prism-like — Prismatic (level tops) — Columnar (rounded tops)

Block-like — Blocky (cube-like) — Blocky (subangular)

Spheroidal — Granular (porous) — Crumb (very porous)

soil structures. Small or granular peds form crumb structures that are easy to cultivate and give the highest agricultural yields. Larger peds form blocky peds, while soils with no peds such as sands or heavy clays have no structure, making them very difficult to cultivate.

The wetting and drying cycle of the soil contributes to structure development, as the soil is forced to swell and shrink. Soil shrinkage occurs as soil water evaporates from pores of increasingly small size so that the soil particles move closer. Soil can shrink vertically leading to changes in the surface elevation, and horizontally through the production of vertical cracks and fissures. Adsorption of water through re-wetting by precipitation and/or irrigation causes the soil to swell. Should the vertical cracks not be aligned when the soil swells and regains its maximum volume, shearing and smearing of the crack faces can take place producing **slickensides**, planes of weakness along which the fissure will reopen with repeated cycles of wetting and drying. Soils with a high proportion of the clay **smectite** are particularly susceptible to shrinkage and cracking when drying.

An important aspect of soil structure is the volume within the soil matrix given to air spaces or **pores**. The interconnectedness of pore spaces influences a soil's drainage ability, the exchange of gases and the penetration within the soil of plant roots and fauna. A stable soil structure will maintain the proportion of air spaces within the soil over time; an unstable structure is one where small particles are moved through the profile and block pores, causing barriers to the movement of water, air and soil fauna. The soil can collapse as the pore spaces become filled up and the bulk density increases. The use of heavy machinery on waterlogged soils can also cause soil structural damage or failure, and significantly reduce porosity.

The soil porosity is the amount of air space within the soil matrix, which can range in size from tiny fissures of less than 1 mm to macropores of more than 300 mm. The size and distribution of pore spaces is linked to texture, through the size and shape of the individual grains and how they are packed into peds. Pores are distributed into two sizes: micropores that are < 0.05 mm (50 μm), and macropores that are > 0.05 mm. Pores are divided into these sizes as they relate to their function in the drainage properties of the soil. Macropores allow the soil to drain quickly in response to gravity. When the macropores are empty, the micropores start to drain. This is a much slower process as water is held under tension. If drainage continues a point will be reached where the larger of the micropores are empty and no more water can be drained.

Macropores may be classed according to their origin (Beven and Germann 1982):

1. Pores formed by soil fauna. These pores can range in size depending upon the size of the burrowing animal, from earthworms to rabbits.

2. Pores formed by plant roots. These pores also have a range of sizes according to the plant that created the channels, from narrow pores by grass roots, to larger structures by tree roots, for example.

3. Cracks and fissures created by desiccation of clay soil.

4. Natural soil pipes created by the erosive action of soil water flow.

ORGANIC MATTER AND HUMUS

The range of organic matter commonly found at the surface of mineral soils in temperate regions is in the order of 4–10 per cent (Brady 1990). Soils with low

slickenside: a smoothed surface within a soil showing parallel striations and grooves

smectite: a clay mineral requiring high ionic concentrations of silica and magnesium for its neoformation (formation in the soil); also called montmorillonite

pores: small voids within rocks, unconsolidated sediments and soils

organic matter, i.e. less than 3 per cent, are classified as mineral soils while peat or organic soils are those with more than 15 per cent organic content (Rowell 1994).

Input of organic matter to a soil comes from leaf litter, dead plant and animals, and the artificial application of slurry wastes. The breakdown and decay of the organic matter creates humus that binds with clay particles to form **clay-humus complexes**. These complexes have immense nutrient and water holding properties and can reduce erosion by binding particles together.

BULK DENSITY AND PARTICLE DENSITY

Bulk density is defined as the mass of a unit volume of dry soil, i.e. both solids and pore spaces, and can be viewed as an indication of the soil structural quality. Bulk density can range from 1.00 g cm^{-3} to 1.80 g cm^{-3} over the range of clay to sand soils. As bulk density is a measure of the soil volume including pore space, its value for a particular soil can change with depth, space and time. Land use is also an important factor: heavy machinery use in agriculture, for example, can compact the soil leading to high bulk densities (e.g. above 1.60 g cm^{-3}), which can restrict root growth.

As bulk density is a measure of the mass of both soil and pores in soil, so the **particle density** is a measure of the solids alone. Typical values of mineral soils are in the range of 2.60–2.75 g cm^{-3}; organic soils have a lower mass per unit volume of 0.1–1.4 g cm^{-3} (Brady 1990).

SOIL NUTRIENTS AND MINERALS

Nutrients are elements essential for plant growth, and are classed into primary and secondary according to their importance. Primary nutrients are carbon (C), hydrogen (H), nitrogen (N) and oxygen (O). Secondary nutrients include calcium (Ca), magnesium (Mg), potassium (K), sulphur (S) and phosphorus (P). There are other elements, termed trace elements, that are required in small amounts (p. 30). Nutrients are sourced from the parent material, from atmospheric or precipitation input to the soil and from artificial fertilizers.

Nutrients are held in the soil by clay-humus complexes. The complexes are negatively charged and thus attract positively charged ions, holding them within the soil, where they are available for uptake by plants. The release of nutrients from the clay particles, by replacing hydrogen ions on plant roots, is termed **cation exchange**. Soils with a high proportion of clay and humus particles are able to retain a large number of minerals and so they possess a large **base** or **cation exchange capacity (CEC)**. A soil with few colloids tends to be nutrient poor as any nutrients within the soil are leached through the profile.

The ability of the soil to retain nutrients is related to its organic content, drainage and CEC, all of which are influenced by texture. The low CEC of sandy soils results in few nutrients being held by the soil, and the good drainage of sandy soils allows leaching of nutrients and elements and the decomposition of organic matter, a source of nutrients. On the other hand, clay soils have a high cation exchange capacity and so can retain nutrients and elements. Drainage is slow due to the small particles leading to reduced leaching and organic matter decomposition. However, clay soils can become waterlogged, and are unsuitable for many crops due to their heavy structure. The best soils for agriculture are clay

---

clay-humus complex: microscopic particles of silicate clays and humus (< 0.002 mm in diameter) with a large surface area for exchange of chemical within the soil solution

bulk density: the mass of material in a unit volume of dry soil

particle density: the mass of soil solids

cation exchange: the exchange of ionic constituents between the soil solution and the clay-humus complex

cation exchange capacity (CEC): the size of the cation exchange complex

loams, soils that possess a high CEC ability but also allow some drainage to prevent waterlogging.

ACIDITY AND ALKALINITY

pH: a measure of the concentration of hydrogen ions in the soil; technically, when measured in soil, it is the negative logarithm of the hydrogen ion activity (meaning the hydrogen ions active in the soil solution and excluding those in organic matter and mineral structure), so a low pH value corresponds to an acid soil with more hydrogen ions in the soil solution, a high pH value with an alkali soil with fewer hydrogen ions

The **pH** index is used to describe acidity and is a function of the hydrogen ion concentration in the soil substrate. Acid soils typically have a pH in the range 3.5 to 7, neutral soils have a pH around 7, and alkaline soils have pH values from 7 to 10 or more. Extreme ranges can reach between pH 2 and pH 11. The acidity level of soils varies spatially as it is influenced by the parent material, moisture regime and climatic situation. Soil acidity can also vary temporally according to the receipt of heavy rains that leach the soil profile. Soil formed from base-poor rocks in cold wet climates, such as Scotland, tend towards acidity. Soils formed from base-rich rocks with a warm temperate climate tend towards neutral or alkaline conditions.

Fertile soils are those with high base levels: if acid ions replace base nutrients on the soil colloids, the soil becomes more acid and the displaced ions are leached through the profile causing the soil to become less fertile. Very acid soils can be toxic to life due to the release of iron and aluminium oxides. Biological activity in decomposing organic matter ceases and peaty soils can develop.

Acid soils can be treated by the introduction of lime ($CaCO_3$) so that the hydrogen ions attached to the soil colloids are replaced with base calcium ions. This solution is only temporary and needs to be repeated regularly to maintain raised alkali levels.

SOIL MOISTURE AND DRAINAGE

vadose (unsaturated) zone: a subsurface zone lying between the ground surface and the water table

capillary zone: a zone in the soil where water may move owing to capillary forces

field capacity: the state in the soil when all the capillary pores are just full

wilting point: the soil water storage at which plants are unable to overcome hygroscopic forces holding water in the soil matrix, fail to take up water through their roots, and therefore lose turgor and wilt

Soil moisture can be defined as the temporary storage of water (from precipitation, snowmelt or irrigation) within the unsaturated zone of the soil. The **vadose** or unsaturated zone is a thin layer in comparison to the soil as a whole, but it is one of the most important facets of the soil profile. Soil moisture is important for soil fauna and plants, provides the solution from which plants take up nutrients and minerals, and controls temperature and erosion.

The lower boundary of the unsaturated zone is the upper boundary of the saturated zone, or water table. In between the two zones is the **capillary zone**.

There are three key concepts when discussing soil moisture: saturation, **field capacity** (FC) and **wilting point** (WP). Each concept relates to a decreasing amount of water in the soil, the actual amount of which changes according to soil texture.

Soil saturation occurs when water occupies all the available pore spaces in the soil. This can occur after heavy rainfall or prolonged wetting, and is the natural state for wetland soils. After gaining saturation, however, the water in most soils starts to drain through the profile by gravity. After some time, (normally between 1–3 days) the macropores have drained, suction forces within the soil balance the gravitation forces, and so the soil ceases to drain. The amount of water held in the soil at this stage is termed the field or storage capacity. The action of plants withdrawing water from their roots and soil evaporation can deplete the water reserve further until the wilting point is reached. This is the point at which there is no water held in the soil that is available for plant use. Note that the soil will still contain some water, but this water is unavailable for plants. This is because water can be held within the soil matrix in two forms:

1.   Adsorbed water, also termed hygroscopic water, is strongly bound to the soil particles and is only a few molecules thick; and

2.   Absorbed water, or available water, is free to move throughout the pore spaces in the soil medium.

When the soil is at field capacity, both adsorbed and absorbed water exists but only the latter is available for plant use. At wilting point, only adsorbed water remains.

The **available water capacity** (**AWC**) of a soil is the difference between moisture content at field capacity and at wilting point. This water is available to plants, and its amount differs according to soil texture. Figure 10.5 shows that the amount of water held in sandy soils at field capacity and wilting point is small, as the coarse texture allows good drainage, and so the amount of water available to plants is small. In addition, the AWC of a clay soil is also small, although for different reasons. Compared to a sandy soil, more water is held at field capacity in a clay soil. However, due to the small size of the particles and the high suction forces the amount of water retained by the soil at wilting point is also high, and so only a small amount is available for plants. The greatest AWC is found in loamy soils.

> *available water capacity*: the amount of water available to plants between wilting point and field capacity; it varies with soil texture

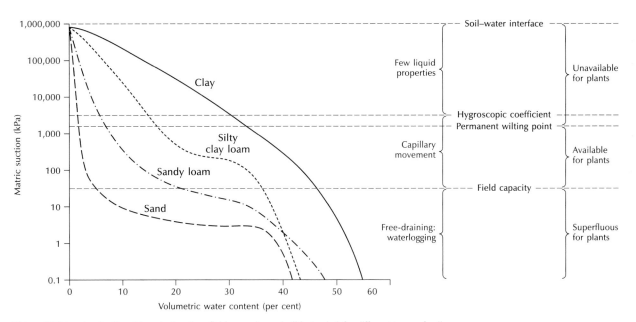

Figure 10.5 Levels of soil moisture (saturation, field capacity and wilting point) for different types of soil

SOIL WATER MOVEMENT

Soil water movement is strongly influenced by soil structure and porosity, in turn dependent upon soil texture. Soil is a complex medium comprising pore spaces of variable size, shape, direction and degree of connectivity, and water flow is constricted by these factors. The interconnectivity of the pore spaces greatly influences the permeability of the soil. Seasonal changes in soil structure and porosity are also important in swelling soils; soil ranges from nearly impermeable when fully wet to very permeable when cracked (Messing and Jarvis 1990).

Sandy soils are very permeable, with large pore spaces, and water can drain through these soils quickly. The tension at which water is retained within the soil is low, however, and so sandy soils have low moisture content at field capacity. At the other extreme, clay soils have small particles and pore spaces, allowing only slow movement of water through the profile. The soil also holds onto water at high tension resulting in relatively high soil moisture levels at field capacity.

10.2 SOIL CLASSIFICATION

SOIL CLASSIFICATION

The most important external factors influencing soil development include the nature of parent material, climate (temperature and moisture levels) and time. As these factors vary widely across the globe, the nature and properties of soil vary widely across the world as well.

The Russian scientist Vasilii Vasielevich Dokuchaev is widely acknowledged as being the first soil scientist as, in the late 1800s, he recognized that soil was a product of its geography: soil could be classified using observable features in the profile and these linked to climate and vegetation zones. The terms **zonal soils**, **intrazonal soils** and **azonal soils** are used to denote soils with particular geographical distribution, defining soils as reflecting the climate and parent matter in which they form, for example. Zonal soil development is dominated by regional climate and vegetation. Azonal soils lack well-developed profiles either because they are young or because substrate or relief limits their development. Found in any climatic zone, they include soils formed on solid rocks, soils formed on unconsolidated materials and alluvial soils formed in active floodplains. Intrazonal soils occur within all climatic zones and reflect the local dominance of a soil-forming factor other than climate. Examples are saline soils, wetland soils and soils formed on limestone.

Many systems have attempted to classify soil types and map their location across the world. One of the most widely used global soil distribution maps is the Soil Taxonomy developed by the United States Department for Agriculture (USDA) that classifies soil types into soil orders and suborders. Another is the FAO–UNESCO Soil Map of the World, first compiled by the FAO in 1961. Other soil classifications exist, particularly at a national level, for example, those by the Soil Survey of England and Wales, and the HOST (Hydrology Of Soil Types) classification, derived by CEH Wallingford (formerly the UK Institute of Hydrology). HOST classifies soil according to drainage ability and other hydrological properties, and is used in the UK flood analysis tool, the *Flood Estimation Handbook* (Institute of Hydrology 1999) (p. 256).

The USDA Soil Taxonomy (Soil Survey Staff 1975; 1992) system is based on the properties of the soil under investigation. These include physical properties, such as colour, moisture, texture, structure and depth, and chemical and mineral properties, such as organic matter, pH and salts (Brady 1990). The system also employs the concept of diagnostic surface and subsurface horizons, referring to surface horizons as epipedons. The classification system is based upon a series of categories, similar to plant classification. Table 10.3 shows the hierarchical nature of the classification. Table 10.4 lists the 12 orders of the USDA Soil taxonomy and Figure 10.6 shows the global distribution of the soil orders.

zonal soils: soils that occur over large areas roughly corresponding to major climatic zones

intrazonal soils: soils occurring within a climatic zone due to the dominance of some factor other than climate, for example, substrate

azonal soils: weakly developed soils found in any climatic zone

	Alfisols	
	Andisols	
	Aridisols	
	Entisols	
	Gelisols	
	Histosols	
	Inceptisols	
	Mollisols	
	Oxisols	
	Spodosols	
	Ultisols	
	Vertisols	
	Rock, sand or ice	

Figure 10.6 Global distribution of different soil orders of the USDA classification

Table 10.3 The hierarchy of categories in the USDA soil taxonomy and their differentiating characteristics

Category	Number of units	Differentiating characteristics[a]
Order	12	Soil-forming process, as suggested by the presence or absence of diagnostic horizons or features
Suborder	64	Subdivisions of orders, the differentiating characteristics of which vary between orders, but include properties associated with wetness, soil-moisture regimes and parent material
Great group	>300	The horizon assemblage of the whole soil profile, base status, moisture and temperature regimes
Subgroup	>2400	Subordinate properties that are still markers of important sets of soil-forming processes
Family[b]	–	Physical and chemical properties that affect a soil's response to management and manipulation for use
Series[b]	19,000 in USA	The same as higher categories but with a much narrower range of properties

Note: [a] For more details, see Soil Survey Staff (1999, Chapter 6)

[b] Essentially pragmatic category whose units are closely allied to interpretive uses of the taxonomic system

> **anthrosol**: a soil with horizons radically altered by human activities, but excluding soils under normal agricultural use

In 1988 the FAO–UNESCO Soil Map of the World introduced a new soil type: the **anthrosol** (Bridges 1997). This category is given to soils that have been profoundly modified by human activities, but does not include soils under normal agricultural practices such as ploughing. The soils are scattered widely and are too small to appear on small-scale maps. They can be defined as soils where the original horizons have been modified or buried because of human actions such as removal of horizons, additions of materials or waterlogging (e.g. paddy fields).

SOIL SURVEY AND LAND EVALUATION

Soil surveys are undertaken to describe, classify and map soil areas. The traditional focus of soil survey has been towards assessing their usefulness for agriculture (Davidson 1992), and in this context the soil is described as the medium in which plants grow. Soil surveys involve sampling the soil at points over a large area using a sampling strategy and digging a soil pit or taking an auger core. General information such as slope, erosion, land use, elevation and bedrock are noted as well as data for each horizon such as depth, texture, colour, stoniness, soil moisture, organic matter, voids, roots, peds and nodules. Maps can be produced using this information to denote the suitability of the soil for various purposes such as agriculture (cultivation, grazing), forestry, housing and other developments, or industrial use.

In England and Wales, for example, the classification for soil suitability for agricultural land is made by grouping soils into five different grades. Grade 1 is excellent quality while Grade 5 is very poor quality. Suitability for forestry use is made using a system of seven different classes, where Class F1 indicates land with excellent flexibility for the growth and management of tree crops while Class F7 is land unsuited for tree crop production (Davidson 1992).

Soil surveys are essential input into land resource inventories that are used for planning and management: an important tool to maximize the best use of land

Table 10.4 Classification of world soils according to the USDA (Soil Survey Staff 1999), and approximate FAO–UNESCO grouping

Soil order	Approximate FAO–UNESCO grouping	Characteristics	Global area (per cent)
Soils with a well developed horizon or with fully weathered minerals resulting from long continued adjustment to prevailing temperature and soil water conditions			
Oxisols	Alisols Ferralsols Nitosols Plinthosols	Ancient, intensely weathered soils of low latitudes. Highly leached (very low base status) with diagnostic subsurface **oxic horizon** of accumulated mineral oxides	7.5
Ultisols	Acrisols Lixisols	Deeply weathered and leached acid soils of equatorial, tropical and subtropical latitudes. Clay accumulation in subsurface horizon. Low base status	8.45
Vertisols	Vertisols	Disturbed and inverted clay soils (with slickensides) of tropical and subtropical zones with high base status. Develop deep wide cracks when dry	2.4
Alfisols	Luvisol	Leached, basic or slightly acid soils of humid and subhumid climates with **argillic (or natric)** diagnostic subsurface horizon of clay accumulation. Medium to high base status	9.65
Spodosols	Podzols	Leached soils of cold moist climates with diagnostic **spodic** B horizon with illuvial accumulation of free sesquioxides of iron and aluminium, and, where undisturbed, an **albic** A horizon. Low base status	2.56
Mollisols	Chernozems Greyzems Kastanozems Phaeozems	Soils of temperate grasslands with diagnostic mollic epipedon (dark, humus-rich and calcium and magnesium-rich). Very high base status	6.89
Aridisols	Calcisols Gypsisols Solochaks Solonetz	Saline and alkaline soils of arid climates. Low organic matter, subsurface horizon of carbonate minerals accumulation or soluble salts	12.2
Soils overlying permafrost			
Gelisols		Soils with permafrost within 1 m of the surface, or gelic materials (frost-churned material or with ice segregation) within 1 m of surface and permafrost within 2 m	8.61
Soils with a high organic matter content			
Histosols	Histosols	Very high organic matter content in upper 80 cm but no permafrost	1.17
Soils with poorly developed or no horizons and capable of further mineral alteration			
Entisols	Arenosols Fluvisols Leptosols Regosols	Very young mineral soils with no distinct horizon development	16.16
Inceptisols	Cambisol	One or more weakly developed horizons with minerals capable of further alteration by weathering	9.81
Andisols	Andosols	Soils with weak horizon development. High proportion of parent materials formed by volcanic activity. High phosphorus retention, available water capacity and cation exchange capacity	0.7

---

oxic horizon: a soil horizon at least 30 cm thick, in which weathering has removed or altered a large portion of the silica and left sesquioxides or iron and aluminium and a concentration of kaolinitic clays

argillic horizon: a mineral subsoil horizon in which clay washed from the overlying surface horizons accumulates

spodic horizon: a soil horizon showing illuvial accumulation of sesquioxides, with or without organic matter, and lying beneath a light-coloured (albic) A horizon

albic horizon: a light-coloured, mineral surface soil horizon from which clay and free iron oxides have been washed out

in terms of quality and suitability. Davidson (1992) describes the work in America in the 1930s to classify risk of erosion in areas of concern, to indicate suitable land uses or agriculture according to the soil properties, and to highlight areas in need of soil conservation schemes.

There is concern in many countries that the spread of industrialization and urban growth occurs at the expense of good quality agricultural land: many countries now have planning rules to prevent the expansion of built-up areas for residential or industrial use on green belt or prime agricultural land. Soil survey can also help to plan water resources, as when identifying areas for irrigation and the type of irrigation required according to the soil properties (e.g. infiltration capacity, soil water retention and so on).

In the developing world, where soil surveys are not available for large areas of land, they are arguably in greatest need. Soil surveys would allow economic assessment of untilled land to assess suitability for conversion to agriculture, making best use of available soil resources. Using a systematic approach based on land or soil survey for planning purposes would allow factors such as soil fertility, drainage and rainfall, to be assessed alongside others such as slope and possibility of failure.

10.3 ENVIRONMENTAL ISSUES WITH SOILS

The degradation of soil occurs on a global scale caused by agricultural mismanagement, overuse, population pressure and overgrazing. Soil degradation is a very real concern now and in the future due to the reduction in fertility and capability that results, and the impact that these effects will have on global food production and security.

Soil degradation takes many forms. It includes erosion by water and wind, biological degradation (e.g. changes in humus content and the loss of animal and plant life), and physical degradation (e.g. loss of structure and change in permeability, with resulting changes in soil drainage). It also includes chemical degradation (e.g. acidification and a loss of soil fertility, alkalization and changes in pH, salinity and chemical toxicity).

Commonly, the complex problem of land degradation is assessed using single approaches or measures. Warren (2002, 449) argues against this approach, stating that land assessment 'is inevitably variable and dynamic because land degradation can only be judged in its spatial, temporal, economic and cultural context'. Warren (2002; Warren *et al.* 2002) points out that the same change in land condition or measured variable (e.g. erosion) may have different consequences in different contexts. Change must be assessed relative to a wide set of components including inputs (nutrients, labour, etc.) their balance with outputs (yield, productivity), spatial extent and the limitations of models used to simulate future scenarios. The evaluation of land degradation must then be placed in the context of a particular component of productivity or potential, its effect now and in the future for a specified group, for example (see Focus 10.4, p. 308).

The link between human health and soils 'has been under-appreciated and under-reported' (Abrahams 2002, 1). One reason for this is the difficulty in establishing definite links between soils and human health. However, as knowledge builds and new risks or discoveries emerge that can impact on human health, there is a need to understand these links so that deleterious effects can be managed. Soils can have either a beneficial or detrimental impact on health in a variety of ways depending upon their mineral, chemical and biological make-ups. Positive links between soils and human health include the antibiotics derived from soil microorganisms, including penicillin and streptomycin amongst many. Soils supply minerals and nutrients for the sustenance of life. Negative links include diseases contracted through skin contact with soils, cancers caused by the inhalation of fibrous minerals or radon gas. In addition, the health of people remote from the soil source can be affected by e.g. the pollution of surface and groundwater by runoff or the erosion of soil particles with adsorbed contaminates, or the volatilization of persistent organic pollutants and their release into the atmosphere. One of the most obvious links between soils and health is the provision of nutrients. Many, if not most, of the essential nutrients and minerals in the human diet are derived from crops or foodstuffs grown from soil. Consequently, deficiencies or excesses of these, or contaminants within the soil that are transferred to humans via consumption (of the plants or animals fed on the plants) can have major negative impacts. Abrahams (2002) gives the example of molybdenum (Mo), an essential micronutrient for which excess or deficiency, linked to soil concentrations, may cause health problems. High soil levels of molybdenum have been linked with gout (a disturbance of uric-acid metabolism that leads to the accumulation of uratia in joints, especially the big toe) in molybdenum-enriched regions within Armenia (World Health Organization 1996) and reduced caries (bone or tooth decay) prevalence (Davies and Anderson 1987), while a high incidence of oesophageal cancer in Transkei, South Africa, has been linked to the deficiency of molybdenum in locally grown produce (Burrell *et al.* 1966).

SOIL SALINITY

Soils are considered salt-affected when there is an excess of salts in the root zone. The term **salinization** is given to the accumulation of salts in soils (p. 39). Soil salts are derived from chlorides and sulphates of sodium, calcium, magnesium and potassium sourced from irrigation water, groundwater and the chemical weathering of soil minerals and bedrock. The **soil solution** is the term given to soil water in which various minerals and salts are dissolved.

Salt-affected soils are found in marine-derived alluvium sediments, coastal locations and hot arid areas where capillary action brings salts to the surface. They are increasingly found in areas where anthropogenic actions, such as the mismanagement of irrigated agriculture and the overexploitation of groundwater causing saltwater intrusions, can cause or exacerbate soil salinity. Most irrigated areas of the world possess salt-affected soils to varying degrees of severity. Generally, leaching processes move mineral and salts downwards through the profile. Their accumulation at the surface occurs in response to evaporative demand and capillary action between surface wetting from precipitation or irrigation. Extraction of water from the soil by plants and evaporation concentrate the soil solution. As water continues to be extracted, salts will be precipitated out of the increasingly concentrated solution.

Layers of precipitate can occur in areas of low, regular rainfall due to the processes of salt release, leaching and precipitation over short vertical distances. Alkaline-earth carbonates are precipitated first (such as calcium or magnesium carbonates). Gypsum minerals (calcium sulphates) are also precipitated and are a

salinization: the accumulation of soluble salts, such as sulphates and chlorides of calcium and magnesium, in salty soil horizons

soil solution: the liquid portion of a soil, that is, soil water plus the materials dissolved in it (solutes)

particular feature of soils in arid and semi-arid areas. Gypsum deposits are commonly found in areas of water evaporation such as former lakebeds or marine sediments (McNeal 1974). Over many years, these layers can become hardened forming impermeable barriers within the soil. Salt layers at the surface can form into crusts of dispersed particles. Rainfall hitting the surface and dissolving the salts at the surface can leach the salts away. This results in high sodium conditions that can cause soil particles to disperse and crusts to form.

High levels of salt and sodium hinder healthy plant growth by causing direct toxicity due to the high sodium, chloride and boron content in the soil solution. Ionic imbalances in the plant are created, which can reduce the availability of water by lowering the osmotic potential, this is known as **physiological drought**. Not only are high levels of salts in soil detrimental to plant growth, but high sodium levels, in relation to low levels of calcium, can also be both physically and chemically detrimental to soil structure.

> **physiological drought**: conditions under which a plant is unable to extract water from the soil, either because the water is held to soil particles at a higher capillary force or osmotic force (or both) than the plant root can exert, or because the water is frozen

Peck and Hatton (2003) examine the problem of soil salinity following the clearance of natural vegetation for dry land agriculture in Australia a major environmental and economic problem where the soil water solution is in the order of 1–100 kg m^{-2} of salts. The challenge is to identify areas at risk of salinization and implement management strategies that are both economically feasible and do not result in the land becoming saline. Such options include replanting natural vegetation on some areas that have been cleared, pumping to lower the water table, and controlling surface and groundwater by constructing ditches to drain the land. In addition to cost, however, these options are constrained by the environmental issue over how to dispose of the saline water. Peck and Hatton (2003) conclude that, while soils and hydrological studies have contributed to the understanding of the salinity problem and its mechanisms, much more research is still needed into the development of effective management strategies.

Saline sodic soils

Should a soil be sodic (i.e. have a dominance of sodium salts), its exchangeable sodium percentage (ESP) will be high. High values of ESP also yield a high soil pH due to the presence of neutral salts. The ESP is the amount of exchangeable sodium expressed as a percentage of the CEC (Rowell 1994). For such soils, the treatment of salinity does not simply mean leaching away salts, as this would cause structural damage through soil slaking, pore blockage and a decrease in permeability of the soil.

Dispersed clay particles move with soil water and form clay 'skins' in pore spaces or block them altogether. The resulting reduction in permeability can be substantial and greatly affect the drainage ability of the soil. A common classification of sodic soil is to use the value of the ESP; soils with an ESP < 15 are classified as non-sodic while those with ESP ≥ 15 are sodic. However, research by Crescimanno *et al.* (1995) shows that soil structural damage can occur below this threshold value (ESP = 15).

Management of salt-affected soils

There are three main management options for saline and sodic soils (Brady 1990):

1. Flushing the soil with water of a low content of soluble salt and improving drainage can eradicate salts in the profile. It will act to leach

and drain away the soil salts. The water used must have a low concentration of soluble salts, particularly sodium.

2. The salt in the soil can be converted into less harmful compounds by the application of certain chemicals. For example, sodic soils are treated with gypsum (calcium sulphate) in order to replace the sodium ion on the clay and colloids with calcium. The sodium is displaced into the soil solution. This approach must be repeated regularly to ensure sodium levels do not increase over time.

3. Control techniques include reduction of evaporation to prevent the upward movement of salts to the soil surface. If the soil is irrigated, enough water must be added to leach the soils downwards and at frequent intervals to prevent the salts migrating upwards when the soils start to dry.

SOIL EROSION AND CONSERVATION

Soil erosion is a worldwide problem (Figure 10.7), and one that has been around for many thousands of years. It has serious ramifications for agriculture, as the finer particles and organic matter that bind the soil and carry nutrients are most easily lost. The result is a reduction in soil fertility. Table 10.5 presents human actions that can cause soil erosion.

Soil erosion not only leads to the reduction of the soil quality at the eroding source but also causes problems at the sink. At the eroding source, the soil is depleted of the finer particles that hold many of the soil nutrients, causing lowered productivity and yields. Away from the eroding surface, eroded particles deposited in rivers can smother aquatic flora and degrade habitats for fauna such as fish. If the sediment particles carry metals or other contaminants these can pollute the riverine environment. A major problem is the **siltation** downstream of structures such as dams that then require costly regular dredging. Aerially transported soil particles can be deposited by way of dust storms, get into machinery and engulf buildings and structures. Aerially transported dust also plays a part in climate change scenarios. Working in New Mexico, Huszar and Piper (1986) found that the cost, effects and impacts of soil erosion were greater outside the farm than inside. They estimated that the annual costs of soil erosion in New Mexico ranged from $260 million to $466 million per year for off-site effects, compared to estimated annual costs of $10 million for on-site effects.

Degraded soils have reduced biological productivity, which can result in reduced cover and an exposed soil surface. A bare soil surface, or one only loosely bound with plant roots, is extremely susceptible to erosion by wind and overland flow. Wind erosion can result in a considerable volume of dry, fine particles being lost from

> **siltation**: the settling out of silt and other fine particles in a lake, reservoir, low-gradient river or estuary

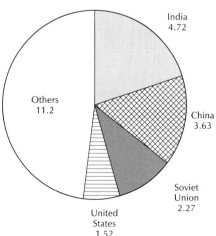

Figure 10.7 Estimated world soil erosion loss. Values are in petagrams per year (Pg yr⁻¹) (see Appendix). *Source:* Data from Brown and Wolf (1984)

Table 10.5 Human activities and their impact on soil erosion

Action	Effect
Removal of woodland or ploughing established pasture	The vegetative cover is removed, roots binding the soil die and the soil is exposed to wind and water. Particularly susceptible to erosion if on slopes
Cultivation	Exposure of bare soil surface before planting and after harvesting. Cultivation on slopes can generate large amounts of runoff and create rills and gullies
Grazing	Overgrazing can severely reduce the vegetation cover and leave surface vulnerable to erosion. Grouping of animals can lead to overtrampling and creation of bare patches. Dry regions are particularly susceptible to wind erosion
Roads or tracks	Collect water due to reduced infiltration that can cause rills and gullies to form
Mining	Exposure of the bare soil

degraded or freshly ploughed fields. Rainfall hitting the soil surface and created runoff in rills and gullies is also a serious mechanism of erosion.

Types of erosion

Erosion occurs after particles have become loose and detached at the surface, and are ready for transportation by water or wind. Particles become detached at the surface when there is little or no vegetation binding the soil or providing a protective barrier, when organic content is low and if the surface dries out.

Water hitting a bare soil surface can detach particles, and the splash effect can transport them away from the point of impact. Water flowing over the surface picks up this loosened material and transports it further. Sheet, rill and gully erosion are all caused by water. Sheet erosion is the erosion of soil over a uniform area, lowering the surface height. Water flowing over the surface may concentrate in small channels called rills. Rill erosion can be substantial. Gully erosion is the formation of large channels by water, which cut deep swathes through the soil, and cannot be removed by ploughing. Water erosion is said to be responsible for some 62 per cent of all land degradation in the USA, and 22 per cent in Europe (Oldeman *et al.* 1991).

Erosion by wind occurs from dry land. The great dust bowl of the American Midwest during the 1930s occurred under drought conditions (Focus 10.1). Wind erosion is a particular problem in arid or semi-arid areas, where erosion, failure of annual rains and overgrazing have contributed greatly to crop failure and famine conditions. Soil tends to have attained wilting point or even lower moisture contents before wind erosion occurs (Brady 1990). In addition to moisture content, the turbulence and strength of the wind is important in determining the size and load of particles it can carry. Particles transported by the wind become an abrasive element to pound into and dislodge other particles, and so the soil structural condition, size of aggregates and exposure of the surface are important factors as well. The dominant movement of soil particles by the wind occurs by **saltation**. This is when the wind lifts the particles into the air,

saltation: the bouncing of air-borne sand grains on hitting the ground

FOCUS 10.1

WIND EROSION: THE NORTH AMERICAN DUST BOWL

A classic example of wind erosion and its devastating effects on land and people is provided by the event in the 1930s on the Great Plains of America – the Dust Bowl, vividly described in John Steinbeck's *The Grapes of Wrath*. Over the period 1900 to 1930, rainfall was plentiful, with average annual precipitation in the order of 500 mm, even reaching 700 mm for some years. During this time, settlers in the mid-western prairies ploughed up the natural grasslands and intensively tilled the land following the prevailing practice in the eastern states, where the soils were heavier textured and rain more reliable. The mass production of tractors enabled vast swathes of land to be cultivated. In the Texas panhandle, land under cultivation increased from 35,000 to 800,000 ha from 1909 to 1929. However, after a few decades of plentiful rainfall, from 1933 precipitation fell to 350 mm and less, and a period of drought ensued. The intensive ploughing had broken up the soil surface, removed the natural vegetation cover,

and caused the degradation of the surface organic layer. Under sustained conditions of high temperatures and low rainfall, the friable soils quickly lost moisture and became exposed as the crops failed. Vast quantities of soil were eroded and deposited as far away as the eastern coastline and into the Atlantic. As the crops failed a mass exodus occurred as people moved out of the area in search of food and work elsewhere. The events in North America during the 1930s have given the name to the *Dust Bowl Syndrome* where environmental destruction is caused by the non-sustainable use of soils or bodies of water as biomass production factors (WGBU 1996) (see Chapter 3). The approach taken is one of maximum yield or return over the short term with no consideration of the long-term effects or repercussions. The implications are not only immediate to the soil, which is degraded, but can also have far reaching impacts, for instance, driving a rural exodus.

letting them bounce along the surface, hitting other particles. The impact of particles moving by saltation on the surface encourages the loosening and break-up of the surface and a continued supply of particles for erosion. When particles are too large for saltation, they roll along the surface in a process termed **creep**. When the particles are very small, they are picked up by the wind and carried aloft in **suspension**. Many particles are transported upwards only a metre or so, but considerable volumes can be lifted into the atmosphere and can be carried great distances.

> **creep**: the rolling and sliding of coarse sand and small pebbles under the force of jumping sand particles and down the tiny crater-slopes produced by impacting particles
>
> **suspension**: the carriage of silt and clay particles by the air

Erosion by water

Research into the causes of soil erosion has led to the development of the Universal Soil Loss Equation (USLE) (Wischmeier and Smith 1978) for sheet and rill erosion by water. This empirical model was derived using approximately 10,000 plot-years of data and can be used to calculate soil loss at any point in a catchment that experiences net erosion.

The equation has been extensively used but also widely criticized. The nature of its 'universalness' has been questioned, especially when applying it to tropical soils (Ellis and Mellor 1995). Nevertheless, the USLE is useful for understanding the principal variables that influence soil erosion. The equation relates erosion to climate, soils, vegetation, topography and management practices. The equation is presented below and explained in Table 10.6 (for more on units see Appendix 1):

Table 10.6 Factors relating to the universal soil loss equation (USLE)

Factor	Description	Derivation
Ecological conditions		
Erosivity of soil, *R*	Rainfall totals, intensity and seasonal distribution. More rain does not necessarily mean more erosion: maximum erosivity occurs when the rainfall occurs as high intensity storms. If such rain is received when the land has just been ploughed or full crop cover is not yet established, erosion will be greater than when falling on a full canopy. Minimal erosion occurs when rains are gentle, and fall onto frozen soil or land with natural vegetation or a full crop cover	The total storm energy and the rain received during the most intense 30 minutes. The annual index is the sum of the index for all storms during the year, and the average over many years is input to the equation
Erodibility, *K*	The susceptibility of a soil to erosion. Depends upon infiltration capacity and the structural stability of soil. Soil with high infiltration capacity and high structural stability that allow the soil to resist the impact of rain splash, have lowest erodibility values	Determined using a plot of land 22 m long on a slope of 9 per cent in continuous fallow. It is expressed as the average soil loss per unit of the *R* factor
Length–slope factor, *LS*	Slope length and steepness influence the movement and speed of water down the slope, and thus its ability to transport particles. The greater the slope, the greater the erosivity; the longer the slope, the more water is received on the surface	Expressed as the ratio of the expected soil loss that is observed for a field of length 22.6 m. S is a slope gradient index, the ratio of the expected soil loss to that observed for a field of specified slope of 9 per cent.
Land use type		
Crop management, *C*	Most control can be exerted over the cover and management of the soil, and this factor relates to the type of crop and cultivation practices. Established grass and forest provide the best protection against erosion, and of agricultural crops, those with the greatest foliage and thus greatest ground cover are optimal. Fallow land or crops that expose the soil for long periods after planting or harvesting offer little protection	The ratio of soil loss from land cropped under current or specific conditions (e.g. crop, stage of growth, etc.), to the loss from tilled land under ploughed, continuous fallow conditions
Soil conservation, *C*	Soil conservation measures that can be undertaken to reduce erosion or slow runoff water. Examples are contour ploughing, bunding, use of strips and terraces	Calculated as the ratio of soil loss under current practice, to the soil loss that would occur if the land were ploughed up and down the slope

Source: Based on discussions in Brady (1990), Sonneveld and Nearing (2003), and elsewhere

$$A = RKLSCP$$

where:

A = the predicted soil loss (t ha^{-1} yr^{-1}),
R = the climatic erosivity or the rainfall erosivity index (MJ ha^{-1} mm^{-1} h^{-1}),
K = soil erodibility (t MJ^{-1} h^{-1} mm^{-1}),
L = slope length
S = slope gradient
C = cover and management
P = erosion control practise.

Focus 10.2 shows an application of GIS-based USLE to a subcatchment of the Lake Victoria basin in central Uganda.

Limitations of the USLE

The empirical nature of the USLE means that its accuracy is limited when applied to conditions or areas outside of those used to derive the model. The

FOCUS 10.2

GIS-BASED USLE IN A SMALL CATCHMENT OF THE LAKE VICTORIA BASIN

A GIS-based application of the USLE has studied soil erosion in a subcatchment (30 km²) of the Lake Victoria basin (LVB), 12 km from Masak town in central Uganda (Lufafa et al. 2003).

The problem
The LVB is an important agricultural area of east Africa (Uganda, Kenya and Tanzania). Owing to recent increases in population and a move from traditional mixed perennial agriculture, to intensive cash crops, severe soil degradation has occurred due to nutrient mining and erosion. As a result, banana yields have fallen severely to 6–17 t ha^{-1} compared to an estimated potential of 60–90 t ha^{-1}, creating food insecurities. An improved understanding of the processes leading to soil degradation is required in order to manage the banana–coffee smallholdings in a sustainable way. Farm observations have shown that erosion is a serious problem but the cost of collecting empirical data has prevented further analyses to assess, for example, the spatial and temporal variability of erosion and dominant forces over wide areas. Thus, predictive models such as the USLE are of immense use in identifying areas of particular concern, and its use within a GIS framework allows prediction for individual cells (micro-areas), allowing the identification

of problem areas more efficiently than traditional lumped methods.

The hilly topography of the study area has a dense network of streams and drains, and the natural vegetation is woodland with papyrus marsh. The climate is tropical wet and dry, with average annual precipitation of 1,218 mm, bimodally distributed with a long rain season between March and June and a short rainy season between August and November. The average annual temperature is 21.5°C. Within the study area, six main land use types were identified: forests, swamp, rangeland, banana, annual crops and banana–coffee intercrop.

Results of the investigation
The predicted soil losses are presented over study area for land use units, terrain units and soil units. In general, the soil loss estimates exceeded the tolerable value (5 t ha^{-1} yr^{-1}) (for more on units see Appendix 1). Lufafa et al. (2003) used four different approaches to parameterize the USLE that gave consistent but different erosion rates, and which are presented as a range of soil losses.

Of the land use units, while the banana–coffee intercrop took up the largest land area, the highest erosion rates were predicted for annual cropland use (65–83 t ha^{-1} yr^{-1}) (Table 10.7). This was followed by

Table 10.7 Annual soil losses computed using the USLE for a subcatchment of the Lake Victoria basin. Employing four different types of parameterization in the USLE generates a range of soil losses

Land use	Land cover (per cent)	Range of soil loss (t ha^{-1} yr^{-1})
Annual cropland	6	65–93
Range land	15	42–68
Banana–coffee	63	36–47
Banana	6	22–32
Forest	1	0
Papyrus marsh	9	0

Source: Adapted from Lufafa et al. (2003)

rangeland (42–68 t ha^{-1} yr^{-1}), banana–coffee intercropping (36–47 t ha^{-1} yr^{-1}), banana (22–32 t ha^{-1} yr^{-1}) and forest and papyrus swamp (both 0 t ha^{-1} yr^{-1}). Lufafa *et al.* (2003) comment that the lower erosion rates computed for the banana monocrop compared to the banana–coffee intercrop are due to the ground cover (mulch) management. Generally, the land cropped to coffee is not mulched. The authors note that groundcover is more important than canopy cover when it comes to erosion control. No erosion was predicted for forest or papyrus marsh, due to their position in the valley bottoms. Overall, soil losses for the area, weighted to land use, were predicted in the range 29–37 t ha^{-1} year^{-1}.

For the terrain units, the most common slope position were backslopes, for which the model predicted the greatest soil loss (32–48 t ha^{-1} yr^{-1}) due to their longer slope length and deeper erodible soils, compared to summits (28–42 t ha^{-1} yr^{-1}) and valley bottoms (0 t ha^{-1} yr^{-1}). Average weighted erosion rates ranged between 20–30 t ha^{-1} yr^{-1}.

For the soil units, the erosion rates of the different soil types generally matched the pattern of the terrain units, due to the link between soil classification and slope position. Chromic Luvisols had the greatest coverage and the highest predicted erosion (41–52 t ha^{-1} yr^{-1}), followed by Petroferric Luvisols (27–39 t ha^{-1} yr^{-1}). Of the soils that occupy the valley bottoms, no erosion was predicted for the Dystric Planosols and very little (3–5 t ha^{-1} yr^{-1}) for the Mollic Gleysols. Weighted average erosion ranged between 33–42 t ha^{-1} yr^{-1}.

small-scale nature of the plot size, the limited range of soil types and the limited geographical location of the data used, mean that the model is ideally applied to similar conditions and therefore not universal in application. In addition, the model output of average annual soil loss is a useful concept in the climatic context in which it was developed, but of limited use in the tropical, semitropical and semi-arid world, where a probability distribution of soil loss would be of more benefit (Lufafa *et al.* 2003). The equation does not take into account the socioeconomic factors that may be influential in soil erosion. More recent research develops physically based simulations of the complex interaction between erosion and its controlling variables.

Sonneveld and Nearing (2003) review the USLE and its shortcomings. They comment that, due to the ease of use and low data demands compared with physically based models, the USLE remains the most popular tool for water erosion hazard assessment. The simple structure of the model allows easy estimates to be made for different scenarios of agricultural practices, soil conservation measures, planning and climate changes by changing the land use under different ecological circumstances. The major shortcomings of the USLE model, however, include overestimation of small soil losses (Nearing 1998) and underestimation of high losses (Keyzer and Sonneveld 1998). The mathematical nature of the equation can also lead to errors whenever one of the factors is imprecise, raising issues about the robustness of the implicit parameter assignments.

In view of these shortcomings, many researchers have attempted to improve the USLE, including the use of fuzzy logic-based modelling (Tran *et al.* 2002) and artificial neural networks (Licznar and Nearing 2003). The most notable attempt is the publication of the Revised Universal Soil Loss Equation (RUSLE) (Renard *et al.* 1998). The RUSLE was devised using new data for rainfall erosivity and consideration of certain erosion processes such as the seasonal variability in K, factors for subsurface drainage, the dependence of the contour P-factor on storm severity and the creation of C sub-factors to account for prior land use, canopy cover and so forth. Sonneveld and Nearing (2003) comment that, as many of the changes of the RUSLE are for specific areas or applications, the new model enables a wider application of the model rather than increased accuracy over the USLE.

Lin *et al.* (2002) report that the USLE is the standard approach in Taiwan to predict soil loss, but that as the USLE was derived with data from small agricultural plots, there is a considerable challenge to scale the USLE to larger areas such as catchments. The USLE should be applied only to areas that experience net erosion. Within the model, erosion from saturated overland flow is not explicitly considered and sediment deposition is not represented. Thus, the application of the USLE to landscape areas rather than small plots is more difficult, particularly as the model does not distinguish areas that experience net deposition from those that experience net erosion. It is difficult to quantify correct values for the six factors for large areas. The most commonly misused parameter is the slope factor due to the complexity in defining slope for entire watersheds. Lin *et al.* (2002) conclude that, as the USLE does not consider gully, riverbed or bank erosion, or sediment deposition, it is optimally applied for land portions where the slope shape is sufficiently steep to ensure net soil detachment and not concave in shape, with slope lengths in the order of 1–100 m.

Erosion by wind

The factors underlying wind erosion have also been examined, and the Wind Erosion Equation (WEQ) was developed in the 1960s (e.g. Woodruff and Siddoway 1965). As with the USLE, this approach is limited in its application due to its empirical nature. Since this approach was put forward, wind erosion research has focused on the development of physically based models examining the interactions between the controlling variables of erosion. However, owing to their complex data needs, these models are also difficult to adapt to new places or climates. The WEQ predicts the potential average annual soil loss using information on the soil erodibility, wind erosivity and land use:

$$E = f(I\ C\ K\ L\ V)$$

where:

E = the erosion per unit area
K = soil erodibility
C = local climate factors
I = soil surface roughness
L = width of the field
V = extent of vegetative cover.

Wind erodibility of soil is similar to soil erodibility for water erosion, and relates to infiltration capacity and structural stability. The local climate factors of importance include wind strength and turbulence, which influence the size of particles that can be dislodged and carried by the wind. The nature and type of precipitation is important, as a wet surface will not suffer erosion by wind. Temperature is also important as it influences the rate at which soil will dry out after rains. A rough soil surface breaks the flow of the wind, slowing it down and reducing its ability to transport or dislodge particles. Surfaces are rough when covered by a crop or vegetation; ploughing to produce large aggregates or clods can give some resistance for bare soils. Windbreaks, such as rows of willow trees, stop the continuous flow of the wind, and decrease velocities. They can also act as sediment traps. The width of the field is important as shorter field widths give a greater roughness to the surface than larger fields and a shorter fetch (the length of uninterrupted wind flow). A network of smaller fields with different land uses or with barriers such as hedges in between, increase surface roughness over short intervals reducing the velocity and therefore erosive power of the wind. The vegetative cover is also important as the denser the cover the more the surface is protected from wind action. In addition, a dense root network acts to bind the soil together giving resistance to destructive forces. Vegetation can also act to buffer the wind, acting as windbreaks, and can trap sediment.

Modelling wind erosion

Wind Erosion on European Light Soils (WEELS) was an international project, started in 1998, to look at sites in England, Sweden, Germany and the Netherlands where serious wind erosion problems occur. The models devised by the project predict soil loss through wind erosion for different management and climatic scenarios.

Böhner *et al.* (2003) enhanced the model's applicability for use over large areas by designing modules incorporating a combination of algorithms and different approaches and using easily collected data for input. Modules on 'wind', 'wind erosivity' and 'soil moisture' cover climatic erosivity, while 'soil erodibility', 'surface roughness' and 'land use' modules cover controlling soil, land and vegetation factors. The model still has limitations: the simplification of the soil-water budget parameters limiting its application to sandy soils, and the non-consideration of suspension losses. The project comprised field measurements at agricultural test sites, as well as wind tunnel experiments. The model was run using a dataset of 29 years (1970–98) for the English site and 13 years (1981–93) for the German site. Owing to the characteristics of wind climates the model results reveal a higher erosion risk for the English site, where the total mean annual soil loss was estimated at 1.56 t ha^{-1}. Greatest erosion was experienced in March with mean monthly values of more than 0.28 t ha^{-1}. For the German test site, the total mean soil loss was less, at 0.43 t ha^{-1} yr^{-1}, with the greatest erosion occurring in April with values in excess of 0.08 t ha^{-1}. To assess the effects of changing crop rotations on erosion risk, a management scenario was performed for the German site. The model was run twice: once with climate and land use data over 1981–93, and again using the same climate data but with land use data from 1971–83. This was done to enable a direct examination of the effect of the intensification of maize cultivation caused by changing market conditions in the 1980s. Because of these land use changes, the areas of bare or barely covered soils in spring increased significantly, and the model results show erosion increased by 50 per cent (0.21 t ha^{-1} yr^{-1}). This result highlights the possibility of managing erosion risk by adjusting land use strategies.

PREVENTING EROSION

As the wind and water equations highlight the most influential factors upon erosion, it follows that control or manipulation of those factors can do the most to reduce or control erosion. Methods to prevent erosion can be mechanical, for instance the use of physical barriers such as embankments or windbreaks, or focus on cropping techniques and soil husbandry. Attempts must be made to reduce surface runoff. This can be done by stopping the flow of water and encouraging infiltration into the soil.

Mechanical methods include the use of **bunding**, terracing and contour ploughing, and shelterbelts such as trees or hedgerows. The key action is to prevent or slow the movement of rain water down the slope, which can have a powerful erosive capability. **Contour ploughing** takes advantage of the ridges formed normal to the slope to act to prevent or slow the downward accretion of soil and water. On steep slopes and those with heavy rainfall, such as the monsoon storms experienced in South-East Asia, contour ploughing is insufficient and **terracing** is undertaken. The slope is broken up into a series of flat stages or steps, with bunds at the edge. In this way, water is ponded in the terrace and drainage controlled from one terrace to another down the slopes. Bunding takes up a relatively large area of space but the use of terracing allows areas to be cultivated that would not otherwise be suitable. In areas where wind erosion is a problem, such as Europe, **shelterbelts** such as tree lines or hedgerows are used. The trees act as a barrier to the wind and disturb its flow. Wind speeds are reduced which therefore reduce its ability to disturb the topsoil and erode particles.

Preventing erosion by different cropping techniques largely focuses on maintaining a crop cover for as long as possible, keeping in place the stubble and root structure of the crop after harvesting, or planting a grass crop. A grass crop maintains the action of the roots in binding the soil, minimizing the action of wind and rain on a bare soil surface. The improved soil husbandry includes improving the soil organic content by growing grass crops and incorporating crop stubble and manure into the soil. Increased organic content allows the soil to hold more water, thus preventing aerial erosion, and stabilizing the soil structure. In addition, care is taken over the use of heavy machinery on wet soils and ploughing on soils sensitive to erosion, to prevent damage to the soil structure. Table 10.8 lists some measures to curb wind erosion.

An understanding of **catenary** and landscape processes is necessary to combat soil erosion. Soil and vegetation cover degradation is prevalent in many areas of Sudano-Sahelian West Africa. An example of a successful programme for combating degradation is described by Roosé *et al.* (1992) and Dugué *et al.* (1993), and was developed for farmers in Yatenga, a north-west province of Burkina Faso (formerly Upper Volta). The programme took into account the multifaceted nature of the problem, by addressing a host of factors such as the diversity of ecological conditions and socio-economic conditions of the farmers, traditional water and soil fertility management, and differences in soil properties (e.g. infiltration rates, organic matter and nutrient balances). The programme combined soil conservation practices with improved management of water resources and soil nutrients, which boosted yields and economic returns. The measures adopted were dependent upon soil type, which in turn varied with catenary position. Farmers were encouraged to first redevelop their individual plots in the main cultivated areas. These are positioned in the central portion of the hillslope, and consisted of the best land of tropical ferruginous soils. Next, the communal areas of the upper catena (gravelly lithosols) should receive attention, and finally the valley bottom (hydromorphic and sandy soils that are

bunds: low banks of earth sometimes created at the edge of human-made terraces. Grass or shrubs may be planted on them

contour ploughing: ploughing along contours, at right angles to the steepest slope

terracing: the construction of a series of flat platforms on slopes

shelterbelts: natural or artificial windbreaks

catenary: pertaining to a soil catena, that is, a linked series of soils running from a hilltop to valley bottom

Table 10.8 List of most common measures to minimize wind erosion risk on the light agricultural soils of Northern Europe

Measures	Remarks
Measures that minimize actual risk (short-term effect)	
Autumn sown varieties	Need to be sown before the end of October to develop a sufficient cover
Mixed cropping	After the main crop is harvested, second crop remains on the field
Nursing or cover crop	More herbicides needed
Straw planting	Unsuitable on light sandy soils
Organic protection layer (e.g. liquid manure; sewage sludge; sugar beet factory lime)	Use depends on availability, and regulations on the use of these products.
Synthetic stabilizers	Unsuitable on peat soils
Time of cultivation	Depends on availability of labour and equipment
Cultivation practice (e.g. minimum tillage; plough and press)	Not suitable for all crop or soil types
Measures that lower the potential risk (long-term effect)	
Smaller fields	Increase in operational time and costs
Change of arable land to permanent pasture or woodland	Loss of agricultural production and farm income
Marling (increasing the clay content to 8–10 per cent)	Suitable material should be available close by
Wind barriers	High investment costs, and loss of production land. Takes several years before it provides full protection. Protection from the shelter reduces with distance

flooded during heavy storms) should be redeveloped. Measures to conserve soil and prevent erosion in the main cultivated areas included stone bunds and rows of vegetation to trap soil particles. The vegetation consisted of groves of trees, and tree lines along roads and farm boundaries, hedges of African vetiver (*Andropogon guyanus*) and shrubs such as *Acacia* and balsam spurge (*Euphorbia balsamifera*). Structures were also built to collect runoff and included artificial ponds and lakes where the topography allowed. The stored water was used for livestock, and irrigation for small gardens of fruit trees and crops. In the upper catena area, the gravelly and sandy hills, soil erosion defences and grasses and shrubs were planted, to help regenerate pasture development. Completely degraded soils were restored using the traditional technique of Zaï, which involves digging holes and filling them with compost to add organic material to the soil and build up fertility. Other measures included redeveloping gullies and valley bottoms, building stone barriers in the villages to prevent gullying and fixing gullies at road-crossing points, and planting orchards and vegetable gardens in the dry season.

SOIL CONTAMINATION

Soils become contaminated when a particular substance within the soil exceeds a safe threshold level. Pollutants arrive in or on soils either as a result of being dumped, as in landfill sites, or by being mobilized through drainage waters, seepage or atmospheric deposition.

Pollutants can be dispersed or concentrated. Dispersed pollutants include the acid compounds in acid rain, or radioactive fallout from events such as the Chernobyl disaster in 1986. Abrahams (2002) uses Chernobyl and its radioactive fallout as an example of how soils and health are linked. Soils over a vast area were contaminated by radionuclides, which could then enter and cause the contamination of food chains, posing a threat to human health. For example, in the UK high concentrations of caesium-137 and caesium-134 deposited on the Welsh and Cumbrian uplands were taken up by heather (*Calluna vulgaris)* and bilberry (*Vaccinium myrtillu*s) and then transferred from the plants into grouse, sheep and honey (via bees), posing threats to human health. The concentrations taken-up and the persistence of the radionuclides are much greater than originally perceived, and as a result, the restricted sale of sheep and their slaughter in parts of the uplands will remain for some decades to come.

An example of concentrated soil contamination is given by Rivett *et al.* (2002), who report the results of a survey of contaminated land and groundwater remediation in England and Wales. The survey of derelict land and other local and national sources suggest that more than 100,000 contaminated sites may exist, although the lack of a national inventory and adequate records mean that this figure may be much larger. Contaminated land is an inheritance from Britain's industrial past, for example from chemical and engineering works, and the high dependence on landfill for waste disposal (over 80 per cent of UK domestic waste ends up in landfill). Smaller activities can also contaminate land, such as the storage tanks of fuel stations, dry-cleaning facilities and power generators. Groundwater underlying contaminated sites may also become contaminated due to the passage of **leachate**. Such contamination has serious impacts in terms of both human health and the economic costs of not only remediation but also the reduced quality of the soil and water resource. As brown-field sites are being increasingly identified for use for human habitation, the remediation of contaminated and hazardous soils has evolved into an important industry.

> *leachate*: the cocktail of chemicals being washed through the soil

Heavy metals

An example of soil contamination can be given by heavy metals. Metals are classified as 'heavy' if their density is greater than 5 or 6 g cm^{-3}, and they are present naturally in certain rocks and soils (Alloway 1990). Heavy metals include lead, cadmium, copper, zinc and nickel. They are of particular concern when they contaminate soil, water and air through human actions. Such actions include the disposal of industrial and domestic waste containing metal compounds in landfill sites (seepage of leachate into soil and groundwater), the spread of sewage and slurry waste over agricultural land, certain fertilizers and atmospheric deposition (e.g. from industrial and vehicular emissions and, though this is not due to human action, active volcanoes).

Only a very small fraction of the metals in soil is exchangeable and easily taken up by plants; the majority is strongly bound to soil constituents. Under conditions of low pH, however, more of the metal fraction becomes available, and so the monitoring of pH in contaminated soils is necessary to ensure it does not fall to levels where metals will be released. Waterlogged anaerobic conditions can also make some metal species become more readily available, by changing their form from insoluble and adsorbed to soluble. Some metals are essential for life and biological activity in small or minute quantities, but harmful in large quantities.

Once the contaminant (metal or other toxic inorganic compound) is introduced into the environment, for example by the application of sludge to input organic matter to the soil, it can be taken up by plants and fauna such as earthworms. These in turn are eaten by organisms higher in the food chain, increasing the levels of contamination stored in the tissues with each trophic step up to the top of the food chain through biomagnification (p. 384). The widespread uptake of such toxic contaminants has led to restrictions being placed on certain foodstuffs such as shellfish, and the introduction of laws regulating the use and disposal of products containing such contaminants.

Heavy metal contamination of soils near mines occurs due to the erosion and deposition of fine-grained tailings and ore by either wind or water, and by particles emitted into the atmosphere during smelting (Focus 10.3). Deposition of heavy metal particles is greatest at the mine site and decreases with distance. Landfill sites are another important source for heavy metal contamination. Landfill sites usually occupy former gravel extraction sites or quarries. Efforts are made to seal the base of the pit with an impermeable layer, such as clay, to prevent leachate from seeping though the pit and into groundwater. Groundwater contamination can occur if the liner becomes cracked or breached, and the resulting contamination plume can move great distances and contaminate water supply aquifers needed for agricultural, industrial and public consumption. When full, landfill sites are covered with topsoil and used for agricultural or other purposes. Plants growing on former landfill sites can uptake metals in their roots and store it within their tissues. Furthermore, owing to the anaerobic conditions within landfill sites, carbon dioxide and methane gases are produced by bacterial respiration, both of which are greenhouse gases.

Sewage sludge is applied to agricultural land as the organic content helps to improve soil texture, and it is a source of nitrogen and phosphate for crops. Problems arise when the sludge contains levels of heavy metals, pathogens or other contaminants, which are hazardous to animal health and toxic to plants. Continued application of slurry over many years can cause a build up of metals

FOCUS 10.3

ITAI-ITAI DISEASE IN JAPAN

Itai-itai disease in Toyama Prefecture, Japan, is an example of the impact of metal contaminated soil on health (Abrahams 2002). Symptoms of the disease include pain in the joints and bones, severe bone decay and serious kidney dysfunction. Autopsies revealed high concentrations of cadmium (Cd) in soft tissues. The existence of the disease was made public in 1955, and in 1968 the Japanese Government officially announced that the cause was cadmium-contaminated food. The blame was placed on the Kamioka mine, which released cadmium-loaded water into the river, which was then used as an irrigation source for rice paddies. The development of the disease cannot be fully explained by excess cadmium levels, however. The severity of the disease in Tomaya Prefecture compared to other regions with high cadmium levels is due to

other contributing environmental factors. The acidic and anaerobic nature of the paddy fields results in relatively high soluble cadmium concentrations compared to well-drained soils, and the solubility of the cadmium leads to its enhanced uptake and bioaccumulation by plants. The malnutrition and restricted diet of the people affected, largely confined to produce grown in the contaminated paddy fields, made them particularly susceptible.

The impact of contaminated soil is shaped not only by the concentration levels, but also by local soil characteristics such as drainage and pH in governing bioavailability, by plant factors such as species, stage of growth and season controlling the rate of uptake, and by the dietary variety and intake of the affected people (Abrahams 2002).

in the soil that may pose problems over long periods with heavy metal cations held on the negatively charged clays and colloids.

Methods to try to reduce the levels of contamination in soil can be achieved by stopping the application of contaminants to the soil in slurry or other forms and to manage those contaminated soils to prevent the metals from becoming available. This is done by maintaining neutral or high pH levels by liming, increasing the organic matter and improving drainage to encourage aerobic conditions. Contaminants can also be physically isolated and contained to minimize movement. Physical barriers such as cement walls or clay lining can be used to isolate and minimize the mobility of the contaminant. Alternatively, injecting or mixing in certain substances with the contaminated soil can cause it to solidify (Mulligan *et al.* 2001).

Biological methods to remediate contaminated land are being developed. Plants can be used to extract metals (or other contaminants) out of the soil, which is termed **phytoremediation**. Certain plants are planted over contaminated areas to **bioaccumulate** toxins and contaminants, and are regularly cropped to remove the accumulated metal or toxin, for disposal away from the site. For example, Pulford and Watson (2002) report the suitability of willow trees (*Salix* spp.) for the phytoremediation of heavy-metal-contaminated land, and Mulligan *et al.* (2001) report that plants such as *Thlaspi*, *Urtica*, *Chenopodium*, *Polygonum sachalase* and *Alyssum* have the capability to accumulate cadmium, copper, lead, nickel and zinc respectively. In addition, the use of bacteria for remediation is termed **bioleaching**. Bacteria such as *Thiobacillus* sp. can be used to reduce sulphur compounds under acidic conditions over a temperature range of 15 and 55°C (Mulligan *et al.* 2001).

phytoremediation: the use of plants to decompose organic chemical wastes and pollutants

bioaccumulation: the selective accumulation of nutrients and toxic substances in the cells and tissues of organisms

bioleaching: the use of certain bacteria to remove heavy metals from contaminated soils

WORLD FOOD SUPPLY

The frequent reports of famine and food shortages in the world media highlight the global problems of providing enough food for the world's population. While 'grain mountains' and food surpluses are common in Europe and America, famine and food shortages seem all too common in other parts of the world.

Major factors underlying this are economic, social and political in addition to physical factors such as soil quality. Food shortages during a drought period can be viewed as a crisis of land degradation resulting from agricultural practices. However, the food crisis can also be viewed as an impact of the global political economy, where the integration of the region into the world economy has led farmers to abandon crops that had formerly provided a highly reliable and drought-resistant food source, to cash crops which are more water intensive and more susceptible to drought (Rocheleau *et al.* 1995).

Global trade arrangements such as GATT and the CAP subsidize agricultural production in western counties to the detriment of many nations in the developing world. Often the import of western foodstuffs into developing nations undermines the local agricultural economy to the extent that farmers cannot sell their produce. In addition, the unequal land distribution in many developing countries results in poorer farmers being pushed onto marginal land, susceptible to flooding and erosion. These factors can give rise to a spiral of poverty (Ellis and Mellor 1995) involving enhanced soil erosion, land abandonment and urban migration.

The problem of food supply and land degradation is complex, and perceived causes differ between 'insiders' and 'outsiders' (Focus 10.4).

FOCUS 10.4

AKAMBA PEOPLE, KENYA

The Akamba people live in the Ukambani region, near Nairobi, Kenya. Their case shows that simple solutions, or policies based on single perspectives of multifaceted problems, result in ineffective responses or, at worst, contribute to the construction of future new crises (Rocheleau *et al.* 1995). Outsiders identified a history of crises, but in doing so, failed to recognize that the critical conditions resulted directly from the policies and interventions of their predecessors in response to previously considered 'crises'. A variety of crises including famine, disease, soil erosion, desertification, fuel wood and energy shortages, overpopulation, poverty, deforestation and overgrazing and the decline of biodiversity were attributed to the agricultural

practices of the Akamba people, rather than those of the European colonial settlers, and reflected global concerns at the time (Table 10.9). The Akamba farmers and herders, however, recounted a different story in which enforced land alienation and market manipulation to protect the settlers' monopoly, together with limits on the mobility of people and herds, have restructured the ecological and spatial order of the region. Ukambani is a predominantly semi-arid region, and only the highest altitude regions permit agriculture without the need for irrigation (Rocheleau *et al.* 1995). Soils are of low fertility and highly erodible. Vegetation comprises dry bush with trees, and, in the higher areas, savannah with scattered trees. Before the Colonial Era,

Table 10.9 One hundred years of 'crisis' in Ukambani

Years	External phenomena	Internalized definition of crisis	Official response
1890–WWI	Rivalry among colonial powers	Identification of need for settlers to have privileged access to land and resources and control of commodity production	Disruption of Akamba land tenure system
	Equation of poverty, high population densities and 'primitive' civilization with disease	Identification of 'irrational' disease prone Akamba cattle rearing processes on crowded reserves	Further land alienation and segregation of Akamba economy
1920s and 1930s	Equation of poor agricultural practices with soil erosion	Identification of destructive Akamba agricultural practices	Forced terracing and production limits
1940s and 1950s	War time and post WWII First World resource needs	Identification of under-production on Akamba reserves: recognition of land hunger as a source of political instability	Forced cash crop production, enclosure and privatization of land
1960s and 1970s	Competition among newly independent states in world commodity markets	Identification of low productivity among Akamba farmers	Further encouragement of cash cropping and land concentration among efficient producers
1980s onwards	Concern about declining soil quality	Identification of 'destructive' agricultural practices as cause of and dam sedimentation river problems	Promotion of intercropping and terracing techniques
	Concern about declining global energy resources	Identification of 'other energy crisis' and Akamba overcutting of trees for fuel wood	Promotion of monocrop or alley cropping, miracle-tree plantings
	Concern about declining global gene pool, endangered species and endangered ecosystems	Identification of the 'threat' posed to national parks and the wildlife therein by area residents	Increased efforts to segregate citizenry from parkland, separate pursuits of conservation and agriculture policies in separate spaces

Source: Adapted from Rocheleau *et al.* (1995)

the Akamba successfully grew crops and kept livestock. Land was a mix of private and common holdings, and flexible patterns of settlement and mobility provided mechanisms for coping with drought. During the early phase of colonization, with the introduction of major infrastructure and contact with European people and livestock, the region suffered epidemics of rinderpest (cattle) and smallpox (humans). Europeans blamed the epidemics on the 'primitive' lifestyle of Akamba people, rather than the introduction of new diseases by the settlers to which the Akamba had no immunity. The loss of cattle and debilitation caused by disease coupled with drought in the 1890s made the Akamba particularly susceptible to the famine over 1897–1901, when over half the population perished. An extensive land seizure and enclosure programme was organized by incoming settlers and, by 1920, two-thirds of the land held by the Akamba had been taken. Land enclosure removed their ability to migrate in search of water or pasture when conditions demanded, and cattle were kept in overcrowded reserves where disease spread easily. These early policies of land seizure resulted in the sedentarization and concentration of the Akamba people that are underlying causes of future crises (Rocheleau et al. 1995).

The disruption of traditional land tenure and land use systems resulted in continuous cultivation of small areas of relatively poor quality land. From initial concerns of overgrazing and soil exhaustion, the focus of the authorities switched to a perceived crisis of underproduction with the subsequent promotion of cash cropping. This led to the replacement of drought resistant sorghum and millet with maize, leading to greater erosion, reduced yields and greater vulnerability to drought; and the ploughing of grazing land, putting increased pressure on common land. Further land-reform policies from the 1940s to Kenyan Independence were implemented to encourage the intensification and development of commercial agriculture. The result for the Akamba was land concentration in the hands of the few, with further poverty, and landlessness for many. Many of those displaced had little option other than to migrate to cities for work or to drier, more vulnerable, fringe areas to try to establish new farms.

Crises attributed to Akamba agricultural practices (Table 10.9) failed to recognize the important part played by the commercial farms run by settlers. For example, deforestation was largely blamed on the Akamba practice of collecting wood for fuel, while the wholesale clearance of large areas to expand commercial farms was overlooked. Loss of biodiversity was attributed to the Akamba rather than the homogenization of the agrarian landscape by the large farms.

Rocheleau et al. (1995) conclude that the focus on one problem at one scale obscures larger-scale processes that operate behind the scenes and form the underlying causes. Owing to this simplistic approach at managing 'crises', previous actions play a part in the development of future crises as the underlying mechanisms are not recognized.

Part of the cause of the unequal supply of food is the high global variation of soil type and productivity as well as the availability of water, technology and fertilizers, and improved plant varieties. Figure 10.8 shows the constraints for agriculture in different regions of the world. Drought and mineral stress are among the major factors affecting productivity; the impact of climate change on precipitation distribution could exacerbate agricultural output significantly on a global scale.

Of the 13 billion ha of land on the major continents, only 25 per cent supports vegetation to provide grazing and 25 per cent could be used for cultivation (Brady 1990). Again, unequal distribution, this time of population density, means that the land area cultivated per person is high in North America, the former USSR and Oceania, and low in Europe, Asia and Africa. Of those areas with low productivity per capita, Europe and parts of Asia have economies strong enough to be able to buy and import food. Populations are increasing rapidly in the regions of Latin America, south and South-East Asia and Africa, and national economies are not strong enough to purchase sufficient food for their populations. However, caution should be taken when thinking of expanding cultivation to new lands. A report by Oldeman et al. (1990) found that overgrazing and poor soil husbandry across the world results in land in

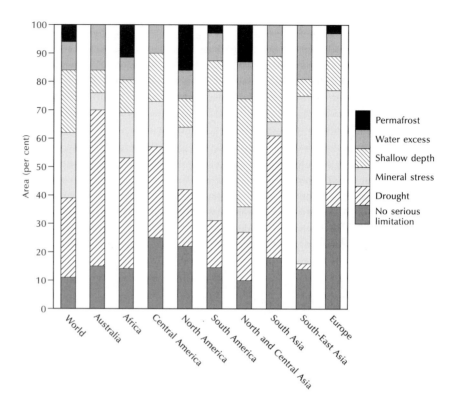

Figure 10.8 Soil-related constraints for agriculture in different regions of the world. *Source:* Adapted from Dent (1980)

cultivation becoming degraded with reduced productivity. Enhanced management of land already in production should be attempted first before new areas are disturbed. Figure 10.9 shows that water erosion is the largest soil problem worldwide caused by human activities. In Africa and Asia, wind erosion is also significant, in addition to the nutrient depletion and poorly managed irrigation practices leading to salinization.

To increase food production, three routes can be taken (Brady 1990):

1.   Untilled land can be tilled and cultivated.

2.   The number of crops grown per year can be increased.

3.   Production on land already under tillage can be intensified.

The first two routes are not options for western countries, as these have already occurred. The only route remaining is the intensification of production, and recent developments such as genetically modified crops are a response to this. In other parts of the world, untilled land can be brought into cultivation, but barriers such as the lack of infrastructure, costs of fertilizers, machinery, transport and sustainable management are real obstacles. The environmental costs of bringing such land into cultivation are also incredibly high. They include not only the potential loss of species and habitat areas, but also the potential impact on climate change through the release of gases from burning forests and the cessation of **carbon sequestration**. Massive soil erosion and increased flooding and landslide risk follow the clearing of slopes for agricultural purposes in tropical areas. The destruction of large pristine habitat areas has seen the reduction of biological diversity and the extinction of many species.

carbon sequestration: the net removal of carbon dioxide in the atmosphere and its long-term storage in live or dead plant tissues (e.g. leaves, branches, stems, roots, litter, peat)

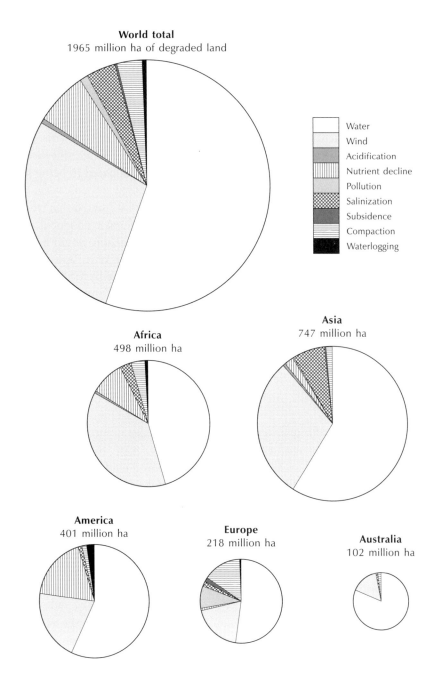

World total
1965 million ha of degraded land

Water
Wind
Acidification
Nutrient decline
Pollution
Salinization
Subsidence
Compaction
Waterlogging

Africa
498 million ha

Asia
747 million ha

America
401 million ha

Europe
218 million ha

Australia
102 million ha

Figure 10.9 Soil degradation (millions of hectares) across the world, caused by human activities. *Source:* Adapted from Oldeman *et al.* (1991)

CLIMATE CHANGE

Soil is a source of many of the greenhouse gases. The decomposition of organic matter produces carbon dioxide, although in small quantities when compared to industrial sources; and methane is produced in wetland and waterlogged soils, such as in paddy cultivation. Denitrification in agricultural soils and under tropical forests is estimated to produce about a third of nitrous oxide content in the atmosphere. Predicting the results of potential climate change are difficult, given the subtle interrelationships between a plethora of environmental and agricultural parameters.

Climate change simulations for a doubled atmospheric carbon dioxide concentration predict regional temperature changes of the range -3 to $+10°C$ and precipitation fluctuations of ±20 per cent. Such meteorological changes will affect rates of evapotranspiration, but the increase in carbon dioxide levels will also change the biological (stomatal) response of plants. Increased carbon dioxide concentrations may reduce the quality of crops, as assimilated nitrogen levels fall and weed growth increases. If temperatures increase in conjunction with atmospheric carbon dioxide, then crops will mature at a faster rate and there is the risk that pests will be prevalent for the entire year as opposed to being killed by low winter temperatures.

The Intergovernmental Panel on Climate Change (IPCC) stated that 'the balance of evidence . . . suggests a discernable human influence on global climate' (MAFF 2000). It is hypothesized by the Panel that this would have an influence on British land cover and soils. The concentration of carbon dioxide is predicted to increase and summer temperatures are set to rise by $1.5°C$ over the next 50 years. The predicted increases in carbon dioxide concentrations are stated as having mainly beneficial effects, such as increasing crop productivity, while an increase in temperature would have implications for water resources. Crops, such as potato, rely on temperature and soil water status to induce each development stage and as such could be affected by potential climatic change. Anthropogenic climate change will affect ecosystems in a variety of ways but the ultimate response will depend on the interactions between climate and other global system components.

The coupled effect of higher temperatures and changes in moisture availability will lead to a range of plant and soil responses (Hillel 1998). In continuously vegetated areas, enhanced levels of atmospheric carbon dioxide can enable photosynthesis leading to sequestration of carbon in organic matter in the soil. In cultivated soil, the reverse effect may occur: the enhanced decomposition caused by higher temperatures can deplete the soil organic matter, releasing carbon dioxide into the atmosphere. This is a particular concern for wetland and peat soils, where desiccation by higher temperatures and changes in moisture availability can lead to the oxygenation of the peat and release of carbon dioxide into the atmosphere, exacerbating the greenhouse effect. Increased rainfall over winter months would likely proliferate the incidence of crop disease, whereas reduced values in spring and summer would have implications on the available water content of a soil. Negative feedbacks would mean that increases in carbon dioxide levels could enhance stomatal resistance and consequently decrease transpiration rates (MAFF 2000). However, this decrease in transpiration will create a warmer canopy, which will amplify rates of water loss.

SUMMARY

Soils are an essential part of global environmental systems and processes. Together with water, soils are immensely important to human societies in the production of crops for humans and grazing plants for animals. This chapter has looked at the basic processes underpinning soil formation and soil properties, and at the impacts of human activities on soil conditions. The mismanagement of soils, from contamination to erosion caused by overgrazing, causes problems that will remain for many generations to come. Soils have an important part to play in climate change, due to their importance in carbon and nitrogen cycles, and the effects of climate change, particularly temperature and precipitation, will impact on land suitability for agriculture, and in those areas already on the edges of overproduction, upon world food supply.

Questions

1 What are the dominant soil-forming processes? How do these differ in different climatic zones?
2 What are the principal causes of soil erosion and how can they be managed?
3 What are the short- and long-term effects of overgrazing and erosion, not only for the soil but also for the local economy as a whole?
4 How can the risk of soil contaminations and pollution be minimized?
5 What can be done to increase food security and yet simultaneously protect vulnerable soil reserves?
6 What might be the impact of climate change on soils and the ability of soils to support agricultural systems?

Further Reading

Brady, N. C. (1990) *The Nature and Properties of Soils*, 10th edn. London: Collier Macmillan.
An excellent introductory text aimed at soil scientists and covering a range of topics. Well worth looking at.

Bridges, E. M. (1997) *World Soils*, 3rd edn. Cambridge: Cambridge University Press.
A very good introduction to all types of soils and soil-forming processes.

Ellis, S. and Mellor, A. (1995) *Soils and Environment*. London: Routledge.
A refreshingly different approach to soils and contemporary environmental issues.

Rowell, D. L. (1994) *Soil Science: Methods and Applications*. Harlow: Longman.
An excellent book covering laboratory experiments and methods used in determining soil properties.

Useful Websites

http://soils.usda.gov/technical/classification/taxonomy/
This is on the site for the United States Department of Agriculture, Natural Resources Conservation Center (NRCC). It contains a downloadable version of the Soil Taxonomy (Soil Survey Staff 1999) with maps and photographs, and lots of other useful information on soils.

http://www.soils.org/
The official website of the Soil Science Society of America. Includes a useful glossary of soil science terms.

http://www.cbr.clw.csiro.au/aclep/asc/soilkey.htm
This site gives the Australian system of soils classification, with maps and colour images of sample profiles.

http://www.soils.org.uk/
The web page of the British Society of Soil Science, has many good links to soil information and online resources.

http://www.fao.org/
The pages of the Food and Agricultural Organization of the United Nations (FAO) give a wealth of information relating to soils, food production and international concerns.

http://www.usda.gov/ *and*
http://www.nrcs.usda.gov/technical/land/
The web page of the United States Department of Agriculture and the associated Natural Resources Conservation Service contains information on soils and resources in an American, but also a world, context.

http://lime.isric.nl/index.cfm?fuseaction=dsp_menu&menuid=2
The International Soil Reference and Information Centre site, hosted at Wageningen University, presents pages giving an introduction to soil science and soil use and misuse.

CHAPTER 11
LANDFORMS

OVERVIEW

Human involvement in the geomorphic system is
extensive, with huge amounts of sediment
derived from human activities being delivered to
the oceans. Humans affect weathering, limestone
landscapes, glacial and periglacial landscapes,
fluvial landscapes, aeolian landscapes and
coastal landscapes. This chapter examines these
human impacts on landforms and land forming
processes.

LEARNING OUTCOMES

Human activities have substantial effects on
landforms and land-forming processes. This
chapter will help you to understand:
- Why historical buildings in many cites are
  succumbing to accelerated weathering
- Why limestone landscapes suffer accelerated
  soil erosion and cave degradation
- How humans alter river regimes and why they
  need to manage rivers
- How frozen areas of the world are susceptible
  to direct human impacts and climatic warming
- How humans disturb beach sediment supply
  and what will happen to coastal landscapes if
  the sea level should rise.

The land surface is fashioned by geological and geomorphic process. Weathering, erosion, transport and deposition are the chief geomorphic processes. They result from chemical and physical instabilities in rocks lying at and near the Earth's surface, from the mining activities of plants and animals, from gravitational forces and from flowing water and wind. Humans interfere with these processes, often to a considerable degree and with disastrous consequences.

11.1 WEATHERING IN CITIES

All rocks weather, but many seem to weather faster in urban environments. Archibald Geikie, who studied the weathering of gravestones in Edinburgh and its environs, found that limestone weathers faster in urban environments than in surrounding rural areas. Recent studies of weathering rates on marble gravestones in and around Durham, England, give rates of 2 microns per year in a rural site and 10 microns per year in an urban industrial site (Attewell and Taylor 1988).

The accelerated weathering of historical buildings in cities, which is caused by pollutants, has considerable economic and cultural costs. Geomorphologists study urban weathering forms, measure weathering rates and establish connections between the two (e.g. Inkpen *et al.* 1994). Based on this research they advise such bodies as the Cathedrals Fabric Commission in an informed way. The case of the Parthenon, Athens, is typical. The Parthenon, a temple dedicated to the goddess Athena, was built 447–432 BC on the Acropolis of Athens, Greece. During its 2500-year history, it has suffered damage. The Elgin Marbles, for example, once formed an outside frieze. Firm evidence now suggests that continuous damage is being caused by air pollution and that substantial harm has already been inflicted in this way. For example, the inwards-facing carbonate stone surfaces of the columns and the column capitals bear black crusts or coatings. These damaged areas are not significantly wetted by rain or rain runoff, although acid precipitation may do some harm. The coatings seem to be caused by sulphur dioxide uptake, in the presence of moisture, on the stone surface. Once on the moist surface, the sulphur dioxide is converted to sulphuric acid, which in turn results in the formation of a layer of **gypsum**. Researchers are undecided about the best way of retarding and remedying this type of air pollution damage.

> **gypsum**: a white or colourless mineral. Hydrated calcium sulphate, $CaSO_4.2H_2O$

Weathering is also damaging St Paul's Cathedral in London, England, which is built of Portland limestone. It has suffered considerable attack by weathering over the past few hundred years. Portland limestone is a bright white colour; before recent cleaning, St Paul's was a sooty black. Acid rain waters have etched out hollows where it has run across the building's surface. Along these channels, bulbous gypsum precipitates have formed beneath anvils and gargoyles, and acids (particularly sulphuric acid) in rainwater have reacted with the limestone. About 0.62 microns of the limestone surface are lost each year, which represents a cumulative loss of 1.5 cm since St Paul's was built (Sharp *et al.* 1982).

Salt weathering is playing havoc with buildings of ethnic, religious and cultural value in some parts of the world. In the towns of Khiva, Bukhara and Samarkand, which lie in the centre of Uzbekistan's irrigated cotton belt, prime examples of Islamic architecture – including mausolea, minarets, mosques and madrasses – are being ruined by capillary rise, a rising water table resulting from over-irrigation, and an increase in the salinity of the groundwater (Cooke 1994). The solution to these problems is that the capillary fringe and the salts connected with it must be removed from the buildings, which might be achieved by more effective water management (e.g. the installation of effective pumping wells) and the construction of damp-proof courses in selected buildings to prevent capillary rise.

11.2 HUMAN IMPACTS ON KARST LANDSCAPES

> **limestone**: a rock containing at least 50 per cent calcium carbonate ($CaCO_3$)
>
> **dolomite**: a rock containing at least 50 per cent calcium–magnesium carbonate ($CaMg(CO_3)_2$), a mineral called dolomite
>
> **anhydrite**: anhydrous calcium sulphate ($CaSO_4$) occurring as a white mineral. Usually found in association with gypsum, rock salt and other evaporite minerals

Karst is terrain in which soluble rocks are altered above and below ground by the dissolving action of water, and it bears distinctive characteristics of relief and drainage (Jennings 1971, 1). It usually refers to limestone terrain, characteristically lacking surface drainage, possessing a patchy and thin soil cover, containing many enclosed depressions and supporting a network of subterranean features, including caves and grottoes. However, all rocks are soluble to some extent in water, and karst is not confined to the most soluble rock types, forming also in sandstone, quartzite and in some basalts and granites under favourable conditions. Widespread areas of karst develop in carbonate rocks (**limestones** and **dolomites**), and sometimes in evaporites, which include halite (rock salt), **anhydrite** and gypsum.

Surface and subsurface karst are vulnerable to human activities. Visitors damage caves, and agricultural practices may lead to the erosion of soil cover from karst areas.

SOIL EROSION ON KARST

Karst areas worldwide tend to be susceptible to soil erosion. Their soils are usually shallow and stony, and, being freely drained, leached of nutrients. When vegetation is removed from limestone soils or when they are heavily used, soil stripping down to bedrock is common. It can be seen on the Burren, Ireland, in the classic karst of the Dinaric Alps, in karst of China, in the cone karst of the Philippines and elsewhere. In Greece, soil stripping over limestone began some 2000 years ago. The **limestone pavement** above Malham Cove (in the north of England) may be a legacy of agricultural practices since Neolithic times, soils being thin largely because of overgrazing by sheep. Apart from resulting in the loss of an agricultural resource, soil stripping has repercussions in subterranean karst. The eroded material swiftly finds its way underground where it blocks passages, diverts or impounds cave streams and chokes cave life.

> **limestone pavements**: extensive benches or plains of bare rock in horizontally bedded limestones and dolomites

The prevention of soil erosion and the maintenance of critical soil properties depend upon the presence of a stable vegetation cover. Soil erosion on karst terrain tends to be higher than in most soil types because features of the geomorphology work together to promote even greater erosion than elsewhere. In most non-karst areas, soil erosion depends upon slope gradient and slope length, as well as the other factors in the Universal Soil Loss Equation (USLE) (p. 299). It also depends partly on slope gradient and slope length in karst terrain but, in addition, the close connections between the surface drainage system and the underground conduit system produce a locally steeper hydraulic gradient that promotes erosive processes. Moreover, eroded material in karst areas has a greater potential to be lost down joints and fissures by sinkhole collapse, gullying or soil stripping. An adequate vegetation cover and soil structure (which reduce erodibility) take on a greater significance in lessening this effect in karst areas than most other places.

HUMANS AND CAVES

Humans have long used caves for shelter, defence, sanctuaries, troglodytic settlements, as a source of resources (water, food, guano and ore in mine-caves) and as spiritual sites. In the last few hundred years, caves have been used for the mining of cave formations (such as stalactites and stalagmites) and guano (especially during the American Civil War), for hydroelectric power generation from cave streams and springs (in China), for storage and as sanatoria and tourist attractions. Humans first occupied caves in China over 700,000 years ago. Many caves are known to have housed humans at the start of the last glacial stage, and several have walls adorned with splendid paintings. Numerous caves in the Guilin tower karst, China, have walls at their entrances, suggesting that they were defended. Mediaeval fortified caves are found in Switzerland in the Grisons and Vallais. In Europe and the USA, some caves were used as sanatoria for tuberculosis patients on the erroneous premise that the moist air and constant temperature would aid recovery. Caves have also been widely used for cheese-making and rope manufacture, as in the entrance to Peak Cavern, Derbyshire, England. Kentucky bourbon from the Jack Daniels distillery relies partly on cave spring water.

Cave tourism is a burgeoning business that is producing degradation of caves as a symptom of Mass Tourism syndrome (p. 58).

MANAGING KARST

Karst management is based on an understanding of karst geomorphology, hydrology, biology and ecology. It has to consider surface and subsurface processes, since the two are intimately linked. The basic aims of karst management are to maintain the natural quality and quantity of water and air movement through the landscape, given the prevailing climatic and biotic conditions. The flux of carbon dioxide from the air, through the soils, to cave passages is a crucial karst process that must be addressed in management plans. In particular, the system that produces high levels of carbon dioxide in soil, which depends upon plant root respiration, microbial activity and a thriving soil invertebrate fauna, needs to be kept running smoothly.

Many pollutants enter cave systems from domestic and municipal, agricultural, constructional, mining and industrial sources. In Britain, 1458 licensed landfill sites are located on limestone, many of which take industrial wastes. Material leached from these sites may travel to contaminate underground streams and springs for several kilometres. Sewage pollution is also common in British karst areas (Chapman 1993).

Limestone and marble are quarried around the world and used for cement manufacture, for high grade building stones, for agricultural lime and for abrasives. Limestone mining spoils karst scenery, causes water pollution and produces much dust. Quarrying has destroyed some British limestone caves and threatens to destroy others. In southern China, many small quarries in the Guilin tower karst extract limestone for cement manufactories and for industrial fluxes. In combination with vegetation removal and acid rain from coal burning, the quarrying has scarred many of the karst towers around Guilin City that rise from the alluvial plain of the Li River. It is ironic that much of the cement is used to build hotels and shops for the tourists coming to see the limestone towers. In central Queensland, Australia, Mount Etna is a limestone mountain containing 46 named caves, many of which are famous for their spectacular formations. The caves are home to some half a million insectivorous bats, including the rare ghost bat (*Macroderma gigas*). The mining of Mount Etna by the Central Queensland Cement Company has destroyed or affected many well-decorated caves. A public outcry led to part of the mountain being declared a reserve in 1988, although mining operations continue outside the protected area where the landscape is badly scarred.

The IUCN (International Union for the Conservation of Nature and Natural Resources) World Commission on Protected Areas recognizes karst landscapes as critical targets for protected area status. The level of protection given in different countries is highly variable, despite the almost universal aesthetic, archaeological, biological, cultural, historical and recreational significance of karst landscapes. Take the case of South-East Asia, one of the world's outstanding carbonate karst landscapes, with a total karst area of 458,000 km^2, or 10 per cent of the land area (Day and Urich 2000). Karstlands in this region are topographically diverse and include cockpit and cone karst, tower karst and pinnacle karst, together with extensive dry valleys, cave systems and springs. They include classic tropical karst landscapes: the Gunung Sewu of Java, the Chocolate Hills of Bohol, the pinnacles and caves of Gunong Mulu and the karst towers of Vietnam and peninsular Malaysia. Human impacts on the South-East Asian karst landscapes are considerable: less than 10 per cent of the area maintains its natural vegetation. About 12 per cent of the regional karst landscape has been provided nominal protection by designation as a protected area, but levels of protection vary from country to country. Protection is significant in Indonesia, Malaysia, the Philippines and Thailand. Indonesia, for instance, has 44 protected karst areas, which amount to 15 per cent of its total karst area. In Cambodia, Myanmar (Burma) and Papua New Guinea, karst conservation is minimal, but additional protected areas may be designated in these countries, as well as in Vietnam and in Laos. Even so, South-East Asia's karstlands have an uncertain future. It should be stressed that the designation of karst as protected areas in South-East Asia is not based on the intrinsic or scientific value of the karst landscapes, but on unrelated contexts, such as biological diversity, timber resources, hydrological potential or archaeological and recreational value. Nor, it must be said, does the conferral of a protected area status guarantee effective protection from such threats as forest clearance, agricultural inroads or the plundering of archaeological materials.

The conservation of karst in the Caribbean is in a similar position to that in South-East Asia (Kueny and Day 1998). Some 130,000 km^2, more than half the land area of the Caribbean, is limestone karst. Much of it is found on the Greater Antilles, with other significant areas in the Bahamas, Anguilla, Antigua, the Cayman Islands, the Virgin Islands, Guadeloupe, Barbados, Trinidad and Tobago and the Netherlands Antilles. Features include cockpits, towers, dry valleys, dolines and caves. Humans have affected the karst landscapes, and the necessity for protection at regional and international level is recognized. However, karst is in almost all cases protected by accident – karst areas happen to lie within parks, reserves and sanctuaries set up to safeguard biodiversity, natural resources or cultural and archaeological sites. Very few areas are given protected area status because of the inherent scientific interest of karst landscapes. At the regional level, 121 karst areas, covering 18,441 km^2 or 14.3 per cent of the total karst, are afforded protected area status. Higher levels of protection are found in Cuba, the Dominican Republic and the Bahamas. Lower levels of protection occur in Jamaica, Puerto Rico, Trinidad and the Netherlands Antilles, and minimal protection is established in the smaller islands.

11.3 HUMAN IMPACTS ON FLUVIAL LANDSCAPES

Fluvial landscapes are dominated by running water. They are widespread except in frigid regions, where ice dominates, and in dry regions, where wind tends to be the main erosive agent.

IMPACTS IN THE PAST

Human agricultural, mining and urban activities have caused changes in rivers. The Romans transformed fluvial landscapes in Europe and North Africa by building dams, aqueducts and terraces. A water diversion on the Min River in Sichuan, China, has been operating ceaselessly for over 2000 years. In the north-eastern United States, forest clearance and subsequent urban and industrial activities greatly altered rivers early in the nineteenth century. A case study from Upper Weardale, England, illustrates well the factors at work in past river modification.

Swinhope Burn is a tributary of the upper River Wear, in the northern Pennines, England. It is a gravel-bed stream with a catchment area of 10.5 km^2 (Warburton and Danks 1998; Warburton *et al.* 2002). Figure 11.1 shows the historical evidence for changes in the river pattern from 1815 to 1991. In 1815, the river meandered with a sinuosity similar to that of the present meanders. By 1844, this meandering pattern had broken down to be replaced by a more-or-less straight channel with a bar braid at the head, which is still preserved in the floodplain. By 1856, the stream was meandering again, which pattern persisted to the present day. The change from meandering to braiding appears to be associated with lead mining. A small vein of galena cuts across the catchment and there is a record of 326 tonnes of galena coming out of Swinhope mine between 1823 and 1846. It is interesting that, although the mining operations were modest, they appear to have had a major impact on the stream channel.

PRESENT RIVER MODIFICATION AND MANAGEMENT

Fluvial environments present humans with many challenges. The degradation of rivers downstream of dams is a concern around the world. Many European rivers are complex managed entities. In the Swiss Jura, changes in some rivers to

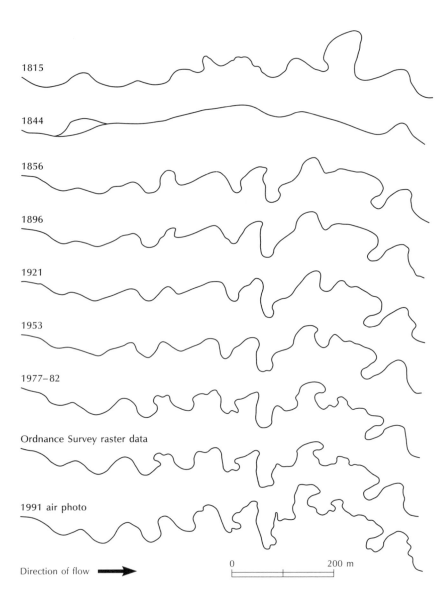

1815

1844

1856

1896

1921

1953

1977–82

Ordnance Survey raster data

1991 air photo

Direction of flow ➡

0   200 m

Figure 11.1 Channel change in Swinhope Burn, Upper Weardale, Yorkshire. The diagram shows the channel centre-line determined from maps, plans and an air photograph. *Source:* After Warburton and Danks (1998)

improve navigation destabilized the channels and a second set of engineering works were needed to correct the impacts of the first (Douglas 1971). Within the Rhine Valley, the river channel is canalized and flows so swiftly that it scours its bed. To stop undue scouring, a large and continuous programme of gravel replenishment is in operation. The Piave River, in the eastern Alps of Italy, has experienced remarkable channel changes following decreased flows and decreased sediment supply (Surian 1999). The width of the channel has shrunk to about 35 per cent of its original size, and in several reaches the pattern has altered from braided to wandering. In England, the channelization of the River Mersey through the south of Manchester has led to severe bank erosion downstream of the channelized section, and electricity pylons have had to be relocated (Douglas and Lawson 2001).

By the 1980s, increasing demand for environmental sensitivity in river management, and the realization that hard engineering solutions were not fulfilling their design-life expectancy, or were transferring erosion problems elsewhere in river systems, produced a spur for changes in management practices.

Mounting evidence and theory demanded a geomorphological approach to river management (e.g. Dunne and Leopold 1978; Brookes 1985). Thus, to control bank erosion in the UK, two major changes in the practices and perceptions of river managers took place. First, they started thinking about bank erosion in the context of the sediment dynamics of whole river systems, and began to examine upstream and downstream results of bank protection work. Second, they started prescribing softer, more natural materials to protect banks, including both traditional vegetation, such as willow, osier and ash, and new geotextiles to stimulate or assist the regrowth of natural plant cover (Walker 1999). River management today involves scientists from many disciplines – geomorphology, hydrology and ecology – as well as conservationists and various user-groups, such as anglers (e.g. Douglas 2000). Thus, in Greater Manchester, England, the upper Mersey Basin has a structure plan that incorporates flood control, habitat restoration and the recreational use of floodplains; while, in the same area, the Mersey Basin Campaign strives to improve water quality and river valley amenities, including industrial land regeneration throughout the region (Struthers 1997).

11.4 HUMAN IMPACTS ON GLACIAL AND PERIGLACIAL LANDSCAPES

GLACIAL LANDSCAPES

Glacial landscapes are products of frigid climates. During the **Quaternary**, the covering of ice in polar regions and on mountain tops waxed and waned in synchrony with swings of climates through glacial–interglacial cycles. Humans can live in glacial and periglacial environments but only at low densities. Direct human impacts on current glacial landscapes are small, even in areas where tourism is popular. Indirect human impacts, which work through the medium of climatic change, are substantial: global warming appears to be melting the world's ice and snow. Over the last 100 years, mean global temperatures have risen by about 0.6°C, about half the rise occurring in the last 25 years. The rise is greater in high latitudes. For example, mean winter temperatures at sites in Alaska and northern Eurasia have risen by 6°C over the last 30 years (Serreze *et al.* 2000), which is why glacial environments are so vulnerable to the current warming trend.

> *Quaternary*: a geological subera spanning the past 2 million years

Relict glacial landscapes, left after the last deglaciation some 10,000 years ago, are home to millions of people in Eurasia and North America. The relict landforms are ploughed up to produce crops, dug into for sand and gravel and are covered by concrete and tarmac. Such use of relict landscapes raises issues of landscape conservation, although knowledge of Quaternary sediments and their properties can aid human use of relict glacial landscapes (Focus 11.1).

PERIGLACIAL LANDSCAPES

Attempts to develop periglacial regions face unique and difficult problems associated with building on an icy substrate (Focus 11.2). Undeterred, humans

FOCUS 11.1

WASTE DISPOSAL SITES IN NORFOLK, ENGLAND

The designing of waste disposal sites in south Norfolk, England, is aided by an understanding of the Quaternary sediments (Gray 1993). Geologically, south Norfolk is a till plain that is dissected in places by shallow river valleys. It contains very few disused gravel pits and quarries that could be used as landfill sites for municipal waste. In May 1991, Norfolk County Council applied for planning permission to create a 1.5 million cubic metres aboveground or 'landraise' waste disposal site at a disused US World War II airfield at Hardwick. The proposal was to dig a 2–4 m deep pit in the Lowestoft Till and overfill it to make a low hill standing up to 10 m above the surrounding plain. The problem of leachate leakage from the site was to be addressed by relying on the low permeability of the till and reworking the till around the edges of the site to remove potentially leaky sand-lenses in its upper layers. In August 1993, after a public inquiry into the Hardwick site, planning permission was refused, partly because knowledge of the site's geology and land drainage was inadequate and alternative sites were available. Research into the site prompted by the proposal suggested that leachate containment was a real problem and that Norfolk County Council were mistaken in believing that the till would prevent leachates from leaking. It also identified other sites in south Norfolk that would be suitable landfill sites, including the extensive sand and gravel deposits along the margins of the River Yare and its tributaries. Landraising in a till plain is also unwelcome on geomorphological grounds, unless perhaps the resulting hill should be screened by woodland. A lesson from this case study is that knowledge of Quaternary geology is central to the planning and design of landfill in areas of glacial sediments.

FOCUS 11.2

PROBLEMS OF DEVELOPMENT ON PERMAFROST

Buildings, roads and railways erected on the ground surface in permafrost areas face two problems (e.g. French 1996, 285–91). First, the freezing of the ground causes frost heaving, which disturbs buildings, foundations and road surfaces. Second, the structures themselves may cause the underlying ice to thaw, bringing about heaving and subsidence, and they may sink into the ground. To overcome this difficulty, the use of a pad or some kind of fill (usually gravel) may be placed upon the surface. If the pad or fill is of the appropriate thickness, the thermal regime of the underlying permafrost is unchanged. Structures that convey significant amounts of heat to the permafrost, such as heated buildings and warm oil pipelines, require additional measures to be taken. A common practice is to mount buildings on piles, so allowing air space below, between the building and the ground surface, in which cold air may circulate. Even so, in ground subject to seasonal freezing, the pile foundations may move, pushing the piles upwards. In consequence, bridges, buildings, military installations and pipelines may be damaged or destroyed if the piles are not placed carefully. Other measures include inserting open-ended culverts into pads and the laying of insulating matting beneath them. In addition, where the cost is justified, refrigeration units may be set around pads or through pilings. Pipes providing municipal services, such as water supply and sewage disposal, cannot be laid underground in permafrost regions. One solution, used at Inuvik, in the Canadian North West Territories, is to use utilidors. Utilidors are continuously insulated aluminium boxes that run above ground on supports, linking buildings to a central system.

The Trans-Alaska Pipeline System (TAPS), which was finished in 1977, is a striking achievement of construction under permafrost conditions. The pipeline is 1285 km long and carries crude oil from Prudhoe Bay on the North Slope to an ice-free port at Valdez on the Pacific Coast. It was originally planned to bury the pipe in the ground for most of the route, but as the oil is carried at 70–80°C, this would have melted the permafrost and the resulting soil flow would have damaged the pipe. In the event, about half of the pipe was mounted on elevated beams held up by 120,000 vertical support members (VSMs) that were frozen firmly into the permafrost, using special heat-radiating thermal devices to prevent their moving. This system allows the heat from the pipe to be dissipated into the air, so minimizing its impact on the permafrost.

Few roads and railways have been built in permafrost regions. Most roads are unpaved. Summer thawing, with the resultant loss of load-bearing strength in fine-grained sediments, and winter frost-heaving, call for the constant grading of roads to maintain a surface smooth enough for driving. Paved roads tend to become rough very quickly, most of them requiring resurfacing every 3 to 5 years. Railways are difficult to build and expensive to keep up in permafrost regions. The Trans-Siberian Railway, and some Canadian railways in the north of the country (e.g. the Hudson Bay railway), cross areas where the ground ice is thick. At these sites, year-round, costly maintenance programmes are needed to combat the effects of summer thawing and winter frost-heaving and keep the track level. The Hudson Bay railway has been operating for over 60 years. For all that time, it has faced problems of thaw settlement along the railway embankment and the destruction of bridge decks by frost-heave. Heat pipes help to minimize thaw subsidence but they are very expensive.

have exploited tundra landscapes for 150 years or more, with severe disturbances occurring after World War II due to exploration for petroleum and other resource development (e.g. Bliss 1990). Permafrost degradation occurs where the thermal balance of the permafrost is broken, either by climatic changes or by changing conditions at the ground surface.

In the Low Arctic, mineral exploration has led to the melting of permafrost. Under natural conditions, peat, which is a good insulator, tends to prevent permafrost from melting. Where the peat layer is disturbed or removed, as by the use of tracked vehicles along summer roads, permafrost melt is encouraged. Ground-ice melting and subsequent subsidence produce **thermokarst**, which resembles karst landscapes. In the Tanana Flats, Alaska, USA, ice-rich permafrost that supports birch forest is thawing rapidly, the forests being converted to minerotrophic (receiving nutrients from minerals dissolved in groundwater) floating mat fens (Osterkamp *et al.* 2000). A hundred years ago or more at this site, some 83 per cent of 260,000 ha was underlain by permafrost. About 42 per cent of this permafrost has been affected by thermokarst development within the 100 to 200 years. The thaw depths are typically 1 to 2 m, with some values as high as 6 m. On the Yamal Peninsula of north-west Siberia, land use and climatic changes since the 1960s, when supergiant natural gas fields were discovered, have led to changes in the tundra landscape (Forbes 1999). Extensive exploration meant that large areas were given over to the construction of roads and buildings. Disturbance associated with this development has affected thousands of hectares of land. The increasing amount of land given over to roads and buildings, together with the associated disturbed land, has driven a roughly constant or increasing reindeer population onto progressively smaller patches of pasture. In consequence, the patches have suffered excessive grazing and trampling of lichens, bryophytes (mosses, liverworts and hornworts) and shrubs. In many areas, sandy soils have been deflated. The human- and reindeer-induced disturbance may easily initiate thermokarst formation and aeolian erosion, which would lead to significant further losses of pasture.

Thermokarst is less likely to develop in the High Arctic, owing to the lower permafrost temperatures and the generally lower ice content. Nonetheless, gully erosion can be a serious problem in places lacking a peat cover. For instance, when snow, piled up to clear areas for airstrips and camps, melts in the spring, the meltwater runs along minor ruts caused by vehicles. In a few years, these minor ruts may be eroded into sizeable gullies. A trickle of water may become a potent and erosive force that transforms the tundra landscape into a slurry of mud and eroding peat. Restoration work is difficult because gravel is in short supply and a loss of soil volume occurs during the summer melt. In any case,

thermokarst: irregular terrain produced by the thawing of ground ice in periglacial environments

gravel roads, although they will prevent permafrost melt and subsidence if they are thick enough, have deleterious side effects. For instance, culverts designed to take water under the roads may fill with gravel or with ice in the winter. In three sites within the Prudhoe Bay Oil Field, studied from 1968 to 1983, blocked drainage-ways have led to 9 per cent of the mapped area being flooded and 1 per cent of the area being thermokarst (Walker *et al.* 1987). Had not the collecting systems, the camps and the pipeline corridors been built in an environmentally acceptable manner, the flooding and conversion to thermokarst might have been far greater. Water running parallel to the roads and increased flow from the culverts may lead to a combined thermal and hydraulic erosion and the production of thermokarst.

Global warming during the twenty-first century is bound to have a large impact on permafrost landscapes, and no effectual countermeasures are available (Lunardini 1996). Much of the discontinuous permafrost in Alaska is now extremely warm, usually within 1–2°C of thawing. Ice at this temperature is highly susceptible to thermal degradation and any additional warming during the current century will result in the formation of new thermokarst (Osterkamp *et al.* 2000). In the Yamal Peninsula, a slight warming of climate, even without the human impacts on the landscape, would produce massive thermokarst erosion (Forbes 1999).

11.5 HUMAN IMPACTS ON COASTAL LANDFORMS

Humans affect erosion and deposition along coasts. They do so through increasing or decreasing the sediment load of rivers, by building protective structures, and indirectly by setting in train climatic processes that lead to sea level rise. Two important issues focus around beach erosion and beach nourishment and the effect of rising sea levels over the next century.

BEACH EROSION AND BEACH NOURISHMENT

To combat beach erosion, especially where it threatens to undermine and ruin roads and buildings, humans have often built sea walls. The idea is that a sea wall will stop waves attacking the eroding coast, commonly a retreating cliff, and undermining a slumping bluff or a truncated dune. Sea walls often start as banks of earth but once these are damaged, stone or concrete constructions usually replace them. Other options are boulder ramparts (also called revetments or riprap) and artificial structures such as tetrapods, which are made of reinforced concrete. Solid sea walls, and even boulder barriers and other artificial structures, are effective and reflect breaking waves seawards, leading to a backwash that scours the beach of material. Such is the demand for countermeasures against coastal erosion that the world's coastline is littered with a battery of artificial structures. Some structures are successful, but the unsuccessful ones stand in ruins. Some have helped to maintain beaches, but others, by promoting eroding backwash, simply worsen beach erosion.

In an effort to prevent beach loss, the dumping of sand or gravel on the shore has become a common practice, mainly in the USA, Western Europe and Australia. Such beach nourishment aims to create a beach formation that 'will protect the coastline and persist in the face of wave action' (Bird 2000, 160).

FOCUS 11.3

BEACH EROSION VERSUS BEACH NOURISHMENT – A DELAWARE CASE STUDY

Nourishing beaches costs money. Any beach nourishment programme has to weigh the economics of letting beaches retreat against the economics of sustaining them. Take the Atlantic coastline of Delaware, eastern USA (Daniel 2001). Delaware's coastline combines high shoreline-property values with a growing coastal tourism industry. It is also a dynamic coastline, with storm damage and erosion of recreational beaches posing a serious threat to coastal communities. Local and State officials are tackling the problem. A comprehensive management plan, called Beaches 2000, considered beach nourishment and retreat. The goal of Beaches 2000 is to safeguard Delaware's beaches for the citizens of Delaware and out-of-state beach visitors. Since the Beaches 2000 plan was published, Delaware's shorelines have been managed through nourishment activities, which have successfully maintained beach widths. Coastal tourism, recreational beach use and real estate values in the area continue to grow. The possibility of letting the coastline retreat was considered in the plan, but shelved as an option for the distant future. One study estimated the land and capital costs of letting Delaware's beaches retreat inland over the next 50 years (Parsons and Powell 2001). The conclusion was that, if erosion rates remain at historical levels for the next 50 years, the cost would be $291,000,000, but would be greater should erosion rates accelerate. In the light of this figure, beach nourishment makes economic sense, at least over the 50-year time period.

Many beach nourishment programmes were implemented at seaside resorts, where beaches are desired for recreational use (Focus 11.3). Recently, the value of a beach in adsorbing wave energy has been realized and nourishment beaches are sometimes used to defend against further cliff erosion or damage to coastal roads and buildings. The key to a successful beach nourishment programme is a thorough comprehension of coastal geomorphology. Before developing and implementing a programme, it is necessary to find out the movement of sand and gravel in relation to the wave regimes and the effects of any artificial structures on the section of shore to be treated. It is also necessary to understand why the beach is eroding and where the sediment has gone – landward, seaward or alongshore. The modelling of beach forms and processes can be helpful at this stage, but an experimental approach, based on accumulated experience, is often more productive. The sediment used to nourish a beach should be at least as coarse as the natural beach sediment and durable, but it may come from any source. More material than is necessary to restore the beach is used, so allowing for onshore, offshore and alongshore losses. It is usually dumped to form a beach terrace, which is worked on by waves and currents to form a natural beach profile, often with sand bars just offshore. The restored beach may be held in place by building a retaining backwater or a series of **groynes**. In some cases, a beach can be nourished by dumping material where it is known that longshore or shoreward drift will carry it to the shore. Nourished beaches will normally still erode and occasionally need replacing. More details of beach management can be found in Bird (1996).

> **groynes**: low walls, usually made of wood, built out into the sea to help prevent beach erosion

THE EFFECTS OF RISING SEA LEVELS IN THE TWENTY-FIRST CENTURY

A current worry is how coastlines will respond to rising sea levels during the present century. Estimates of sea level rise are 10 to 15 cm by the year 2030, accelerating to 30 to 80 cm by 2100 (Wigley and Raper 1993). Inevitably, submerging coastlines, presently limited to areas where the land is subsiding, will become widespread and emerging coastlines will become a rarity. Broadly speaking, low-lying coastal areas will be extensively submerged and their high- and low-tide lines will advance landwards, covering the present intertidal zone. On steep, rocky coasts, high- and low-tide levels will simply rise and the coastline stay in the same position. It seems likely that the sea will continue to rise, with little prospect of stabilization. If it does so, then coastal erosion will accelerate and become more prevalent as compensating sedimentation tails off. The rising seas will reach, reshape and eventually submerge 'raised beaches' created during the Pleistocene interglacials. Forms similar to those found around the present world's coasts would not develop until sea level stabilized, which would presumably occur either when the measures adopted to counterbalance increasing greenhouse gases worked, or else when all the world's glacier, ice sheets and snowfields had melted, occasioning a global sea level rise of more than 60 m (Bird 2000, 276).

Specific effects of rising sea levels on different types of coasts are summarized in Figure 11.2. Cliffs and rocky shores were largely produced by the tolerably stable sea levels that have dominated over the last 6000 years. Rising sea levels will submerge shore platforms and rocky shores, allowing larger waves to reach cliffs and bluffs, so accelerating their erosion on all but the most resistant rocks (Figure 11.2a). Some eastern British cliffs are retreating about 100 cm yr^{-1} and this rate will increase by 35 cm yr^{-1} for every 1 mm rise of sea level (Clayton 1989). Cliff notches will enlarge upwards as the rising sea eats into successively higher levels. Rising sea levels are also likely to increase the occurrence of coastal landslides and produce new and extensive slumps, especially where rocks dip towards the sea. The slump material will add to sediment supply for beaches, perhaps in part compensating for the rising sea level. The rise of sea level by 1 to 2 mm per year over the last few decades has caused beach erosion in many places around the world. Accelerating sea level rise will greatly exacerbate this problem. The seaward advance of prograding beaches will stop, and erosion set in (Figure 11.2b). Where the beach is narrow, with high ground behind it, the beach may rapidly disappear unless nearby cliff erosion provides enough replenishment of sediment. Beaches fronting salt marshes and mangals will probably be eroded and over-washed. Beaches ahead of sea walls will be eroded or be removed by the scour resulting from the reflection of incident waves. Beaches will persist wherever the supply of sand and shingle is sustained, or where additional material is provided by cliff erosion or increased sediment load from rivers. Most present beaches will probably be lost as sea levels rise, but on coastal plains with coastal dunes, new beaches may form by the shoreward drifting of sediment up to the new coastline, along the contour on which submergence stops.

Salt marshes, mangals and intertidal areas will all be submerged beneath rising sea levels (Figure 11.2c). Small cliffs on the seaward margins of salt marshes and mangrove terraces will erode faster than at present. Continued submergence will see the seaward and landward margins move inland. In low-lying areas, this may produce new salt marshes or mangals, but steep-rising hinterlands will cause a narrowing and perhaps eventual disappearance of the salt marsh and mangal zone. The loss of salt marshes and mangals will not occur in areas where sediment continues being supplied at a rate sufficient for a depositional terrace to persist. And modelling suggests the salt marshes of

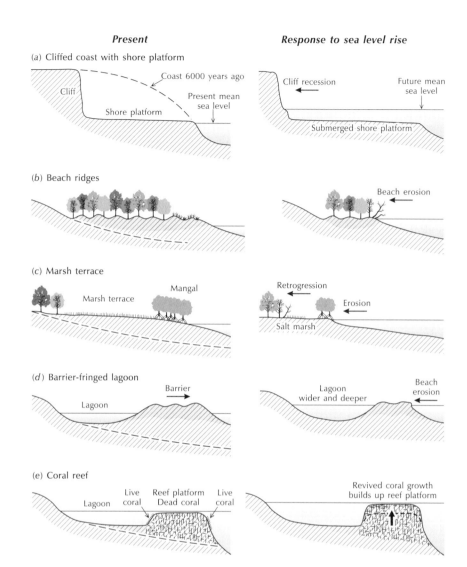

Figure 11.2 Coastal changes brought about by a rising sea level. See text for further details. *Source:* Adapted from Bird (2000, 278)

mesotidal estuaries, such as the Tagus estuary in Portugal, do not appear vulnerable to sea level rise in all but the worst-case scenario, with several industrialized nations not meeting the terms of the Kyoto Protocol (Simas *et al.* 2001). Inner salt marsh or mangal edges may expand inland, the net result being a widening of the aggrading salt marsh or mangrove terrace. Intertidal areas – sandflats, mudflats and rocky shores – will change as the sea level rises. The outer fringe of the present intertidal zone will become permanently submerged. As backing salt marshes and mangals are eroded and coastal lowland edges cut back, they will be replaced by mudflats or sandflats, and underlying rock areas will be exposed to form new rocky shores.

Estuaries will generally widen and deepen as sea level goes up, and may move inland. Coastal lagoons will become larger and deeper, and their shores and fringing swamp areas suffer erosion (Figure 11.2d). The enclosing barriers may be eroded and breached to form new lagoon inlets that, with continued erosion and submergence, may open to form marine inlets and embayments. New lagoons may form where rising sea levels cause the flooding of low-lying areas behind dune fringes on coastal plains. They may also form where depressions are flooded, as water-table rises promote the development of seasonal or permanent lakes and swamps. Wherever there is a supply of replenishing sediment, the deepening and enlargement of estuaries and lagoons may be countered.

Corals and algae living on the surface of intertidal reef platforms will be spurred into action by a rising sea level and grow upwards (Figure 11.2e). However, reef revival depends upon a range of ecological factors that influence the ability of coral species to recolonize submerging reef platforms. In addition, the response of corals to rising sea levels will depend upon the rate of sea level rise. An accelerating rate could lead to the drowning and death of some corals, and to the eventual submergence of inert reef platforms. Studies suggest that reefs are likely to keep pace with a sea level rise of less than 1 cm yr^{-1}, to be growing upwards when the sea level rise falls within the range 1–2 cm yr^{-1}, and to be drowned when the sea level rise exceeds 2 cm yr^{-1} (Spencer 1995).

SUMMARY

Humans have a great impact on the geomorphic system. Rocks weather faster in cities, causing severe damage to some historic buildings. Limestone landscapes suffer from soil erosion in areas where agricultural practices lead to overgrazing and caves and cave features (stalagmites and so on) may be damaged by the direct (physical damage) and indirect (changing cave atmosphere) effects of cave tourists. Humans dam, divert and canalize rivers. They may be contributing to the melting of the world's ice, which is causing severe problems in periglacial regions where the permafrost is thawing. Along coastlines, humans disrupt erosion and deposition regimes directly, by increasing or decreasing the sediment load of rivers and building protective structures, and indirectly by inadvertently driving climatic change in a direction leading to rising sea levels.

Questions

1 Discuss the problem of accelerated weathering in cities.

2 Explain why cave tourism may lead to cave degradation, and describe remedial measures.

3 To what extent do human activities affect river regimes?

4 Examine the problems of building houses, roads and railways in permafrost regions.

5 Outline the major changes likely to occur along coasts if the sea level rises over the next century and beyond.

Further Reading

Bird, E. C. F. (2000) *Coastal Geomorphology: An Introduction*. Chichester: John Wiley & Sons.
A highly readable and excellent text on coastal forms and processes, with information on management problems.

Ford, D. C. and Williams, P. W. (1989) *Karst Geomorphology and Hydrology*. London: Chapman & Hall.
An excellent book on karst.

Useful Websites

http://daac.gsfc.nasa.gov/DAAC_DOCS/geomorphology/GEO_HOME_PAGE.html
Geomorphology from Space: A Global Overview of Regional Landforms is an out of print 1986 NASA publication edited by Nicholas M. Short, Snr and Robert W. Blair, Jnr. Designed for use by the remote sensing science and educational communities to study landforms and landscapes, it is a gallery of space imagery consisting of 237 plates. This site offers a free online version of the publication.

French, H. M. (1996) *The Periglacial Environment*, 2nd edn. Harlow: Addison-Wesley Longman.
The best recent account of periglacial landforms and processes.

Gillieson, D. (1996) *Caves: Processes, Development and Management*. Oxford: Blackwell Publishers.
A superb book on subterranean karst with chapters on management.

Huggett, R. J. (2003) *Fundamentals of Geomorphology*. London: Routledge.
An introduction to all aspects of geomorphology.

Thorne, C. R., Hey, R. D. and Newson, M. D. (1997) *Applied Fluvial Geomorphology for River Engineering and Management*. Chichester: John Wiley & Sons.
Not a starting text but worth a look.

http://hum.amu.edu.pl/~sgp/wel.htm
Useful links put together by the Association of Polish Geomorphologists.

http://main.amu.edu.pl/~sgp/gw/gw.htm
The Virtual Geomorphology site. Very useful with many links.

4

HUMANS AND THE BIOLOGICAL ENVIRONMENT

CHAPTER 12
SPECIES AND COMMUNITIES I:
LAND COVER AND BIOTIC IMPACTS

OVERVIEW

Before the eighteenth century, humans adapted to the landscapes in which they lived. During the Great Transformation of the last two centuries (p. 5), they radically altered many of those landscapes, and especially the plant communities that they supported, converting forest and grassland to cropland and built-up areas. Habitat fragmentation is a major side effect of land cover change. Broad tracts of natural and semi-natural habitats have been broken into smaller and isolated blocks. The fragmentation of habitats occurs locally, but has cumulative effects that are regional in scale. A knock-on effect of this habitat fragmentation is a reduction in biodiversity. As with habitat fragmentation, so with the loss of biodiversity: it occurs locally but cumulates over regional and global scales. The Great Transformation also saw humans exploiting the natural world to an unprecedented extent. Conversion of land to agriculture and silviculture has promoted the expansion of some species that formerly had smaller populations, making them newly abundant. Bourgeoning trade between far-flung places encouraged both the accidental and deliberate spreads of exotic species, some of which have become invasive and cause enormous management problems.

LEARNING OUTCOMES

This chapter should help you to understand:
- What biodiversity is and how it varies across the planet
- How humans alter biodiversity by destroying habitats and breaking them into smaller and more isolated blocks
- How humans alter biodiversity by encouraging the rise of newly abundant species
- How humans alter biodiversity by aiding and abetting the spread of exotic invasive species.

12.1 INTRODUCING BIODIVERSITY

The tremendously rich biological diversity of the Earth is always changing. Since the mid-twentieth century, it has fallen fast owing to human activities.

WHAT IS BIODIVERSITY?

genetic diversity: the mix of genes or genetic variability (expressed as the number and relative abundance of alleles) in a population, or species, or other group of organisms

habitat diversity: the number of different habitats in an area

species diversity: the number and variety of species living in an area

allele: a form of a gene at a locus on a chromosome

Biological diversity, or biodiversity, is the collection of genes, species, communities and ecosystems that form the biosphere. Biodiversity therefore embraces **genetic diversity**, **habitat diversity** and **species diversity**. Genetic diversity is the sum of genetic characteristics of populations, subspecies, species or group of species, often measured as the number and abundance of **alleles** (different forms of a particular gene). Habitat diversity is the number of habitats within a given area. Species diversity can mean species richness (the total number of species), species evenness (the relative abundance of species) and species dominance (the most abundant species).

Estimates of the total number of species living on the Earth today vary enormously, ranging from 3 million to 100 million. The number of known species is about 1,400,000. Therefore, the low estimate of 3 million would mean that 46 per cent of all species are known to date; the figure drops to a mere 1.4 per cent for the high estimate of 100 million. Total species diversity figures disguise enormous differences between groups of organisms, with insects and plants being by far the most numerous groups on the planet (Table 12.1).

BIODIVERSITY GRADIENTS AND HOT-SPOTS

diversity gradients: changes in biodiversity along environmental gradients, such as the climatic gradients associated with increasing altitude and latitude

diversity hot-spots: areas or regions with exceptionally high biodiversity

Total diversity also disguises two important geographical diversity patterns: (1) altitudinal and latitudinal **diversity gradients**: and (2) **diversity hot-spots**. Many more species live in the tropics than live in the temperate regions, and more species live in temperate regions than live in polar regions. In consequence, a latitudinal species diversity gradient slopes steeply away from a tropical 'high diversity plateau'. Virtually all groups of organism exhibit the diversity gradient. An example is the species richness gradient of American mammals, where the diversity falls from a tropical high of about 450 mammal species to a polar low of about 50 mammal species. Species diversity also diminishes with increasing altitude. Superimposed on latitudinal and altitudinal trends are diversity hot-spots. These are areas where large numbers of endemic species occur (Figure 12.1). Most hot-spots occur in tropical forests and in Mediterranean biomes. These hot-spots contain 20 per cent of the world's plant species on 0.5 per cent of the land area. All are under intense development pressure.

BIODIVERSITY CHANGE

Biodiversity responds to changes in the physical and biological environment in a web of complex ways. Many scientists think the recent steep fall in

Table 12.1 The diversity of life on Earth

Group	Diversity (number of species)	Group	Diversity (number of species)
Viruses	100,000	Animals	
Monerans (bacteria and cyanobacteria)	8500	Sponges	5000
Protists (single-celled organisms)	60,000	Cnidarians (includes jelly fish, sea anemones and stony corals)	9000
Fungi	100,000		
Plants		Flatworms	13,000
Mosses	12,000	Rotifers	2000
Liverworts	10,000	Nematodes	12,000
Hornworts	100	Molluscs	110,000
Whisk ferns	Several	Annelids (segmented worms)	12,000
Club mosses and scale trees	1000	Arthropods	1,000,000
Horsetails	15		(including 500,000 insects)
Ferns	12,000	Brachiopods	330
Conifers	550	Echinoderms	7000
Cycads	100	Chordates	45,000
Gnetophytes	70	Vertebrates	44,000
Flowering plants	235,000	Cartilaginous fish	850
		Bony fish	20,000
		Amphibians	3900
		Reptiles	6000
		Birds	9000
		Mammals	4500

Sources: Various (see Margulis and Schwartz 1998)

global levels of biodiversity disquieting. Their worry is that biodiversity decline is an irremediable process that will undermine the basis of human existence.

A handful of factors seem to be the chief determinants of biodiversity change at a global scale: land cover change, changes in atmospheric carbon dioxide concentrations, nitrogen deposition and acid rain, climatic change and species exploitation and species exchange (the accidental or deliberate introduction of plants and animals to ecosystems outside those in which they live). These factors have individual effects, but they also have synergistic effects. For example, a high availability of nitrogen would augment the effects of elevated atmospheric carbon dioxide levels on ecosystems. Likewise, the effects of biotic exchanges are likely to be greater when they occur at the same time as land use changes. Similarly, predictions suggest that the combination of climate change and habitat destruction can be disastrous (Travis 2003).

The drivers of biodiversity change fall into four main categories: land cover, species exploitation and exchange, climate and environmental chemistry

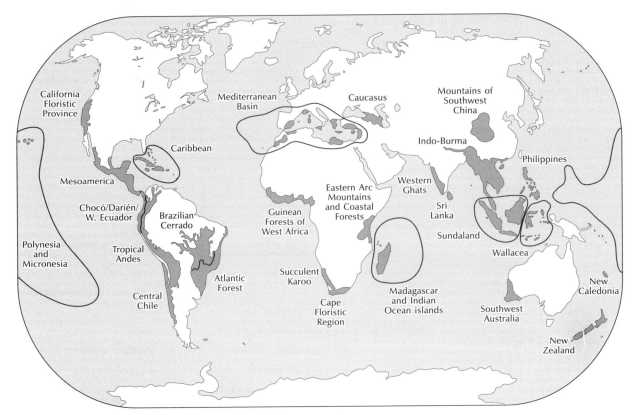

Figure 12.1 Diversity hot-spots – regions of high endemism containing many species threatened with imminent extinction. The circled areas enclose (1) the Mediterranean Basin and (2) four island hot-spots – the Caribbean, Madagascar and the Indian Ocean islands, Sundaland and Wallacea. Notice that most hot-spots occur in the tropics and in Mediterranean-type climates

(Figure 12.2). The present chapter will address the impacts of habitat fragmentation, which results from land cover change, the arrival of newly abundant species, the effects of invasive exotic species and the overexploitation of species. The next chapter will look at chemical and climatic impacts of life, and will give an overview of the drivers of biodiversity change.

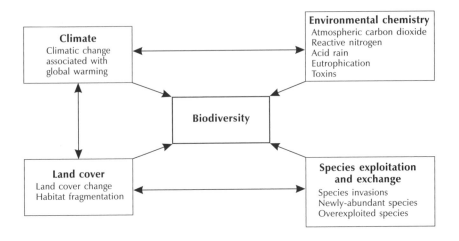

Figure 12.2 Drivers of biodiversity change and the main interactions between them

12.2 FRAGMENTED SPECIES

Humans change land cover. They clear large blocks of vegetation, bit by bit, leaving several smaller and more isolated blocks. This process, known as **habitat fragmentation**, has three linked components (Figure 12.3; Bennett 1999, 13):

> **habitat fragmentation**: the breaking up of large areas of habitat into smaller and more isolated parcels

- Habitat loss – an overall loss of the original habitat in the landscape

- Habitat reduction – a reduction in the size of blocks of habitat accompanying subdivision and clearing

- Habitat isolation – the increasing isolation of habitat blocks as new land uses occupy the intervening landscape.

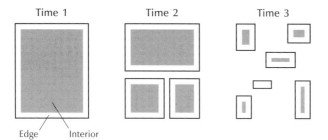

Figure 12.3 Components of habitat fragmentation. The diagram shows that as time progresses, fragmentation causes (1) a general loss of habitat, (2) a reduction in size of surviving habitat fragments and (3) an increased isolation of surviving habitat fragments

In detail, five processes cause land cover change – perforation, dissection, fragmentation, shrinkage and attrition (Forman 1995, 406–15). Forests, for example, are being perforated by clearings, dissected by roads, broken into discrete patches by felling, and the newly created patches are shrinking and some of them disappearing through attrition. Habitat fragmentation is the overall effect of land cover change. It is an extremely significant process for many animals and plants and poses a global environmental problem.

Habitat fragmentation affects many species and communities living in fragmented landscapes in several ways. The key effects are a reduction in the total amount of original habitat; a continued decrease in the size of surviving habitat blocks; the increasing isolation of surviving habitat blocks; a change in the composition of the original habitat; a change in shape of surviving habitats; and the creation of sharp boundary zones (**ecotones**) across habitat edges. In simple terms, a formerly extensive cover of, say, forest, fragments into a landscape comprising **patches** of original forest separated by **corridors** of various kinds (road, railways, trails, powerlines and so forth), all sitting in a **matrix** of a new habitat such as wheat field (Figure 12.4). The original forest would also have contained patches and corridors – natural clearings and rivers for instance – but human activities tend to produce patchier and more corridor-rich landscapes.

> **ecotones**: transitional zones between two ecosystems
>
> **patches**: fairly uniform (homogeneous) landscape areas that differ from their surroundings (e.g. woods, fields, ponds and houses)
>
> **corridors**: strips of land that differ from the land to either side (e.g. roads, hedgerows, fencerows and rivers)
>
> **matrix**: the background ecosystem or land use type in which patches and corridors are set (e.g. forest, arable and residential)

The changes that occur as landscape fragments have many consequences for animals and plants. In brief, species are lost; the composition of floral and faunal communities alters; and changing ecological processes affect animals and plants. The reasons for these changes become clear in the following discussion.

Figure 12.4 Fragments of forest vegetation in a farmland matrix at Naringal East, south-western Victoria, Australia. Photograph by Andrew Bennett

PATCH METRICS

Patch size

Patch size affects the occurrence of animal and plant species. Larger patches tend to house more species than smaller patches, an observation accounted for by the species–area curve (Basics 12.1). There are three explanations (Connor and McCoy 1979):

- A small patch contains a small 'sample' of the original habitat and is less likely, therefore, to contain all the species found in a larger patch

- Habitat diversity tends to be lower in small patches, so affecting the number of species that a small patch can support

- Species in small patches tend to have lower populations than those in larger patches, with the result that fewer species can maintain viable populations in small patches. Species with large home ranges are more sensitive to this process than species with smaller home ranges.

Some species are restricted to large patches while other species are restricted to small patches. Species restricted to large patches are far more common than species restricted to small patches, partly owing to differing 'minimum area requirements'. Some species require a larger area of continuous habitat in which to live than others, the smallest area necessary being called the 'minimum area'. In a study of bird species in sixteen small and isolated woodlands in North Humberside, England, four species out of forty-nine recorded had minimum area requirements (McCollin 1993). These species were the great tit (*Parus major*) and spotted flycatcher (*Muscicapa striata*), with minimum area requirements of about 1 ha, and the jay (*Garrulus glandarius*) and turtle dove (*Streptopelia turtur*), with minimum area requirements of about 4 ha. Chaffinch (*Fringilla coelebs*), blue tit (*Parus caeruleus*), robin (*Erithacus rubecula*), wren (*Troglodytes troglodytes*) and blackbird (*Turdus merula*) occurred in all 16

BASICS 12.1

SPECIES–AREA RELATIONSHIPS

Count the species in increasingly large areas, and the species diversity will increase. This, the species–area effect, is a fundamental biogeographical pattern that applies to both mainland species and island species.

Figure 12.5 is a species–area curve for anole lizards (*Anolis* spp.) living on selected Caribbean islands.

A log–log line expresses the relationship between the number of species, S, and island area, A:

$$\log S = \log c + z \log A$$

where c is the intercept value (the number of species recorded when the area is zero), and z is the slope of the line. For the Caribbean example, the equation is:

$$\log S = -0.93 + 0.50 \log A$$

The z-value (0.50) indicates that, for every 1-km^2 increase in island area, an extra 0.5 anole species will occur. A power function also describes the species–area relationship as:

$$S = cA^z$$

or, in this example:

$$S = 8.7A^{0.5}$$

As a rule, the number of species living on islands doubles when habitat area increases by a factor of ten. For islands, values of the exponent z normally range from 0.24 to 0.34. In mainland areas, z-values

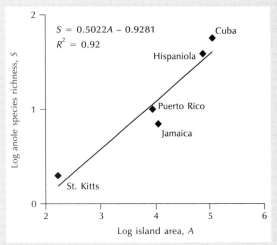

Figure 12.5 Species–area curve for anole lizards on selected Caribbean islands

normally fall within the range 0.12 to 0.17. This means that, as area increases, the number of species on islands increases at roughly twice the rate of the number of species on mainlands. The difference may be partly attributable to relative isolation of islands, which makes colonization more difficult than on mainlands.

---

woodlands, which suggested that their minimum area requirements were less than 0.73 ha, the area of the smallest wood.

Species restricted to small patches are uncommon, but they do exist. Gray squirrels (*Sciurus carolinensis*) in New Jersey, USA, are common in small woodland patches, but scarce or absent in large patches, probably because small patches do not house the owl predators that large patches do. In New Zealand, large (up to 8 cm diameter) carnivorous kauri snails (*Paryphanta busbyi*) are more likely to survive in small patches because in large patches they are prey to wild pigs.

Focus 12.1 is a case study showing the effect of patch size, forest fragmentation and other factors on forest-floor herbs in North Carolina.

Interestingly, latitude and longitude affect species–patch-size relationships. Bird species in small woods within agricultural landscapes of the Netherlands, the United Kingdom, Denmark and Norway displayed species–area relationships (Hinsley *et al.* 1998). Species richness across all woods declined with increasing latitude, as did the proportion of resident species; the proportion of migrant species rose. In addition, the slopes and intercepts of the species–area curves declined with increasing latitude. So, not only were there fewer species available to colonize woods at higher latitudes, but the gain of species richness for a unit increase in wood area was smaller than at lower latitudes.

FOCUS 12.1

FOREST PATCHES AND FOREST HERBS, BLUE RIDGE PROVINCE, NORTH CAROLINA

The effects of forest fragmentation on cover-forest herbs in the French Broad River Basin, Buncombe and Madison Counties, North Carolina, which lie in the Southern Blue Ridge Province, USA, reveal the complex repercussions of habitat fragmentation on ecosystems (Pearson *et al.* 1998). Fourteen patches of closed-canopy, mesic (not too wet, not too dry), cove-forest (forest growing in topographic hollows) lying in the altitudinal range 600 to 920 m were sampled with 4 ha study plots, each of which was divided in 1 ha subplots. The plots were set down in patches ranging from 5 ha to more than 10,000 ha. There were eight small (<25 ha) plots, three isolated (> 200 ha) plots and three large plots within continuous forest. A control on the environmental characteristics was achieved by limiting the investigation to closed-canopy, mesic, deciduous forest. Mesic forests were defined as communities on north-facing or sheltered slopes with one of the following trees dominant in the canopy: yellow poplar or tulip-tree (*Liriodendron tulipifera*), sugar maple (*Acer saccharum*), American basswood (*Tilia americana*), yellow buckeye (*Aesculus octandra*), bitternut hickory (*Carya cordiformis*) and sweet birch (*Betula lenta*). American beech (*Fagus grandifolia*), northern red oak (*Quercus rubra*), and red maple (*Acer rubrum*) were commonly, but not always, present. Small stands of rosebay rhododendron (*Rhododendron maximum*) and eastern hemlock (*Tsuga canadensis*) were avoided. Seventeen herb species, selected as good representatives of cove-forest communities, were recorded as cover and density estimates.

The survey revealed that the coverage and density of herb species were greater in large patches (>200 ha) than in small patches (<10 ha). Eight of the seventeen species displayed a greater abundance in large patches. The eight species were wild ginger (*Asarum canadense*), false goatsbeard (*Astilbe biternata*), toothwort (*Cardamine diphylla*), yellow mandarin (*Disporum lanuginosum*), spotted mandarin (*Disporum maculatum*), bloodroot (*Sanguinaria canadensis*), foam flower (*Tiarella cordifolia*) and bellwort (*Uvularia grandiflora*). Ant-dispersed spotted mandarin and bellwort were more likely to be absent from small patches than from large patches. Patch size and isolation did not affect wind-dispersed species, such as ferns and composites. Hairy chickweed (*Stellaria pubera*) was more abundant in small patches.

Several factors may explain the species–area relationships of herbs in the cove-forests. First, habitat fragmentation is apt to disrupt the dynamics of herb populations in smaller patches, possibly leading to the local extinctions. Six species that were more likely to occur in larger patches had limited seed-dispersal abilities that may prevent them from colonizing isolated patches. Of these, wild ginger, spotted mandarin, bellwort, foam flower and Canada violet (*Viola canadensis*) are dispersed by ants; while false goatsbeard and foam flower produce many small seeds that fall close to the parent plant. Good dispersers, such as the maidenhair fern (*Adiantum pedatum*), rattlesnake fern (*Botrychium virginianum*) and broad-leaved goldenrod (*Solidago flexicaulis*), were unaffected by patch size. A second environmental factor that might help to explain the species–area relationships is habitat differences between small and large patches. Organic matter content was the only soil factor that varied significantly with patch size. Indeed, humic matter affected the coverage and density of many species. This is because humic soils have higher moisture retention, improved aeration and tilth and more nutrients. Plainly, if small patches have less humic matter than large patches, then this will have knock-on effects in the herb community. Disturbance may affect the species–area relationships. Although there were no signs that the patches had been cultivated over the last 75 years, some small patches were adjacent to land currently or formerly in agricultural use, which renders them more prone to being used for grazing and as a source for forest products than is the case for the larger plots. Such disturbance might have affected herb populations by increasing mortality rates and by degrading the habitat. Therefore, although forest patch-size does go some way towards explaining the cover and density of herb species, habitat fragmentation, habitat quality and human disturbance may have had an impact on the species–area pattern.

Patch shape and edge effects

Smaller patches usually have a higher ratio of perimeter to area than larger patches. For this reason, smaller patches are more vulnerable to edge effects than are larger patches (Figure 12.6). Obviously, large patches contain more 'interior' than 'edge'. Indeed, there is a minimum patch size below which there is no 'interior', the whole patch being an 'edge'. Even 12 years ago, research revealed

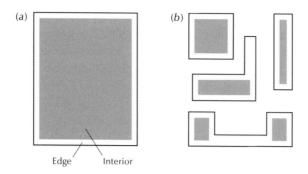

Figure 12.6 The effects of habitat fragment shape and size on edge effects. Large fragments with a low ratio of perimeter to area (a) contain proportionally less edge than small fragments with a high ratio of perimeter to area (b). Some small fragments may be all edge, as may narrow linkages between fragments

that many surviving woodland fragments in England and Wales lay in the range 1 to 5 ha (44.0 per cent by number, 9.6 per cent by area) and therefore were predominantly 'edge' habitat (Spencer and Kirby 1992). This section will consider two aspects of edge effects: edges and interior species, and edges as ecological traps.

Edge species and interior species

Ecologists commonly distinguish between edge species and interior species. Studies commonly reveal the prevalence of edge species in small patches and interior species in large patches. Interior species live in the core of a habitat. They actively avoid the habitat edges if they are able to meet their resource needs within their territories or home ranges. English woodland examples include the great spotted woodpecker (*Dendrocopos major*) and nuthatch (*Sitta europaea*). Edge species use a habitat edge. Two types of edge species are recognized, the first of which are intrinsically edge species, and the second of which are ecotonal species (McCollin 1998). Ecotonal species occur near the edge because the edge habitat suits them. They are not dependent on adjacent habitats for food, shelter or anything else. Intrinsic edge species live near edges because the adjacent habitat provides resources. For instance, in highly fragmented agricultural landscapes, bird species living in woodland edges next to open country depend upon food resources offered by farmland. Examples are the rook (*Corvus frugilegus*) and carrion crow (*Corvus corone corone*), which feed mainly on grain, earthworms and their eggs and grassland insects, with the crow also taking small mammals and carrion; and the starling (*Sturnus vulgaris*) feeds on leatherjackets and earthworms in the upper soil layers of pasture (McCollin 1998).

Edges as 'ecological traps'

Species diversity in edge habitats tends to be high, and frequently includes a number of exotic species. Edge habitats also typically support a higher diversity of herbivores and predators than adjacent habitats: the 'rich pickings' of food supplies attract the herbivores, and the abundance of herbivores lures the predators. For instance, many game birds commonly live at higher densities in edges than in patch interiors and provide a banquet for predators. Hawks, cats, canines and other predators often centre their foraging in edge habitats. However, predation rates vary with edge type. Three types of edges in a bottomland forest lying along the Roanoke River in North Carolina, USA, had different levels of predation during the 1996 breeding season (Saracco and Collazo 1999). The three edge types were: (1) a forest–farm edge, which has a sharp exterior edge; (2) a forest–river edge, which has an abrupt interior edge;

and (3) a levee–swamp edge, which has a gradual interior edge at the boundary of the two dominant communities in the floodplain – cypress-gum swamps and coastal plain levee forests. Predation rates of northern bobwhite (*Colinus virginianus*) eggs and clay eggs were significantly higher along forest–farm edges than along the other two edges, where the predation rates were roughly the same. Taken with higher avian predator abundance on forest–farm edges, the pattern of egg predation indicated that avian predators exerted more predation pressure along these edges, a finding consonant with other studies where agricultural encroachment into forested landscapes may have a deleterious effect upon breeding birds.

The 'ecological trap hypothesis' purports to explain the richness of species in edge habitats. It has several interesting aspects. For example, an elevated rate of herbivory in edge habitats puts pressure on edge plants, some of which survive the herbivore onslaught by being less palatable or less sensitive to trampling. Predation and scavenging rates may also alter in edges, to the detriment of some species (Focus 12.2).

CORRIDORS

Corridors serve several ecological roles. They are at once habitats, conduits and filters. As habitats, corridors of all kinds tend to have a high species diversity with edge and generalist species dominating. As conduits, they provide routes for wildlife and for humans. The rate of movement along corridors depends upon such factors as the species moving, the type of corridor, the corridor width and length, the number of 'narrows' and gaps, the number of entrances and exits, the curviness, the patchiness, the degree of crisscrossing with other corridors and the strength of any environmental gradients present. As filters, corridors permit some species to cross but bar others. The filtering process may create separate patches on either side of the corridor, each with a different species composition.

Roads

Roadsides, and even roads themselves, provide habitats that are different to surrounding habitats. They act as wildlife refuges for some species. In particular, they often support 'edge' species from surrounding forest. They are also potential avenues of movement for various groups of animals and plants through what otherwise might be unpleasant terrain. Grassy roadsides provide routes for grassland species across forested and intensive agricultural regions. The length of the routes can be enormous: the Interstate Highway System in the USA has created about 70,000 km of potential movement corridors. Movement may take place on the road surface and on the vehicles that move along it, in the open space above the road and in the roadside habitat.

Vehicles move a surprisingly large number of plant seeds along roads. In a survey of car-wash flora from the Shell Canberra Superwash, Australia, 259 plant species were identified, of which 19 were not native to, naturalized in, or cultivated in Canberra or the nearby New South Wales tablelands (Wace 1977). Mud and soil clinging to vehicles also transports fungal spores. The cinnamon fungus (*Phytopthora cinnamoni*) spread through the forests of southern Australia in this manner. Seed transport by vehicles leads to a large variety of pioneer and adventive plants in roadsides and in particular the immediate road edges.

FOCUS 12.2

THE DECLINE OF THE AMERICAN BURYING BEETLE

The American burying beetle (*Nicrophorus americanus*) is a member of the carrion beetle family Silphidae (Plate 11). It once lived throughout eastern and central North America, in thirty-five US states and southern fringes of Ontario, Quebec and Nova Scotia in Canada (Figure 12.7). East of the Appalachians, its range was shrinking fast by 1923; west of the Appalachians, the decline occurred later. In the Midwest, the decline appears to have begun at the centre of the range and moved outwards, with populations surviving at peripheral sites since 1960. The American burying beetle now lives in just four states – Nebraska, Rhode Island, Oklahoma and Arkansas – and its abundance has declined throughout its range since the 1980s.

Efforts to account for the American burying beetle's decline have pinpointed habitat fragmentation as a possible culprit. Large expanses of the beetle's natural habitat have suffered fragmentation, which has had two effects. First, it has altered the original species composition of the predator–scavenger community in which the American burying beetle lives. The increase in edge habitat created by fragmentation has boosted the occurrence and density of species that compete with the beetle for carrion – crows, racoons, foxes, opossums, skunks and other vertebrate predators and scavengers. Second, it has reduced the reproductive success of indigenous species upon which the beetle preys and upon which it depends for optimal reproduction. Other factors may also have hastened the American burying beetle's decline: artificial lighting has decreased populations of insects active during the night; land use change has tended to isolate patches of the beetle's preferred habitat; and the possibility that the genetic characteristics of the species have led to reduced rates of reproduction. Biologists are making strenuous efforts to understand the causes of the American burying beetle's decline and to develop a recovery plan.

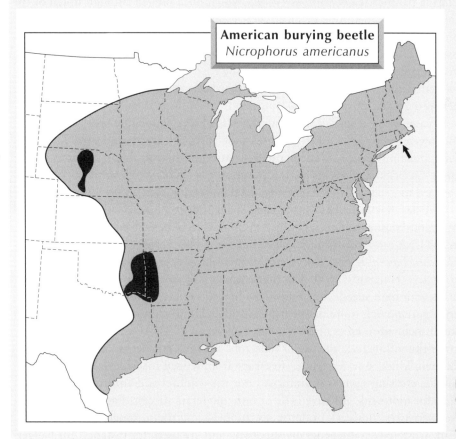

Figure 12.7 The historical and current range of the American burying beetle. *Source:* After Kozol *et al.* (1994)

In general, animals make little use of the open road and roadside, both of which harbour predators and other hidden dangers (such as cars and lorries). Some frogs and snakes are dispersed by 'hitching a ride', though this is not a common process, a documented case being the establishment of the spotted grass frog (*Limnodynastes tasmaniensis*) at Kununurra, in north-western Australia, some 1800 km from its south-eastern Australia homeland (Martin and Tyler 1978). Such predators as the red fox (*Vulpes vulpes*), wolf (*Canis latrans*), dingo (*Canis familiaris dingo*), cheetah (*Acinonyx jubatus*) and lion (*Panthera leo*) use tracks and lightly used roads, especially at night. They use them as clear pathways, uncluttered with vegetation, along which to move and hunt. Animals do not generally use metalled road surfaces as conduits, though moose (*Alces canadensis*) sometimes amble along them at night (and charge at cars). Bats use open spaces above roads as flight paths and foraging spaces (e.g. Crome and Richards 1988).

Roads act as barriers or filters to animals that would cross them. Almost every animal, from spiders and beetles to kangaroos and deer, is a potential road-crosser. In wetlands, roads pose barriers to the free movement of aquatic animals, and may divide and isolate populations. The chief components of roads that deter would-be crossers are: the bare road surface itself; the altered roadside habitat; and the noise, movements, emission and lights that are part of road traffic (Bennett 1991). On wide roads with much traffic, all three elements conspire to create a daunting and formidable barrier to wildlife. The Florida panther (*Puma concolor coryi*), which once ranged through much of the south-eastern USA, lives in south-west Florida in national and state parks and on nearby private lands. With a population of about 70 adults, the Florida panther is among the most critically endangered animals in the world. Fragmentation of the panther's habitat by roads is in part a cause of its reduction that, in 1992, was so severe that biologists gave the subspecies between 24 and 60 years until extinction.

In cases where a species will not cross a road, the population splits into two, the interbreeding rates on either side of the corridor change, and two subpopulations evolve. **Genetic differentiation** in the two subpopulations may ensue. In Britain, roads separating common frog (*Rana temporaria*) populations have led to genetically differentiated subpopulations (Reh 1989).

genetic differentiation: the splitting of a gene pool leading to the evolution of different subspecies or species

Vehicles may kill animals that cross roads (Focus 12.3). Road kill takes an enormous annual toll on wildlife. A million vertebrates die every day on roads in the USA (and not through natural causes). There is even a book called *Flattened Fauna: A Field Guide to Common Animals of Roads, Streets and Highways* (Knutson 1987). The highest road kill rates occur on two-lane main roads with high speeds, but multi-lane highways have a greater ecological impact, removing more native habitat and creating barriers that many animals are disinclined to cross or even go near (Forman 1995, 168). Techniques to lower the toll include the use of reflectors, mirrors, repellents, bait, fencing of different kinds, one-way gates (to escape from fenced roadsides), lighting, wildlife crossing signs (for motorists, not the wildlife) and animated warning signs for motorists. All have little or only moderate success. Far more successful are overpasses (bridges), underpasses (tunnels) and the seasonal closure of a road. Underpasses are generally successful and are used, for instance, for badgers (*Meles meles*) in Great Britain and mountain goats (*Oreamnos americanus*) in Montana, USA.

FOCUS 12.3

DEATH ON THE A83 MOTORWAY, FRANCE

The A83 motorway, which runs north-north-west through the Département de Vendée to Nantes, opened in October 1994. A study of a 68.3 km section between Montaigu and Fontenay-Le-Comte from April to November 1995 (Lodé 2000) revealed high road kill rates of vertebrates. Thirty-three weekly surveys recorded 2266 road-killed individuals from 97 species, including some uncommon or rare species. The road kills comprised about 29 per cent amphibians, 1 per cent reptiles, 27 per cent birds and 43 per cent mammals. Rodents bore the heaviest toll (27 per cent),

but their predators (falcons, owls and carnivores) also suffered heavy losses (about 22 per cent). The common toad accounted for 16.6 per cent of the vertebrate deaths. Road-killed rare and uncommon species included the following mammals: barbastrelle bat (*Barbastrella barbastrellus*), lesser horseshoe bat (*Rhinolophus hipposideros*), otter (*Lutra lutra*), genet (*Genetta genetta*); and the following birds: bluethroat (*Luscinia svecica*) and rock sparrow (*Petronia petronia*); and an amphibian, the midwife toad (*Alytes obstetricans*).

Trails

Some plants use trails as conduits. The invasion of cheatgrass (*Bromus tectorum*) over huge, dry areas of north-western North America took place largely along cattle trails and railroad corridors. Many terrestrial mammals make trails within their home range, which they use for foraging and so forth. Humans are inveterate trail-makers. They use their trails for movement – on foot, on horseback, on a motorcycle or other kind of vehicle.

Powerlines

Electricity transmission lines, gas lines, oil lines and dikes tend to be fairly straight with sharp boundaries, and to have a fairly constant width over which disturbance or maintenance is evenly distributed. They favour edge and generalist species. In a forested Tennessee landscape, in the USA, almost all the birds in powerline corridors were edge species (Anderson *et al.* 1977). Just two forest interior species – the scarlet tanager (*Piranga olivacea*) and the red-eyed vireo (*Vireo olivaceus*) – ranged into narrow powerline corridors.

   In the Mojave Desert, California, a study explored the relationships between linear right-of-ways (paved highways and transmission powerlines) and raven (*Corvus corax*) and red-tailed hawk (*Buteo jamaicensis*) populations (Knight and Kawashima 1993). Ravens were as common along highway transects as they were along powerline transects, but were more abundant along highway and powerline transects than they were along control transects. Raven nests were more abundant along powerline transects than along highway and control transects. Both red-tailed hawks and their nests were more abundant along powerline transects than along highway and control transects. Ravens used power poles as nest sites more than expected based on availability, but did not use them as perch sites more than expected. Red-tailed hawks used power poles for both nesting and perching more than expected based on availability. Ravens appear to be more abundant along highways due to the carrion produced by automobiles. Ravens are facultative scavengers and road kill is a ready meal. Red-tailed hawks hunt live prey and are not interested in road carrion. Both ravens and red-tailed hawks appear to be more common along powerlines owing to the superior perch and nest sites that powerlines afford. Powerline towers are tall and may give predators a wide field of vision. Nests built in tall towers may benefit from greater cooling, owing to the increased openness around the nest, while the beams and latticework offer sturdy nest site anchors to guard against dislodgement by extreme winds.

Powerlines act as strong filters, largely because they make a loud 'buzzing hum' that deters would-be crossers, especially in wet weather. To some crossers they are a hazard. In south-central Nebraska, USA, spring migrating sandhill cranes (*Grus canadensis*) may crash into powerlines. A study marked alternating spans of powerline with 30 cm diameter yellow aviation balls (Morkill and Anderson 1991). The number of cranes flying over marked and unmarked powerline spans did not differ, but cranes reacted more often to marked than unmarked spans, mainly by gaining altitude or by changing the direction of flight. More dead cranes were found under unmarked than marked powerline spans.

Hedgerows and other wooded strips

Hedgerows and windbreaks are line corridors dominated by edge species living at high densities. Forest interior species are normally present, albeit in low numbers. Width is a key factor in determining the species richness and abundance of many species living in hedgerows and windbreaks, though vertical structure also exerts a major influence of bird species diversity. Hedgerows support a large variety of game birds. If a hedgerow is associated with a wall, fence, ditch or soil bank, then even greater species richness is encouraged. A ditch attracts amphibians and reptiles, while the sunny side of a soil bank encourages drought-tolerant species. Some of the species harboured by a hedgerow or windbreak may frequent adjacent fields where they eat crops. Conversely, many birds that live in fields use hedgerows for perching or for foraging. The high density of animals in hedgerows draws in predators from the surroundings.

A variety of wildlife uses hedgerows and other wooded strips as conduits. The high number of road kill victims found where a woodland strip is broken by a road attests to this fact. Not all species will cross hedgerow gaps. The dormouse (*Muscardinus avellanarius*) is an arboreal (tree) habitat specialist that is averse to crossing even narrow gaps in hedgerows, though they are prepared to move across grass fields (Bright 1998). Plants may also move along wooded strips, though the evidence is patchy. A recent study of remnant and regenerated hedgerows in Tompkins County, in the Finger Lakes region of central New York State, USA, indicated that forest herbs colonize hedgerows from source areas (Corbit *et al.* 1999).

Stream and river corridors

riparian: along the edge of a river

Stream or **riparian** corridors play a starring role in many ecosystems, exerting a big influence over animals and plants. Many species rely on stream corridors for food and water, for shelter, for travel and rest and for reproduction. The varied habitats and excellent food base (water plants, herbs and shrubs bearing berries and seeds, leafy foliage) foster high species diversity. A reason for this richness of wildlife and wildlife habitats is the fact that riparian ecosystems receive water, nutrients and energy from upstream ecosystems. These upstream additions allow greater species richness and help to maintain a roughly constant supply of resources (Harris 1984, 142). Permanent lakes and rivers support many amphibians and aquatic birds and mammals. They also contain fish and other aquatic organisms that form the seat of several food webs.

The chief habitats in stream corridors are riverbank and floodplain. Changing water levels, changing soil moisture levels, and erosion and deposition of sediment characterize riverbank habitats. Floodplains are often rich in wetland habitats, different types being dictated by the frequency and the degree of inundation – marsh or bog, shrub swamp or thicket, forested swamp, vegetation of rarely flooded levees, ridges and hummocks. All these floodplain habitats share periodic flooding, poor soil drainage, occasional surface deposition of sediment and nutrient-rich soils. In dry regions, such as Arizona, riparian corridors sometimes form 'linear oases' (thin green lines), which are supported primarily by groundwater, and contain rare species. In tropical grasslands, biodiverse 'gallery forest' meanders across the plains, providing water, food and shade for many species in a grassland matrix.

Many mammals use river corridors as conduits. Mountain lion (*Felix concolor*), bobcat (*Lynx rufus*), grizzly bear (*Ursus arctos horribilis*) and black bear (*Ursus americanus*) are known to move many kilometres along river corridors. If river meanders are present, people and animals (e.g. egrets, kingfishers and river otters) may move directly between meanders, rather than follow the circuitous channel. Hawks and related birds of prey commonly migrate along windward edges of ridges, where updrafts of air facilitate gliding.

Some exotic (non-native) plants use river corridors to spread. In Britain, floods disperse Japanese knotweed (*Reynontria japonica*) rhizomes; and Indian balsam (*Impatiens glandulifera*), introduced from the Himalayas as a garden plant, has spread along riverbanks. In the western United States, two exotic species – tamarisk or salt cedar (*Tamarix hispida rubra*) and Russian olive (*Elaeagnus angustifolia*) – have colonized widely along rivers and cause serious problems in local ecosystems. The tamarisk is a very invasive shrub-tree from Eurasia. Government agencies planted it across the western United States in the early 1900s in an effort to control soil erosion. The Russian olive is a native tree of Europe and Asia. Settlers introduced it to North America some 150 years ago, since when it has colonized widely along rivers where it is having a severe impact on native birds and fish (e.g. Dixon and Johnson 1999). Similarly, purple loosestrife (*Lythrum salicaria*), originally introduced into North America from Europe in the early 1800s as an ornamental plant and a contaminant of ship ballast, has spread with waterborne commerce, and has invaded many wetlands causing severe problems in some areas (Mullin 1998).

NETWORKS

Tree networks and life

Corridor networks, tree networks and circuit networks all influence animals and plants. The influences of corridor and circuit networks are the most studied, but tree networks play a role in determining the distribution of species in landscapes and deserve further investigation. Some studies hint at relationships between species distribution and stream networks. Some relationships are broad, as in seasonally migrating mammals using river networks to go from lowlands to uplands. Examples in the Grand Tetons of North America are elk (*Cervus canadensis*), mule deer (*Odocoileus hemionus*) and moose (*Alces americana*), which migrate to subalpine meadows in early summer and back in autumn to winter feeding areas in the plains. Other relationships are more specific and relate species distribution to stream order. Plants in first-order and second-order streams are usually different from plants in high-order streams (e.g. mosses and ferns in headwaters). Because terrain elevation changes with stream order, floodplain vegetation also changes. In Europe, riverine hardwood forests of big river plains contain similar species and genera that often recur in different geographical regions. Typical species are common hawthorn (*Crataegus monogyna*) or black hawthorn (*Crataegus nigra*), ash (*Fraxinus excelsior*) or narrow-leaved ash (*Fraxinus angustifolia*), fly honeysuckle (*Lonicera xylosteum*) or perfoliate honeysuckle (*Lonicera caprifolium*), pedunculate oak (*Quercus robur*) or holm oak (*Quercus ilex*), hoary willow (*Salix eleagnos*) or common osier (*Salix viminalis*) (Schnitzler 1994).

Some animals are more abundant near streams. Racoons (*Procyon lotor*) and opossums (*Didelphis virginiana*) are predators of forest songbird eggs and nestlings. In Missouri, USA, abundances of racoons and opossums were related to, among other things, stream density (Dijak and Thompson 2000). More generally, a sequence of terrestrial mammal species occupying similar functional niches occurs along different sized streams according to the river-continuum concept (Vannote *et al.* 1980). In the western Cascades of North America, a series of carnivorous, amphibious mammals with similar ecological roles eat prey of different sizes, live in streams of different orders, and occur at different elevations in a drainage basin (Figure 12.8). The northern water shrew (*Sorex palustris*) and marsh shrew (*Sorex bendirei*) live in the headwater and lower-order streams, the mink (*Mustela vison*) lives in slightly higher-order streams, the otter (*Lutra canadensis*) lives in middle-order streams, and the grizzly bear (*Ursus arctos horribilis*) leaves in higher-order streams.

Corridor and circuit networks

Many network properties affect the distribution of abundance of animals and plants. Location within a circuit, like location within a river network, is

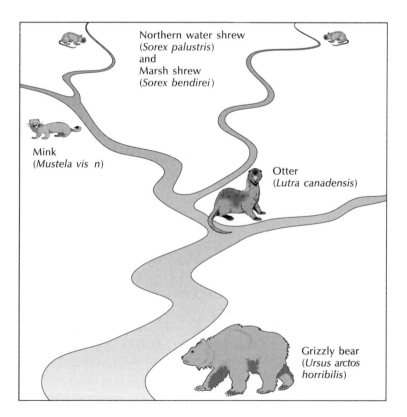

Figure 12.8 The association of different-sized carnivorous mammals with stream order and typical food-particle size according to the river-continuum concept. *Source:* Adapted from Harris (1984)

sometimes a significant property and determines the species present. In Brittany, France, the carabid beetle fauna differs in various parts of the hedgerow network. Some species live near the periphery, some nearer the centre, and some in wide corridors within the network (Burel and Baudry 1990). Similarly, while single corridors tend to favour edge species, networks that involve the proximity of two corridors (at right angles or an acute angle) may be able to support some patch species, even in the absence of a patch. This appears to be the case for kangaroos using roadside strips in Western Australia (Arnold *et al.* 1991). But the main thrust of research into landscape networks concerns connectivity and its effects upon population dynamics, which is studied using mathematical models and in the field.

Network connectivity

Of all network properties, connectivity is of particular consequence for animals and plants. It affects the viability of metapopulations by determining how easy or difficult moving between resource patches is for animals. A mathematical simulation suggested that two- or three-patch metapopulations are doomed to extinction, no matter how much movement there is between patches, when all local populations (subpopulations) are below a minimum viable population size (Wu *et al.* 1993). That finding has management implications: given a set of scattered small populations, augmenting individual populations might be more advisable, rather than trying to bolster migration between patches by maintaining or by building corridors. Another simulation showed that, when at least one subpopulation is larger than the minimum viable population size, there is a critical size for that subpopulation above which the metapopulation as a whole will persist and below which it will collapse. A third simulation indicated that, when a metapopulation comprises two or more patches, metapopulation dynamics and persistence were strongly affected by two factors: (1) the pattern of patch connections; and (2) the spatial position of the populations above the viable minimum level. All three simulation results suggested that both the number of connections between patches and the magnitude of movements along them are decisive for overall patch connectivity. And the magnitude of migration between patches is positively related to the minimum size of the above-minimum-viable-population subpopulation in both the two-patch and three-patch metapopulation systems due to a population sink effect.

A battery of field investigations that explore the effect of connectivity on species distributions and abundance, support theoretical studies (e.g. Focus 12.4).

MOSAICS

Studies of landscape mosaics are in their infancy. At least two lines of enquiry have emerged. The first considers population dynamics in landscapes with emphasis on movement between patches, corridors and matrixes. The main techniques employed in these studies are observation by capture or radio-tracking or mathematical models. Radio-tracking has proved an effective tool to study animals' movements. Small radio-transmitters may be fitted to a range of animals – mice, geese or lions – which enables them to be tracked with relative ease. Radio-tracking is helping studies of foraging within home ranges, dispersal out of home ranges and migration between habitats. A key idea in these studies is landscape structure and connectivity. The second line of enquiry focuses around

FOCUS 12.4

EASTERN CHIPMUNKS IN CANADA

Eastern chipmunks (*Tamias striatus*) were studied in a patch-and-corridor network set in 200 ha of farmland at Manotick, about 20 km south of Ottawa, Canada (Bennett *et al.* 1994). The eastern chipmunk is a small, diurnal, burrow-dwelling member of the squirrel family. In the Ottawa region, it is active from April to October, hibernating in winter when snow covers the ground. It is a native woodland species, but persists in farm landscapes where some wooded vegetation remains. In the study area, they lived in a farmland mosaic of patches and corridors. The patches were small woodland remnants, with sugar maple (*Acer saccharum*), white ash (*Fraxinus americana*), eastern white cedar (*Thuja occidentalis*), basswood (*Tilia americana*), and white elm (*Ulmus americana*) being the dominant species. Fields used for pasture and for corn (*Zea mays*) and oats (*Avena sativa*) crops surrounded the patches. Vegetated fencerows that

occupied narrow uncultivated strips of land along fence-lines between fields connected the patches. These fencerows provided a corridor network linking the woodland patches (Figure 12.9). The vegetation in the fencerows was variable, ranging from long grasses to shrubs and vines, or to woodland strips of mature trees. In addition, there were stone walls, rocks and fence rails within some fencerows. Trapping and radiotelemetry in a separate study showed that the chipmunks confined themselves to wooded or shrubby vegetation, and were seldom recorded in, or moving across, fields of pasture or crops.

In the present study, chipmunks were trapped in four woods and eighteen fencerows in four trapping sessions between May and September 1989. They were assigned to one of three residence statuses valid for the season studied. 'Resident' chipmunks were recorded in the same fencerow or wood during two or

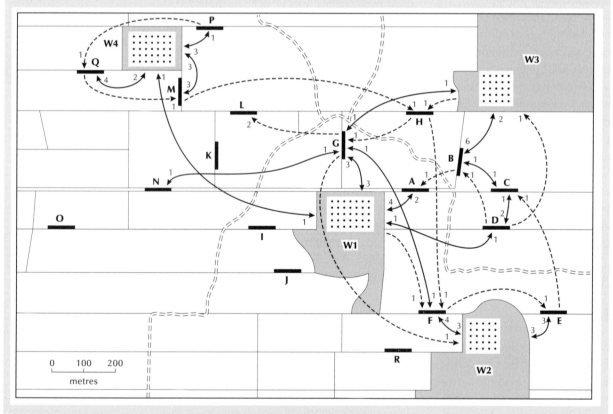

Figure 12.9 Fencerows provide a corridor network linking woodland patches. Eastern chipmunks (*Tamias striatus*) were studied in a patch-and-corridor network set in 200 ha of farmland at Manotick, about 20 km south of Ottawa, Canada. Trapping sites are labelled W1–W4 (woods, where trapping sites are shown by circles) and A–R (fencerows). Arrows show the direction of chipmunk movement, and not the route taken. Heavy solid lines show two-way movement, heavy dashed lines show one-way movement, and the digits indicate the number of recorded directional movements between two landscape elements. Fencerows are depicted by thin solid lines, and streams by thin dashed lines. *Source:* After Bennett *et al.* (1994)

more trapping sessions. 'Temporary resident' chipmunks were recorded during only one session, even if trapped several times. 'Transient' chipmunks were trapped in a particular fencerow or wood only once. In all, 530 captures of 119 chipmunks were recorded during the study. They resided in all four woods, and were trapped in fourteen of the eighteen fencerows. Resident individuals lived within and along many fencerows, which enhanced the continuity of the resident population between woods. These residents preferred fencerows with tall trees and a woodland structure. Transient chipmunks seem to use the fencerow network as a pathway across the farmland. They were more abundant in fencerows with high linear continuity (preferring fencerows without too many gaps) and with 'woody' habitat attributes (preferring tall trees, small trees and tall shrubs). Chipmunks never used grassy vegetation, which, like the surrounding farmland, appears to be an inhospitable habitat.

the relation between vegetation and landform and landscape properties, the main tools used in these investigations being GIS, DTMs and statistical analysis.

Landscape structure and connectivity

Landscape mosaics are specific arrangements of patches, corridors and matrixes. The arrangement of these three elements affects the ease with which animals and plants may move through a landscape. In its turn, the ease of movement affects population dynamics. **Landscape resistance** describes the ease of movement. It is a mosaic property determined by the structural landscape properties, including the degree of connectedness, that impede movement. There is no standard way of measuring landscape resistance, though several landscape factors contribute to it, including high fragmentation of patches, gaps in corridors, a lack of corridors between patches and the presence of main roads. In southern Holland, landscape resistance affects the movement of woodland birds, including the nuthatch (*Sitta europaea*) (Harms and Opdam 1990; see also Knaapen *et al.* 1992). In this study, built area, glasshouses and busy roads increased landscape resistance, while wooded vegetation decreased it.

> *mosaic*: the landscape pattern produce by a particular combination of patches, corridors and matrix
>
> *landscape resistance*: the ease with which an animal or plants can move through a landscape

An important point to bear in mind is that the resistance offered by a landscape to one species, or group of species, may not be the same as the resistance offered by the same landscape to another species. So, if the spatial structure of a landscape should change, species will fare differently (Henein *et al.* 1998). For instance, species with small home ranges will respond differently than species with large home ranges. Indeed, many wide-ranging species (caribou, tigers, black bears and vultures) are sensitive to the arrangement of regional landscapes, and commonly use two or more landscapes that need to be close together (Forman 1995, 25). In addition, habitat generalists may find it easier to move through heterogeneous landscapes created by habitat fragmentation than may habitat specialists (e.g. Mabry *et al.* 2003).

Field investigations

Most of the empirical data identifying landscape spatial patterns (connectedness) that interfere with individual movements, concern walking animals with low powers of dispersal. For flying animals, the distance between patches is in almost all cases a straight-line distance between points. This metric does not account for the behavioural characteristics of species that depend on landscape patterns for their movements. The presence of the short-toed tree creeper (*Certhia brachydactyla*), for instance, seems to depend on the spatial structure of agricultural landscapes. In western France, around the city of Roz, two adjacent

bocage (French): a landscape of small, enclosed fields with woodland

polder: an area of low-lying land reclaimed from sea, lake or river, and protected by dikes

and contrasting rural landscapes – long-established **bocage** and a recent **polder** – differ in grain size, in the quality of linear landscape elements delimiting fields and in their history (Clergeau and Burel 1997). The bocage landscape consisted of a mosaic of fields with an average size of less than 2 ha comprising permanent pasture (15 per cent), temporary grassland (20 per cent), maize fields (20 per cent) and scattered small woodlands, with an old hedgerow network. The polder landscape was reclaimed between 1850 and 1930, and since 1980 all the polder fields have been used for market gardening. In the recently reclaimed polder, tree creepers are present only in linear rows of 'well-connected' trees that are long enough to support the home range of the short-toed tree creeper. Colonization of this recent polder landscape from the bocage, which serves as a source of dispersers, required connections between the bocage hedgerows and the planted dikes.

A consideration of the structure and composition, and particularly the connections between corridors and the matrix, in cultivated landscapes has helped to explain seasonal fluctuations in wood mouse (*Apodemus sylvaticus*) populations (Ouin *et al.* 2000). A study conducted in the polders of the Mont Saint-Michel Bay, western France, used GIS to look into the seasonal dispersal of wood mice from hedges to crops at field and landscapes scales. Ninety per cent of the area was under intensive agriculture, with wheat (*Triticum aestivum*), maize (*Zea mays*), peas (*Pisum sativum*) and carrots (*Daucus carotta*) as the main crops. The semi-natural habitats (hedges and grassy linear habitats) were distributed over a dike network. The results suggested that the summertime drop in hedgerow populations of wood mice resulted from a movement into the crops. Hedgerows serve as a source of wood mice in spring, and the rate at which they colonize fields in summer depends on the quality of the crops, as reflected in the crop cover and seed availability in fields and in landscapes.

A study of butterfly species richness in fragmented calcareous grassland near the city of Göttingen in Germany showed that the landscape context plays a role (Krauss *et al.* 2003). Figure 12.10a shows that the richness of butterfly species in the grassland fragments increases with increasing habitat area, for specialists more than for generalists. However, in Figure 12.10b, butterfly species richness is unrelated to an index of habitat isolation. Figure 12.10c indicates a connection with landscape context in that generalist butterfly richness increases with landscape diversity.

12.3 NEWLY ABUNDANT AND INVASIVE SPECIES

The species composition of different places is always changing owing, in the short-term, to emigration, immigration and local extinctions, and in the longer term to speciation. Human activities affect all these processes, with the possible exception of speciation. Two important human impacts are the emergence of superabundant species and intercontinental invasions of exotic species.

NEWLY ABUNDANT SPECIES

The populations of some species have grown far more abundant than they were in historical times. The growth of these populations often results from

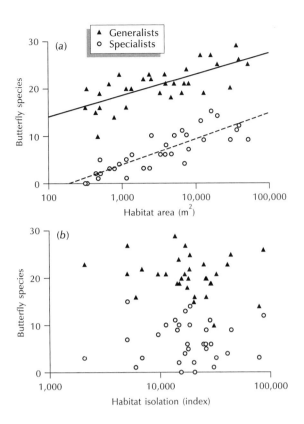

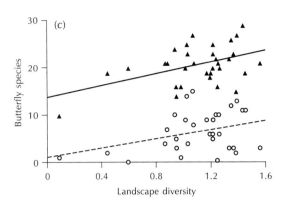

Figure 12.10 Relationships between the number of specialist and generalist butterfly species in grassland fragments and three landscape measures. (a) Butterfly species richness versus habitat area. (b) Butterfly species richness versus an index of habitat isolation. (c) Butterfly species richness versus landscape diversity. *Source:* After Krauss *et al.* (2003)

environmental changes brought about by humans. No word or phrase serves to describe such newly abundant species. Conover (2002, 125) proposed the term 'anthropogenic abundant' species, or 'AA' species for short, but newly abundant species seems preferable. He pointed out that many species today regarded as 'common' are newly abundant species that have benefited from human activities. In North America, the conversion of vast areas of forest to farms, fields, pastures and residential areas has led to the superabundance of mourning doves (*Zenaida macroura*), American robins (*Turdus migratorius*), mockingbirds (*Mimus polyglottus*), cowbirds (*Molothrus ater*) and red-winged blackbirds (*Aeglaius phoeniceus*). Also, coyotes (*Canis latrans*) became superabundant after humans had decimated populations of cougars (*Puma concolor*) and gray wolves (*Canis lupus*), which had kept coyote numbers in check through competition and predation. The newly abundant species themselves can cause environmental change, much of which is often degradational or destructive. An excellent example of a newly abundant species in North America is the white-tailed deer (Focus 12.5).

INVASIVE EXOTIC SPECIES

Exotic, introduced, invasive, non-indigenous, alien or non-native are names given to animal or plant species living outside their normal range. 'Biological pollution' is another name sometimes given to such species. Invasions by exotic species of animals and plants are a current and grave environmental problem, and one of the foremost causes of biodiversity loss.

FOCUS 12.5

THE WHITE-TAILED DEER AS A NEWLY ABUNDANT SPECIES

The white-tailed deer (*Odocoileus virginianus*) is a long-legged and fast-moving cervid (Figure 12.11). Males weigh between about 70 and 140 kg and can run at speeds up to about 58 km per hour. The white-tailed deer ranges from southern Canada to Central America, including mountain regions. It is not common in deserts, but it does occur in parts of the Great Basin Desert of Utah and Colorado, the upper, eastern part of the Sonoran Desert of Arizona, and upper regions of the Chihuahuan Desert of New Mexico and southern Texas. Over the last twenty years or so, the populations of white-tailed deer expanded rapidly. This expansion resulted partly from habitat changes brought about by agriculture and silviculture, and partly from the ending of overhunting. In many areas of the eastern USA, high densities of white-tailed deer have altered forest composition. Deer browse on the seedlings and saplings of such palatable trees as aspen, sugar maple and cherry, which find it difficult to regenerate. In consequence, unpalatable species, including American beech, have come to dominate the heavily browsed forests. Slow growing and palatable trees – the eastern hemlock and northern white cedar, for example – may disappear from stands where they have been growing for centuries.

But the impacts of deer browsing go further: they also affect plants in the understorey and herb layer, which are open to repeated browsing. Some lilies and orchids and others that cannot tolerate deer browsing require protecting by building deer-proof fences. Where deer densities are very high, all tree and understorey species can disappear, leaving an open, park-like forest with ferns the sole survivor in the understorey. These radical vegetation changes affect ecosystem processes; they change nutrient cycles, they change the quantity and quality of plant litter available to decomposers, and they change the soil microenvironment. Moreover, the changes in the understorey vegetation affect the bird community in the forests. Birds such as black-and-white warblers, black-throated green warblers and wild

Figure 12.11 White-tailed deer (*Odocoileus virginianus*) in the Smokey Mountains, USA. Photograph by Pat Morris

turkeys, which depend on the understorey vegetation for nesting or foraging, are uncommon in areas stripped of understorey plants. High deer populations also suppress squirrel and rodent populations through competition for acorns.

What is an exotic species?

It is not always clear whether a species is native or exotic. Several criteria can be identified indicative of a native species and several criteria suggestive of an exotic species. For example, one criterion of exoticness is a species that lies outside its natural range owing to human activity. A criterion of 'nativeness' might be a species that occurred in its natural range before the arrival of Europeans. Such criteria are problematic, however. The first European to visit the Hawaiian islands was Captain James Cook in 1800. However, Polynesians had already introduced species to the island before the arrival of Europeans. Similarly, some species spread to new areas without the aid of humans: are these native or exotic? And how are native and exotic species identified in Europe, Africa and Asia, where humans have lived for thousands of years but evidence of early introductions is lacking? Such problems as these can be resolved philosophically by accepting that there are several criteria that are typical of native species, and a corresponding set of criteria that are typical of exotic species, but that none of these criteria is either necessary or sufficient to label a species as either native or exotic (Woods and Moriarty 2001). A further complication is that some researchers use the term 'invasive species' to signify an introduced species that causes, or is likely to cause, economic or environmental harm or harm to human health.

The impact of exotic species

Not all exotic species are successful. This needs saying to temper the long list of detrimental effects discussed below. For example, although many of the mammals introduced to New Zealand have thrived, others – bandicoots, kangaroos, racoons, squirrels, bharals, gnus, camels and zebras – failed to become established. However, populations of exotic species may grow to higher levels than in their natural range and cause severe problems for the native animals, plants and ecosystems. They become so abundant because the diseases, parasites, competitors or predators with which they co-existed in the homeland do not check their numbers. The European starling (*Sturnus vulgaris*) had become one of the most abundant birds in North America within about a century of its introduction. Globally, exotic species have contributed hugely to the biodiversity loss of the last few centuries. They are responsible for 42 per cent of reptile extinctions, 25 per cent of fish extinctions, 22 per cent of bird extinctions, and 20 per cent of mammal extinctions (Cox 1999). The reasons for this heavy toll are varied.

Exotic species, especially if they should become abundant, may cause several problems for native species: outcompete them for food and shelter; heavily predate them; adversely alter their habitats; precipitate cascade effects in their communities; reduce their genetic integrity by interbreeding with them and pass on exotic diseases.

Exotic competitors sometimes oust native rivals. The rose-ringed parakeet (*Psittacula krameri*) on Mauritius threatens the native Mauritius parakeet (*P. echo*) with extinction because it can exclude it from nest cavities. It is a potential danger to native species in the UK, where three main colonies exist in south-west London, south-east London and the Isle of Thanet. The population of rose-ringed parakeets in the UK grew from a few in 1969, to 500 in 1983, to 5700 in 2002. In Australia, feral goats endanger the survival of the yellow-footed rock wallaby (*Petrogale xanthropus*) and the brush-tailed rock wallaby (*P. penicillata*) by competing for food and water.

Exotic predators often have profound impacts on native prey species, especially on islands where the native prey species have evolved in the absence of native predators and tend to lack the ability to defend themselves and an innate sense of caution. The vulnerability of island species to introduced predators is borne out by the fact that since the 1600s, some 90 per cent of reptile extinctions, 93 per cent of bird extinctions and 81 per cent of mammal extinctions have occurred on islands. Exotic rats, cats and mongooses have played havoc with island native reptile and bird populations. In the Galápagos Islands, native species include endemic reptiles (iguanas and the giant tortoise) and birds, including Darwin's finches. In the 1800s, the first settlers introduced feral cattle, donkeys, horses, goats, pigs, rats and dogs. Fishermen and visitors later introduced other species. The exotic herbivores overgrazed vegetation, causing the demise of many native plants and prompting the spread of exotic plants. Giant tortoise (*Geochelone nigra*), green turtle (*Chelonia mydas*) and other reptile eggs were prey to the pigs. Galapagos or dark-rumped petrels (*Pterodoma phaeopysia*) and baby tortoises were prey to black rats (*Rattus rattus*). The native rice rats (*Oryzomys spp.*) also suffered in competition with the exotic black rat. Feral dogs ate iguanas (*Conolophus* spp.), Galápagos fur seals (*Arctocephalus galapogoensis*), boobies (*Sula* spp.) and Galápagos penguins (*Spheniscus mendiculus*). Eradication campaigns have purged exotic species from some of the Galápagos Islands: feral goats from several islands; feral cattle from San Cristobal, Santa Cruz and Santa Maria; feral dogs from San Cristobal and Santa Maria (by shooting and poisoning). And they have reduced exotic rat populations on a couple of the smaller islands.

Sometimes, an exotic species alters the native habitat in a way that is unfavourable to the native species. Exotic carp have aided the decline of waterfowl in many wetlands, partly through competition, but partly by increasing the turbidity of the water during spawning and feeding, and so lowering the invertebrate populations upon which the native water birds feed. They also destroy aquatic vegetation used by native birds for food, cover and nesting. Modern humans are an exotic species. In many places they have altered habitats in such a way as to advertently or inadvertently attract exotic species. In south Florida, USA, some 2 million people live on a strip of land that formerly supported a continuous stand of pine trees. The developers and residents replaced the native forest with hundreds of colourful exotic plants collected from around the tropics. Some 1000 insect and 900 exotic plant species now have free-ranging populations in Florida, while some 25,000 plant species are cultivated but have not yet escaped. The new plants and insect communities make a habitat favourable to tropical birds, several species of which, including many parrots and parakeets, have established populations. The importation of exotic plants for ornamental purposes poses a problem in many developed regions of the world (Focus 12.6). The worst invasive plants seem to have several characteristics in common: they are large, grow fast, grow as clones, regrow fast after being damaged, are able to re-establish themselves from roots or rhizome fragments and often produce profuse numbers of seeds.

Exotic species may affect native species indirectly by setting in train a cascade of community changes. An exotic predator might give another exotic herbivore a competitive edge over a contending native herbivore by preferentially preying on the native herbivore, which may be less well adapted to competing with predators than the exotic herbivore. Likewise, exotic plants may not be able to compete with native plants unless an exotic herbivore overgrazes the native species, which may lack adaptations to guard against herbivory. With the native plants reduced by the exotic herbivore, the exotic plant can then gain a foothold and spread. As an example of this sort of cascade effect, take the case of the feral pig in Hawaii. Feral pigs on Hawaii found difficulty in establishing themselves because their diet lacked protein. After the introduction of the earthworm, a ready source of protein, the pig population soared. Once abundant, the pigs caused a decline in such native plants as ferns, this decline permitting a rise in

FOCUS 12.6

INVASIVE ORNAMENTAL SPECIES IN GREAT BRITAIN

In recent years, plant collectors have been searching the globe for new varieties of ornamental plants to satisfy the garden trade in Great Britain. This collecting has been responsible for the importation of several thousand species, of which around 900 have become established in the wild. Many species of plant, and notably those with a naturally vigorous growth strategy, have become problematic to nature conservation. Japanese Knotweed (*Fallopia japonica*) and Himalayan balsam (*Impatiens glandulifera*), for example, exhibit extreme competitiveness and are now common sights on riverbanks and in neglected areas of Britain. Other more recent introductions such as New Zealand pygmyweed or Australian swamp stonecrop (*Crassula helmsii*), parrot's feather (*Myriophyllum aquaticum*) and floating pennywort (*Hydrocotyle ranunculoides*) are now present in the British countryside and exhibit distinctly invasive growth strategies. At present *Crassula helmsii* is by far the most widely distributed of these three species in Britain and as such has been the focus of much research regarding the plant's impact and subsequent control.

the abundance of some exotic plants. A similar kind of cascade effect occurred on the Pacific island of Guam, where the native fauna evolved in the absence of snakes, the island being too remote for reptiles to reach. However, brown tree snakes (*Boiga irregularis*), an efficient nocturnal predator (Figure 12.12), arrived sometime during World War II, carried in military vehicles and equipment shipped in from New Guinea. Unable to adapt to predation by the brown tree snake, extinctions occurred in thirteen of twenty-two native bird species, two of three native bat species, and four of ten native lizard species (Table 12.3). Of the nine surviving bird species, seven are rare; and the surviving mammal species – the Marianas fruit bat (*Pteropus mariannus*) – is endangered. The depredations caused by snake predation might not have arisen were it not for the introduction of various exotic animals, including musk shrews (*Suncus murinus*), green anoles (*Anolis carolinensis*), black drongos (*Dicrurus macrocercus*) and curious brown skinks (*Carlia fusca*). These exotic species arrived shortly after the brown tree

Figure 12.12 Brown tree snake (*Boiga irregularis*) in a logging camp, Papua New Guinea. Photograph by Pat Morris

Table 12.3 Status of native species on the island of Guam, Pacific Ocean

Common	Uncommon	Rare	Extirpated
Birds			
Yellow bittern (*Ixobrychus sinensis*)	Pacific reef-egret (*Egretta sacra*)	White-tailed tropicbird (*Phaethon lepturus*)	Wedge-tailed shearwater (*Puffinus pacificus*)
		Common moorhen (*Gallinula chloropus*)	Mallard (*Anas platyrhyncos*)
		Brown noddy (*Anous stolidus*)	Micronesian megapode (*Megapodius laperouse*)
		White or fairy tern (*Gygis alba*)	Guam rail (*Rallus owstoni*)
		Mariana grey swiftlet (*Aerodramus vanikorensis*)	White-browed rail (*Poliolimnas cinereus*)
		Mariana crow (*Corvus kubaryi*)	White-throated ground-dove (*Gallicolumba xanthonura*)
		Micronesian starling (*Aplonis opaca*)	Marianas rose crown fruit-dove (*Ptilinopus roseicapilla*)
			Guam Micronesian kingfisher (*Halcyon cinnamomina*)
			Nightingale reed-warbler (*Acrocephalus luscinia*)
			Guam broadbill or flycatcher (*Myiagra freycineti*)
			Rufous fantail (*Rhipidura rufifrons*)
			Cardinal honeyeater (*Myzomela cardinalis*)
			Bridled white-eye (*Zosterops conspicillatus*)
Lizards			
Blue-tailed skink (*Emoia caeruleocauda*)	Moth skink (*Lipinia noctua*)	Oceanic gecko (*Gehyra oceanica*)	Snake-eyed skink (*Cryptoblepharus poecilopleurus*)
Mourning gecko (*Lepidodactylus lugubris*)	Mutilating or stump-toed gecko (*Gehyra mutilata*)	Rock or pelagic gecko (*Nactus pelagicus*)	Azure-tailed skink (*Emoia cyanura*)
			Slevin's skink (*Emoia slevini*)
			Micronesian or spotted-belly gecko (*Perochirus ateles*)
Mammals			
		Marianas fruit bat (*Pteropus mariannus*)	Pacific sheath-tailed bat (*Embollonura semicaudata*)
			Little Marianas fruit bat or Guam flying fox (*Pteropus tokudae*)

snake, for which they provided a ready source of prey. Had these species not come to Guam, the brown tree snakes might not have become abundant enough – there are up to 13,000 snakes per square mile on Guam – to cause such havoc to the native fauna.

The interbreeding of exotic and native species reduces the genetic integrity of the native species. In Britain, native red deer (*Cervus elaphus*) and exotic sika deer (*Cervus nippon*) from the Far East, tend to interbreed, often producing a hybrid zone where they do so. The fear is that such hybridization may imperil the red deer populations. Interbreeding of introduced mallards with native species in New Zealand, Hawaii, Australia and Florida in the USA threatens the integrity of native duck populations. In the case of Florida, mallards (*Anas platyrhynchos*) used to occur as wild, winter migrants. Recently, some mallards – probably from those released in backyards and by hunting clubs – stay year-round and have started breeding in the wild. These new resident mallards are interbreeding with the endemic Florida mottled duck (*Anas fulvigula fulvigula*). This hybridization poses a considerable threat to the future of the Florida mottled duck. A botanical example from the British Isles is the native bluebell (*Hyacinthoides non-scripta*) that hybridizes with the introduced Spanish bluebell (*Hyacinthoides hispanica*). Most of the bluebells in London are such hybrids. Introduced species carrying diseases may decimate native populations. In Queensland, Australia, a virus introduced by pet fish seems to have led to a sharp fall in frog populations. In 1998 and 2002, native North Sea seal species – mainly harbour seals and grey seals – suffered population crashes when harp seals (*Phoca groenlandica*) from the Arctic Ocean introduced the distemper virus. Fortunately, the native species have staged a comeback.

SUMMARY

Biodiversity takes in the diversity of genetic material, the diversity of species and the diversity of habitats. It is unevenly distributed around the world, with a couple of dozen or so, mainly tropical and Mediterranean-type climate, diversity hot-spots superimposed upon a tropical high to polar low biodiversity gradient. Biodiversity is always changing, but the recent biodiversity nose-dive is worrying. The chief drivers of current biodiversity change are changes in land cover, species exploitation and exchange, climatic change and changes in environmental chemistry. Habitat fragmentation is a large part of land cover change. It leads to habitats being lost, shrinking and becoming increasingly isolated. The size, shape and degree of connection between habitat fragments affects the distribution and abundance of many species, as well as community processes. Particularly important is the proportion of habitat edges to habitat interiors, with edge effects more noticeable in smaller and irregularly shaped habitat fragments. Corridors of various kinds – road, railways, trails, powerlines, hedgerows, rivers and so on – act at once as conduits, as barriers and as filters. Linkages between habitat fragments appear to help in maintaining species diversity, and a network of well-connected fragments seems to fare better than a network of ill-connected fragments. 'Landscape context' as measured, for example, as landscape diversity, also affects biodiversity. Human activities have altered the species composition of many places: witness the emergence of newly abundant species, such as the white-tailed deer, and the introduction of invasive exotic species, such as efficient predators on formerly prey-friendly islands.

Questions

1 Why does biodiversity vary from place to place?
2 Why does habitat fragmentation affect some animal and plant species more than others?
3 How important are wildlife 'corridors' to species conservation?
4 Try to identify the traits of a successive invasive species.
5 Examine the range of steps taken to overcome the overexploitation of species.

Further Reading

Bennett, A. F. (1999) *Linkages in the Landscape: The Role of Corridors and Connectivity in Wildlife Conservation*. Gland, Switzerland and Cambridge, UK, IUCN.
An excellent book looking at the detrimental effects of habitat fragmentation on biodiversity and ways of combating it through landscape linkages. Contains many examples and a discussion of conservation strategy.

Elton, C. S. (1958) *The Ecology of Invasions by Animals and Plants*. London: Chapman and Hall.
Old, but a classic and well worth having a look at.

Huggett, R. J. and Cheesman, J. E. (2002) *Topography and the Environment*. Harlow: Prentice Hall.
Chapter 6 contains many examples of the effects of landscape elements (patches, corridors and matrices) and landscape structures (networks) on species and communities.

Hunter, M. L. (2001) *Fundamentals of Conservation Biology*, 2nd edn. Oxford: Blackwell Science.
Covers issues of habitat loss and overexploitation.

Jeffries, M. L. (1997) *Biodiversity and Conservation*. London and New York: Routledge.
An excellent basic text.

Quammen, D. (1996) *The Song of the Dodo: Island Biogeography in an Age of Extinctions*. London: Hutchinson.
An excellent, readable and entertaining introduction to biogeographical ideas.

Useful Websites

http://www.centerforplantconservation.org;
www.mobot.org
A useful site at the Center for Plant Conservation, Missouri Botanical Garden.

http://www.iucn.org
The International Union for the Conservation of Nature and Natural Resources (IUCN), the foremost organization in the field of conservation.

http://www.rbgkew.org.uk
The world's largest botanical garden – the Royal Botanic Gardens, Kew.

http://www.unep-wcmc.org.uk
The United Nations Environment Programme World Conservation Monitoring Centre (UNEP-WCMC)

http://www.wwf.org
The World Wildlife Fund (WWF), the leading conservation organization. Massive number of links.

http://www.zsl.org
Website of the Zoological Society of London. Includes information on its conservation programmes.

CHAPTER 13
SPECIES AND COMMUNITIES II: CHEMICAL AND CLIMATIC IMPACTS

OVERVIEW

The increasing reliance on fossil fuels during the twentieth century has led to rising levels of atmospheric carbon dioxide. Coupled with rising levels of other 'greenhouse' gases, the changing chemistry of the atmosphere has seemingly triggered a warming trend that, unless humans adopt remedial measures, will continue through the present century and beyond. The climatic changes stimulated by global warming will affect animal and plant species. Some species appear able to weather the changes by making small adjustments to their behaviour. The climatic changes imperil other species, for instance those that live at high altitudes and high latitudes, those that are poor dispersers and those that have a small gene pool. The same climatic changes will force many communities to change. Biomes may expand or contract, or disassemble and reassemble elsewhere, probably with a different set of species. Other human-induced chemical changes in the environment disrupt natural biogeochemical cycles. Humans have boosted the amount of reactive nitrogen in ecosystems, which has both beneficial and detrimental effects. In industrialized regions, the burning of fossil fuels has produced acid rain, the effects of which on animals and plants are negative. The release of nitrates and phosphates from agricultural and domestic sources produces high nutrient levels in lakes, so accelerating the natural process of eutrophication and harming aquatic wildlife. Many human-made toxins escape into the environment causing local pollution incidents and, more worryingly, contamination in ecosystems far removed from the source areas. All these human-made drivers of global change, and those discussed in the previous chapter, act together. Their integrated effects are complex, but scientists are beginning to understand them.

LEARNING OUTCOMES

After reading this chapter, you should be able to understand:

- Which kinds of species are safe in a climatically changed world
- Which kinds of species are at risk in a climatically changed world
- The likely changes in the world's biomes brought about by the warming trend
- The effects of changing environmental chemistry on the living world, and in particular the effects of reactive nitrogen, acid rain, nitrates and phosphates and environmental toxins
- How the chief drivers of biodiversity change – land cover, species exploitation and exchange, climate and environmental chemistry – act in concert to affect the world's biomes.

The planetary environment is ever changing. At present, important trends arising from human activity appear to be global warming (a switch to a hotter and more humid world) and a changing chemistry of the environment. Like habitat fragmentation, the spread of invasive exotic species and overexploitation of wildlife, these climatic and chemical changes threaten biodiversity.

13.1 GREENHOUSE SPECIES

natural selection: The process by which the environmental factors sort and sift genetic variability and drive microevolution

Species become adapted to the environment in which they live by processes of **natural selection**. If that environment should change, the species must re-adapt, move elsewhere, or perish. The fate of many species during the twenty-first century, as the world warms and habitats shrink or vanish, is worrying. It seems that extinction will continue apace and global biodiversity will drop, but which species are the most vulnerable to global warming?

'SAFE' SPECIES

The safest species are migratory birds, insects and mammals that can track their preferred climatic zone by adjusting the timing or destination of migration to fit annual variations in climate. However, much wildlife stays put and cannot respond swiftly to climatic changes. For such sedentary species, global warming is likely to encourage a slower response that involves shifts of species' ranges towards the poles or to higher altitudes. The shift would occur, not through changes in the movements of individuals, but from changes in the balance between local extinctions and colonization events at the northern and southern range boundaries. With net extinction at the southern boundary or net colonization at the northern boundary, the range would shift northwards.

range shift: a change in the distribution (range) of species
range contraction: the shrinking of a species range
range expansion: the enlargement of a species range

Studies of such **range shift** generally focus on single species or on a portion of the species' range. For instance, in the British Isles, climatic change during the twentieth century led to a decline and **range contraction** of many of the UK's butterfly species, but an increase in the numbers and **range expansion** of some of the more common species. The increase in abundance and range of commoner species seem largely a response to generally warmer summer weather, where an abundance of habitat remains for the species concerned both within and outside their current range. Species currently expanding their range in the UK include small skipper (*Thymelicus sylvestris*), Essex skipper (*Thymelicus lineola*), large skipper (*Ochlodes venatus*), orange tip (*Anthocharis cardamines*), brown argus (*Aricia agestis*), comma (*Polygonia c-album*), speckled wood (*Parage aegaria*), marbled white (*Melanargia galathea*), gatekeeper (*Pyronia tithonus*) and ringlet (*Aphantopus hyperantus*). Confined to southern England in the mid-twentieth century, the comma butterfly has expanded its range northwards since the 1970s, and is now in the north-east. The white admiral butterfly (*Ladoga camilla*), a shade-tolerant woodland species, also expanded its range because coppicing ceased and woodlands became shadier.

An ambitious study, which used data on 35 non-migratory European butterflies, furnished the first large-scale evidence of poleward shifts in complete species' ranges (Parmesan *et al.* 1999). The results indicated that 63 per cent of the butterflies studied shifted their ranges to the north by 35–240 km during the twentieth century, and just 3 per cent shifted their ranges to the south during the same period. These range shifts are some 5–50 times greater than the colonization distances achieved by comparable butterflies in single colonization events. This fact suggests that the northwards extension seems to result from a sequence of new populations each giving rise to further colonizations. To be sure, the speckled wood butterfly (*Pararge aegeria*) in England, Wales and Scotland has established at least one, and usually several, new populations in each of the 10-km grid squares to the north of its historical range since 1915 (Figure 13.1; Figure 13.2). Not all species show this northward range expansion: several species seem to have a stable range, and the range of the lesser purple emperor (*Apatura ilia*) has pulled back southwards.

SPECIES AT RISK

Six kinds of species appear at high risk of extinction under global warming: peripheral species, geographically localized species, highly specialized species, species with poor dispersal ability, species sensitive to climate and genetically impoverished species.

Peripheral species

Populations of animals or plants that are at the contracting edge of a species range – **peripheral populations** – are at increased risk of extinction compared with those near the centre. The same is true of **disjunct subpopulations** that may

> **peripheral population**: a population lying at the edge of a species range
>
> **disjunct subpopulations**: populations of the same species separated geographically from each other

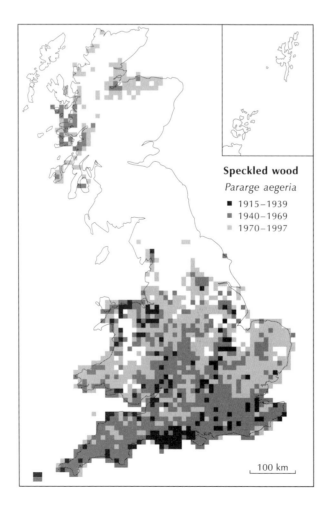

Figure 13.1 The speckled wood butterfly (*Pararge aegeria*) moves north. *Source:* After Parmesan *et al.* (1999)

Figure 13.2 Speckled wood butterfly (*Pararge aegeria*) in Lock's Copse, Isle of Wight, England. Photograph by Pat Morris

---

bioclimate: the climatic conditions that affect living things

adaptive radiation: the diversification of many species from a single founding species

historical distribution: a species range at some time in the past

---

have adapted to a local **bioclimate**. Some researchers argue that, because peripheral populations are prone to extinction, efforts to conserve them are a waste of time. A counterargument suggests that some peripheral populations may survive in isolated refuges (biogeographers tend to call them refugia) that later, after environmental conditions change again, provide a source population for an expanded range and subsequent **adaptive radiation**. In short, today's peripheral population could be the core of tomorrow's species range. Indeed, peripheral populations survive more frequently than do core populations when species undergo dramatic reductions in their range (Channell and Lomolino 2000). The present California condor (*Gymnogyps californianus*) population is peripheral to its **historical distribution**, which ranged from California to New York (Snyder and Snyder 2000). It is not possible to predict how the condor population will change in the future, but it seems sensible to protect the peripheral population. Equally, it may be folly to protect species in those parts of their range where they are common, as the grey wolf (*Canis lupus*) is in Alaska, and not to protect it in those parts of its range where it is rare, such as in Idaho.

Geographically localized species

Many currently endangered species live in alarmingly small and isolated habitats. The golden lion tamarin (*Leontopithecus rosalia*) is a squirrel-sized primate that lives in lowland Atlantic Coast rainforest, Brazil, just north of Rio de Janeiro (Figure 13.3). Lowland Atlantic forest is one of the most endangered rainforests in the world. It once covered 1 million km² along the Brazilian coastline. Today, a mere 7 per cent of the original forest remains, putting the golden lion tamarin under enormous threat. An estimated 1000 golden lion tamarins live in the wild, some 220 in the Poço das Antas Biological Reserve, 130 in the Fazenda União Biological Reserve, 400 in several reintroduced forests, and about 250 scattered outside protected areas. Additionally, some 480 golden lion tamarins live in 150 zoos around the world. During the 1990s, a combination of relocating tamarin families from vulnerable areas into the protection of the Poço das Antas Biological Reserve, and reintroducing captive-bred tamarins into the wild, has put the little

Figure 13.3 Golden lion tamarin (*Leontopithecus rosalia*) in Jersey Zoo, Jersey. Photograph by Pat Morris

primate on the road to recovery. Predators, probably 3–6 kg weasel-like tayra (*Eira barbara*), destroyed all but one of fifteen tamarin study groups between 1996 and 2000. Fortunately, golden lion tamarins soon recolonized some of the territories occupied by the extinct groups. Even so, for the first time in 17 years, some areas with suitable habitat in the reserve no longer contain golden lion tamarins and the population size in the reserve has dropped by 36 per cent.

Highly specialized species

Species that have tight **habitat requirements** are notably vulnerable to changes in their habitats. Many species have a close association with only one other species. An example is the Everglade snail kite (*Rostrhamus sociabilis plumbeous*), a crow-sized bird of prey that feeds exclusively on the large and colourful Florida apple snail (*Pomacea paludosa*) in Florida wetlands. Since about 1950, swamp drainage, the choking of water bodies by water hyacinth (*Eichhornia crassipes*), which makes the snail difficult to see, and adverse effects of pesticides and other pollutants, has already robbed the kite of its prey in some areas. Once just 30 snail kites lived in Florida, but there are now about 300 and efforts to clean up water sources and habitats may ensure a continued growth of the population. In North Dakota, USA, the piping plover (*Charadrius melodus*) is a small shorebird that nests only on bare sandbars or gravel on islands in rivers or shorelines of alkali lakes. The construction of reservoirs on the Missouri River has reduced the sandbar habitats available to the plover, while wetland drainage, contaminants and cattle grazing has reduced its habitat in alkali wetlands. Habitat loss is much more likely to endanger a specialized species than it is a generalist species such the mourning dove (*Zenaida macroura*) that nests successfully on the ground or in trees in the countryside or in urban areas.

> **habitat requirements**: the environmental conditions required for a particular species to thrive. Habitat generalists have broad habitat requirements; habitat specialists have very specific habitat requirements

Poor dispersers

Many trees have heavy seeds that do not travel far from their parents. Plants with limited **dispersal** abilities have knock-on effects along the food chain. Some birds, mammals and insects are closely associated with specific forest trees,

> **dispersal**: the spread of organisms into new areas

which cannot migrate rapidly. These species would be hard pressed to survive. The endangered Kirtland's warbler (*Dendroica kirtlandii*), for example, only breeds in and nests upon the ground under jack-pine (*Pinus banksiana*) forest on well-drained sandy soil in north-central Michigan, USA. But not any jack-pine forest will do – it has to be young (5- to 23-years old), secondary growth forest that emerges in the aftermath of a forest fire. The region is the only place in the world where Kirtland's warbler breeds. With more heat and less rain caused by global warming, the jack pines would be unable to remain dominant and perhaps even die out in the region. Even if they should manage to migrate northwards, they would then be growing on less well-drained soils unsuitable for the ground-dwelling warbler, which may find itself without any suitable nesting sites (Botkin *et al.* 1991). Therefore, this species will likely be a casualty of global warming.

Climatically sensitive species

climatically sensitive communities: communities that respond quickly to changes in climate, and particularly to changes in temperature and water budgets

global amphibian decline: the somewhat mysterious loss of amphibian biodiversity that started in the 1970s

Species in **climatically sensitive communities** are vulnerable to global warming. Communities in this class include wetlands, montane and alpine biomes, Arctic biomes and coastal biomes. Wetlands will dry out, Arctic regions and mountaintops will warm up and coastal biomes will be flooded. Animals and plants in these regions will have to cope with the most rapid changes.

The drying up of many wetlands, owing to altered rainfall patterns, would have dire consequences for moisture-loving amphibians (frogs, toads, salamanders and newts). Indeed, a **global amphibian decline** is a current concern amongst biologists, with many writers speaking of an alarming rate of loss since the 1970s (e.g. Houlahan *et al.* 2000). Indeed, all six native UK amphibian populations are dwindling. In Australia, fourteen species are in sharp decline, and four have become extinct over the last 30 years. In places, entire populations have almost completely vanished in a short space of time. Many of the declines occur in areas of minimal human contact. Nonetheless, the evidence for a global decline in amphibian populations tends to be either undependable or derived from short-term studies at local scales. It is very difficult to test the idea of global decline for two reasons. First, amphibian populations tend to fluctuate without much assistance from environmental change, making the detection of time trends complicated. Second, it is tricky to infer global trends from local and regional information. With these problems in mind, a recent study set out to assess large-scale temporal and geographical variations in amphibians using data from 936 populations (Houlahan *et al.* 2000). Looking at the global scale, the results showed relatively rapid declines from the late 1950s and early 1960s to the late 1960s, followed by a reduced rate of decline to the end of the twentieth century. Amphibian populations fell in Europe, including the UK, and North America during the 1960s, but only in North America did they continue declining from the 1970s to the late 1990s. It would seem, therefore, that while large-scale trends show considerable variability in time and space, amphibian populations are indisputably declining and have been doing so for several decades.

High latitudes will see greatest rises in temperature, which could have serious repercussions for Arctic wildlife. The polar bear (*Thalarctos maritimus*), the world's largest land carnivore, weighing up to 700 kg, faces threats from global warming and from such human activities as hunting (Figure 13.4). The worldwide polar bear population comprises some 22,000 to 27,000 individuals

Figure 13.4 Polar bear (*Thalarctos maritimus*). Photograph Andrew E. Derocher

split into twenty populations of varying size (ranging from a few hundred to a few thousand) in Alaska, Canada, Russia, Norway and Greenland. Some 15,000 of the bears live in Canada. The bears occupy most of their original range and live at a density similar to that before the industrialized era. Nonetheless, evidence suggests that global warming could deny the bears the huge expanses of Arctic sea ice where they feed on seals, particularly ringed seals (*Phoca hispida*) and bearded seals (*Erignathus barbatus*), and other prey (young walruses, beluga whales, fish and seabirds) during the winter. Indeed, global warming may already be affecting polar bears. In Hudson Bay, Canada, a study by the Canadian Wildlife Services shows that the polar bear population is declining. Ice on the bay now melts an average of three weeks earlier than it did in the mid-1970s. Consequently, polar bears need to withdraw further inland to replenish their reserves of fat by feeding on seal pups, which live on the ice. By 2050, the ice-free period during the summer could increase from 60 days to 150 days, so further curbing the feeding period for the bears. The body condition of the bears is likely to deteriorate and reproductive success to falter, triggering local extinctions. The Arctic marine ecosystem would feel the loss of polar bears, who are the top predators. The worse-case scenario predicts no sea ice at all by 2080, which would cause the total extinction of polar bears unless they were able to adapt and secure an alternative food source.

A 3°C warming at high altitudes would require animals and plants to shift their ranges upwards by about 500 m. Animal and plant populations on mountains may retreat upwards as the climate warms, but eventually some of them will have nowhere to go. This would probably happen to the pikas (*Ochotona* spp.) that lives on alpine meadows on mountains of the western USA (Figure 13.5). Pikas feed on grasses and do not range into boreal forests. As the boreal forest habitat advances up the mountainsides, so the pikas will retreat into an ever-decreasing area that may one day vanish. This process is already happening to Edith's checkerspot butterfly (*Euphydryas editha*), which lives in western North America (Parmesan 1996). Records of this species suggest that its range is shifting upwards and northwards, and extinctions have occurred at some sites.

Figure 13.5 Pika (*Ochotona princeps*) in the Rocky Mountains, western USA. Photograph by Pat Morris

> **genetically impoverished species**: species with a reduced gene pool owing to population reduction and isolation

Genetically impoverished species

Species forced to a set of small populations often have a smaller gene pool than the more widespread parent population. Biologists believe that such **genetically impoverished species** are normally less well equipped genetically to adapt to climatic change. However, the Mauritius kestrel (*Falco punctatus*)

Figure 13.6 Mauritius kestrel (*Falco punctatus*) in the Black River Conservation Aviaries, Mauritius. Photograph by Pat Morris

(Figure 13.6), which habitat destruction and pesticide contamination reduced to a single breeding pair in the wild by 1974, has recovered through a captive breeding programme. Despite having a smaller gene pool than museum specimens collected before 1974, the recovered Mauritius kestrel is thriving and has started exploiting gardens and other habitats created by humans. It is not clear whether many other species would be able to recover with such a small gene pool.

13.2 GREENHOUSE COMMUNITIES

COMMUNITIES UNDER PRESSURE

Community composition and structure change in the face of **environmental perturbations**. During the next century, global warming will perturb many communities, and will affect the species composition of communities and the geographical location of biomes.

> **environmental perturbation**: a change in any environmental factor that disrupts the 'balance' with a community

Biome and ecotone shifts: case studies

Global warming fuelled by a doubling of carbon dioxide levels could produce large shifts in the distribution of **biomes**. Each biome will tend to respond individually, rather than all biomes moving en masse. Arctic tundra biomes should shift northwards under a warming climate, for example. Of course, other factors – including habitat loss, habitat fragmentation and the invasion of alien species – also drive biome change. Nonetheless, the climatic changes likely to occur over the next century will induce big changes in the placement of biomes and the zones of transition (ecotones) between them.

> **biome**: a community of animals and plants occupying a climatically uniform area on a continental scale

Arctic tundra biomes

Boreal (northern) tree species will probably invade the shrub tundra along its southern boundary, and shrubs will invade the true tundra (M. D. Walker *et al.* 2001). The Arctic tundra biome as a whole is in danger, but species extinctions may be rare. Expansion of boreal forest into tundra may be pegged back by the position of the Arctic weather front, which makes it difficult for trees to establish themselves to the north of the latitudinal tree-line. Alpine tundra is likely to fare far worse under a warming climate. Indeed, some models predict the total loss of that biome. Palaeoecological evidence from the Pleistocene epoch (between 2 million to 10,000 years ago) tempers these gloomy predictions. The position of the altitudinal **tree-line** and the overall diversity of alpine tundra have held fairly constant throughout the Pleistocene with changes being only localized. Even so, smaller fragments of alpine tundra will likely disappear, along with rare species and habitats sensitive to the fragmentation caused by road-building, forestry and so on.

> **tree-line**: the altitudinal and latitudinal limits of tree growth

Boreal forest biomes

These should expand northwards under a warming climate, but agricultural landscapes and managed temperate forests should invade their southern fringes (Chapin and Danell 2001). Overall, the area of boreal forest should stay roughly the same as at present, but about a third of present boreal forest will probably advance into Arctic tundra and another third be converted to other biomes

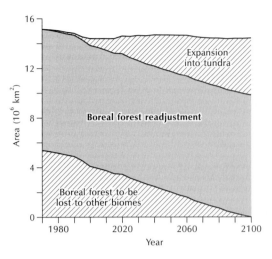

Figure 13.7 Boreal forest change.
Source: After Chapin and Danell (2001)

(agricultural and forest biomes) (Figure 13.7). Increased outbreaks of fires, outbreaks of pathogens and insect swarms, and conversion to agriculture along the southern margins of the biome should promote an increase in species and habitat diversity. They should also lead to habitat fragmentation that will encourage the colonization of non-boreal species. Worryingly, the changes would also reduce the regional albedo (the reflectivity of a surface), so boosting the climate warming at the tundra–boreal forest border through a positive feedback mechanism. Single tree-species dominate large areas of boreal forests, and are the foundation of many community and ecosystems processes. For this reason, the gain or loss of just one species of tree may have large regional repercussions.

Temperate grassland biomes

Global warming may severely affect these for several reasons (Sala 2001). Agriculture will likely eat up a sizeable portion of the temperate grassland area. Overgrazing may well exacerbate grassland biodiversity loss. Changes in water availability associated with global warming will probably affect grassland biomes, which are water-limited ecosystems that depend on a seasonal drought. The predicted changes in water availability will benefit some species and harm others, but act to the detriment of temperate grassland biodiversity as a whole. The changing hydrological conditions should alter the competitive balance of grassland species, which in turn will produce changes in species abundances and local extinctions. The effects will probably vary from region to region, as the changes in the water balance are unlikely to be uniform.

Savannah (tropical grassland) biomes

In a warming world, habitat loss, habitat fragmentation and an intensification of land use will probably affect savannah biomes (B. Walker 2001). New species from neighbouring biomes are unlikely to invade the savannahs, although alien species may. Alterations in the seasonal distribution of rainfall will be the main climatic driver of change. Seasonal shifts in rainfall distribution could well lead to species loss and their replacement by aliens. A change in the relative abundances of species and plant functional types (particularly in the relative abundances of woody plants and grasses) will greatly affect savannah biodiversity.

Mediterranean-type biomes

Around the world, Mediterranean-type biomes are exceptionally biodiverse and are havens of endemic species, especially endemic plant species. Global warming threatens them all (Mooney *et al.* 2001). Mediterranean-type biomes have a history of landscape conversion, habitat destruction, habitat fragmentation and the invasion of alien species. Climate warming will probably aggravate these problems, with changes in the annual and year-to-year patterns of precipitation being notably important. Temperature changes could have material effects, especially at lower elevations. A changed frequency of fires would also have a big impact on biodiversity. Overall, the biodiversity of Mediterranean-type biomes may well be hit the hardest of all biomes over the present century.

Desert biomes

These contain species adapted to surviving difficult conditions (Huenneke 2001). Many species have reserves on which to draw when the 'going gets tough': rodents and harvester ants have seed caches; reptiles have eggs or resting stages. However, these reserves could be vulnerable to changing rainfall regimes, especially to wetter conditions and increased fire frequency. Single species often dominate large areas of desert. Should climate disrupt such species, then changes in the desert biome are likely. Equally, some desert species are endemic to local and specialized habitats; these species would be susceptible to change. Riparian and aquatic habitats are the most threatened parts of desert biomes. They are habitats where pressure of recreational use and livestock grazing will make themselves felt most. In short, desert life will probably respond dramatically to climatic and land use change.

Temperate forest biomes

The effects of global warming upon temperate forests will differ in the Northern and Southern Hemispheres (Armesto *et al.* 2001). Southern temperate forests tend to be isolated and occupy small areas. They harbour mainly endemic species assemblages that contain representatives of ancient families and genera that have one or a few living species. These unique forests are highly susceptible to global warming and their conservation is a priority. South American temperate forests, for example, may lose much of their biodiversity over the next hundred years, partly owing to human exploitation and land use change. Northern temperate forests vary in composition from region to region, as do their disturbance histories and potential human impacts. For these reasons, the response of northern temperate forests to climatic warming will vary from place to place and generalizations are tricky. More so than southern temperate forests, which have little room for southward expansion, northern temperate forests could spread into boreal latitudes.

Tropical forest biomes

Climatic warming is unlikely to affect these (Dirzo 2001). The only places that may see significant climate-induced changes are biomes at mid-elevations where the incidence of typhoons or changes in the degree of seasonality may alter. However, deforestation and forest fragmentation do pose an enormous threat to tropical forest biodiversity.

Biome and ecotone shifts: general patterns

As the world warms up and climates change over the next hundred years and beyond, species within biomes may move in the same general direction. However, that does not mean that entire communities will move as one. Quite the contrary, species move at different rates in response to climatic change. The result is that communities disassemble by splitting into their component species and losing some species that fail to move fast enough or cannot adapt. Mathematical models are helpful in understanding the possible changes in communities as the world warms up. Clearly, as the previous section showed, profound changes in the composition and geography of communities would occur. Although biome response to changing climates is likely to be individualistic (each species 'does its own thing'), some general patterns seem to emerge.

Spatial shifts of biomes should produce discrete 'change' zones and 'no-change' zones (Figure 13.8) (Neilson 1993). Such changes alter the pattern of vegetation. Large, relatively uniform biomes reduce in size, while new biomes, which combine the original biome and the encroaching biomes, emerge. This may happen as the boreal forests move northwards into the southern stretches of tundra biomes, and temperate forests biomes move northwards into the southern fringes of the boreal biomes. Drought followed by infestations and fire would probably affect biomes and ecotones alike, if climate were to change rapidly. Increased frequency of fires in Mediterranean-type biomes would almost certainly reduce biodiversity there.

Ecotones would be especially sensitive to climatic change, whether it was slow or fast. Models suggest that the pattern of ecotone change would be sensitive to **water stress** (Figure 13.9). With climatic warming but no accompanying water stress, habitats should not fragment as they would with water stress, but should display a wave of high habitat variability as the ecotone gradually tracks the climatic shifts. Under extreme drought stress, the entire landscape should fragment to a high degree, as every variation in topography and soil would become important to site water-balances and to the survival of different species. The ecotone would disappear for a while and would reappear at a new location. It would not visibly shift geographically, as it would with no water stress, but rather would disassemble and then re-establish itself elsewhere later, probably with a different set of species.

> **water stress**: a lack of soil water during at least one season

Implications for conservation

The projected shifts of biomes and their component species give managers of nature reserves a quandary. Should they protect reserves in spite of the likely effects of climatic change? Should they allow the reserves to lose species and readjust to the new climatic regimes? Or should they adopt a different approach

Figure 13.8 Geographical shifts in biomes induced by global warming. Notice the discrete 'change' and 'no change' zones. New biomes, which combine the characteristics of the original biomes and the encroaching biomes, emerge. *Source*: After Neilson (1993)

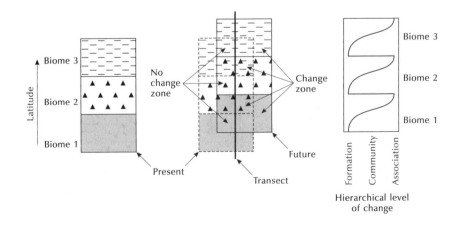

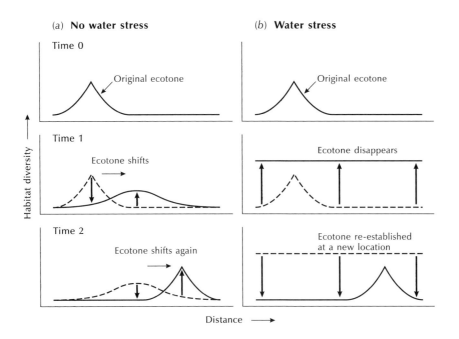

(a) **No water stress**   (b) **Water stress**

Figure 13.9 Ecotone changes induced by global warming. (a) Changes with no water stress. (b) Changes with water stress. *Source*: After Neilson (1993)

that extends conservation to areas outside nature reserves? The MONARCH Report (Modelling Natural Resource Responses to Climate Change) for the UK Climate Impacts Programme tackled this crucial question (Harrison *et al.* 2001). It gave predictions about how the changing climate is likely to modify the character of the British and Irish countryside up to 2080 (Focus 13.1). It concluded that nature conservation would need to evolve with the changing

FOCUS 13.1

KEY FINDINGS OF THE MONARCH REPORT FOR TERRESTRIAL SPECIES AND COMMUNITIES

For terrestrial environments, the key predictions of the changing bioclimatic space (the area that experiences climate conditions suitable for particular species to live) of selected species and communities were:

Species
Plant species displayed a varied response. For example, the ranges of globeflower (*Trollius europaeus*) and trailing azalea (*Loiseleuria procumbens*) may shrink and decline while the range of sea purslane (*Halimione portulacoides*) may expand. Of bird species, the capercaillie (*Tetrao urogallus*) could lose 99 per cent of its bioclimatic space by 2050, and the snow bunting (*Plectrophenax nivalis*) could lose more than half of its bioclimatic space by 2050. On the other hand, the turtle dove (*Streptopelia turtur*), yellow wagtail (*Montacilla flava*) and reed warbler (*Acrocephalus scirpaceus*) may all enlarge their potential ranges. The large skipper butterfly (*Ochlodes venata*), which in the year 2001 occupied England, Wales and south-west

Scotland, may expand its range, invading most of Ireland and Scotland by the 2050s.

Communities
Montane heath communities – which comprise dwarf willow, lichens, mosses, sedge and herbs – occur in the windswept uplands of the Pennines, the Lake District and Wales. Their vegetation is already susceptible to various human activities and they seem to be the most vulnerable habitat. Upland oak woodland and upland hay meadows in the Pennines and Lake District are also likely to come under pressure from climatic change. Beech woodlands may suffer from drought in the southern part of their range, such as in parts of Kent, Surrey and the Chilterns. Some wetlands – raised bogs, wet heaths and coastal dune slacks – may profit from the expected increase in winter rainfall. Other wetlands may fare badly where summer rainfall is likely to decline, a fall of up to 110 cm being possible in south-east England.

climate by taking a more forward-thinking and adaptable approach. To accommodate species' movements and displacements, emphasis should extend to wildlife management in the wider countryside, and not stick solely to nature reserves. And new conservation techniques should be employed, including the establishment of special corridors to allow wildlife to move.

13.3 SPECIES, COMMUNITIES AND CHANGING ENVIRONMENTAL CHEMISTRY

REACTIVE NITROGEN: FROM DEARTH TO GLUT

Human-induced change in the nitrogen cycle (p. 31) is one of the foremost drivers of current biodiversity change. Nitrogen is crucial to life. The genetic character of all living things, and the enzyme proteins that impel the metabolic mechanism of every living cell, depend upon nitrogen. Nitrogen is an abundant element in the ecosphere. About 78 per cent of the atmospheric mass consists of nitrogen. However, atmospheric nitrogen is a triple-bonded gas (N_2), which means that three bonds join the two nitrogen atoms, making the molecule almost inert and unavailable to most organisms. For most organisms to utilize atmospheric nitrogen, the triple bonds must be broken and the single atoms of nitrogen thus formed be bonded chemically with oxygen or hydrogen, or both, or with carbon. **Nitrogen-fixing processes** bond nitrogen with oxygen and hydrogen. Nitrogen-assimilating processes bond nitrogen with carbon. Breaking the triple bonds demands energy. Under natural conditions, unique microorganisms that have evolved the necessary metabolic pathways to create biologically active forms of nitrogen – ammonia, amines and amino acids – the building blocks of proteins and nucleic acids, do the nitrogen-fixing. A few free-living bacteria and blue-green algae, as well as symbiotic bacteria associated with the roots of such crop plants as clover and such trees as alder, are also nitrogen-fixers. Some atmospheric nitrogen combines with oxygen at high temperatures associated with, for example, lightning to form reactive, oxidized forms of nitrogen. When burnt, nitrogen in fossil fuels also produces nitrogen oxides.

Before humans started to manipulate biogeochemical cycles, nitrogen fixation by living things was the dominant supplier of new reactive nitrogen to the ecosphere. The amount of reactive nitrogen passing though the atmosphere and biosphere was quite small and nitrogen was a limiting factor. In consequence, strong competition of many life forms in a nitrogen-limited environment shaped the evolution of biodiversity and food webs. Over the last two hundred years, humans have come to dominate the production of **reactive nitrogen** (Figure 13.10). At the end of the nineteenth century and start of the twentieth century, humans used guano and nitrate deposits from Pacific islands and the deserts of South America to boost agricultural production in Europe and elsewhere. Then in 1913, the Haber–Bosch chemical process for converting atmospheric nitrogen to ammonia was invented to provide an almost unlimited supply of reactive nitrogen (Smil 2001). Since about 1965, human nitrogen-fixation has exceeded natural terrestrial nitrogen-fixation. This development has caused huge changes in the flux and storage of nitrogen in water and air that have far-reaching environmental consequences (Galloway and Cowling 2002; Mattson *et al.* 2002;

nitrogen-fixing processes: processes involving nitrogen-fixers (a few free-living bacteria and blue-green algae, as well as symbiotic bacteria associated with the plant roots) that manufacture biologically active forms of nitrogen – ammonia, amines and amino acids

reactive nitrogen: nitrogen in a form that is available to living things

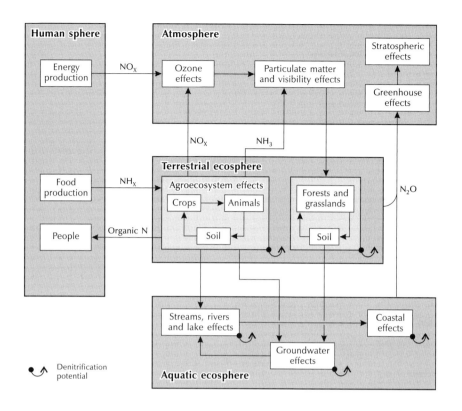

Figure 13.10 The cascade of human-produced reactive nitrogen. *Source*: After Galloway and Cowling (2002)

Vitousek *et al.* 2002). Denitrifying processes (processes that convert reactive nitrogen back to non-reactive forms) cannot keep pace with the creation of new reactive nitrogen, which is therefore accumulating in the air, soils and water. The accumulation in the environment affects humans and ecosystems, both beneficially and detrimentally (Table 13.1).

ACID RAIN

What is acid rain?

Robert Angus Smith (1817–84) is the 'father of acid rain'. He first noted the phenomenon of acid rain in the *Memoirs and Proceedings of the Manchester Literary and Philosophical Society* of 1852, in a paper called 'On the air and rain of Manchester'. He was probably the first person to campaign for smokeless fuels.

Rainwater is normally a weak acid – **carbonic acid** – with a mean pH of around 5.6. Acid rain is even more acid in reaction, normally owing to human contributions of sulphur dioxide, oxides of nitrogen and carbon dioxide to the atmosphere. Any rain below pH 5.6 is 'acid rain'. That means that most of Europe and eastern North America receives acid rain throughout the year. Technically speaking, acid rain is a misnomer. The correct term is **acid precipitation**, which encompasses acid snow, acid hail and acid fog, as well as acid rain. However, most writers use acid rain as an umbrella term for all forms of acid precipitation reaching the surface through **wet deposition**. A related process, called **dry deposition**, is the fallout of the oxides of nitrogen and sulphur as dry gases or absorbed on other aerosols such as soot or fly ash. Once dry

carbonic acid: a weak acid with the chemical formula H_2CO_3

acid precipitation: acid rain, acid snow, acid hail, acid dew and acid fog. Scientists usually use the term 'acid rain' to mean acid precipitation

wet deposition: the deposition on surfaces of dissolved substances and particles by any form of precipitation

dry deposition: the deposition on surfaces of dry gases and particles in the atmosphere

Table 13.1 Beneficial and detrimental effects of reactive nitrogen

Beneficial effects	Detrimental effects
Human health	
Increased yields and nutritional quality of foods	Respiratory and cardiac disease (on exposure to high ozone concentrations and fine particulate matter)
	Nitrate and nitrite contamination of drinking water – 'blue-baby syndrome' (methaemoglobinaemia) and certain kinds of cancer
	Blooms of toxic algae
Ecosystems	
Increased productivity of nitrogen-limited natural ecosystems	Ozone-induced injury to crops, forests and natural ecosystems and predisposition to attack by pathogens and insects
	Acidification and eutrophication effects on soils, forests and freshwater ecosystems
	Eutrophication and hypoxia in coastal ecosystems
	Nitrogen saturation of soils in forests and other natural ecosystems
	Biodiversity losses in terrestrial and aquatic ecosystems and invasions by nitrogen-loving weeds
	Changes in abundance of beneficial soil organisms that alter ecosystem functions
Indirect effects on societal values	
Increased wealth and wellbeing of human populations in many parts of the world	Significant changes of land use
	Regional hazes that decrease visibility at scenic vistas and airports
	Depletion of stratospheric ozone by N_2O emissions
	Global climate change induced by N_2O emissions and the formation of ozone in the troposphere
	Damage to useful materials and cultural artefacts by ozone, other oxidation and acid deposition
	Long-distance transport of reactive nitrogen, causing inimical effects in countries far from emissions sources, and increases in background concentrations of ozone and fine particulate matter, or both

NB: The magnitude of the reactive nitrogen flux commonly determines whether effects are beneficial or detrimental. All the effects are interlinked though biogeochemical circulation pathways of reactive nitrogen

Source: Adapted from Galloway and Cowling (2002)

deposition makes contact with moisture on the ground, it produces acidic water of much the same composition as wet deposition. Wet deposition and dry deposition are normally lumped together as acid rain. In Britain, around two-thirds of the acid precipitation falls as dry deposition.

What causes acid rain?

Emissions of oxides of nitrogen – nitric oxide (NO) and nitrogen dioxide (NO_2) – and sulphur dioxide by industry, electricity generation, households and vehicles are the main causes of human-induced acid rain. At the global and continental scale, volcanic emissions of the gases are also sometimes significant. Once airborne, some of the gases and particles come back down to the ground without water being involved (dry deposition). The rest of the gases oxidize in the presence of sunlight (photoxidation) to produce nitric acid and sulphuric acid, either in gas or in liquid phase reactions, the latter being a more effective producer of acids. The rate of conversion depends upon the intensity of sunlight, the presence of ammonia and the concentration of heavy metals in airborne particulate matter (which act as catalysts in the conversion process). When in a watery environment (such as clouds), the acids and other chemicals in the atmosphere dissociate into positively or negatively charged particles – ions. For instance, sulphuric acid in solution is a mix of positively charged hydrogen ions (cations) and negatively charged sulphate ions (anions). Acid rain derived by wet deposition is thus a cocktail of nitrate and sulphate ions. (For more discussion of these and other atmospheric pollution processes, see Chapter 8.)

The longer that the original emissions are airborne, the more likely it is that chemical reactions will produce nitric and sulphuric acids. Transport of emissions over 500 km, known as **Long Range Transport of Atmospheric Pollution (LRTAP)**, is most favourable to the formation of a strong acid rain. Until about 30 to 40 years ago, acid rain tended to be a local phenomenon that affected the area near the source of emissions. To reduce pollution at ground level, tall smokestacks were constructed. For example, in 1972, the International Nickel Company (INCO Limited) of Canada added a 381-metre superstack to its nickel smelter complex at Sudbury in Ontario. The tall stacks policy on smelters and electricity power stations, combined with higher exit velocities of the emissions, solved the local pollution problem by forcing the pollutants higher into the atmosphere. However, the pollutant stayed in the air for longer, so enhancing their conversion to acid rain via the LRTAP mechanism, and spreading the acid rain problem further to other regions. Figure 13.11 shows present problem areas and potential problem areas of acid rain.

> **Long Range Transport of Atmospheric Pollution (LRTAP)**: the transmission of atmospheric pollutants over 500 km. Some pollutants may travel worldwide

What are the effects of acid rain on life?

To understand how acid rain affects living things, it is necessary to study what it does to their habitat.

Acid rain can increase the acidity of soils. However, the effect varies with rock type and the ability of the soil to buffer incoming acid input (that is, its acid neutralizing capacity). In soils formed on rocks containing few bases, hydrogen and aluminium ions replace calcium and magnesium ions. This leads to a progressive **soil acidification** and a **mobilization** of metals in the soil. On soils formed in rock that is rich in bases, calcium and magnesium ions released in weathering may replace those lost through leaching. The result is that the soil buffers the effects of the acid rain and no soil acidification takes place.

> **soil acidification**: the loss of bases from a soil, leading to a higher acidity (lower pH)
>
> **mobilization**: the rendering of metals in the soil into a soluble, and therefore transportable, form

Soil acidification depletes base cation stocks in soils, so depriving trees of essential nutrients. This puts the trees under stress and they become more vulnerable to pests, disease, climate and weather. If the rain is below pH 4.2, clay–humus complexes in the soil begin to release aluminium. The aluminium damages root systems, reduces the rate of tree growth, encourages the development of abnormal cells, promotes a premature loss of leaves and needles, causes the necrosis of leaf tissue, favours the leaching of nutrients from leaves and leads to crown dieback and a loss of biomass. These effects can affect large areas of forest. In areas of Europe and North America that suffer from acid rain, forest soil acidity has increased fivefold to tenfold since around 1950 or later. In Ontario and Quebec in Canada and Vermont in the USA, the progressive dieback of sugar maples since 1980 has had substantial economic repercussions –

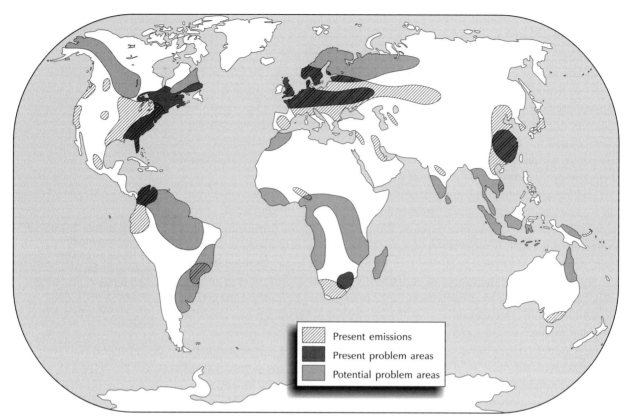

Figure 13.11 Acid rain: present and potential problem regions. The potential problem sites include rapidly industrializing regions and areas with sensitive soils

Quebec is the source of 75 per cent of the world's maple sugar. In Germany, *Waldsterben* (forest death) – now usually referred to as 'forest decline' (Schulze *et al.* 2002) – had affected at least a third of the country's forests by 1980, and half the country's forests by 1990 (Figure 13.12). Indeed, in central Europe, the

Figure 13.12 Visual signs of forest decline in forests of the Fichtelgebirge, Germany. Photograph by Ernst-Deflef Schulze

curtailing of natural forest regeneration compounds the damage to existing trees. Failure of the central European forests to regenerate would cause gigantic economic disruption in a region dominated by the forest industry. However, acid rain is not the sole culprit for this decline of European forests – other factors, including land use, tree harvesting processes, drought and fungal attacks, play a role (Focus 13.2). Nonetheless, Bernhard Ulrich, who first identified the processes by which acid rain might damage trees, argued that these other factors would operate fully only once soil acidification had rendered a forest vulnerable. The latest thinking is that dieback of German forests, which is often observed at high altitudes, may result from ozone pollution.

Acid rain alters the chemical composition of lakes and groundwaters. Lakes naturally become more acid with time, but acid rain accelerates the process. Accelerated lake acidification is widespread in areas affected by acid rain where the lake waters lack a buffer. In Sweden, there are some 90,000 lakes acidified to some extent, 4000 of them severely so, as well as 100,000 km of rivers. Some groundwaters are more susceptible to acid rain than others are. The susceptibility

FOCUS 13.2

SOIL ACIDIFICATION IN BRITAIN

Approximately one third of the land surface of Britain is 'upland'. The uplands lie in the north and west of the country and possess a complex geology and cool wet windy climate. Soils are variable but generally acidic, of low base status with predominately peats and podzols. Vegetation is mainly upland grass and herb species, heather, bracken and scrub forest, with land use of sheep grazing and raising of game animals such as grouse and red deer. Following the timber shortages during World War I, the Forestry Commission was established in 1919 and acquired vast tracts of upland moorland in order to establish plantations for commercial timber production. The majority of the trees planted were conifers, as the harsh conditions preclude the profitable silviculture of broadleaf tree species.

One of the investigations at the Plynlimon study catchment in central Wales was to assess the impact of large-scale upland afforestation upon water resources and quality. The study catchment comprises the headwaters of the Severn and Wye, considered very similar climatically and physiographically, but differing in land use. Grassland dominates the Wye, whereas forest covers 60 per cent of the upper Severn: a Forestry Commission plantation commenced in the 1930s with Sitka spruce (*Picea sitchensis*) being the dominant species.

One of the most important research findings from the studies at Plynlimon was that forested streams are more acidic than their grassed counterparts are. An acidic stream environment is typical of the British

uplands. The limited buffering potential of the upland soils make them particularly sensitive to any changes in the acidity of the incoming precipitation. Causes of the changes in water chemistry are complex and result from ploughing and draining the soil and the development of the tree root and canopy systems. Forest canopies capture a greater volume of sea salt and artificial pollutants than moorland vegetation: high altitude forests may also collect greater quantities of solutes from cloud or mist, which is higher in concentration of solutes than rainfall. On acid organic rich soils, the increased anion inputs from these sources can further the acidity of drainage waters. The large surface areas of the forest canopy makes it an effective scrubber of dry pollutants from the air stream, which are then flushed from the canopy by storm events. Tree metabolism also plays a part, with studies showing neutralization of acidification by younger trees while older trees have more of an acidifying effect. The uptake of nutrients by the trees promotes soil acidity, and changes in soil hydrology though draining and ploughing mean that water is in contact with the soil for shorter periods of time, minimizing any neutralizing effect of remaining bases in the soil. In addition, the tree litter is in itself acidic and not easily decomposed as the Sitka spruce is not native to Britain and specific decomposing bacteria are absent. Not only does this prevent the recycling of leaf nutrients back into the soil but in time creates an acid, rich, deep humus layer in the soil.

depends on how the 'acid load' compares to the base cation content of the aquifer, the soil depth, soil texture, aquifer size and mineral composition, and the water available for recharge. In Europe, the most sensitive groundwaters occur at high latitudes and high altitudes where soils are thin and weathering rates low. The least sensitive groundwaters occupy deep-soil agricultural areas.

The accelerated acidification of lakes and rivers has drastic impacts on aquatic biology, at least in the areas worst affected by acid rain. Acid waters kill fish, but not so much through the direct effects of acidity as through the changes in water chemistry that the increased acid induces – heavy metals such as aluminium, mercury, manganese, lead and zinc attain lethal concentrations. In Sweden, there were 4000 lakes with no fish by the end of the 1970s. In southern Norway, about 80 per cent of lakes and streams were technically dead, or nearly so, with fish eliminated from 33,000 km^2 of lakes by the mid 1980s. Similarly, on the other side of the Atlantic, thousands of lakes in the eastern USA were too acid to support fish, while in Ontario in Canada, 300 lakes had waters with a pH less than 5 (which is the level at which fish begin to die), and another 48,000 lakes were threatened. Viable fisheries demand a pH of at least 4.6, though the acidity tolerance varies with species. Pike (*Esox lucius*) is the least sensitive to acidification, with adult mortality rates not affected until the pH drops below 4.0; small mouth bass (*Micropterus dolomieu*) is the most sensitive, with adult mortality rates affected below pH 5.5.

EUTROPHICATION

Eutrophication is the enrichment of water bodies by nutrients, mainly nitrates and phosphates. It occurs naturally, taking thousands of years to produce a more productive lake. Artificial or human-induced eutrophication is an accelerated version of the natural process. It affects thousands of lakes around the world and is a form of water pollution brought about by the excessive accumulation of plant nutrients.

Lake ecosystems overloaded with nutrients display several symptoms that result from deterioration in the water quality and disruption of the ecological balance (e.g. Skei *et al.* 2000). The symptoms include higher primary production, an excessive growth of algae and other aquatic plants, and a shift in the algal community to a larger fraction of large, freshwater blue-green algae. The blue-green algae are unpalatable to herbivorous zooplankton and they can produce toxins that are inimical or even fatal to the health of humans, other mammals, fish and birds. The toxins either pass along the lake food chain, or enter organisms directly by contact with, or ingestion of, the algae. Algal species also cause fish deaths by physically clogging or damaging gills, which causes asphyxiation. Another possible symptom of eutrophication is wild swings of dissolved oxygen concentrations between day and night. Low oxygen levels, caused by the respiration of the increased lake phytomass, may kill invertebrates and fish. This process is more acute when decaying algal blooms further reduce the oxygen content of water. The growth or decay of benthic (bottom-dwelling) mats of macro-algae can also lead to the deoxygenation of sediments. Ultimately, eutrophication reduces lake biodiversity by spurring the proliferation and dominance of nutrient-tolerant plants and algal species, which displace more sensitive species of higher conservation value, so changing the structure of the lake ecosystem.

The changes associated with eutrophication are commonly undesirable and conflict with other water uses, including water supply (e.g. algae clogging filters in treatment works), livestock watering, irrigation, fisheries, navigation, water sports, angling and nature conservation. The changes can also make the lakes less desirable to humans by increasing turbidity and producing discolouration, unpleasant odours, slimes and foam formation. In short, artificial eutrophic lakes often become less economically beneficial and less aesthetically attractive to humans.

Nutrients that enter lakes to cause artificial eutrophication come from point sources and from diffuse sources. Point sources include sewage effluent. Diffuse sources include run-off from agricultural land. The relative contribution of point and diffuse sources polluting nutrients varies in different drainage basins. In the UK, the average percentage contributions of phosphorus entering surface waters are: agricultural sources 43 per cent; human and household waste (including detergent) sources 43 per cent; industrial sources 8 per cent; and background sources 6 per cent (Morse *et al.* 1993). For nitrogen, inputs to fresh waters come principally from diffuse sources, particularly agriculture, although point sources (usually urban wastewater) may make a material contribution in many regions. In England and Wales, 70 per cent of the total nitrogen input to inland surface waters comes from diffuse sources: in order of decreasing importance, from agriculture, precipitation and urban runoff. The remaining 30 per cent of the total nitrogen comes from sewage effluent and industrial discharges.

Artificial eutrophication became a 'hot environmental issue' in the mid-1960s. Lake Erie was one of the first lakes where artificial eutrophication was recognized (Focus 13.3). Even so, environmental issues tend to wax and wane with changing scientific fashions. By the mid-1970s, interest in eutrophication had faded to some extent. However, this particular form of nutrient pollution remains a serious water-quality problem worldwide and is still the subject of research. In the UK, for example, the United Kingdom Eutrophication Forum (UKEF) identifies research and disseminates information on eutrophication for the conditions that can occur in UK waters. Research continues elsewhere. A recent study looked at lake eutrophication at the urban fringe in the Seattle region of Washington, USA (Moore *et al.* 2003). Three levels of local residential development defined three types of lake: 'undeveloped' lakes, with no residential development; 'sewer' lakes, with houses serviced by sewer systems and mainly occurring near urban centres; and 'septic' lakes, with one or more houses with active septic tanks and mainly occurring along the urban–rural fringe (Plates 12 and 13). The findings showed that septic lakes were more eutrophic than sewer lakes and undeveloped lakes, suggesting that septic systems account for high levels of eutrophication in lakes along the urban–rural fringe, and indicating a need for proactive management to minimize the eutrophication in those lakes.

ENVIRONMENTAL TOXINS

Human-made persistent organic pollutants

Chemical pollution affects all groups of species, from large Arctic carnivores to small insects. Persistent organic pollutants (POPs) have a carbon-based molecular structure and often contain highly reactive chlorine. They are mostly toxic synthetic chemicals produced many decades ago. They do not break down easily

FOCUS 13.3

EXPERIMENTAL LAKES AREA

During the 1960s, Lake Erie underwent rapid artificial eutrophication, owing to its overloading with nutrients from human sources, and particularly with phosphates from detergent-rich sewage. The phosphates in Lake Erie and other lakes triggered the excessive growth (blooms) of algae, which in turn altered the water quality. The algal blooms led to oxygen depletion and the death of fish. Many native fish species disappeared, and species more resistant to the new conditions replaced them. Masses of rotting, stinking algae fouled beaches and shorelines. The problem caused grave concern and prompted much research. For instance, the University of Manitoba set up the Experimental Lakes Area (ELA) in 1968 to investigate the problem. Between June 1969 and May 1976, eutrophication was virtually the sole focus of whole-ecosystem experimental studies at the ELA.

Scientists at the ELA used small, natural lakes as experimental 'laboratories'. They added various combinations of nutrients to the lakes to determine which macronutrients – carbon, nitrogen and phosphorus – were instrumental in producing artificial eutrophication. After several years' research, they discovered that phosphorus was the key nutrient in accelerating eutrophication. ELA Lake 226 was the site of the most visually spectacular experiment. A plastic divider curtain divided the lake into two roughly equal portions. Scientists added carbon and nitrogen to one half of the lake, and carbon, nitrogen and phosphorus to the other half. For eight years in succession, the half receiving phosphorus developed eutrophic algal blooms, while the half receiving just carbon and nitrogen did not. After only two years, the experiment in this lake convinced even the sceptics that phosphorus was the culprit. The result was the institution of a multi-billion dollar phosphate control programme within the St Lawrence Great Lakes Basin. Part of this programme involved legislation to control phosphates in sewage, and to remove phosphates from laundry detergents.

and persist in the environment where they cause pollution. Worryingly, they dissolve in fat and tend to accumulate in living tissue. POPs include such toxic substances as DDT, dieldrin, PCBs and dioxins that are a serious health threat to animals and humans. Sea currents and winds carry them north from their sources in industrial and agricultural regions. In Arctic ecosystems, **biomagnification** takes place and their concentration increases in the higher echelons of food chains. Concentrations become particularly high in marine mammals, which form the basis of the traditional Inuit diet, and recent findings record unacceptably high levels of human-made environmental toxins in the Inuit population of Greenland. In parts of East Greenland, all the population had toxin burdens high enough to cause concern. The source of the toxins is polar bears, seals and whales. The Arctic Monitoring and Assessment Programme (AMAP) concludes that, unless Greenlanders change their eating habits, such health effects as reduced fertility, genetic damage and deformities in children will result. To date, Greenland is the only place in the world where the levels of environmental toxins in humans have set alarm bells ringing.

> **biomagnification**: the increasing concentration of certain substances as they move up a food chain

Many other human-made chemicals cause problems in wildlife. The 'horror stories' of the defiling and poisoning effects of some chemicals, even those already banned, are very worrying. Fish suffering from chemical pollution are vulnerable to hormone disruption, causing eggs to be produced in testes and disrupted production of sperm. In 1980, dicofol (a pesticide related to DDT) was accidentally spilt into Lake Apopka, Florida, USA. Four years later, the lake's alligator population had fallen by 90 per cent, the successful hatching rate fell and the population has remained low. Male alligators now have reduced levels of male hormones and females have raised levels of female hormones. Some amphibians exposed to atrazine, which is a very persistent herbicide, grew extra testes and ovaries. Aquatic mollusc reproduction is disrupted by tributyltin (TBT), a chemical used mainly to protect ships' hulls and as an antibacterial agent on sportswear, a disinfectant and a flame retardant. Some animals accumulate the breakdown products of TBT, leading to high concentrations in such top predators as dolphins, tuna and sharks.

Heavy metal pollutants

Several **heavy metals** produced by industrial processes pose a threat to wildlife. Such metals include zinc, nickel, chromium, mercury, lead, cadmium, antimony, gold, platinum, silver, bismuth, arsenic, selenium, vanadium, thallium and manganese. They can be stored in body tissues, some permanently. A little arsenic administered to a human each day will eventually accumulate to a fatal dose, as many a writer and reader of murder mysteries knows. The body burden is the content of heavy metals in a human body. For a 70-kg person, the average burden is about 8 mg of antimony, 18 mg arsenic, 30 mg cadmium and 150 mg of lead.

Heavy metals are often subject to **bioaccumulation** and biomagnification. The events at Minamata in Japan spotlighted the injurious effects of mercury (Focus 13.4). Mercury causes pollution problems elsewhere too. In the Alto Paraguay River Basin, Brazil, gold mining uses large amounts of mercury, some of which enters the air and the rivers running into Pantanal, a wildlife reserve (Hylander *et al.* 1994). The mercury content of local catfish (*Pseudoplatystoma coruscans*) was above the limit for human consumption, and significantly above the natural background level. Mercury content in bird feathers also indicated biomagnification. However, soil and sediment samples contained no significant accumulations of mercury, suggesting that organisms more readily absorb the mercury originating from the gold-mining process than they do mercury naturally present in soil minerals. Ditches along busy roads are liable to pollution by heavy metals. In Louisiana, United States, animals and plants living in ditches have accumulated cadmium and lead (Naqvi *et al.* 1993). The cadmium level in the red swamp crayfish (*Procambarus clarkii*) was 32 times that in the water, and

> **heavy metals**: metals with a relatively high atomic weight
>
> **bioaccumulation**: the processes by which certain animals and plants store chemical pollutants in their cell tissues. Fish tissues, for example, concentrate low levels of DDT

FOCUS 13.4

MERCURY POISONING AT MINAMATA

In 1956, a strange illness afflicted first birds, cats and then people, mainly fishermen and their families, in the Japanese coastal town of Minamata. Birds lost their coordination and fell from the sky, cats went mad, running in circles and foaming at the mouth. Humans displayed a set of progressively worsening symptoms, starting with fatigue, irritability, headaches, a numbness in arms and legs and difficulty in swallowing, and culminating in blurred and restricted eyesight, poor hearing, a lack of muscle coordination, inflamed gums and diarrhoea. Some 43 people died, 111 were severely disabled and 19 babies were born with

congenital defects. The source of the disease was a plastics factory on the bay that used an inorganic form of mercury in its production process. The mercury issued from the plant in water effluent and flowed into the bay, where microbial action converted the inorganic mercury to methyl mercury, an organic form of the metal that readily enters the food chain and passes through fish and shellfish to other animals and humans, damaging brain cells in the process. The Minamata disease occurred later in Niigata City in Japan, near a caustic soda factory in Canada, in Jirin City in China where methyl mercury was discharged into a river, and elsewhere.

the lead level 12 times that in the water, giving bioaccumulation factors of 5.1 and 1.7, respectively.

Heavy metals in sediment cores from lakes record human releases of these elements over the last few centuries. In Norway, surface sediments revealed significantly elevated concentrations of mercury, antimony, bismuth, cadmium, arsenic, nickel, copper, lead and zinc, which appear to have come from atmospheric deposition (Rognerud and Fjeld 2001). For most of these elements, the most elevated levels occurred in lakes in the south of the country and at lower altitudes. Smelters raised the levels of copper and nickel locally.

13.4 DRIVERS OF GLOBAL CHANGE: INTEGRATED EFFECTS

The chief drivers of biodiversity change act singly and in concert. It is impossible to know exactly how these drivers will alter the world's biomes during the current century. Nonetheless, models using various scenarios give an indication of the possible directions and degrees of change induced by human impacts. An interesting model predicted changes in ten major biomes up to the year 2100 (Sala *et al.* 2000; 2001). The biomes were Arctic tundra, alpine tundra, boreal forest, temperate grassland, savannah, Mediterranean-type, desert, northern temperate forests, southern temperate forests and tropical forests. The model also predicted changes in lakes and in streams. The chief drivers of biodiversity change were assumed to be land use change, climate change, nitrogen deposition, biotic exchange and a rise in atmospheric carbon dioxide concentrations. Three models predicted changes in these driving variables during the present century: a business-as-usual climate model, a potential vegetation change model, and a land use change model. The researchers estimated the magnitudes of change in the drivers from the model results. Finally, they estimated the sensitivity of each biome to a unit change in the drivers. Figure 13.13 summarizes their findings.

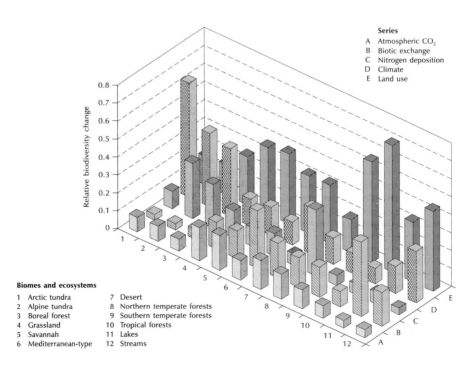

Figure 13.13 Biome responses to drivers of biodiversity change for the year 2100. *Source*: Based on data in Sala *et al.* (2000)

Biomes and ecosystems

1	Arctic tundra	7	Desert
2	Alpine tundra	8	Northern temperate forests
3	Boreal forest	9	Southern temperate forests
4	Grassland	10	Tropical forests
5	Savannah	11	Lakes
6	Mediterranean-type	12	Streams

Series

A Atmospheric CO_2
B Biotic exchange
C Nitrogen deposition
D Climate
E Land use

Notice that biomes differ enormously in their response to different drivers of change. A single factor – land use change – affects tropical forests and southern temperate forests, with other factors playing a minor role. A different single factor – climate change – affects Arctic tundra. In contrast, several factors, all with moderate to large effects, conspire to affect Mediterranean-type ecosystems, temperate grasslands and savannahs. Low to moderate impacts of all drivers affect deserts and northern temperate forests. Land use change, biotic exchange and climate change will likely cause the greatest changes of biodiversity in freshwater ecosystems. Land use change is particularly effective in these ecosystems because human activity tends to focus around lakes and streams, and human-applied nutrients tend to collect in aquatic environments. Biotic exchange also affects lakes and streams significantly, especially in waters with introduced fish stocks and with discharged ballast water (water used as ballast in ships and potentially able to carry harmful organisms around the world).

Part of the study of potential biodiversity change involved predicting the impact of all drivers of change on the ten terrestrial biomes. Predictions for the year 2100 used three cases: no interactions between the drivers; antagonistic interactions between the drivers, and synergistic interactions between the drivers (Figure 13.14a, b, c).

1.   In Case 1, where the drivers of change do not interact, Mediterranean-type ecosystems and temperate grasslands experience the largest changes in biodiversity. This is mainly because several drivers in these ecosystems have moderate to high potential impacts. On the other hand, Arctic, alpine and desert ecosystems are least affected, chiefly because the drivers have low to moderate impacts in those ecosystems.

2.   In Case 2, where drivers interact antagonistically, the biomes changing the most and the least alter radically. In practice, total biodiversity change in this case is equal to the change resulting from the driver expected to have the largest effect, and is the maximum of the effects of all the drivers. Under this assumption, tropical and temperate forests and Arctic ecosystems suffer the greatest biodiversity change. This is because they respond to the change of a single driver – land use change for tropical and temperate forests and climate change for Arctic tundra.

3.   In Case 3, synergistic interactions occur between the drivers and the total biodiversity change is the product of the changes brought about by each driver. Under this assumption, the synergistic interaction between the drivers augments the difference among the biomes. Mediterranean-type ecosystems change the most and Arctic ecosystems the least. However, the change in the Arctic ecosystems is just 2 per cent of that of the change in the Mediterranean-type ecosystems. That contrasts with Case 1 results where the minimum biome change is 60 per cent of the maximum biome change.

The above study predicts that differences in biodiversity change in the world's biomes will depend upon the way in which the drivers of change interact. However, some patterns are common to all three cases. First, Mediterranean-type ecosystems and temperate grasslands seem destined to experience the greatest

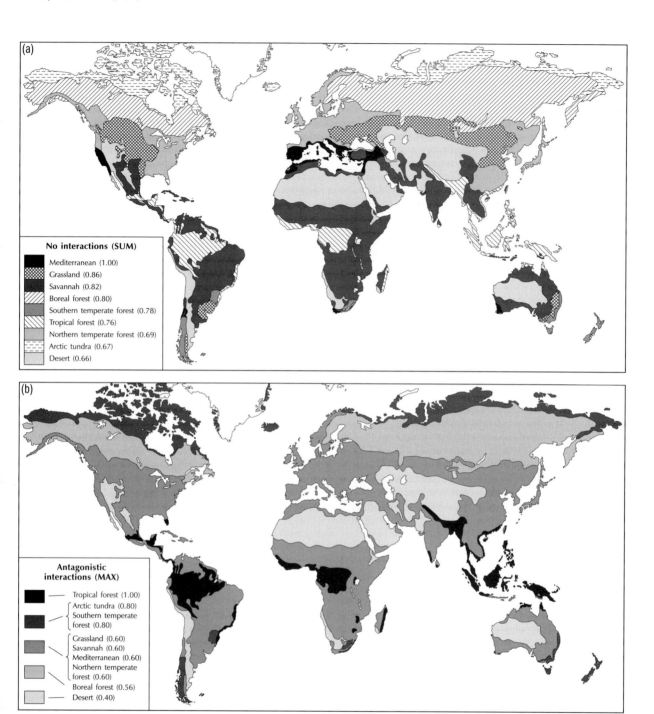

proportional biodiversity change by the year 2100, while savannahs are likely to experience moderate biodiversity change. Deserts and northern temperate forests seem likely to suffer a low to moderate proportional change in biodiversity. Against these shared patterns, the predicted changes in tropical forests and southern temperate forest vary a great deal between cases. With antagonistic interactions between the drivers of change, these biomes display the greatest proportional change in biodiversity; while with no interactions and synergistic interactions between the drivers of change they show the least proportional

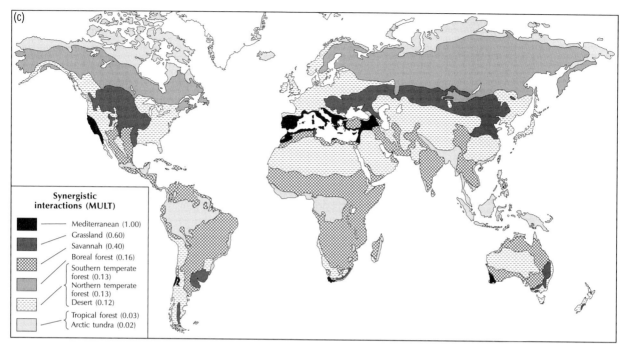

Figure 13.14 Maps showing combined effects of drivers of biodiversity change on biomes. (a) Predictions for no interactions between the drivers, with the total change computed as the sum of the effects for each driver. (b) Predictions for antagonistic interactions between drivers, i.e. the driver expected to have the greatest effect, with the total change calculated as the maximum effects of all the drivers. (c) Predictions for synergistic interactions between drivers, with total change computed as the product of changes resulting from each driver. *Source*: Reprinted with permission from O. S. Sala *et al.* (2000) 'Global biodiversity scenarios for the year 2100' (*Science* 287, 1770-74). Copyright © 2000 American Association for the Advancement of Science.

change in biodiversity. Note that the predicted biodiversity changes are expressed as proportional, rather than as absolute, values. In actuality, the total number of species lost in a biome would depend upon the number of species currently present. Accordingly, tropical forests stand to lose far more species than, for instance, Arctic tundra.

It is interesting to note that biodiversity is rather responsive to drivers of change in all biomes. Land use is the most effective driver of biodiversity change, but the role of different drivers varies greatly among the biomes: some biomes are predominantly affected by just one factor – land use change or climatic change – while others are affected by most drivers. Further complications will arise from changes in the relative importance of the different drivers as the century progresses (Sala *et al.* 1999). Land use is likely to have the speediest rate of change, but should attain a peak value fastest and then decline when conversion to agricultural land is complete. The relative importance of the other drivers will grow throughout the century, but some will grow faster than others. The relative importance of atmospheric carbon dioxide levels and nitrogen deposition are likely to grow second fastest after land use change. Biotic exchange comes third in the list as land use change, carbon dioxide increases and nitrogen deposition rates magnify its effects. Finally, climate will change the least rapidly as it requires changes in atmospheric carbon dioxide levels and other gases and particulates to occur first.

SUMMARY

Global warming, the changing chemical cocktail in the atmosphere and its knock-on effects in soils and water, and environmental toxins threaten biodiversity. Mobile species, such as most birds and butterflies, with broad tolerance of environmental conditions are the least likely to suffer from rising temperatures. Species that live at the margins of a contracting range, species that live in small areas, species that are highly specialized, species that are disinclined to move and species that have impoverished gene pools are most at risk from rising temperatures and associated changes in weather. This will affect all the world's biomes, but to varying degrees. The Arctic tundra and Mediterranean-climate biomes are exceptionally vulnerable, tropical forest biomes the least vulnerable. Boreal forest biomes should move northwards but occupy about the same amount of land. Biomes may move bodily northwards or southwards, creating zones of change where one biome moves into another, as when boreal forests advance into tundra biomes, and zones of no change where, for instance, the southern part of a biome moves into the northern part. Alternatively, biomes may simply disassemble and reassemble elsewhere with a different set of species. Humans have changed the chemistry of the environment, with some alarming repercussions. They have increased the amount of reactive nitrogen in ecosystems, the results of which are many and varied, some being beneficial and some detrimental. They have created 'acid rain', which adversely affects soils, forests, lakes, rivers and their associated wildlife, and accelerates the weathering of buildings. They have encouraged the accelerated build-up of nutrients, mainly nitrogen and phosphorus, in lakes (eutrophication), which has inimical effects on lake ecosystems. They have also released toxins, causing one-off environmental disasters near the site of release. Some toxins spread around the world and accumulate in remote places, such as the Arctic tundra. All the drivers of biodiversity change act together. Land use change is the most potent driver of biodiversity change, but the role of different drivers varies among biomes.

Questions

1 Why are some animals and plant species more vulnerable to a warmer climate than others?

2 How are present biomes likely to respond to a warming global climate?

3 Explain how humans have altered the nitrogen cycle. What effects have these alterations had?

4 Why does 'acid rain' pose a threat to forests and to lakes?

5 Examine the causes and effects of artificial eutrophication.

6 Why are so many environmental toxins a cause for concern?

Further Reading

Hinchliffe, S. (2003) *Understanding Environmental Issues*. Chichester: John Wiley & Sons.
A good source of examples.

Lomborg, B. (2001) *The Sceptical Environmentalist: Measuring the Real State of the World.* Cambridge: Cambridge University Press.
Always worth hearing a good counterargument to the consensus view.

Park, C. C. (2001) *The Environment: Principles and Applications.* London: Routledge.
An excellent text, including much material on the topics covered in this chapter.

Sala, O. E., Chapin III, F. S. and Huber-Sannwald, E. (2001) 'Potential biodiversity change: global patterns and biome comparisons'. In F. S. Chapin III, O. E. Sala and E. Huber-Sannwald (eds) *Global Biodiversity in a Changing Environment: Scenarios for the 21st Century* (Ecological Studies 152), pp. 351–67. Springer: New York.
A good overview of the changes likely to occur in the world's chief biomes over the next century, though not an easy read.

Snyder, N. F. R. and Snyder, H. (2000) The California Condor: A Saga of Natural History and Conservation. San Diego: Academic Press.
An excellent case study of biology and conservation with superb photographs. A pleasure to read.

Useful Websites

http://www.mobot.org and
http://www.centerforplantconservation.org
Center for Plant Conservation, Missouri Botanical Garden.

http://www.iucn.org
International Union for the Conservation of Nature and Natural Resources (IUCN).

http://www.rbgkew.org.uk
Royal Botanic Gardens, Kew.

http://www.unep-wcmc.org.uk
United Nations Environment Programme World Conservation Monitoring Centre (UNEP-WCMC).

http://www.wwf.org
World Wildlife Fund (WWF).

http://www.zsl.or
Zoological Society of London.

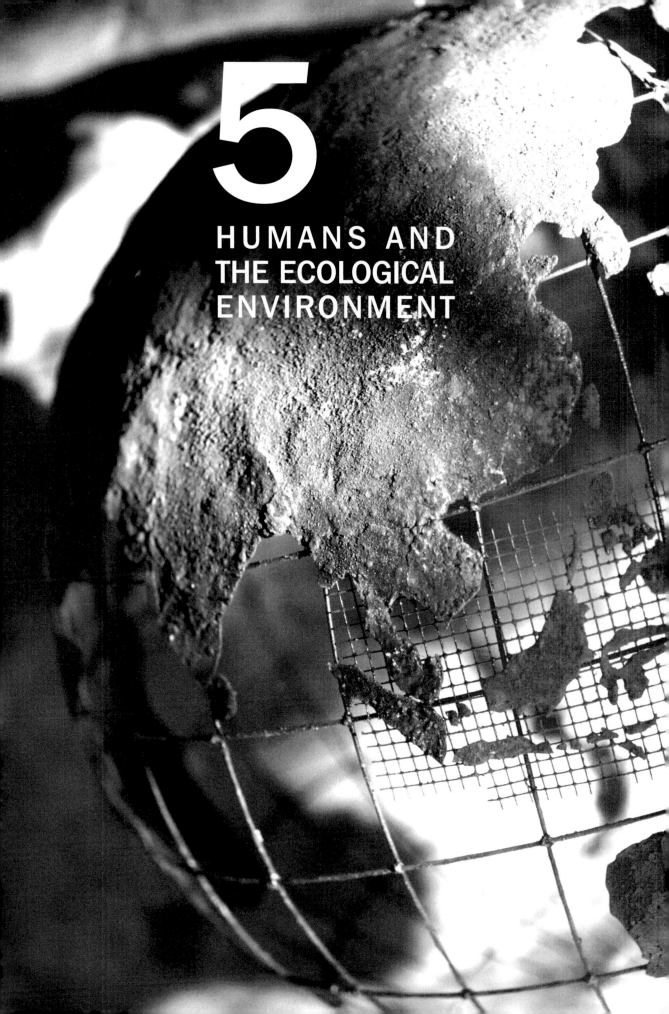

5

HUMANS AND THE ECOLOGICAL ENVIRONMENT

CHAPTER 14
LOCAL, REGIONAL AND GLOBAL ECOSYSTEMS

OVERVIEW

The environmental syndromes introduced in Chapter 3 operate over many different scales, each with associated symptoms as local, regional and/or global impacts. An appreciation of spatial scale is therefore a critical prerequisite for a thorough understanding of the environmental syndromes examined in this book. This is one of the reasons why geography has such an important role to play in this task. Scale applies equally in terms of understanding the principal drivers of a particular problem and in appreciating the processes through which direct and indirect impacts have a bearing on the environment and its peoples. It follows, therefore, that part of the solution has also to involve action at different scales to be effective.

This chapter will explore issues of scale further through selected examples of environmental syndromes. In some cases, it is useful to broaden out the analysis from the specific syndromes themselves, to consider the wider implications of processes. Since urbanization is now such an important global driving force, this is one area considered in some detail. Specifically, the chapter identifies and discusses the range and nature of impacts as they relate to the four environmental media covered in Part 3 of the book.

LEARNING OUTCOMES

By the end of this chapter, you should feel able to:

- Identify and outline the impact of selected syndromes on local, regional and global ecosystems and on the environment as a whole
- Appreciate the role of spatial scale operating within specific environmental systems and syndromes
- Summarize the range of impacts and related modification of physical processes associated with urbanization
- Appreciate the interlinkages between the different spheres of the physical and human environment.

14.1 INTRODUCTION

The impacts of humankind's alteration of the physical environment and of the processes that operate in the physical environment can have far reaching effects on the Earth's ecosystems at every scale. This chapter investigates the nature of these impacts with a range of examples that are related to environmental syndromes. It goes on to consider the impacts of broader drivers, such as urbanization, both on ecosystems and the other spheres covered in this book – air, water, soil and landforms.

Central to the discussion is the idea of local, regional and global ecosystems. Here the term ecosystem is used in a traditional sense but also with a wider meaning. On one side, there is the notion of an ecosystem as a community of plants and animals living in balance with the physical environment. However, the widespread influence of humans across the globe makes it increasingly difficult to identify 'natural' ecosystems; instead, these exist (in different degrees of harmony) with an environment that has varying degrees of human manipulation. It is more useful, therefore, to consider the term ecosystem as including humankind and its artificial environments so that all of the relevant interlinkages are considered. In some situations, humans have been so successful at removing plants and animals that some 'human ecosystems' exist in an (almost) entirely human-constructed, managed and maintained environment. However, this is something of a misleading notion since the materials for this artificial environment originally came from somewhere, and the water, food and energy necessary to maintain it (and the humans within it) must still be brought in from outside. Getting these materials, moving them, transforming them and using them causes a whole series of localized impacts of different degrees of severity which in turn have regional or global influence (or both). The utilization, development and sink syndromes in Chapter 3 hark back to this central idea and show how considering something at face value is always problematic.

In terms of communities of plants and animals, the impacts of humans at local and regional scales can be enormous and destructive. However, this is not necessarily always an unhappy relationship and some species groups have adapted so well that some commentators propose that entirely new and apparently successful ecosystems have emerged. The human impacts on the global ecosystem are also seen as regional consequences of global changes, such as global warming. These impacts are combined with, and not always easy to disentangle from, the effects of natural climatic changes associated with, for instance, ENSO events, the variability in solar output, and large volcanic explosions.

The syndrome approach attempts to facilitate a better understanding of problems and therefore, in time, will hopefully lead to effective solutions. The final chapter of this book will touch on some of the ways and means by which this last part of the equation is being approached, and what the prospects for the future therefore may be.

14.2 UTILIZATION SYNDROMES

This group of syndromes encompasses some of the key issues associated with our use of and impacts on a range of global resources. From Chapter 3, it can be seen that examples include the Katanga syndrome, the Scorched Earth syndrome, the

Overexploitation syndrome, and the Rural Exodus syndrome. Another utilization syndromes covered elsewhere in this book is the Mass Tourism syndrome (Chapter 3).

ENVIRONMENTAL DEGRADATION THROUGH DEPLETION OF NON-RENEWABLE RESOURCES: THE KATANGA SYNDROME

The Katanga syndrome refers to the severe environmental impacts associated with excessive mining of non-renewable resources. This activity can be viewed as comprising two broad and related groups, the first associated with the release of toxic substances such as mercury, and the second associated with changes to landforms, including the consumption of energy associated with the removal of very large quantities of material (WBGU 1996). The latter may take place either to obtain the material itself (for example limestone for construction or industrial uses) or for what the material contains (for example in the case of harvesting precious minerals from parent rock). When carried out on a large scale, especially using mechanized techniques, such activity can give rise to local landscape and ecosystem destruction. With the release of toxic substances into air and water, it can also cause impacts that are longer-term and which affect a wider geographical zone.

It is increasingly common for industrial and post-industrial countries to allow for some form of re-habilitation of the natural landscape and ecosystems following a mining operation. This may involve, for example, the removal, storage and subsequent reestablishment of the original topsoil as the medium for re-landscaping and the reintroduction of indigenous plants and animals. Of course, it must be said that this rebuilding of the natural landscape is an imperfect process. Nevertheless, it is a welcome step in the right direction where the needs and costs of subsequent restoration are considered before any exploitative activity is undertaken. The purpose of carrying out such activities is not just for the protection of ecosystems; it may also be necessary to ensure that modifications of natural landforms do not lead to stored up problems for the future, for example through creating slope instability, subsidence or pollution. Unfortunately, many countries in the world do not currently take such steps.

Another symptom of the Katanga syndrome relates to the mass movement of people. In areas where large mines are established, considerable flows of people can result. There may be a major influx of workers (principally young men), but also a counterflow made up of those that have been displaced by the activity (for example farmers and other residents). The resultant degradation of **social capital** in an area can take as long, if not longer, as the environment to fully recover once the mines are exhausted and the developers move on.

Focus 14.1 gives a case study discussion of the Katanga case itself. Other examples cited by WBGU (1996) include Carajás in Brazil, associated with the production of iron ore and aluminium, and the Gulf States, associated with oil production (risks of oil spills, fire and contamination).

> *social capital*: the inherent wealth of communities of people, represented for example by sets of social values and knowledge networks

THE SCORCHED EARTH SYNDROME

The Scorched Earth syndrome refers to the serious environmental, as well as social, impacts associated with military activities (WGBU 1996). In common with other syndromes, these can be seen to operate over a number of spatial scales with the local focus of the activity often being associated with extensive

FOCUS 14.1

THE KATANGA CASE

The area known as Katanga covers some 500,000 km² in the southern part of the Congo in Africa. It borders Angola to the south-west and Zambia to the south-east. Rich in natural mineral resources, including diamonds, copper and cobalt, the area became a very important mining region, supplying global markets with the inputs to manufacturing processes and other industrial and commercial activities. Although mining has a long history in the region, the natural resources were particularly savagely exploited in the nineteenth and twentieth centuries, largely through Belgian colonial interests. Over time a certain reliance on mining developed, and even by 1989 copper mining was providing half of the national revenue of the Congo and almost all of its foreign currency. This, in turn, gave rise to rapid development; but this was uneven in both spatial and social terms. Some commentators estimate a loss of 50 per cent of the native population in the first 30 years of colonial rule (Renton 2003). Working conditions in the mines were extremely poor and the workers low paid, despite the profits being made by the mining companies, such as the Union Minière du Haut Katanga. In 1960, the new republic of the Congo was formed and the old mining companies nationalized. However, the area has suffered from continued social, economic and political instability ever since, with struggles against **neocolonialism** and internal fighting between different political factions. This internal unrest has led to a large population displacement, and associated poverty and famine. The realization of the remaining wealth of the mines has been hampered by other problems too. The cost of extracting the remaining mineral ores is more expensive than in the past, due to the mix of copper and uranium present within them. The toxicity of these elements means that sophisticated modern machinery is required to produce a suitable final product to sell. A final issue relates to the need to deal with the legacy of past mining activity. The metal dust that was stockpiled into large mounds over the years, often near to large settlements, now poses an urgent and significant disposal problem.

---

neocolonialism: where one or more foreign nations exert an influence over a host nation through economic power and social influence

and permanent damage – in many cases quite literally scorched earth. Where conflicts involve traditional low-cost, high-impact weaponry, impacts may be largely restricted to the immediate vicinity of the fighting itself. However, with the use of high-tech weapons, and those related to chemically and biologically based threats, there is an increasing potential for wider-scale global damage and longer-term damage to the areas in which the weapons have been used. Recent conflicts in Iraq and Kuwait have also clearly indicated the central importance of natural resources within the military strategy of both sides of the conflict. These can be highly valued goals, both in terms of controlling the inherent wealth of an area and also as a means by which to demonstrate economic and environmental sabotage. In the context of rising concerns over the depletion of global natural resources and the impacts of the pollution that this has already caused, this seems particularly catastrophic. However, for the same reason, it is hardly a surprising military tactic either.

WGBU (1996) note the existence of large zones of contaminated land at the sites of former areas of conflict. These are associated with a range of environmental symptoms such as biodiversity loss due to toxic substances, soil degradation and landform destruction. The human impact in terms of the destruction of settlements and infrastructure and the related movement of peoples away from sites of conflict are also extremely marked in these syndromes.

THE OVEREXPLOITATION SYNDROME

WGBU (1996) see the Overexploitation syndrome as encompassing two main processes. The first involves the conversion of natural ecosystems and the second the overexploitation of terrestrial and marine biological reserves to the point at which they are no longer able to regenerate. In marine environments, the advent of highly advanced technological methods and their extensive application has led to enormous advances in the efficiency of the process of fishing. By the 1990s, all 17 of the world's major fishing grounds were already at or already beyond their capacity to regenerate (WGBU 1996).

In terrestrial ecosystems, the classic example is forest clearance, which is associated with extreme local impacts such as loss of habitat, loss of biological diversity and instability of slopes and soils. At local and regional scales, the removal of protective vegetation cover can cause acute susceptibility to natural disasters, such as flooding and landslides. Globally, forest clearance can incur the penalty of increased carbon dioxide emissions (e.g. where forest is cleared by burning) as well as the loss of capacity for carbon removal and storage. The human cost of this overexploitation is just as far-reaching, from the loss of native homelands through to the development of an over-reliance on unsustainable agricultural systems. Even though these issues are nationally and internationally recognized, and systems are in place to try and counteract unsustainable trends and negative activities, the complexity and far-reaching nature of the processes surrounding individual examples of the Overexploitation syndrome makes them very difficult to overcome. This syndrome is exemplified by clear-felling of tropical rainforests and associated land use changes such as have occurred in Brazil, Malaysia, Indonesia and Borneo (WGBU 1996).

The following sections give some specific examples of current and historical overexploitation activity. Later in the chapter, the numbers of species that are presently under threat from the collective influence of all of the environmental syndromes are examined.

Overexploitation in the past

Overhunting may have caused the extinctions of elephant birds and giant lemurs in Madagascar, giant kangaroos in Australia, moas in New Zealand and megaherbivores in the Americas. More recently, the commercial exploitation of marine mammals, fish, big game and birds has caused the extinction of several species and the near-extinction of several others. The cases of the passenger pigeon and sea otter illustrate this point.

Passenger pigeon

The graceful passenger pigeon (*Ectopistes migratorius*) once thrived in North America (Plate 14). On 1 September 1914, 'Martha', the last passenger pigeon and named after the wife of George Washington, died at 1.00 pm at the age of 29 in the Cincinnati Zoological Garden. The rapid extinction of such a flourishing population resulted from three circumstances. First, the pigeons were very tasty when roasted or stewed. Second, they migrated in huge and dense flocks, perhaps containing two billion birds, that darkened the sky as they passed overhead. Third, they gathered in vast numbers, perhaps several hundred million birds, to roost and to nest. A colony in Michigan was estimated to be 45 km long and an average of around 5.5 km wide. Professional hunters – pigeoners – formed large groups to trap and kill the pigeons. They felled nesting trees to collect the squabs (young, unfledged pigeons). The clearing of forest for farmland and the expansion of the railroad made huge tracts of the pigeon's range accessible to humans and the persecution intensified. By about 1850, the catching and marketing of passenger pigeons employed several thousand people. Every means imaginable was used to obtain the tasty bird. Special firearms, cannons and forerunners to the machine gun were built. In 1855 alone, one New York handler had a daily turnover of 18,000 pigeons. In 1869, 7,500,000 birds were captured at one spot. In 1879, a billion birds were captured in the state of Michigan. The harvesting rate disrupted breeding and the population size began to plummet. Even by 1870, large breeding assemblies were confined to the Great Lake States, and the northern end of the pigeon's former range. The last nest was found in 1894. The last bird was seen in the wild in 1899.

The sea otter

A population of around 200,000 sea otters (*Enhydra lutris*) (Plate 15) once thrived on the kelp beds lying close to shore from northern Japan, through Alaska, to southern California and Mexico (Duggins 1980). In 1741, Vitus Bering, the Danish explorer, reported seeing great numbers of sea otters on his voyage among the islands of the Bering Sea and the North Pacific Ocean. Some furs went to Russia, and soon this new commodity became highly prized for coats and hunting began. In 1857, Russia sold Alaska to the United States for $7,200,000. This cost was recouped in forty years by selling sea otter pelts. In 1885 alone, 118,000 sea otter pelts were sold. By 1910, the sea otter was close to extinction, with a worldwide population of fewer than 2000. It was hardly ever seen along the Californian coast from 1911 until 1938.

The inshore marine ecosystem changed where the sea otter disappeared. Sea urchins, which the otters ate, underwent a population explosion. They consumed large portions of the kelp and other seaweeds. While the otters were present, the kelp formed a luxuriant underwater forest, reaching from the seabed, where it had a purchase, to the sea surface. With no otters to keep sea urchins in check, the kelp vanished. Stretches of the shallow ocean floor changed into sea urchin barrens, which were a sort of submarine desert. Happily, a few pairs of sea otters had managed to survive in the outer Aleutian Islands and at a few localities along the southern Californian coast. Biologists took some of them to intermediate sites in the United States and Canada where strict measures protected them. With a little help, the sea otters staged a comeback and the sea urchins declined. The lush kelp forest grew back and many lesser algae moved in, along with crustaceans, squids, fish and other organisms. Grey whales migrated closer to shore to park their young in breaks along the kelp edge while feeding on the dense concentrations of animal plankton.

Current overexploitation

In many parts of the world, people are eating wild animals and moving them further toward extinction. These species need saving to preserve both biodiversity and human supplies of animal protein. It is ironic that the world's biodiversity hotspots have the greatest number of threatened species, as well as the highest numbers of malnourished and poor people. Bushmeat is a staple diet in these biodiverse areas. In the Côte d'Ivoire, people harvest about 80,000 tons of wild meat each year, at a value of $200 million. In the Amazon Basin, the harvesting figure is about $175 million-worth of wild meat a year.

Over the last thirty years or so, hunting has probably lain behind the extinctions of the Alagoas currasow in north-eastern Brazil, and Miss Waldron's red colobus monkey in forested zones of Ghana and the Ivory Coast. Last recorded in 1988, Alagoas currasow (*Mitu mitu*) is extinct in the wild, although there are a few captive specimens. Not sighted since 1978 despite extensive searches, scientists declared Miss Waldron's red colobus monkey (*Procolobus badius waldroni*) extinct in autumn 2000. It was the first known primate extinction since 1800.

Today, hunting imperils about a third of the threatened mammals and birds worldwide. The mammals most at risk from overexploitation are big and reproduce slowly, such as larger antelopes and elephants. The birds most at risk from overexploitation are generally big and conspicuous, such as pheasants and megapodes (ground-dwelling, Australasian birds).

THE RURAL EXODUS SYNDROME

The Rural Exodus syndrome can be seen as environmental degradation caused by the loss of the traditional land use practices that have proved a sustainable means of survival for native populations for many centuries (WGBU 1996). Being developed over generations, these practices represent the collective cultural consciousness of a people's relationship with their environment. Such experience can enable subsistence, even prosperity, in the face of extremely harsh conditions. Techniques such as terrace based agriculture, sustainable irrigation and wind erosion control are just some of the invaluable systems affected.

The drivers for the processes of change behind the Rural Exodus syndrome are inextricably linked to other environmental syndromes such as the Favela and Asian Tiger syndromes. In these cases, the young and able are attracted to what is perceived as a better way of life in urban and industrial centres. With the inevitable changes to traditional practices and rural society this brings, there may additionally be the introduction and adoption of a range of techniques brought in from elsewhere, which prove to be unsuitable and ultimately unsustainable. Examples cited by WGBU (1996) include that of the development of the Karakoram Mountains in northern Pakistan. Here, the establishment of communications networks facilitated large movements in people and goods. As there became fewer people available to work the intensive agricultural systems, the cultivated land area reduced (despite the use of semi-mechanized methods) and could no longer be adequately managed and maintained. In the worst cases, this led to the enforced abandonment of large areas of previously productive agricultural land (as a result of landslides and avalanches), thereby having implications for the provision of food not just for the former food producers themselves but for the country as a whole. The symptoms of this syndrome include: loss of biodiversity, migration, food insecurity and severe soil erosion (WGBU 1996). As with the Overexploitation syndrome, soil erosion can then lead to wider impacts in lowland areas, where increasing silting-up of watercourses and flooding can occur as a result. Where this then affects the remaining agricultural areas of the same (or neighbouring) nation, the knock-on effect in terms of the supply of food can be even more severe.

14.3 DEVELOPMENT SYNDROMES

Chapter 3 outlined the specific development syndromes identified by WGBU (1996): the Favela syndrome and the Urban Sprawl syndrome. Other development syndromes include: the Aral Sea syndrome (landscape destruction due to large-scale projects); the Green Revolution syndrome (use of inappropriate technologies); the Asian Tiger syndrome (rapid economic development) and the Major Accident syndrome. Since the city is fast becoming the preferred form of human settlement the world over, it is useful to review how this has had, and is likely to continue to have, impacts on the processes underpinning our understanding of physical geography. Chapter 15 looks further towards the future of cities and the prospects for the urban environment. This section will consider the implications of the wider process of urbanization itself.

THE ISSUE OF GLOBAL URBANIZATION AND ITS IMPACTS

An increasing proportion of the world's population now live in what is described as a built-up environment. Built-up environments are characterized as areas of concentrated human activity and the infrastructure associated with that activity, that is, buildings for various purposes and structures for movement and communication. The construction of the infrastructure to support human activity has, over time, transformed the 'natural' landscape and altered the physical and chemical processes associated with it. Some outcomes of this include the development of specific meteorological and climatological characteristics, such as **urban heat islands** and altered hydrological characteristics, such as river channel modification, through direct intervention or indirect outcomes. The concentration of human activity, which is increasingly associated with polluting activity, has also had an influence on the quality of the air and water in and around built-up areas. This issue is true of cities in the developed and the developing world, but the richer the city the more resources it demands and the more waste it produces. Hardoy *et al.* (1999) summarize the regional impacts from rural interactions with urban areas as:

> **urban heat island**: the phenomenon of elevated temperatures within some large built-up areas relative to adjacent non-built-up areas caused by differences in surface cover and heat sources

- The expansion of the built-up area – reshaping land surfaces and natural water networks, changing land cover, moving materials and tapping water resources

- The demand from city residents and enterprises for products that are directly or indirectly generated from resources outside the urban zone (e.g. forests, watersheds, farmland). See Focus 14.2 for an example

- The disposal of solid, liquid and gaseous waste to land, air and water outside the city limits.

So, the areas around towns and cities have become both a source of inputs (materials and energy) to them, and the means of disposing the outputs (waste products) that they produce when transforming the materials into something else. This is the basis of the concept of industrial ecology and encompasses the idea of towns and cities having an ecological footprint (see Chapter 4). The ecological footprint of an average industrialized urban dweller is far greater than that of the average agrarian rural dweller, but there are important variations between countries and areas depending on total wealth, the distribution of that wealth, and the history and longevity of development in a particular place. The ecological footprint of the Lower Fraser Basin (in which the Canadian city of Vancouver is situated), for example, has been estimated to be around 20 times its land–surface area (Rees, cited in Hardoy *et al.* 1999). However, the burden each

FOCUS 14.2

ENERGY AND URBANIZATION – THE CASE OF FIREWOOD

Human settlements require fuel to support industrial and domestic activity. In developing nations, where electricity is still a relative luxury, there is a heavy reliance on solid fuel for the provision of energy for cooking and bathing. A report on India's environment, published in the 1980s, contained some startling facts about the demand for firewood from its urban inhabitants and the increasingly large distances over which it was necessary to bring their supplies (CSE 1986). This report notes that during 1981–82 there was around 600 tonnes of firewood brought into Delhi by railroad every day, most of it sourced to the area of Madhya Pradesh over 1000 km away. This is in addition to the use of wood resources within the city

district itself and the ready availability of other forms of liquid and solid fuels. Another Indian city, Bangalore, which at the time had about half the population that Delhi had, consumed in the region of 400,000 tonnes of commercial firewood a year. This corresponds to a daily consumption double that of Delhi. Although 85 per cent of the supply was sourced to farms and woodlands within a radius of around 150 km, the remainder was transported a distance of between 500 and 1100 km in order to reach the city. The sustainability of this demand and the implications of it for the ecosystems of the surrounding area clearly gives some food for thought (Hardoy et al. 1999).

person puts on the environment is not equal. Indeed, it has been calculated that the average ecological footprint of the poorest 20 per cent of Canadians is only a quarter of that for the wealthiest 20 per cent (Wackernagel and Rees cited in Hardoy et al. 1999).

In the developing world, the overall ecological footprint of cities may, at present, be smaller than for developed countries. However, other environmental issues have severe implications for both the cities themselves and their **hinterlands**. Focus 14.3 gives an overview of the regional impacts of the city of Jakarta in Indonesia.

> **hinterland**: the zone of influence around a settlement, for example for work, markets or services

Urbanization itself, then, is a key driver of environmental change. Changes can be direct and indirect in nature as well as local to global in scale. However, the idea of an Urban Sprawl syndrome perhaps goes beyond an analysis of the impacts of towns and cities. To some, 'urban sprawl' implies a certain speed of change and/or a type of development that is somehow 'out of control', and which is consuming larger and larger proportions of the Earth's surface. Some of the most marked examples of this have occurred in the United States where some cities, famously Los Angeles, have become almost impossible to negotiate on foot and necessitate the use of some form of mechanized transport to make the long trips between work, leisure and home. The idea of urban sprawl is considered further in the next section.

Given the importance of the city and its ability to influence natural form and process, it is useful to consider the impacts of urbanization in the different environmental spheres covered in this book, namely: air, water, land and life. Before that, however, it is first useful to consider the nature and development of cities themselves.

THE DEVELOPMENT OF THE CITY

The city has evolved as the focus for human activity in the modern world. The development of humankind, a highly social animal, has involved a movement towards increasingly large communities and the adaptation of our environment

FOCUS 14.3

THE REGIONAL IMPACT OF JAKARTA

Jakarta, capital of Indonesia, is a world megacity. It has undergone very rapid development from a settlement of 1.4 million in the 1950s to 11.5 million in 1995 (WRI 2003). In the 1980s, the population was growing at a rate of around 10 per cent per year because of large influxes of people from other regions in the country; it is expected that the population will reach an incredible 21.1 million by the middle of this century. The process becomes more understandable when put in the context of average wages in the urban area having been around 70 per cent higher than average wages elsewhere in Indonesia (Clarke *et al.* 1991). The urban region associated with Jakarta now stretches some 120 km west to east and 200 km north to south, with population densities as high as 12,435 people per square kilometre (WRI 2003). Over time, it began to be recognized that this growth was having severe direct and indirect impacts on the environment in and surrounding the emerging city. Like many of the globe's major cities, Jakarta is located within an agricultural area and its expansion has led to both the loss of good quality land and the increased marginalization of agricultural activity to more upland regions. Since many of these areas are intrinsically less suitable for agricultural activity, this has, in turn, led to soil erosion and reduced productivity. This erosion then leads to other impacts downstream through the silting of rivers and modification of the water system (for example, flooding in the rainy season and low flows in the dry season, which exacerbates water quality problems) (Hardoy *et al.* 1999). Jakarta Bay has been associated with mercury poisoning episodes as well as industrial and agricultural pollution. Much of the development has been so rapid (in common with many other developing cities – see the Favela syndrome, p. 63) that it has outpaced the capabilities of the authorities to provide essential infrastructure, leading to the development of informal 'kamgpung' developments (UNCHS and WB 1993).

In the city itself, there are further problems to contend with: WRI *et al.* (1996) report ambient particulate matter concentrations exceeding world health standards over 150 times per year and respiratory tract infections killing around 12 per cent of the population, a rate which is twice that experienced in rural areas. The adequate and sustainable supply of safe water is a further challenge. Less than half of the urban population has piped water supplies, many using private wells that tap directly into the groundwater supplies of the city. In the early 1990s, it was noted that the groundwater table was falling by between 30 cm and 70 cm a year for the preceding 15 years (McBeth 1995).

It is also important to note that despite these seemingly hopeless statistics, the authorities in Jakarta are making clear efforts to overcome the environmental and social burden associated with the development of their capital city. Examples include the Kampung improvement programme and the adoption of a spatial planning system in the early 1990s (WRI 2003). It is true that there is still some way to go, but developments in understanding the causes and consequences of developmental problems, as well as schemes to help monitor and manage the environment, is providing a useful basis for ongoing improvement.

(For more information, see World Resources Institute (2003), Hardoy *et al.* (1999), Douglass (1989).)

---

to support these communities. Our inventiveness and drive now enables this to be done on scales that have never before been possible in the history of our species. Urbanization is such a powerful idea that it is catching on almost everywhere. The speed and extent of urbanization is actually quite difficult to measure and monitor, but this is an essential task if the process is to be managed in a sustainable way.

The concept of urban sprawl was introduced in the previous section. One interesting study has attempted to determine the extent of urban sprawl in the USA using night-time satellite imagery (Sutton 2003). Sutton (2003) measures per capita land use consumption (through comparing the population of areas with the extent of their built-up area, as represented by different light intensities) for different urban zones in the US, and attempts to identify areas that exhibit the most sprawl. These areas may, in turn, be associated with higher rates of the negative aspects of this sprawl, such as the loss of open space, traffic congestion and pollution. The results highlight just how complex the issue has become, since although inland US cities such as Atlanta and Kansas City have high sprawl rates, eastern and western seaboard cities such as Los Angeles show remarkably low rates, even though they undoubtedly suffer the negative consequences associated with sprawl. Sutton (2003) attributes this to the larger 'functional region' (i.e. catchment areas), a mismatch between work and residential characteristics (necessitating longer travel times), high land costs and the degree of globalization associated with a city.

To truly understand the growth and development of urban areas it is necessary to gain an appreciation of the human spatial processes that are traditionally the realm of human geography. Human spatial processes are to some extent the geographical expression of human values and beliefs as they influence behaviour. Since behaviour is modified by a range of physical (economic and technological) and socio-cultural parameters, our characteristics and situation influence what we do, when we do it and how we do it, which in turn shapes the fabric and form of the cities in which we live. The wider human perspectives of this are considered in Chapter 15.

THE IMPACT OF CITIES ON AIR

In view of the ever-increasing numbers of people living in an urban environment in both the developed and developing world, the interest in and research into urban climatology has grown rapidly over recent years. The impacts of urbanization on climate can be summarized into five interlinked drivers of change (Bridgman *et al.* 1995):

- The replacement of natural land cover types, such as grass, soil and tree cover, by artificial land covers like concrete, tarmac/asphalt and glass. These materials have different properties that affect the amount of solar radiation that is reflected and absorbed

- The replacement of non-solid, rounded structures of natural vegetation with hard, angular built forms. These have impacts on the movement and strength of wind

- The addition of artificial heat to the environment from mobile and stationary human-made sources

- The modification of natural processes of channelling and disposing of water and the reduced infiltration of water into soil

- The addition of a range of contaminants to the atmosphere.

Taken together, the impacts are complex and interconnected. They can affect the amount of incoming solar radiation, the influence of solar radiation on patterns of temperature, circulation and wind characteristics, and cloudiness and moisture budgets. Figure 14.1 provides a diagrammatic representation of how radiation and energy flux characteristics differ between a generalized urban and a generalized rural case.

Looking in a little more detail, Arnfield (2003) traces developments in understanding in two key areas: urban atmospheric turbulence and urban temperature. The first key point made is that urban areas exhibit enormous heterogeneity, with different types of land cover each with distinct radiative,

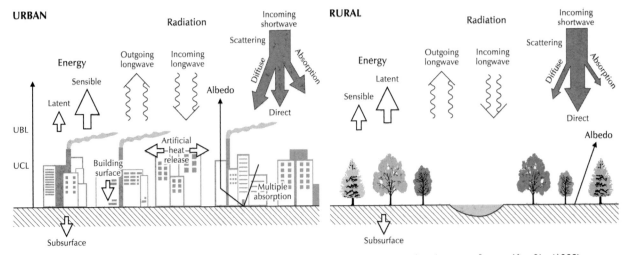

Figure 14.1 Characteristics of radiation and energy flux processes in an urban compared to a rural environment. *Source:* After Oke (1988)

thermal, aerodynamic and moisture-related properties. These essential differences tend to become greater with decreasing scale so that at the localized scale there can be dramatic changes in climate (e.g. windiness and temperature) over very small distances. As might then be expected, the complexity of interactions between the patchwork of land covers is also highly influential. Characteristics are generalized in terms of the urban canopy layer (building to ground) and the urban boundary layer (above roof level to the point at which the influence of the presence of the city is no longer distinct from its surroundings) (Arnfield 2003). The influence of the form and structure of the city makes it difficult to model meteorological processes, especially at very localized scales, which makes the study of related processes, such as air quality, challenging too. This potential for highly varying characteristics in specific localized areas (e.g. urban street canyons) is also important for understanding measured parameters in these locations, which is one of the reasons why there is a particular need for guidelines to promote the consistency in the siting of meteorological and air quality monitoring stations in the urban environment.

It has been recognized for some time that there is a clear temperature excess associated with the urban environment (e.g. Oke 1982), but recent research indicates that the nature of these urban heat islands and the processes that form them actually suggest a number of different types exist (Arnfield 2003). The nature of urban heat islands would also appear to differ depending on whether it is the surface or the air that is being considered and the measurement technique that is being employed. As well as the influence of the properties of the materials that make up the land cover of urban areas, there are also sources of anthropogenic derived heat (from vehicles, heating and people themselves), which can vary widely between cities and according to the time of day and the season. Urbanization has an influence on the amount of solar radiation received at ground level, with air pollutants being cited as some of the most influential factors. A reduction of solar energy input of 33 per cent has been cited for Hong Kong over a 35-year period (with no change in cloud cover), an influence that balances out any urban heat island effect in this area (Stanhill and Kalma 1995). In some instances then, taken at the whole city level, energy balances in urban areas may be no different to those in rural areas.

Arnfield (2003) notes that much recent research has confirmed hypotheses about the nature of urban heat islands such as those proposed by Oke (1982), where urban heat island intensity is:

- Inversely proportional to wind speed and cloud cover

- Greatest under anticyclonic conditions and in the warmest part of the season

- Proportional to urban population and extent

- Greatest at night and may not exist at all during the day.

The location of cities, whether inland or coastal, their latitude, and their orientation are also likely to impact upon the degree to which the effect is observed and therefore the likely modification of environmental processes.

Poor air quality is another major symptom of urbanization. Focus 14.4 describes the development of urban air quality issues in the UK into the 1990s. This shows the interlinkages between the drivers of urbanization as well as the role of scientific understanding and government policy in shaping the impacts that result.

FOCUS 14.4

URBAN AIR QUALITY IN THE UK AND ITS DEVELOPMENT INTO THE 1990s

Urban air quality is not a new concern and its degradation has been noted as far back as the thirteenth century in the UK (Brimblecombe 1987). However, it was the Industrial Revolution in the late eighteenth and nineteenth centuries, and the associated expansion in the use of coal for energy in homes and industry, that resulted in an increasingly severe problem (NSCA 1996). The growth and concentration of industry and population in the newly formed cities in the north, such as Leeds, Liverpool and Manchester, as well as in major cities elsewhere, was instrumental in the development of both general levels of poor air quality and the occurrence of smog episodes. Smogs, the combination of elevated smoke and sulphur dioxide (SO_2) concentrations, were recorded in many towns and cities and had important consequences for human health and the natural environment. For example, in January 1931, smog covered Manchester and Salford for nine days, during which there were 592 deaths from respiratory disease (NSCA 1986). However, the best known and most influential incident, the Great London Smog, occurred much later, between the 5th and 8th of December 1952. During this period, smoke and SO_2 concentrations reached a peak of around 6000 and 4000 mg m^{-3} respectively, and led to some 4000 excess deaths (Brimblecombe 1987). Within four years, public concern, media pressure and the findings of the subsequent Beaver Committee into the event had led to the first Clean Air Act of 1956 (Elsom 1992).

The Act represented a consensus of developing ideas on the control of air quality that included the use of smokeless zones and smokeless fuels and the principle of 'dilute and disperse'. A number of smokeless zones had already been established with mixed success using powers obtained from private Acts of Parliament, including in Manchester in 1952 and Salford in 1954 (MACCANC 1982). However, the use of such zones was to become an important and successful aspect of national air quality control policy. The Clean Air Act also included some industrial sources not already covered by the series of Alkali Acts, the first of which had been introduced as early as 1863. Together, the Clean Air and Alkali Acts formed the basis of the UK's air quality control framework.

Domestic and industrial emissions of smoke and SO_2 continued to be important well into the 1960s, until alternative sources of energy to coal became widely available (Wood et al. 1974). As well as the influence of changing fuel usage in homes and industry, the legislative measures that enabled the designation of smokeless zones also played a role in improving air quality in selected urban areas. Indeed, by the 1970s some eight million domestic and industrial premises were covered by smoke control orders (Elsom 1992). Urban smoke and SO_2 pollution has been greatly reduced, with a few notable exceptions where the ready availability of coal as cheap fuel has lead to its continued use (QUARG

1993a). These exceptions include traditional coal mining areas, such as in south and west Yorkshire, and areas in Northern Ireland. Sulphur dioxide has persisted as an occasional urban problem due to the large number and variety of sources involved, and due to the pollutant being less easily detectable by the public than is the case with a number of other pollutants such as smoke (Elsom 1992). Industrial SO_2 emissions are still a concern at regional, national and international scales, partly due to the legacy of policies that encouraged the use of increasing chimney heights to disperse pollutants over long distances. Sulphur dioxide and associated reaction products have been shown to be a cause of acidification in waters and ecosystems as far away as northern Europe as well as in sensitive areas in the UK itself (CLAG 1994).

Oxides of nitrogen (NO_x), another contributor to acid deposition, are also released as combustion products from fossil fuels. However, unlike SO_2, the primary source of emissions of NO_x is road transportation (Gillham et al. 1994; Salway et al. 1996). This has meant that this pollutant has grown in importance as a local issue due to the unprecedented increase in road transport usage since the 1960s. The number of vehicles registered in the UK grew from 8.5 million in 1960 to 25.2 million in 1994 (DoT 1996). Road transport is also an important source of other pollutants in the urban environment, carbon monoxide (CO), Volatile Organic Compounds (VOCs) and particulate matter (PM), with the latter predominantly associated with diesel engines (QUARG 1993b). Although urban areas generally no longer experience high concentrations of black smoke, as they did earlier in the century, emissions of particulate matter less than 10 microns in aerodynamic diameter (PM_{10}) have become an important issue due to associated risks to human health (QUARG 1996; COMEAP 1995a). More recent research and the associated policy responses are becoming centred on the importance of controlling even smaller particle sizes (for example, $PM_{2.5}$).

As well as transportation and combustion related activities, other sources of pollutants can also be seen to have an important role in determining modern air quality. A number of VOCs, for example, are emitted from industrial processes that, over time, have become increasingly complex in nature. Some of these organic species are highly toxic and even at very low concentrations can have detrimental effects on humans and the environment. Significant quantities of VOCs are also evolved from the widespread use of solvents for domestic, commercial and industrial purposes (Passant 1993; Gillham et al. 1994; Salway et al. 1996). Many of these VOCs are also contributors to the formation of tropospheric ozone, which itself is the cause of many occurrences of poor air quality in the UK (Derwent 1995a; Derwent 1995b; Marlowe et al. 1995).

THE IMPACT OF CITIES ON WATER

The same characteristics of the urban environment that have been highlighted as determinates of urban climate distinctiveness also have an important role in explaining the impacts of urbanization on water. The influence of urbanization can be seen in water inputs, water flows and water quality. The schematic diagram in Figure 14.2 nicely illustrates the range of impacts in terms of the hydrological cycle. In summary, these are (Bridgman et al. 1995):

- Amount of precipitation input. Since urban pollutants can provide more condensation nuclei on which water droplets can form it is not uncommon to find increased precipitation in and around urban areas

- Evaporation. Urban surfaces store water more efficiently than natural ones and therefore, given the right conditions (in particular, warm surfaces, unsaturated air and sufficient air movement), evaporation can greatly increase

- Interception. Urban structures and materials can act to intercept precipitation before it reaches ground level. For example, urban housing has systems of drain pipes that are specifically designed to remove water as efficiently as possible

- Transpiration. Urban areas tend to have smaller amounts of vegetation than corresponding rural areas and therefore transpiration losses are generally lower in urban than rural areas

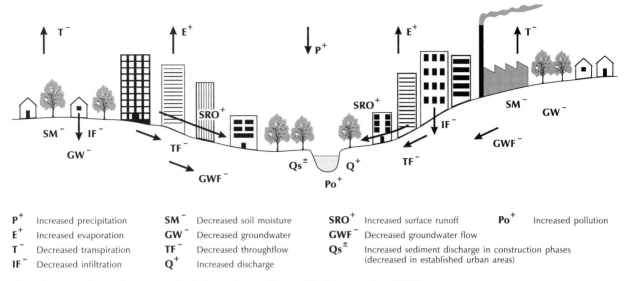

Figure 14.2 The relationship between urbanization and water. *Source*: After Bridgman *et al.* (1995)

- Surface water. Landscaping and engineering can either encourage the formation of surface water or promote its rapid dispersal

- Soil and groundwater. Soil moisture in urban areas can be severely depleted due to the effects of urban land cover and drainage systems. These same processes can have implications in terms of depleting groundwater reserves. Furthermore, the contamination of water reaching these water reserves can also cause problems

- Runoff. Owing to the tendency for increased input to the urban hydrological system and the reduced influence of the processes that naturally remove water, runoff can be greatly increased. Where precipitation events are high intensity this can give rise to natural and artificial water removal systems becoming overwhelmed, leading to flooding. This is made particularly problematic where urbanization has occurred on flood plains.

The above list gives some indication of the expected influence of urban environments on the hydrological system, but the reality is much more complex than this simple list suggests. To get a full appreciation of the impacts of urbanization, it is useful to look at how the urban climatic characteristics interact with urban hydrological characteristics. Of course, none of these issues can be viewed in isolation from the impacts of humans within these human-oriented environments.

Following on from the discussion of urban climatology, the urban heat island effect also drives some changes in the urban water budget of the city. Again, this is a complex issue interrelated to land use and human behaviour, as well as the undisturbed characteristics of environmental processes themselves. For example, a period of low precipitation may lead to artificial watering of public and private greenspace that then modifies the water balance of the city as a whole. Even allowing for 'natural' evapotranspiration processes in the urban environment, there will be some alterations in view of the make-up of areas surrounding a particular greenspace, for example through the passage of warm air from built-up surfaces onto vegetation, which then increases evapotranspiration rates (Grimmond and Oke 1986).

The presence of impervious land cover that makes up the larger proportion of urban land use types is also influential in terms of the impact of rainfall events. To prevent the build up of water in flat and depressed areas, elaborate drainage networks have been established to channel water into streams and rivers that take the water through and out of urban areas. Without these networks, there would be regular and dramatic flood events. Even with well-designed drainage systems, storm events can still lead to localized flooding. This problem is becoming marked in areas where entire catchments are becoming urbanized and the existing drainage networks can no longer cope with the excess surface flow produced. Given that recent studies are also suggesting the character of rainfall events are becoming more intense (New et al. 2001), the issue will continue to challenge urban planners for some time to come. In the USA, a study has suggested that between 1910 and 1996 daily average precipitation has increased by around 10 per cent. The increase in rain days and the number of extreme events accounts for more than half of this trend (Karl and Knight 1998). In the UK, Osborn et al. (2000) analysed data from over 100 stations between 1961 and 1995. Their findings revealed a trend towards reducing rainfall during the summer and increasing rainfall during the winter periods.

Old et al. (2003) document all of these characteristics in relation to their recent research into the river flow and sediment transport in the highly urbanized River Aire catchment in northern England. Their investigations of the impact of a summer 2001 rainfall event revealed a change in discharge from 0.45 to 34.6 cubic metres per second, in just 15 minutes, and a suspended sediment concentration that rose from 14 through to 1360 milligrams per litre. Clearly, this water volume and sediment load has implications for areas downstream.

In addition to issues associated with the amount of rainfall and associated runoff, another influence of urbanization is its impact on the quality of water. Burton (2003) traces the history of water pollution in the Mersey Basin in north-west England over 200 years of urbanization. He notes that, before the industrial revolution, anecdotal evidence suggests that freshwaters throughout the catchment of the River Mersey were relatively clean but, by the 1990s, the estuary itself was considered one of the most polluted in Europe. The industrial revolution, which brought first cotton and then chemical industries, was also associated with population expansion. People were attracted from many parts of the UK and Ireland and the area was associated with considerable wealth, prosperity and opportunity. Even by the beginning of the nineteenth century, fish species were still present in the major rivers of the catchment, but the cities of Manchester and Liverpool had rising problems of poor drainage, frequent flooding, polluted water and poor sanitation due to overcrowding and very poor living conditions (Burton 2003). Sewage was an acute problem for the new cities (Figure 14.3) and, even though extensive public works resulted in an impressive maze of sewers and drains, this wastewater was not treated in any way and was instead discharged to rivers and subsequently the sea. The problem was, therefore, not really resolved but simply transferred elsewhere with devastating effects on aquatic life in the areas downstream of the cities. With the advent of chemical industry, particularly the alkali works, there were further problems of

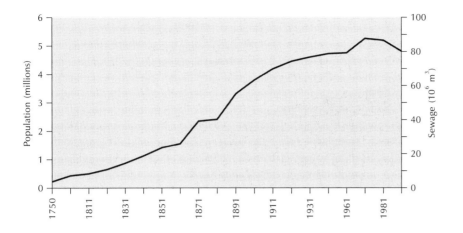

Figure 14.3 Trends of population and sewage generated in the Mersey Basin in north-west England since 1750. *Source*: After Burton (2003)

water quality with some waterways being so grossly polluted that entire tracts of land were uninhabitable. During the latter half of the twentieth century, strict controls on industry and sewerage outflows and work carried out by local organizations led to a turnaround in the quality of rivers in the Mersey basin. By 2000, 80 per cent of catchments were found to support fish (Mersey Basin Campaign 2000) and the dissolved oxygen and biochemical oxygen demand levels were much improved.

A second example from the UK demonstrates the critical role of urbanization on water-based ecosystems. Preston *et al.* (2003) note 35 per cent of the 65 species of aquatic plants present in the River Cam in Cambridge in the 1660s have since been lost. The loss of species is attributed to pollution from sewage and land use changes from riparian pastures to suburban open spaces. Thus, even a relatively small city, with a population that increased from just over 6000 people in the 1750s to just under 70,000 by the 1950s, could not escape widespread and far-reaching effects. Even after several decades of improvements, the long-term impacts of urbanization are not eradicated.

As areas of the UK were seen to be undergoing an improvement in water quality, urbanization was again causing problems in other parts of the world. A study of the water quality of Shanghai in China, for example, showed a strong relationship between water quality, land use and urbanization between 1947 and 1996 (Ren *et al.* 2003). Shanghai is located in the Chanjiany (Yangtze) river basin, which is a fertile agricultural area as well as one in which considerable urban and industrial development has taken place. Today, the extent of the city is estimated to cover some 280 km², more than twice the size of its built-up area during the late 1940s. In common with many other cities exhibiting the characteristics of urban sprawl, the rate of land usage by the city grows decade on decade, with the 1980s and 1990s showing the largest rate of annual land consumption at around 6.5 km² per year. This compares to around 3 km² per year prior to the country's cultural revolution (1966–76, when there was very little development). Ren *et al.*'s (2003) work suggests that up to 94 per cent of the degradation of water quality in the city can be explained by changes to industrial land uses.

It is being increasingly recognized that even in an urban environment, water is a multi-functional resource (Figure 14.4). Nowhere is this recognized more than in the urban areas of the Netherlands where a very high importance is placed on the protection of urban water resources, just as they are in the rural environment

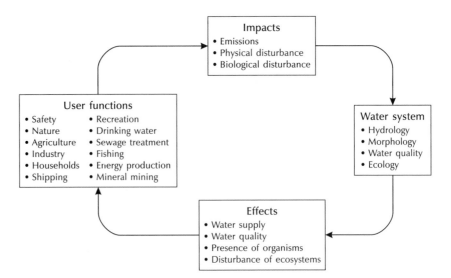

Figure 14.4 Components of the water system. *Source*: After van Beek (1998)

(Geldof and van de Ven 1998). Dutch researchers have invested considerable effort in trying to find solutions to the problems of the negative aspects of urbanization on the hydrological cycle, for example through developing systems of water management that can store and re-use stormwater runoff and therefore place less pressure on other parts of the water system.

THE IMPACT OF CITIES ON LAND

Implicit to the above discussion is the issue of the modification of natural landscapes in the urban environment. Accordingly, urbanization has an influence on the geomorphology of the city in terms of processes occurring within catchments, river systems and estuaries. Other impacts can be felt through the effects of development on slope stability, soil erosion and even the creation of new landforms, such as with the stockpiling of waste products.

The actual process of urban development influences the land in many ways. For example, the excavation of sites for the construction of buildings and infrastructure modifies natural and semi-natural processes both during construction and post construction. During construction, land is very susceptible to erosion especially where the nature of the development (such as commercial, industrial or high density residential projects) means that large areas must be cleared with little retention of existing vegetation or landforms (Douglas 1983; Bridgman *et al.* 1995). Land is at such a high premium in urban areas that development sometimes takes place in areas that are not well suited to the construction of permanent buildings or residences, such as floodplains or steep sided valleys. People who live and work in these areas are often the least prepared to cope with the inevitable natural disasters that occur.

The stability of slopes in a particular area is determined by many different factors, including the topography of an area, its drainage characteristics, the underlying bedrock, soil type, the frequency and severity of earthquakes, the quantity of rock and other materials above the surface and human activity (Douglas 1983). In Rio de Janeiro, development of the city up the steep sided hillsides has led to very severe problems. Here, the natural topography, coupled with deeply weathered rocks and heavy rainfall means that any removal of vegetation and modification of hillsides greatly increases the chances of movement. Disturbance of slopes high up on a hillside, for example associated with the development of favelas, coupled with planned developments at the base of hillsides where slopes are interrupted at critical points, as happened for example in Copacabana district, puts everyone living on the hillside at risk (Douglas 1983).

Where the developments themselves pose a potential environmental risk, such as industries with a high polluting capability, flood events can pose particular problems and the effects can be felt much further away than the immediate vicinity of the facility itself. To counter these issues, artificial means of controlling river channels and otherwise protecting developments on natural floodplains are required. In the past there has been much emphasis on hard engineering solutions to these types of problems. There is an important debate occurring between planners, developers, hydrologists and engineers into just how much humankind can rely on its 'command and conquer' approach to these issues, particularly in view of the likely impacts and implications of issues such as climate change.

Urban river channels can be seen to undergo four stages of development along-side the development of urban areas (Bridgman *et al.* 1995):

- Pre-urban state – natural or largely natural in form, perhaps with modification in relation to agricultural practices

- Early urban state with land degradation and high rates of sedimentation associated with construction

- Established urban state with surface sealing and higher runoff coupled with sediment deficiency and the addition of contaminants. Channel enlargement and degradation of banks and beds may occur

- Stabilized or engineered state which may result in complete modification of the original channel characteristics (as in Paris, Vienna or Sydney) and/or the addition of completely artificial channels (such as Los Angeles) to cope with extreme seasonal requirements.

As was pointed out in earlier chapters, large concentrations of people create large concentrations of waste materials. The traditional response has been to dispose of unwanted waste materials from humans and their activities through using the natural environment as a recycling system. For most of the solid, non-biological waste produced, the response has been to dispose of this on land. Figure 14.5 demonstrates the processes of land degradation that can occur due to the increased input of waste products. The Australian cities of Sydney and Melbourne generate around 3 to 4 million tonnes of residential waste every year and, if compacted into a 1 km² zone, this would correspond to an added 2–3 metres depth every year (Bridgman *et al.* 1995). In many well-established cities worldwide, space for this waste is literally running out. Landfill sites have important implications for water movements and there are other considerations too. In particular, the decomposition of this waste over time can lead to the build up of waste gases, notably methane. Since methane is an important greenhouse gas, this therefore contributes to climate change. In the high concentrations that can build up due to waste decomposition, methane not only produces an unpleasant odour but, in combination with other gases from landfill sites, can

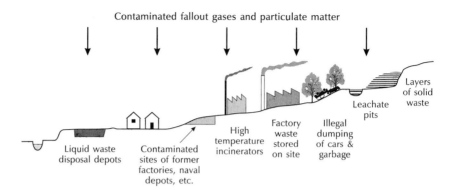

Figure 14.5 Urbanization and its impacts on land. *Source*: After Bridgman *et al.* (1995)

actually be explosive. Just such an occurrence happened in the UK in 1986 where a part of a house in Loscoe, Derbyshire, was destroyed by landfill gas from a nearby landfill site (Mather *et al.* 1996). Many waste disposal sites must therefore have a system of vents and be monitored to ensure that waste gases are not allowed to build up. Where waste materials include industrial wastes, the potential for contamination of waterways, groundwater and the air is greater still. In these situations, careful monitoring is required to ensure that toxic chemicals do not leach out to contaminate the land and soil around the site, especially where these areas are used for agricultural purposes as this can lead to toxic chemicals being brought into the human food chain and building up in natural ecosystems.

As well as burying unwanted waste below the surface, there are examples of where waste has been stockpiled above the surface of the ground. Where this is done for domestic refuse, the impacts usually are limited to bad odours, encouraging vermin (rats and birds) and posing a fire risk. However, occasionally, this can give rise to a major disaster, as happened in Turkey in 1993. Here, the combination of explosive gases and the instability of a 50 m high refuse dump resulted in an avalanche of refuse that buried 11 houses and killed 39 people (Wilson 1993 cited in Mather *et al.* 1996). In Aberfan in Wales, two days of unusually heavy rain in October 1966 led to a rapid mudflow of saturated tip material dumped over a quarter of a century earlier. The mudflow buried the village school and devastated the local community, all the more so since the instability of the tip had been noted on at least three earlier occasions in 1939, 1944 and 1962 (Douglas 1983).

THE IMPACT OF CITIES ON ECOLOGY

Some of the implications of urbanization on ecosystems have already been discussed since they are closely related to the other drivers of change in the urban environment. This section will bring together these ideas and look at the implications of human settlement in terms of ecosystems as a whole.

An interesting example of the extent of the impact of urbanization on biological systems is given in Bridgman *et al.* (1995), who trace the change in ecosystems through the history of the development of Sydney, Australia. The estimated changes in natural vegetation types (Table 14.1) are attributed to land cover change and the associated modification of physical and chemical processes. Taken together, these modify the original ecosystems of the area, leading to another cycle of change.

Table 14.1 Estimated vegetation change in Sydney 1788 to 1998

Vegetation Type	1788 (estimated) area (thousand ha)	1988 per cent remaining
Forests		
Blue gum	11	0.9
Turpentine-ironbark	35	0.5
River flats	19	3
Woodlands and scrubs		
Cumberland plain	107	6
Castlereagh	19	32
Eastern suburbs scrub	8	0.2
Sandstone formations	162	85
Wetlands		
Estuarine	6	29
Freshwater	3	18

Source: After Bridgman *et al.* (1995)

Open space, particularly greenspace, can be very limited in some cities. Simpson (1996) notes the differences in open space and greenspace in different European Cities. In Milan in northern Italy, 58 per cent of the city area is classified as open, but only 10 per cent of that area is considered to be greenspace. In contrast, Paris in France has only 23 per cent of its area classified as open space but some 92 per cent of this area is greenspace. The importance of greenspace, particularly natural greenspace, within built-up areas as a refuge for urban wildlife and for human health and well-being is gaining increasing recognition. One response to this is to try and promote a minimum provision of natural greenspace, as is happening in the UK (Pauleit *et al.* 2003).

Plainly, the change of land cover through the process of urbanization removes, sometimes permanently, the habitat that supports species within a particular ecosystem. The degree to which individual species are able to adapt to such changes depends on a number of factors such as their dependence on particular food sources and the range over which they operate. The species that survive (and sometimes thrive) in urban environments have modified their patterns of activity and form part of a modified urban ecosystem that is distinctly different from a non-urban counterpart. Remaining habitats become increasingly isolated and the existence of large tracts of impassable urban area around each effectively marks them out as islands. For this reason, ideas borrowed from island biogeography have been used to help research and explain the processes that occur within and between patches of similar habitat type as a means of finding ways in which remaining biodiversity within them can best be preserved (Bridgman *et al.* 1995). These urban habitat islands are not only inherently weaker due to their size and large amount of edge, they are more likely to be subject to threats as a result of chemical, physical and biological pressures coming from the surrounding area. As Douglas (1983) points out, this is not just an inter-urban issue, and the modified urban ecosystems have an effect on the surrounding surviving pre-urban ecosystems. Savard *et al.* (2000) use birds as an indicator species to assess the impacts of urbanization on ecosystems (Table 14.2) and help in the more effective management of remaining ecosystems. They suggest the following approaches to enhancing bird abundance:

Table 14.2 The relationships between degree of urbanization and reproductive characteristics of the European starling

	Urban centre	Urban periphery	Countryside
Laying date	6th April ± 1 day	4th April ± 2 days	1st April ± 2 days
Clutch size	5 ± 0	5.6 ± 0.9	6 ± 0.5
Number of eggs hatching	2.8 ± 2	5 ± 0.8	5.7 ± 0.5

Source: After Savard *et al*. (2000)

- Restore bird habitats around the periphery of urban areas

- Identify and consolidate green corridors linking areas of habitat with the surrounding habitats

- Increase the volume and diversity of vegetation within the city; enhance parks; increase structural diversity of vegetation; plant tree species that provide suitable food and shelter within the city

- Promote night-time blackout of tall buildings to promote migration

- Encourage the sensitive management of parks and private gardens, including distributing appropriate guidance.

Another reason for modification is the competition caused by both the deliberate and accidental introduction of alien species, from ornamental plants through to animals kept as pets. These may provide direct competition or predatory risk for native species, or affect species indirectly through their roles as vectors for diseases or other competitive species. In Australian cities direct impacts have been caused through the introduction of carp, house sparrows, rabbits, cats and blackberry (Bridgman *et al.* 1995).

Although most of the implications of urbanization for plant and animal communities are negative, some species can profit from the creation of what is effectively a new habitat or foraging ground for them. For example, the black-headed gull (*Larus ridibundus*) has, over time, adapted from being an estuary dwelling species to one that thrives in an urban setting, making use of food scraps and other waste, particularly during the harsh winter months (Douglas 1983). There is increasing evidence of new communities of species developing in areas that have been cleared by dereliction or are made hostile to indigenous communities through land contamination or other forms of pollution. Nevertheless, this does not always outweigh the loss of native communities.

14.4 SINK SYNDROMES

The final part of the jigsaw is to consider syndromes associated with the efforts of humans to use the environment as a means of disposing of their waste products. The Smokestack syndrome has been considered in some detail elsewhere in this book, both in terms of broad characteristics (Chapter 3) and the physical processes that are associated with the syndrome (Chapter 8). A case study example of dioxins is given in Focus 14.5. The section goes on to consider the Contaminated Land syndrome. Finally, another sink syndrome identified by WGBU (1996) is the Waste Dumping syndrome. Here harmful substances are concentrated into one area so that they can be sealed off as a form of management

FOCUS 14.5

DIOXINS AND THE ENVIRONMENT

The term dioxin stands for polychlorinated dibenzo-p-dioxin, where pairs of benzene rings are joined with two oxygen atoms. Furans (dibenzofurans) are usually also considered under the umbrella term 'dioxin' as they are chemically similar. There is a range of sources for these substances, related to the by-products of human activity. In particular, sources are industrial processes (metal smelting, the paper and pulp industries and plastic industries) and waste disposal (municipal, domestic and industrial waste incineration). They are very long-lived in the environment. Once they have entered the bodies of humans and animals, they are very effectively stored. Human exposure is possible through a number of routes but food is the most important source. This is because dioxins resist the normal removal processes and therefore tend to build up in the food chain over time. Even for populations living near to a waste incinerator, estimates still suggest that 98 per cent of dioxin (PCDD/PCDF) exposure is through diet, especially fat-containing animal products, fish and shellfish (Llobet *et al.* 2003).

These substances are known carcinogens and there are no safe levels. At low levels they are known to affect human immune systems and hamper liver and thyroid functions. At higher concentrations, dioxins cause cancer, birth defects and reduced fertility. Although dioxins are still produced from the incineration of waste and some industrial processes, a number of alternatives have been identified for new products, and pressure continues to be placed on organizations to ensure that the use of chlorine free materials are promoted and used. In a recent study of dietary intake of dioxins in the Catalonia region of Spain, the relative proportions of intake from different food groups were 31 per cent seafood, 25 per cent dairy products, 14 per cent cereals, 13 per cent meat and 20 per cent all other remaining food groups (Llobet *et al.* 2003). The overall intake was estimated to be below the Tolerable Daily Intake (TDI) value of 1–4 pg TEQ per kg per day* that was established by the World Health Organization in 1998. The encouraging news here is that this study, in keeping with the majority of results from other studies in Europe, suggests that the intake of dioxins is now lower than in the 1990s (Llobet *et al.* 2003).

*One pg (picogram) is 10^{-12} grams (or one million millionth of a gram). TEQ (Toxic Equivalent Quantity) is a measure that allows information about the quantity and toxicity of a range of toxic substances to be combined, and therefore enables an assessment to be made about overall harmfulness. TEQ is calculated through multiplying the mass of each toxic substance with the equivalent TEF (Toxic Equivalency Factor) for that substance. The TEF shows how harmful each substance is compared to the most toxic dioxin, which is given a value of 1. The specific example here shows that the Tolerable Daily Intake (TDI) value of 1–4 pg TEQ per kg per day is the equivalent mass of the most toxic substance but this could actually be made up of larger quantities of less toxic substances.

(i.e. the opposite of the Smokestack syndrome's 'dilute and disperse' rationale). This strategy becomes problematic as these areas are a constant threat to soil and groundwater quality – a so-called environmental 'time bomb', where natural processes make them an environmental disaster waiting to happen.

THE CONTAMINATED LAND SYNDROME

The Contaminated Land syndrome, like many of the syndromes noted in this chapter, has a long history. Contaminated land represents the legacy of past industrial activity in many parts of the world, sometimes where the activity itself is long gone. The symptoms of the syndrome include a loss of biodiversity in affected areas, loss or disruption of soil processes leaching of pollutants from sites and their subsequent contamination of soil, water and air, and significant health hazards (WGBU 1996). There are a number of hot-spots in both the developed and developing world which have been particularly identified in terms of the Contaminated Land syndrome. These include the Saxony–Anhalt area in Germany, the Manchester–Liverpool–Birmingham triangle in the UK, Wallonia in Belgium, Katowice in Poland, Pittsburg in the US and Cubatao in Brazil. One of the most famous examples of the Contaminated Land syndrome is Love Canal (Focus 14.6).

FOCUS 14.6

THE CASE OF LOVE CANAL

Love Canal is one of the landmark cases of contamination causing serious long-term effects on the natural environment and its human inhabitants. The canal itself, located near Niagara Falls in New York, is a 20 m wide 3 m deep trench, around 1 km in length, constructed in the 1890s initially for a hydropower scheme. With the failure of the scheme, the canal later became a waste disposal site for toxic chemicals during the late 1940s and early 1950s. Later in the 1950s, the land was sold on for development and became the site for a school and residential homes, the residents within which had no idea about the previous history of the site. By the 1970s, residents had become suspicious of the strange health issues associated with the area, including chromosome damage and unusually high rates of cancer. Eventually the concerns were taken on by the relevant authorities and the United States Environmental Protection Agency carried out an extensive survey which revealed the extent of the human health problems in the area around the canal. As a direct result, by 1981 over 500 families were moved out of the area. Although the level of risk for those remaining is difficult to ascertain, it is important that this is a known risk and one that people have information to make an informed judgement about (Mather *et al.* 1996).

IMPLICATIONS FOR GLOBAL ECOSYSTEM CHANGE

This discussion on environmental syndromes has highlighted some of the issues of scale when considering both the causes and the effects of environmental problems. Ultimately, the impacts that are felt at local and regional scales aggregate up to further global scale change, and this in turn will then play its role in affecting further change at a more localized level. In view of this, it is useful to end with a consideration of the current state of the world's ecosystems, which can be considered an important indicator of the rate and nature of change that is currently taking place across the globe as a whole. The Pilot Analysis of Global Ecosystems (PAGE) provides a broad scale analysis of current issues in relation to five of the world's most important ecosystem resources: agroecosystems, forest ecosystems, grassland ecosystems, freshwater ecosystems and coastal ecosystems.

AGROECOSYSTEMS

Agroecosystems by their very definition are highly modified and managed. Cropland and managed pasture have been estimated to cover some 28 per cent of the Earth's land area, with 31 per cent devoted to cropland. The land area of permanent cropland is currently expanding at a rate of around 2 per cent per annum compared to fairly stable patterns in annual cropland and a rate of expansion of pastureland equivalent to around 0.3 per cent per annum (Wood *et al.* 2000). The stable patterns in global annual cropland mask increased intensity of harvesting within the land area that it occupies, 1.38 billion hectares. The area of irrigated agricultural land in 2000 was around 270 million hectares, expanding at a rate of 3.3 million hectares per year (1.6 per cent), but at the same time a land area of 1.5 million hectares per year was lost due to salinization processes (Wood *et al.* 2000). Wood *et al.* (2000) also indicate some critical gaps in knowledge about these ecosystems, for example, what patterns of changes in agricultural practices these land cover changes are associated with and therefore what the true dynamics of change might really be.

FOREST ECOSYSTEMS

Matthews *et al.* (2000) suggest that global forest cover has been reduced by between 20 and 50 per cent since pre-agricultural times. Although forest areas in the developed world have increased since 1980, almost all is under some form of management and supports reduced biodiversity levels. The developing world, which has just over half of all forest resources, lost 10 per cent of forested land area during the same period. However, the actual loss may be greater since many mixed agricultural–forest uses are not recorded as a loss of forest area. Almost 40 per cent of tropical forest remains largely intact and undisturbed by human activity, but the rate of tropical deforestation is still estimated to be in the region of 130 thousand kilometres per annum. The loss of forest land cover is due to a number of factors including transfer to agricultural land and logging for wood supply. However, natural forest fires are also an important driver of change.

The degree to which the forest ecosystem has been degraded and fragmented can be estimated by looking at statistics describing the rate of expansion of the global road network. Matthews *et al.* (2000) see the opening up of previously intact forested areas as an important prerequisite to the further intrusion of human activity, through settlements, mining, logging, hunting and poaching. They call for improved information on the extent and location of the global road network, better information on forest condition, for example in relation to agricultural uses and the potential for forest fires, and better mapping and classification of resources generally as an aid to understanding ecosystem dynamics.

GRASSLAND ECOSYSTEMS

Grasslands are found in all regions of the world (with the exception of Antarctica and Greenland) and cover 40 per cent of the Earth's surface. At the national scale, the largest tracts of grassland area are found in North America (the United States and Canada), Australia, Russia and China but the largest proportions of grassland within individual countries are all found within sub-Saharan Africa (White *et al.* 2000). The sub-Saharan region has the largest overall land area covered by grassland, some 14.5 million square kilometres. Grasslands are most common in semi-arid zones (28 per cent), humid (23 per cent), cold (20 per cent) and arid zones (19 per cent) with temperate grassland types having experienced a heavy loss to agricultural land uses. As with forested areas, White *et al.* (2000) note that differences in classification techniques presents considerable difficulty in fully assessing rates of change and the character and condition of grasslands across the globe. It is hoped that improvements in the resolution of satellite imagery will partially help in being able to better monitor changes in the future.

FRESHWATER ECOSYSTEMS

The value of freshwater to global ecosystems and the human population cannot be overstated. The PAGE assessment of freshwater ecosystems revealed that there have been considerable human modifications to freshwater systems that have far-reaching implications. These include (Revenga *et al.* 2000):

- Large-scale dam construction – of the 227 largest rivers in the world, 60 per cent are strongly or moderately fragmented as a result of the construction of dams, channel modification and diversion and canal development. Large dam projects are currently under development on the Yangtze, Tigris, Euphrates and Danube. Such developments are known to have dramatic effects on freshwater ecosystems. This is the basis of the Aral Sea syndrome (WGBU 1996).

- Land use change – over half of the world's wetlands have been lost to agricultural or urban uses since the beginning of the twentieth century.

- Groundwater extraction – the quality and quantity of groundwater is threatened as in excess of 1.5 billion people worldwide rely on its direct extraction.

These, together with pollution, overexploitation (an increase in freshwater fishing and farming), the introduction of alien species and habitat degradation are seen as drivers for the loss of freshwater species. The World Wildlife Fund has identified 53 freshwater ecoregions for priority conservation. One positive observation was that the population of North American wetland bird species has increased over recent times (Revenga *et al.* 2000). The PAGE group called for increases in the monitoring of fluvial systems and for this to include biological information as well as chemical information.

COASTAL ECOSYSTEMS

The Earth has been estimated to have 1.6 million kilometres of coastline and 19 per cent of the area within 100 km of the coast (excluding Antarctica and water bodies) has been classified as altered, usually by agricultural or urban uses (Burke *et al.* 2000). There have been recorded losses of a number of coastal habitats, particularly mangroves (ranging between 5–80 per cent between different countries), sea-grasses and coral reefs. The near-sea area of the coastline of many countries is under pressure from heavy trawling activity. Statistics cited by Burke *et al.* (2000) suggest that trawling is carried out over more than half of the continental shelf area of the 27 countries for which data were available. Burke *et al.* (2000) call for improved recording of information by all nations to help better measure and monitor the extent of human use of coastal resources.

ENDANGERED SPECIES

The loss of global ecosystems to land use change and other forms of modification and stress is clearly one primary cause of increasing vulnerability of individual species to extinction. Table 14.3 traces the change in numbers of species of different types that have been classified as being under threat in 2000 and 2003. The table shows that, although there has been no improvement to the numbers of species under critical threat of extinction (in fact a considerable worsening of figures for the most part), there are some positive indications. For instance, the numbers of fish, mammals, reptiles, insects and molluscs all showed reductions in the number of species classified as vulnerable in 2003.

Table 14.3 Numbers of species classified by the International Union for Conservation of Nature and Natural Resources as critical, endangered or vulnerable in 2000 and 2003

Group	Critical		Endangered		Vulnerable	
	2000	2003	2000	2003	2000	2003
Mammals	180	+4	340	−3	610	−1
Birds	182	0	321	+10	680	+1
Reptiles	56	+1	74	+4	161	−3
Amphibians	25	+5	38	−1	83	+7
Fishes	156	+6	144	0	452	−8
Insects	45	+1	118	0	392	−3
Molluscs	222	+28	237	+6	479	−5
Plants	1014	+262	1266	+368	3311	+553

Source: International Union for Conservation of Nature and Natural Resources (2003)

SUMMARY

This chapter has outlined a number of development, utilization and sink syndromes of environmental change, using examples from different geographical locations with contrasting features. Throughout this discussion, the importance and influence of drivers and impacts has been identified, as have the scales over which the syndromes operate. In view of the increasing populations living in built-up areas, considerable emphasis has been placed on the specific impacts of urbanization on the environmental media looked at in this book: air, water, landforms and life. The chapter ends with a consideration of the state of the world's ecosystems. Chapter 15 goes on to explore the issue of human impacts and trends for the future in more detail.

Questions

1 How does the environmental syndrome approach help in understanding the environmental impacts of human activity?

2 Which of the environmental syndromes referred to in this chapter are in evidence in your country? Where possible, try to give some specific examples.

3 Using the IUCN website, list the top 10 endangered species in 2003. Which of the environmental syndromes referred to in this chapter are likely to have been an influence on the current status of these species?

4 Draw up a table that summarizes the key differences between rural and urban areas in terms of the physical processes and characteristics of air, water, land and life systems.

5 Using the suggested websites as a guide, produce a summary of the environmental problems associated with two megacities, one from the developed world and a second from the developing world.

Further Reading

Many interesting books explore the impact of urban areas on the natural and semi-natural environment. Two of the more recent of these are recommended for follow up reading.

Bridgman, H., Warner, R. and Dodson, J. (1995) *Urban Biophysical Environments* (Meridian Australian Geographical Perspectives). Melbourne: Oxford University Press.
This is a very accessible and readable introduction to the impacts of urbanization on key processes associated with physical geography. It uses a range of examples from Australia to illustrate the ideas described in the book, but the ideas and concepts are generic to any urban area.

McCall, G. J. H., de Mulder, E. F. J. and Marker, B. R. (1996) *Urban Geoscience*. Rotterdam: A. A. Balkema.
A collection of essays about different aspects of the urban–environment relationship. The book is oriented towards the geologist but provides a good grounding for the physical geographer.

For further reading on the nature and characterization of environmental syndromes the reader is directed to:

WBGU – Wissenschaftlicher Beriat der Bundesregierung Globale Umweltveränderungen (German Advisory Council on Global Change WGBU) (1996) *World in Transition: The Research Challenge Annual Report 1996*. Internet reference http://www.wbgu.de. Last accessed December 2003. English version of this website is available through a link on the homepage.

Useful Websites

Some recommended follow up websites are as follows:

http://www.iucnredlist.org
International Union for Conservation of Nature and Natural Resources. Interesting statistics and specific information about threatened species, plus a searchable database.

http://www.wri.org
World Resources Institute. Examples and data about global resource use and the state of the world ecosystems.

CHAPTER 15
PERSPECTIVES FROM THE HUMAN SPHERE

OVERVIEW

The last chapter examined some of the far-reaching effects of the activities of humans in shaping the environment and modifying the physical geography of the world. This chapter looks towards the future and considers the progress that has been made, and that will continue to be made, in understanding environmental issues and syndromes. In doing so, the overall value of a syndrome approach to categorizing, modelling and tackling environmental issues will be considered. Given that people are the ultimate driving force of change, it is vital to understand future trends in terms of population, urbanization and resource use. As might be expected, determining future patterns in drivers is by no means a straightforward task, and there is much intellectual effort invested in understanding the societary processes that underpin them. In this endeavour, the different dimensions of human geography research (with links to allied disciplines such as psychology, sociology, economics and governance) are an essential foundation. But just as society influences the environment, the environment and associated physical processes also, in turn, influence society. Hence, for a complete understanding of the human sphere, it is essential to appreciate the physical sphere. The process of economic globalization, driven in large part by a complex ever-shifting landscape of patterns of supply and demand for goods and services, is one example of this. Recognition of the symptoms of environmental syndromes and their interrelationship with human society and economy is the crux of the ethos of sustainable development. The chapter will end with a reflection on the future and prospects for 'global management' and its delivery through a framework of local and global activity.

LEARNING OUTCOMES

At the end of this chapter, you should feel able to:
- Outline a number of projected trends in human activity and explain their uneven global patterns
- Appreciate some of the interlinkages between future human activity and potential future environmental change
- Consider the role of global management in helping meet the environmental challenges of the twenty-first century.

This chapter draws together the ideas that have been presented in the previous chapters through a survey of trends and prospects for planet Earth. This is done particularly from the perspective of the human sphere, considering projected changes in population, urbanization and resource use. It ends with a view on the utility of the syndrome approach for understanding and solving global issues and prospects for global management.

15.1 TRENDS FOR THE FUTURE

To get an overview of the likely future of human–environment interactions, it is useful to review some of the key projected trends in global activity. The following sections describe estimated trends in population, urbanization, energy and resource use, pollution and quality of life. In considering these trends, it is important to realize that each is associated with a degree of uncertainty in terms of the key assumptions that support the particular scenarios being presented. This uncertainty is then magnified where groups of trends are considered and their implications for the global environment are hypothesized.

GLOBAL POPULATION TRENDS AND PROJECTIONS

Figure 15.1 shows the world's total projected population to 2050. The diagram shows the steady growth in numbers of people worldwide from 2.6 billion in 1950 (increasing by some 37.8 million per year or 1.47 per cent) to 6.3 billion (increasing by some 73.4 million per year or 1.16 per cent) in 2003 (US Bureau of Census 2003). By 2050, it is estimated that the global population will reach an incredible 9.1 billion. This sheer volume of people is expected to place further pressures on global resources.

However, as can be seen from Figures 15.2a and 15.2b, it is also true that, overall, the global rates of population growth are exhibiting a marked downward

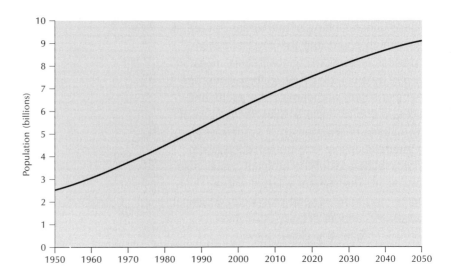

Figure 15.1 Total global population totals 1950 through 2050. *Source:* After US Bureau of Census (2003)

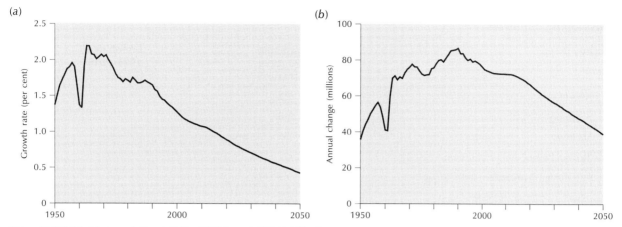

Figure 15.2 Total global population statistics 1950 through 2050 (a) World population growth rates and (b) world annual population change. *Source:* After US Bureau of Census (2003)

trend. This rate reduction is due to a number of developing countries now beginning to experience a stabilization of their populations, as has been observed over the past half century or so in the developed world (WRI *et al.* 1996). Although the reduction is encouraging, there are some developing nations still exhibiting considerable population expansion. These countries tend to be associated with high levels of poverty, relatively poor opportunities for women and high rates of migration both within and between nations, all of which act to further exacerbate the problem (WRI *et al.* 1996). This rapid growth will then, inevitably, place pressure on already stretched resources in some parts of the world and is likely to lead to further environmental degradation before improvements are possible. All the signs are, though, that eventually the global population will stabilize. Nevertheless, the WRI *et al.* (1996) warn against complacency since, for the projected growth characteristics to be realized, work that has already been initiated in terms of promoting reduced levels of fertility (such as through the provision of family planning services, better education and health care, and the encouragement of economic opportunities for men and women alike) must be continued.

URBANIZATION

A large proportion of the world population growth discussed in the previous section is anticipated to either occur within urban areas themselves or to expand urban population growth through providing a basis for migration to cities. 'A gigantic urban revolution is under way today: in 40 years the equivalent of 1000 cities, each with three million inhabitants will have to be built' (Bindé 1998, 499). This process is set to continue with 50 per cent of the world's population now living in urban areas which will rise by 2020 to an estimated 60 per cent (WRI *et al.* 1996).

Annual urban population growth rates are not uniform across the globe. Table 15.1 gives an idea of the different degree of growth observed between 1970 and 1995 compared to that anticipated in 1995 to 2015. This table clearly shows the accelerated rates in the least developed countries. In some places, it is anticipated that as much as 90 per cent of future population growth will be associated with

Table 15.1 Comparison of the different rates of urban population growth between developed and developing countries

Country Group	Growth rate 1970–1995 (%)	Growth rate 1995–2015 (%)
Least developed	5.1	4.6
All developing	3.8	2.9
Industrialized	1.1	0.6
Low Human Development Index Group	4.1	3.7
Medium Human Development Index Group	3.9	2.8
High Human Development Index Group	3.3	1.7

Source: After UNCHS (2001)

Note: the Human Development Index refers to a measure of the average level of GDP per capita, educational attainment and life expectancy at birth

cities (WRI *et al.* 1996). It is also interesting to consider further the characteristics of the urban areas themselves that are likely to result. Figure 15.3 shows a comparison of the likely distribution of different sized settlements in the years 2000 and 2003. These data suggest that there will be an increase in the number of people living in megacities of 10 million people or more, compared to numbers living in smaller settlements of 500,000 to one million and settlements of less than 500,000 people.

Bindé (1998) draws together some key challenges associated with the inexorable rise of the city:

1. The first key challenge is the growing impact of big cities on the environment, which involves problems of land consumption, pollution and impacts on rural areas. The land consumption issue is illustrated by Cairo, which consumes 600 ha of agricultural land each year, even though only 5 per cent of Egypt's land area is agriculturally productive.

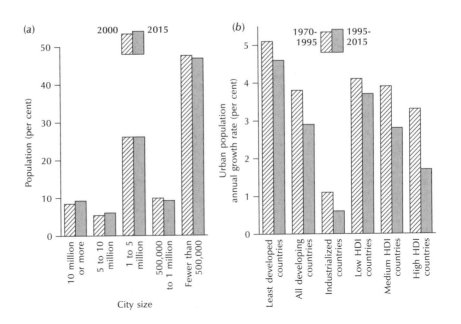

Figure 15.3 The distribution of the global population across different sized cities as in the year 2000 and expected by the year 2015. *Source:* After UNCHS (2001)

New York has expanded in size by a massive 61 per cent over the last twenty-five years despite a population rise of a mere 5 per cent. In terms of the cost of pollution, the OECD estimates the cost of local pollution as costing nations between 1 and 10 per cent of total GDP.

2. The second key challenge relates to the ultimate ongoing sustainability of the urban way of life as a whole. What some see as the predatory nature of the relationship between the city and its hinterland, established from pre-industrial times, cannot be sustained into the future. In western society, even people who do not consider themselves urban dwellers generally have a strong link with urban areas for work and leisure. However, issues are now most pressing in the developing world where, according to the World Bank, the population of cities is expanding at an incredible 65 million inhabitants per year. This raises questions about how adequate housing and living conditions could possibly be provided within cities that are already developing large informal settlements due to rates of growth outstripping the rate at which provisions can be made for the in-migrants. Although the rate of urbanization in the developed world is more manageable, it is also true to say that the average western urban dweller is a far greater burden than their developing world counterpart, for example one urban American consumes the same as 20 urban Bangladeshi.

Bindé (1998) relates three series of problems on which to focus attention: water, transport and energy. However, there are also other issues related to air quality, materials use, biodiversity, human health and quality of life. There is also the important issue of how cities can adapt to changing environments, in particular climate change.

The impacts of urbanization on local, regional and global ecosystems, which were outlined in Chapter 14, are clearly going to continue to be important well into this century, and finding the means by which this growth and inevitable impact can best be managed is an important goal for geographers.

ENERGY AND RESOURCE USE

By 1994, about 1.2 billion people in the developing world did not yet have safe water supplies and 3 billion lacked proper sanitation facilities (Livernach and Rodenburg 1998). Despite this remaining need, in the 40 years prior to 1990 there had been a fourfold increase in freshwater extraction from fluvial and groundwater systems in response to the needs of agriculture, industry and urban areas (WRI *et al.* 1996). Focus 14.3 noted the impact this was having on groundwater supplies in Jakarta. WWF (2003), reporting some of the outcomes of the World Water Forum in Japan, note that should the current rates of freshwater consumption continue, half the world's projected population would live in water-stressed catchments in twenty years' time. This has implications not just for humankind but also for the health of ecosystems at a range of spatial scales. The rise in water consumption was seen to be one particular issue, with per capita increases expected as well as general population growth. Another issue was the expected continued degradation of quality of the remaining freshwater supplies. The control of water quality and adequate provision of water services is clearly a high priority for the future and one that is receiving a large amount of investment the world over, an estimated US$90 billion worth per year (WRI *et al.* 1996; WWF 2003).

This issue of water availability has implications for future food provision. An estimated 67 per cent of water supplies are used as an input into agricultural systems (WWF 2003), so increased pressure is likely to have an impact on the capability of these systems to provide sufficient food for the future population (WRI *et al.* 1996). The most recent suggestions are that it will be possible to provide sufficient food for the projected rise in population but the equitable distribution of food may still be an issue (see p. 422).

Much effort has been invested in trying to determine likely future patterns of energy demand and consumption. Increases in the region of 34–44 per cent between the mid-1990s and 2010 and 54–98 per cent by 2020 seem likely (WRI *et al.* 1996). Highest increases in demand are expected in Asia and Latin America, where it is anticipated that demand will be met predominantly by non-renewable fossil fuel sources rather than by renewable energy. Under current scenarios, renewable sources are likely to be associated with only 2–4 per cent of total global energy supply (WRI *et al.* 1996). However, some researchers have noted the potential for further use of renewables in the future. Fischer and Schrattenholzer (2001), for example, describe a high economic growth, low emission scenario that relies on a 15 per cent supply of global primary energy from bioenergy sources by 2050. Another important source could be wind power, and there is increasing interest from national and supranational organizations in the potential for onshore and offshore wind farm developments. The European Commission, for example, consider that it is possible to increase European use from the 1997 contributions of 4450 Megawatts (MW) to as much as 40,000 MW by 2010, much of this supplied by offshore developments that are not associated with as many conflicts with other land uses as onshore wind farms (Gaudiosi 1999). Although there is much still that could be done, promotion of wind energy sources and government assistance for associated wind energy establishments has already seen a fivefold increase in the installed grid-connected capacity across Europe between 1990 and 2000 (McGowan and Connors 2000). In terms of global potential, up to 60,000 MW could be provided by 2010, compared to 1997 levels of 7200 MW (Gaudiosi 1999).

Another important source of energy demand is associated with transport systems. Research into the global patterns of energy demand for transport suggests that US 'automobile cities' consume eight times more energy per capita in private-vehicle-based systems, compared to some public transport oriented cities in the Far East, such as Hong Kong (Kenworthy and Laube 1999). Careful land use planning can help to counteract some of these issues for the future.

POLLUTION

Although there is some potential for switching energy supplies to more renewable sources and an increasing amount of effort being placed in developing and promoting means of energy conservation, there is still the prospect of increasing fossil fuel usage in the future. In view of this, it is inevitable that there will be some local and regional air pollution consequences (WRI *et al.* 1996). In the developed world, air-quality management is now well developed and many air-quality problems are broadly manageable. Nonetheless, there are still some issues that will be important in this century, such as urban air quality associated with nitrogen dioxide and particulate matter and more widespread issues associated with particulate matter, tropospheric ozone and a number of hazardous organic compounds. In the developing world, urban air quality is still a very challenging problem for numerous pollutants. These will all have an influence on global issues. The biggest impact of increased energy demand, however, is the expected impact of use on greenhouse gas emissions, globally anticipated to increase by another 30 to 40 per cent by 2010 (WRI *et al.* 1996) and the subsequent impact of these on global change.

Many of the issues associated with predicted climate change have been examined elsewhere in this book, but one issue that has not yet been considered in any detail is that of associated sea level rise (see Chapter 8). The predicted rises (Figure 7.15b) will put coastal ecosystems, already under considerable pressure from development, under further risk (p. 418); this is also in the context of around 60 per cent of the world's population living within 100 km of the coast (WRI *et al.* 1996). Some 58,000 km^2 of the Atlantic and Gulf coastlines of the United States lie below the 1.5 metre contour and may therefore be vulnerable (Titus and Richman 2001). There are also some specific areas where the vulnerability to sea level rise is expected to be particularly felt, as in the South Pacific region where many island nations are threatened (Mimura 1999). Given the low-lying nature of these islands, which include Tonga, Fiji, Samoa and Tuvalu, all are likely to see increased flooding and inundation, with other environmental impacts expected to include beach erosion, saltwater intrusion and the disruption of the communications and other infrastructure.

QUALITY OF LIFE

The UNFAO have classified 27 nations as having either low or critical food security index values (WRI *et al.* 1996), and although trade in food across the world is expected to double over the 20 years between 1990 and 2010, this is unlikely to satisfy demand everywhere. As a result, the UNFAO fear that the number of undernourished people, for example in sub-Saharan Africa, could rise from 175 million to some 300 million by 2010 (WRI *et al.* 1996).

Across the globe, almost 200 million children are either moderately or severely underweight (Iyengar and Nair 2000). The migration of populations from rural to urban centres and the resultant pressure that this places on limited resources is an important factor in explaining poor quality of life and elevated levels of malnutrition. Deficiencies in vitamin A, iron, iodine, calcium and folic acid are compounded by the effects of pollutants caused by human activity, such as environmental lead, arsenic and other heavy metals entering the food chain. These deficiencies then make individuals more susceptible to disease and the impacts of other pollutants in air and water. It is also interesting to compare statistics showing the causes of death in under-5s between the developed and developing world (see Figure 15.4, p. 428).

Another issue related to quality of life and the likely expansion of global megacities is the question of how these cities are going to accommodate the expected growth in population. It has been noted that, due to the high rates of growth and in-migration, many developments occur as informal urbanization. This type of development may then take place in areas that are not suitable and which have a high potential for the occurrence of natural disasters. Such disasters are more likely to occur in some cities since people become forced to carry out

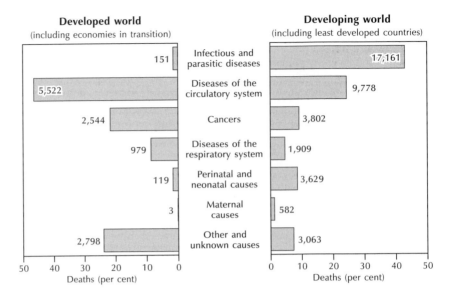

Figure 15.4 Distribution of 10.4 million deaths among the under-5s in developed and developing countries, World Health Report, 1995. *Source:* After Iyengar and Nair (2000)

planned and unplanned development in increasingly unsuitable areas, such as floodplains and steep-sided slopes (Hamza and Zetter 1998). Sometimes the developments themselves can form the basis of the risk, for example through the potential for fires and health hazards in slum developments.

These issues are certainly not, however, only restricted to the developing world. For example, it is common knowledge that much of the development in the Californian basin is on tectonically unstable land. In Europe, the desire to protect remaining areas of landscape with high agricultural and ecological value is creating ever-increasing problems of where to develop new housing. In these areas, direct population pressure is not the chief issue, but instead problems of supply and demand associated with uneven development.

15.2 ANALYSIS OF THE SYNDROME APPROACH

This book has made considerable use of the syndrome concept of global change as a basis of characterizing a number of human–environmental relationships, and has shown a syndrome approach might provide a starting point from which to build an understanding of physical geography from a human perspective. The WGBU (1996) argue that their view of a holistic Earth system is an essential prerequisite to elucidating key processes influencing global change and their subsequent management. By modelling relationships in particular syndromes, they argue, it is possible to gear research activity towards solving identified problems. In doing this, the process of research would go through a number of stages: initial problem analysis; development of guiding principles; identification of responsible bodies and instruments; implementation and decision-making and risk management analysis. To facilitate this approach, the WGBU have proposed a framework for problem analysis that enables a series of relationships and connections to be built which reflect the complexity, interactivity and inherently multi-scaled nature of global change issues (Lüdeke *et al.* 2002). An example of the global trends and interrelationships associated with the Favela syndrome is shown in Figure 15.5. This shows the complexity of the problem but also maps

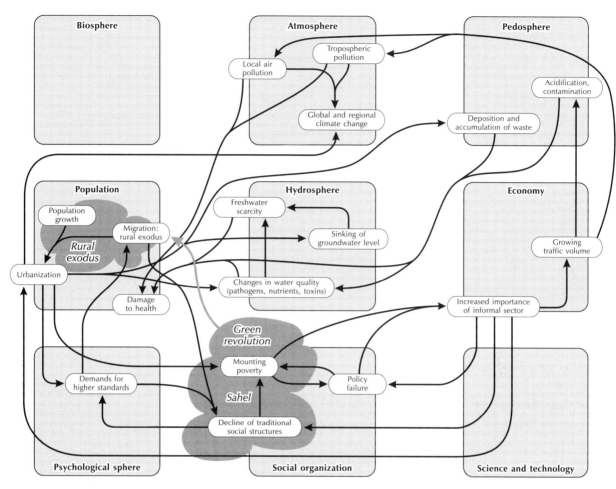

Figure 15.5 Network of relationships for the Favela syndrome. *Source:* After WGBU (1999)

out where changes are likely to result in impacts, so that these can be fully considered.

The value of the syndrome approach has been recognized by a number of other organizations as well as the WGBU. For example, in a report recently produced for the European Commission, the environmental syndromes approach has been used as a means of identifying syndromes relevant from a European perspective (Pfaul *et al.* 2000). Table 15.2 shows some of the key issues from a western European and southern European perspective.

Of course, the syndrome approach does not have universal acceptance. For example, Krings (2002) provides a critical review of the Sahel syndrome as presented by the WGBU in 1996, contending it provides an erroneous picture of the central causes of the syndrome. He argues that population pressure and associated agricultural intensification and the development of so-called 'vicious circles of land degradation' (in particular marginal agrarian areas with few alternative means of livelihood for the subsistence resident population) omits an important consideration, that of the power-relations in the area that help determine patterns of land use and misuse. This sort of comment and discussion on the nature and workings of syndromes is important to help further develop understanding.

Table 15.2 Environmental syndromes from a European perspective, examples of the western and southern European regions

Environmental issue	Associated syndrome
Western Europe	
Trans-boundary air pollution	Smokestack
Loss of biodiversity and landscape degradation	Dust Bowl; Mass Tourism; Aral Sea; Urban Sprawl
Depletion of fish stocks	Overexploitation
Industrial pollution	Waste Dumping; Contaminated Land
Transportation of dangerous goods	Contaminated Land
Water quality	Aral Sea
Glacier melting	Smokestack
Traffic	Smokestack
Southern Europe	
Shortage of fresh water	Overexploitation; Rural Exodus
Drought, desertification and soil erosion	Sahel; Rural Exodus; Aral Sea
Deforestation (fires)	Overexploitation; Major Accident
Hazardous waste disposal	Waste Dumping; Contaminated Land
Loss of biodiversity	Sahel; Overexploitation; Dust Bowl
Mass tourism and coastal vulnerability	Mass Tourism
Natural disasters	
Marine pollution and fish stock depletion	Overexploitation
Agro-industries and genetic engineering	Dust Bowl; Aral Sea

Another issue relates to the need to balance an analysis of problems with a similar analysis of solutions. Where successful management and control at a range of scales can be characterized and analysed in a similar way to the environmental problems, it may provide the basis for replicating positive aspects of human–environment interactions to other global issues.

PROSPECTS FOR GLOBAL MANAGEMENT

The syndrome approach, then, provides one mechanism through which effective global management can be promoted. However, there is still a considerable way to go to meet the challenges of global change. As Schellnhuber (1998) points out, although the UN Framework Convention on Climate Change (Kyoto agreement) has been a celebrated example of successful international cooperation on a critically important environmental issue, the agreed measures would only result in a reduction of global warming by around 0.1°C at the end of the twenty-first century under the business-as-usual scenario. He argues that only a reduction in greenhouse gas emissions nearer 80 per cent by 2050 would give sufficient climate stabilization to negate the need for considerable adaptation measures and reduce the likelihood of any non-linear environmental 'surprises' such as alterations to monsoon events, El Niño events and the oceanic 'conveyor belt' (Broecker 1995; Schellnhuber 1998). Given the current withdrawal of the US (the largest emitter of total and per capita greenhouse gases) from the Kyoto agreement, it has to be said that, unfortunately, the likelihood of achieving this level of stabilization looks increasingly remote. Nevertheless, there is much international consensus elsewhere on the need to provide meaningful practical measures to try to manage and control global change (under sustainable development) as well as adapting to it. This is the spirit of Agenda 21 and the 'think global act local' mantra which has gained so much widespread support.

Table 15.3 identifies a number of 'top missions' for global managers for the twenty-first century. Focuses 15.1 and 15.2 give two different perspectives on tasks that can and are being undertaken in the pursuit of these goals. It is increasingly recognized that further scientific research will be necessary to meet the grand challenges ahead and to underpin global management activity in the quest for solutions. In the words of Kofi Annan, Secretary-General of the United Nations 'It is impossible to devise effective environmental policy unless it is based on sound scientific information' (April 2000 Millennium Report to the United Nations General Assembly quoted in WRI (2003b)). The relationship between humankind and its environment is complex but understanding this is an essential goal for research endeavour. Although our knowledge is growing, there are still many gaps to fill, both in terms of measuring and monitoring and also in terms of our understanding of the drivers and processes for change in which geographers, both physical and human, should have an important input.

Table 15.3 Top missions for global managers in the twenty-first century

Basic development targets	Core environmental factors	Grand challenges
Security	Climate	Risk management
Health	Water	Resource efficiency
Food	Soils	Renewable energy
Habitation	Bio-potentials	Sustainable land use
Mobility	Mineral resources Energy	Re-inventing cities and the countryside Genetic protection

Source: After Schellnhuber (1998)

FOCUS 15.1

FILLING DATA GAPS – THE MILLENNIUM ECOSYSTEM ASSESSMENT PROJECT

In the light of global environmental concerns, the World Resources Institute (WRI) has initiated a full scientific assessment of the world's major ecosystems. The Millennium Ecosystem Assessment, which published its framework for assessment in 2003, is a large international effort to provide the basis for improved decision support in the use of ecosystem resources. It focuses on finding the best use of ecosystem services (direct food, fuel and fibre, and indirect climate and disease regulation, aesthetic and cultural roles) without irreversible ecosystem degradation, which would then threaten the future provision of ecosystem services and, ultimately, human wellbeing (WRI 2003).

FOCUS 15.2

RE-INVENTING THE CITY – CITIES AS SOLUTIONS

The UN Centre for Human Settlements (1999) argues that the negative impacts of cities on the environment are exaggerated and, rather, cities have been fundamental in providing economic security to nations and have underpinned improvements in living standards across the globe. They cite patterns of family size and a slowing of population growth as positive outcomes from the adoption of an urban way of life. Other advantages of cities are given as:

- Higher densities of people results in lower cost per household in the provision of vital infrastructure and services such as healthcare

- Concentration of production and consumption gives rise to the potential for increased efficiency in the use and re-use of resources, for instance through recycling schemes

- Urban dwellers demand less land than rural dwellers. In most countries, although urban dwellers are in the majority, they take up only around 1 per cent of the surface area of their nation

- Concentration of home and work places gives rise to the potential for economies of scale in energy use

- Concentrations of home and work places gives rise to the potential for less travel and the use of more sustainable forms of travel such as public transport systems

- Cities are centres of cultural development and expression.

Undoubtedly, there are some powerful arguments in favour of the city as a valuable form of human settlement. However, the real value of what the UN Centre for Human Settlements describes is not in denying the negative aspects of urbanization on the environment and the influence it has in key physical and chemical processes underpinning physical geography, but rather in the *potential* that the city holds in providing a better future. It is for this reason that many researchers, urban planners and environmental managers are working to try to guide the development of the sustainable city.

SUMMARY

People are the ultimate driving force of current environmental change. Future trends of population, urbanization, energy and resource use, pollution and quality of life are essential to issues of conservation and sustainability. Predicting trends in these factors is no easy task and requires a deep understanding of the societary – as well as biological, physical and chemical – pressures that drive them. In this area of research, physical geography comes face to face with human geography and allied disciplines (psychology, sociology, economics and governance). The two groups should depend on each other: physical geographers need to be aware of the socio-cultural and economic drivers of change studied by human geographers; equally, human geographers need the physical geographers' knowledge of drivers in the natural world. The process of economic globalization, which is chiefly driven by complex and ever-changing patterns of supply and demand for goods and services, is one example of this. Recognition of the symptoms characterizing the environmental syndromes and their interrelationship to human society and economy forms the foundation of a sustainable approach to development. Although not without its critics, the syndrome approach offers a useful means of categorizing, modelling and tackling environmental issues, and a productive device through which to promote effective global management in a framework that spans local and global activity.

Questions

1 Compare and contrast different views of the city.
2 Outline the expected trends in population growth, urbanization and health in the twenty-first century.
3 What role might the physical geographer play in global management?

Further Reading

Alverson, K. D., Bradley, R. S. and Pedersen, T. F. (eds) (2003) *Paleoclimate, Global Change and the Future.* Berlin and Heidelberg: Springer.
A synthesis of ten years of Past Global Change (PAGES) research.

Blowers, A. and Hinchliffe, S. (eds) (2003) *Environmental Responses.* Chichester: John Wiley & Sons and The Open University.
An informative read.

Brown, L. C. (2003) *Plan B: Rescuing a Planet in Stress and a Civilization in Trouble.* New York: W. W. Norton.
Another scheme for saving the world. Well worth reading.

Hinchliffe, S., Blowers, A. and Freeland, J. (eds) (2003) *Understanding Environmental Issues.* Chichester: John Wiley & Sons and The Open University.
An interesting set of offerings.

Hinchliffe, S. and Woodward, K. (2000) *The Natural and the Social: Change, Risk and Uncertainty.* London: Routledge and The Open University.
Delves more into the human side of the environmental equation. Develops some of the ideas raised in Physical Geography.

Lomborg, B. (2001) *The Skeptical Environmentalist: Measuring the Real State of the World.* Cambridge: Cambridge University Press.
An alternative argument that should be read.

Mannion, A. M. (1997) *Global Environmental Change: A Natural and Cultural Environmental History*, 2nd edn. Harlow: Longman.
An ideal book for students.

Morris, R., Freeland, J., Hinchliffe, S. and Smith, S. (eds) (2003) *Changing Environments.* Chichester: John Wiley & Sons and The Open University.
Prosecutes an interdisciplinary approach. Excellent.

WGBU (Germany Advisory Council on Global Change) (1996) *Annual Report World in Transition: The Research Challenge.* Berlin: Springer. [Available at www.wgbu.de/wbgu_jg1996_engl.html]
If you have not already had a look at this document, now is a good time to do so.

Useful Websites

Some useful web resources for further reading and research can be found at the following locations:

http://www.census.gov/ipc/www/
US Bureau of Census. Investigates trends and projections for different countries and regions of the world as well as characteristics and trends of the US population itself. This site also includes a world population projection 'clock' http://www.census.gov/cgi-bin/ipc/popclockw

http://www.un.org/esa/population/publication/six billion/htm
UN Department of Economic and Social Affairs – Population Division. Data and analysis reflecting on the world passing the six billion population mark.

http://www.unfpa.org/swp/swpmain.htm
United Nations Population Fund (UNFPA) State of the World Population 2003. Online report.

http://www.wri.org
World Resources Institute. Examples and trends about global, regional and national resource use and the state of the world ecosystems.

http://www.unece.org/stats/trend/trend_h.htm
UNECE Trends in Europe and North America: The Statistical Yearbook of the Economic Commission for Europe 2003. Country profiles for trends in a wide range of sectors, including population, households, education, employment, health, energy, crime and environment. Also lists further Internet resources.

http://www.unchs.org
United Nations Centre for Human Settlements. Includes news, campaigns, publications and events.

APPENDIX 1:
UNITS OF MEASUREMENT

Investigating the environment inevitably involves measuring physical quantities such as length, time, mass, area, volume, density and temperature. Most physical quantities can be expressed in terms of four selected quantities – mass (M), length (L), time (T) and temperature (θ). There are several systems of arbitrarily chosen standard units in which the magnitude of these quantities are expressible. This book uses the *Système Internationale d'Unités* (SI).

BASE UNITS

The base units, and their abbreviations, in the SI system are:

Quantity	Unit	Symbol
length	metre	m
mass	kilogram	kg
time	second	s
temperature	Kelvin	K
electric current	ampere	A
luminosity	candela	cd
amount of substance	mole	mol

Kilogram and metre may be spelt kilogramme and meter.

DERIVED UNITS

Several units may be derived form the base units. Examples include area, volume, density, velocity, acceleration, force, pressure, heat energy and heat flux.

Force

The SI derived unit is the Newton, N, defined as the force necessary to impart and create acceleration of 1 metre per second per second (1 m/s^2) to a mass of 1 kg.

Pressure and stress

The SI derived unit of pressure and stress is the pascal, Pa, defined as the pressure exerted by a force of 1 N evenly distributed over 1 m^2. Atmospheric pressures are conventionally expressed in bars or millibars (1 bar = 10^5 Pa = 10^5 N m).

Work and energy

The derived SI unit for all forms of energy is the joule, J, defined as the energy needed to displace a force of 1 N through a distance of 1 m. The joule is sometimes called the Newton-metre, N m. The calorie, cal, is still occasionally used in biology and ecology (1 cal ≈ 4.18 J; 1000 cal = 1 kcal = 1 Cal).

Power

The derived SI unit is the watt, W, defined as the power required to equal the rate of working of 1 J/s.

MULTIPLES AND SUBMULTIPLES

A large number of multiples and submultiples are used with metric units (Table A.1). The SI system recommends that only the multiples and submultiples of a thousand are used (10^3, 10^{-3}, 10^6, 10^{-6}, 10^9, 10^{-9}, etc.), and discourages the use of the prefixes hecto, deca, deci, centi and myria. However, these 'banned' prefixes can be useful and are sometimes used.

UNITS COMMONLY USED IN PHYSICAL GEOGRAPHY

Length

The basic unit of length in the SI system is the metre. In geography, the kilometre is the commonest multiple. A variety of submultiples are employed (Figure A1. Note that the micrometre, μm, is often called a micron).

Area

Area has the dimensions of length squared (L^2). The basic unit of area in the SI system is the square metre, m^2. Commonly used multiples and submultiples are

Table A.1 Metric multiples and submultiples

Prefix	Symbol	Scientific notation	Decimal notation	Description
exa	E	10^{18}	1,000,000,000,000,000,000	–
peta	P	10^{15}	1,000,000,000,000,000	–
tera	T	10^{12}	1,000,000,000,000	trillion
giga	G	10^9	1,000,000,000	billion
mega	M	10^6	1,000,000	million
myria	my	10^4	10,000	ten thousand
kilo	k	10^3	1000	thousand
hecto	h	10^2	100	hundred
deca (deka)	da	10^1	10	ten
deci	d	10^{-1}	0.1	tenth
centi	c	10^{-2}	0.01	hundredth
milli	m	10^{-3}	0.001	thousandth
micro	μ	10^{-6}	0.000,001	millionth
nano	n	10^{-9}	0.000,000,001	billionth
pico	p	10^{-12}	0.000,000,000,001	trillionth
femto	f	10^{-15}	0.000,000,000,000,001	–
atto	a	10^{-18}	0.000,000,000,000,000,001	–

Note: The terms billion and trillion (and billionth and trillionth) here follow the US definitions, which have become widely accepted. In the UK, a milliard is a thousand million (10^9), and a billion is a million million (10^{12}). Similarly, in the UK, the terms millardth and billionth denote a thousand millionths and a million millionths, respectively.

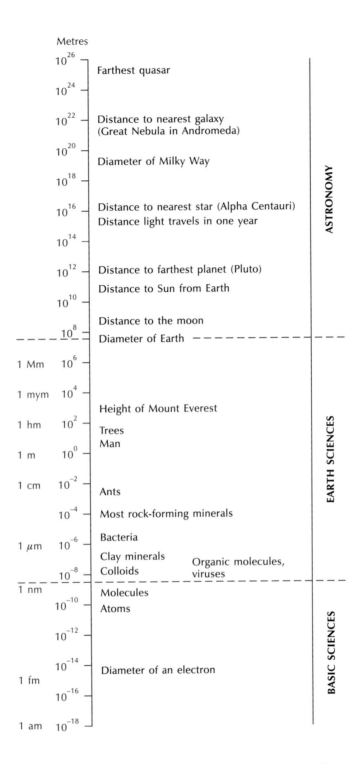

Figure A1 The length dimension

the square kilometre, km², and the square centimetre, cm². In land measurement, the area, a, and hectare, ha, are frequently adopted as units (1 a = 100 m²; 1 ha = 10,000 m²).

Volume

Volume has the dimension of length cubed (L^3). The basic unit of volume in the SI system is the cubic metre, m³. Commonly used multiples and submultiples are the cubic kilometre, km³ and the cubic centimetre, cm³ (sometime written cc).

Fluid volumes are often expressed as litres, l, or millilitres, ml (1 l = 1 dm³ = 1 cm³).

Mass

Mass is the quantity of matter that a given object contains. The basic SI unit of mass is the gram (sometimes spelt gramme). In physical geography, commonly used multiples and submultiples for small quantities of mass are the milligram, mg, and microgram, µg (Figure A2). Intermediate masses are expressed in kilograms, kg, or tonnes, t (1 t = 1000 kg = 1 Mg). For the enormous stores of materials in the ecosphere, high multiples are helpful, including megatonnes, Mt, teragrams, Tg, and gigagrams, Gg. For clarity, it is worth denoting the type of mass being referred to. So, when describing the amount of carbon in a store, g C

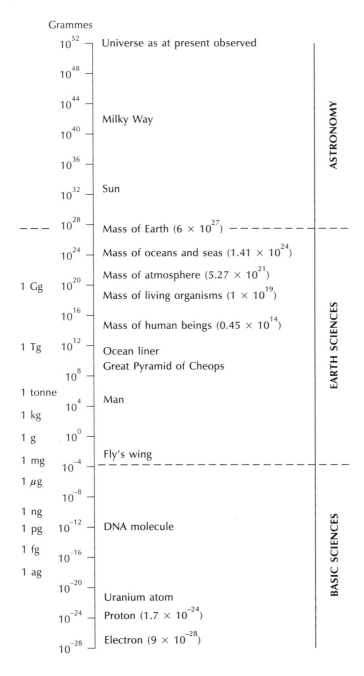

Figure A2 Mass in the environment

can be used as a shorthand for grams of carbon, and Gt C for gigatonnes of carbon.

It is hard to visualize vast quantities of materials stored in the environment. To put some figures in perspective, it might be helpful to think of blocks of ice. A cubic metre of ice has a mass of 1 t (assuming its density is 1 g/cm³). A block of ice 100 m long, 100 m wide, and standing 100 m high would weigh 10^{12} g = 1,000,000 t = 1000 Gg = 1 Tg. A cubic kilometre of ice would weigh 10^{15} g = 1,000,000,000 t = 1,000,000 Gg = 1000 Tg = 1 Pg.

Density

This is mass per unit volume, unit area, or unit length; for example, kg/m³.

Concentration

This is the amount of a given substance in a unit amount of another substance. For material in solution, it may be expressed as the number of moles of solute in a litre of solvent. Solute concentrations may be expressed as grams per litre, g/l. When concentrations are very small, it is common to use one of the following expressions: parts per thousand (ppt), parts per million (ppm) and parts per billion (ppb). Parts per million means the weight of solute in a million parts of solution. To convert g/l to ppm, multiply by 1000. For example, 0.99 g/l = 990 ppm. If the concentration is measured on a volumetric basis, as is commonly the case with atmospheric constituents, then this fact should be denoted by writing the concentrations as parts per million by volume (ppmv). In geochemistry, concentrations are commonly expressed as milligrams per kilogram (mg/kg).

To denote concentrations of chemical species, the letter p (from the German potenz, meaning power) may be used. Everybody is probably familiar with the expression pH, which means the concentration of hydrogen ions. The letter p may be used to signify the concentration of any substance; for example, pCO_2 stands for the concentration of carbon dioxide.

Mass flow

This is defined as either mass per unit item or volume per unit time. River discharge, for instance, is commonly measured as cubic metres per second (m³/s), also styled cumecs, while the flow of suspended sediment in a river is measured as tonnes per year (t/yr). It is occasionally necessary to measure the density (intensity) of mass flow. This is the mass flow rate per unit area, and would be expressed in such units as kg/m²/s.

Energy flow

This is defined analogously to mass flow. For instance, heat flow rate may be expressed as J/s (= W). The density of heat flow rate would be expressed as J/m²/s (= W/m²). In meteorology, the Langley, ly, was once widely used (1 ly = cal/cm² ≈ 4.18 J/cm²).

NB. The use of more than one solidus (/) in an expression is nowadays generally frowned upon. However, in most systems of interest to physical geographers, density of mass (and energy) flow has traditionally been written in a form such as kg/ha/yr. This kind of expression is unambiguous enough and seems acceptable. Pedantic scientists would prefer to see the less elegant alternatives – kg/(ha yr) or kg ha^{-1} yr^{-1}.

APPENDIX 2:
DATING ENVIRONMENTAL MATERIALS

In the last 30 years, and even the last decade, our understanding of environmental change over time periods for which we have no human records has been revolutionized by developments in scientific dating techniques. Human records still give the most precise dates – no scientific dating method will tell you that the tsunami struck at 3.30 am on 27 October – but the precision and accuracy of other methods has been considerably refined, and the number of techniques available expanded.

Knowing when an event happened or a sedimentary unit was laid down is fundamental to the understanding of environmental change. We need to understand rates of change in order both to interpret what happened in the past and to predict what may happen in the future. Dating also allows records of change that are not actually in the same place to be tied together. For example, it is important when producing a climate model to know what happened in low latitudes when glaciers expanded in the high latitudes; the only way to associate two geographically separate records is to date both and then make informed comparisons.

The types of dating method available can be divided into relative, absolute and correlation methods. Relative dating can tell you that one unit or landform is older than another. Absolute dating gives an age in years. Correlation connects different units or landforms using some event that is recorded at those different places, such as deposition of volcanic ash layer from a particular eruption. If the common event can be absolutely or relatively dated, this age can then be applied to all the sites at which it is identified.

One problematic aspect of dating is that it is much easier to date periods of deposition than of erosion or of cessation of deposition, both of which are recorded as gaps in the sedimentary record. There is a wide range of dating methods that can be used for different types of deposit. However, absolute dating of erosion has until recently relied either on the relative dating of weathering or soil development on the exposed surface, or on getting a limiting age by dating more recent material which has been deposited on top of that eroded surface.

Three main questions help you to choose the right dating method:

- What technique will work on the material which you have available?

- What is age range over which you need to date?

- How much money do you have available to spend?

The first two are discussed below for a variety of individual methods and age ranges and are illustrated in Figure A3. Finance is obviously dependent on the project in hand. Some methods are more expensive than others, depending on the level of technology and the time taken to calculate an age.

The following is not a comprehensive list of dating techniques, but covers those most used by physical geographers.

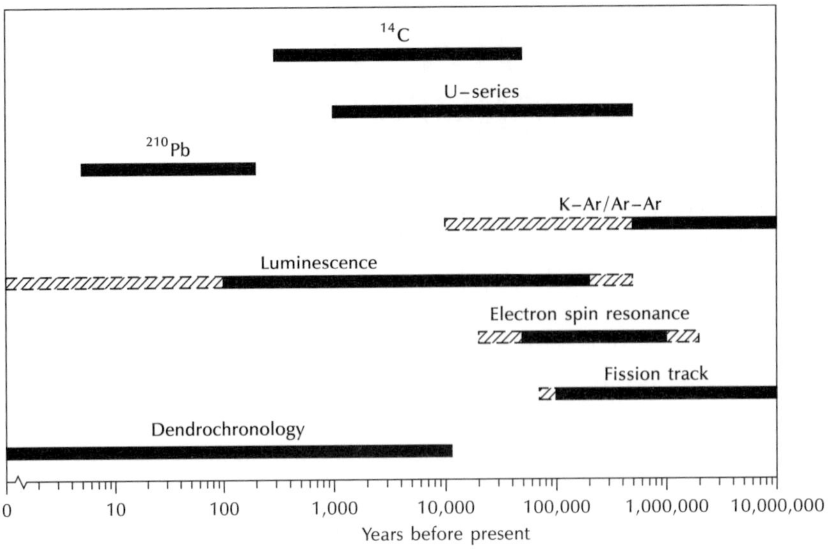

Figure A3 The age ranges over which a variety of absolute dating methods are effective. Filled bars indicate the time ranges over which a method is usually applicable; hatched bars indicate where it may be applicable in favourable circumstances

RELATIVE DATING

Superposition

The most fundamental way of distinguishing ages is known as the law of superposition: if one deposit is found on top of another, the lower one is usually older and the upper one younger (Figure A4). There may be gaps in the record due to erosion or lack of deposition during a time interval, but this would not affect a simple older/younger age being assigned. Obviously, there are situations in which superposition does not work in a simple way – for example where tectonic activity has displaced units relative to one another – but it is a good initial working hypothesis

Weathering and soil development

Weathering and soil development both occur over time, so the degree to which they have occurred can be used to indicate how long a surface has been exposed. It may be possible to create calibration curves in a particular environment by dating a range of soils or weathered profiles at different degrees of development. Clearly, this would only work within a limited geographical area since both soil development and weathering are so dependent on climate and substrate. Relative or even calibrated ages obtained in this way are unlikely to be very precise, but the technique can be useful where no others can be used – a particular problem where eroded surfaces are involved.

Lichenometry

Lichens are among the first living things to colonize exposed surfaces or deposited sediments. They provide a way of dating the stabilization of a surface (Noller and Locke 2000). The concept behind lichenometry is very simple: that lichens grow with age, so there is a relationship between the size of the lichen and its age (and thus the age of the surface on which it is growing). Different species of lichen grow at different rates and are more or less suitable for dating because of their form (Innes 1985).

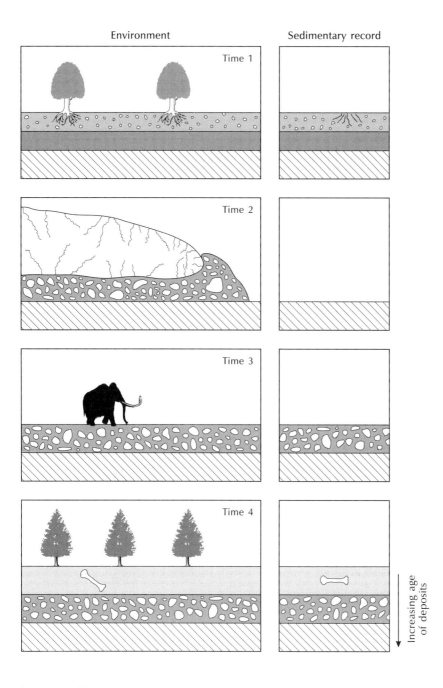

Environment

Sedimentary record

Time 1

Time 2

Time 3

Time 4

Increasing age
of deposits

Figure A4 The most basic law of dating is that of superposition: in general, younger material is found above older. A sequence of events as shown here (a landscape which undergoes glaciation, then a postglacial warming through permafrost to forest) will lead to the deposition of different kinds of sediment, but the oldest will be at the base and the youngest at the top. There may be gaps in the sequence caused by erosion (for example, glacial stripping of some units, as at Time 2) or by a break in deposition. This can become complicated by, for example, tectonic activity leading to folding or faulting, so that units move relative to one another, or by the emplacement of volcanic rocks within the sequence rather than at the surface. However, superposition is the first working hypothesis when trying to interpret the timing of a sequence of sediments or rocks

The basis of lichenometry is as a relative dating tool – that a surface with larger lichens is older than one with smaller lichens. This is only applicable within a restricted area – lichens respond to their ecological environment, and there are considerable differences in growth rate in different areas (Lowe and Walker 1997). The method can be used to give absolute ages for a surface if a calibration curve is constructed. In order to do this, lichens are measured on a wide range of known age surfaces, either human constructions like houses, gravestones and walls, or natural surfaces which have been dated using other methods (Innes 1985). The age of other surfaces in that area can then be determined by measuring the sizes of particular species of lichen.

This is a useful method to apply to eroded surfaces, and has been widely applied to glacial moraines. It is also fairly cheap and simple to apply (although a good sampling strategy is critical). However, it can be difficult to obtain a precise age. The growth rate of lichens is not only dependent on their age, but on the changing environments in which they are growing (Innes 1985). Lichen growth rates will alter according to the local climate, pollution, snow cover and overshadowing by vegetation, among other factors. There is also an issue about colonization – this may occur some time after the surface is exposed (Innes 1985). It has also been noted that active glacial moraines may have lichens already growing on them (Lowe and Walker 1997). Dates are thus not likely to be very precise.

The age range over which lichenometry can be applied varies according to the environment. Species like *Rhizocarpon geographicum* can continue to grow for thousands of years in a cold, dry climate (Lowe and Walker 1997; Noller and Locke 2000). However, in more humid climates, an upper limit of growth may be reached within 500 years.

ABSOLUTE DATING

Radiocarbon

Radiocarbon dating is the best-known and most widely used scientific dating method for the late Quaternary (Taylor 1997). It is based on the radioactive decay of ^{14}C (carbon-14), hence its other name, ^{14}C dating. ^{14}C is created by the action of cosmic rays in the atmosphere (Trumbore 2000), where it rapidly oxides first to carbon monoxide (^{14}CO), then to carbon dioxide ($^{14}CO_2$), diffuses through the atmosphere and enters the global carbon cycle (Taylor 1997; Trumbore 2000).

There are three **isotopes** of carbon, which occur in different proportions in nature. The most common is ^{12}C (carbon-12), which makes up around 98.89 per cent of all carbon. ^{13}C (carbon-13) makes up around 1.11 per cent; ^{14}C makes up less than a billionth of one percent of the carbon found on Earth (Trumbore 2000). ^{12}C and ^{13}C are both stable isotopes – they are not radioactive. ^{14}C is radioactive (hence 'radiocarbon') and has a **half-life** of 5730 ± 40 years. This means that half of any amount of ^{14}C will have radioactively decayed to form the stable isotope ^{14}N (nitrogen-14) in 5730 years.

All three isotopes of carbon are found in living tissue, as all form part of the carbon cycle. This is why the decay of ^{14}C can be used as a dating method. When a plant or an animal is alive, it takes in ^{14}C (along with ^{12}C and ^{13}C), for example, plants take it in as carbon dioxide (CO_2) during photosynthesis. When an organism dies, the ^{14}C that is within the tissue undergoes radioactive decay and is not replaced by new ^{14}C. Thus, in order to tell how long it is since an organism died, you measure the remaining amount of ^{14}C in the wood, bone, shell, etc. and compare that with the amount of ^{14}C that there would be in a living organism, and the amounts of the stable forms of carbon. The maximum age that can be obtained by radiocarbon dating relates to the smallest amount of ^{14}C that can be detected. In theory, ages can be obtained of up to around 75,000 years if large enough samples are available, but in practice you need to be very careful about ages above around 40,000–50,000 years, since any contamination with younger carbon would form a much higher proportion of the final results obtained (Taylor 1997). For example, a contamination of only 0.1 per cent modern carbon in an infinitely old sample could yield an age of around 55,000 years (Bradley 1999).

isotopes: isotopes of an element have the same chemical properties, but differ from each other in the number of neutrons in the nucleus, so have different atomic masses. They may behave differently in their nuclear properties, so one isotope of the same element could be radioactive while another is stable

half-life: the half-life ($t_{\frac{1}{2}}$) of a radioactive isotope is the average time it takes for half of the atoms present to undergo radioactive decay. This does not depend on the amount that was there to begin with, so if an isotope had a half-life of 5 years, the initial quantity would be reduced to half after 5 years, to a quarter after 10 years, to an eighth after 15 years, etc.

Radiocarbon ages cannot be directly compared with those from other dating methods, including historical records. There are two main reasons for this:

1. The half-life of ^{14}C that was originally used (5568 years) has now been re-evaluated to 5730 years. The older 'Libby' half-life is still used in the determination of dates. This is because a large number of radiocarbon dates had already been published when the re-evaluation was made, and it was important to avoid confusion by having consistency between the generations of dates (Aitken 1990).

2. One main assumption originally made for radiocarbon dating was that ^{14}C concentration has remained constant in different parts of the carbon cycle ('reservoirs') over time (Taylor 1997). This, however, is not the case, due both to human and natural effects. The amount of ^{14}C in the atmosphere was severely affected by two recent human effects: the industrial revolution, which released large amounts of 'old' carbon through the burning of fossil fuel, and atmospheric nuclear testing. These two effects together mean that radiocarbon dating cannot be used for very recent material, and has an effective minimum age for dating of around 300 years (Taylor 1997). There are also longer term, natural variations in ^{14}C levels in the atmosphere, which relate, at least in part, to the level of cosmic rays which reach the atmosphere and create ^{14}C (Taylor 1997).

Because of these issues, radiocarbon ages need to be calibrated in order to make them comparable with other techniques and with solar/calendar years. It is critical to discover whether a ^{14}C date has been calibrated before making any such comparison (Bartlein *et al.* 1995). In order to create data sets through which radiocarbon ages can be calibrated, comparisons are made between radiocarbon ages and ages obtained on the same samples using other methods that do not suffer from the same deviations from calendar years (see, for example, Stuiver *et al.* 2003). Dendrochronology is a particularly good way of making such comparisons, since it will give an age to a one-year precision and the wood may be dated by radiocarbon. Currently, dendrochronology provides radiocarbon calibration over the last 11,000 years (ORAU 2004). This calibration has been extended to 24,000 years using a variety of records including marine **varves** and uranium-series dating of corals (Stuiver *et al.* 1998; 2003).

> **varves**: annually laminated sediments that were deposited through water. They are usually deposited in lakes, but may also be marine

Radiocarbon dating can be carried out through counting of the radioactivity of the samples. This is termed conventional radiocarbon dating. During the 1970s, advances in mass spectrometry technology allowed the development of accelerator mass spectrometry (AMS) radiocarbon dating. In this technique, the amount of ^{14}C is measured directly. This is a faster and potentially more precise technique, which can work with smaller sample sizes, but is more expensive and requires the use of a nuclear accelerator (Taylor 1997). Although it is very useful, therefore, conventional techniques are still widely used. It was hoped that AMS dating would also mean an increase in the maximum age of samples that could be dated. However, there is still a severe practical limitation due to micro-contamination (Taylor 1997).

Uranium series

Uranium series dating depends on the radioactive decay chains that start with isotopes of uranium, in particular uranium-238 (^{238}U) and uranium-235 (^{235}U)

and end with stable (non-radioactive) isotopes of lead (Pb) (Schwarcz 1997; Lowe and Walker 1997):

$$^{238}U\rightarrow^{234}Th\rightarrow^{234}Pa\rightarrow^{234}U\rightarrow^{230}Th\rightarrow^{226}Ra\rightarrow\ldots\,^{210}Pb\rightarrow^{210}Bi\rightarrow^{210}Po\rightarrow\,^{206}Pb$$
$$^{235}U\rightarrow^{231}Th\rightarrow^{231}Pa\rightarrow\ldots\,^{207}Pb$$

The uranium decays, forming thorium (Th) and protactinium (Pa), which as they are created undergo radioactive decay themselves. After millions of years, the levels of all the parts of the decay chain in a closed system will reach what is termed secular equilibrium – that is, the daughter isotopes will be decaying at the rate at which they are supplied by the decay of the parent isotope (Ku, 2000). Before this point is reached, however, it is possible to determine how long it is since the system was formed and the decay chain began. In order to calculate this, we need to know three key things (Schwarcz 1997):

1.   The half-lives of the isotopes involved.

2.   The starting point of the system – that is, how much there was of each isotope present when the mineral to be dated formed.

3.   That the system has remained closed since its formation – i.e. that it has not gained or lost any of the isotopes from an outside source, but only through the radioactivity of the isotopes that were present to begin with.

The age is then determined by using the ratios between reasonably long-lived members of the decay chain, such as the $^{230}Th/^{238}U$ or $^{231}Pa/^{235}U$ ratios (Bradley 1999).

The first of the three key factors is satisfied, as the half-lives of the different isotopes have been determined. The second could potentially pose more difficulties, but is not usually a problem for minerals that have formed from solution. Uranium ions are very soluble in water whereas those of thorium and protactinium are not (Schwarcz 1997). This means that minerals that form from solution, in particular calcium carbonates, may have small amounts of uranium present when they form, but will not contain thorium or protactinium. This gives a known starting point, with the parent isotope present but not its daughters.

The materials to which U-series dating is most applicable are thus organic and inorganic carbonates. This includes coral, which obtains its carbonate content from water, and inorganically precipitated carbonates such as stalagmites and stalactites and soil carbonates (Ku 2000). Often, the precipitated material is not pure, but rather contaminated with some sedimentary detritus when it forms. In this case, there may be daughter isotopes present in the mixture at the time of formation, thus invalidating the assumption of zero thorium and protactinium at formation. The first way of combating this is through careful sampling and avoidance of obviously contaminated material. After this, the problem can be detected and may be corrected for through the presence of ^{232}Th, which has a very long half-life and is not a daughter product of any other isotope.

The third key factor is that the systems must have remained closed since they were formed. This means that, for example, coral that has recrystallized will not be datable, nor will other samples which have exchanged radioisotopes with external systems (e.g. groundwater) since they were formed (Ku 2000).

U-series ages are calculated in two main ways: by alpha spectrometry, which measures the radioactive decay as it occurs; and by TIMS – thermal ionization mass spectrometry (Schwarcz 1997). TIMS is faster, requires smaller samples and gives better precision, as it is a direct measurement of the isotopes present. However, there are more facilities available for alpha spectrometry at present.

The age range and precision of the technique depends on the ratio of isotopes chosen and the measurement technique used. At best, using TIMS, it can obtain ages up to around 500,000 years, with a precision of 1 per cent (Schwarcz 1997). Using alpha spectrometry, the limit is around 350,000 years, and the precision 5–10 per cent (Schwarcz 1997).

Lead-210

Lead-210 (^{210}Pb) is a radioactive isotope that forms part of the decay chain of uranium-238 (^{238}U). It has a half-life of around 22 years (Faure 1986). Models have been created that allow you to calculate the amount of ^{210}Pb which would have been derived from the atmosphere and the geology and would have been present in the sediments when they were laid down (Noller 2000). The decay from this starting point can then be used to date the sediments (and sometimes ice in glaciers and icecaps). The technique has been applied most widely to lake sediments.

Lead-210 dating is a good complement to radiocarbon dating (Roberts 1998) as it covers a time period younger than that for which radiocarbon is useful: the optimum time range is 5–100 years, and it can be applied to sediments up to around 200 years old (Noller 2000).

Potassium–argon and argon–argon

Potassium–argon (K–Ar) and argon–argon (Ar–Ar) dating, like uranium-series, use an accumulation clock of radioactive decay, as opposed to the decay clock used by radiocarbon dating (Walter 1997). In the study of Quaternary environments, K–Ar and Ar–Ar dating are applied most frequently to volcanic material such as lava flows and tephra (volcanic ash). They are based on the decay of potassium-40 (^{40}K) to argon-40 (^{40}Ar). Potassium has three naturally occurring isotopes: ^{41}K, ^{40}K and ^{39}K, the relative amounts of which are very constant (Renne 2000). Of these isotopes, only ^{40}K is radioactive, decaying in two different ways, to form either ^{40}Ca or ^{40}Ar, with a half-life of 1250 million years (Renne 2000).

As with uranium series dating, in order to produce a valid date, it is necessary to know the half-lives of the isotopes involved, the starting point of that system, and for the system to be a closed one, with no inputs or outputs from other sources once it has formed. In the case of K–Ar, the initial starting point of the system is not usually a problem. Argon will diffuse out of an igneous or metamorphic rock as it cools after forming, meaning that the effective starting point is with only the parent isotope, ^{40}K, present (Walter 1997).

In order to apply K–Ar dating, a sample is split into two subsamples ('aliquots'), one of which is used to measure the potassium, and the other the argon. Total potassium is usually measured using flame photometry or atomic absorption, and the amount of ^{40}K is then calculated from the general isotopic composition, given above (Walter 1997). The level of ^{40}Ar is then determined using mass spectrometry.

Ar–Ar dating is another way of determining the ^{40}K–^{40}Ar ratio. In this case, samples are placed in a nuclear reactor before measurement and bombarded with high-energy neutrons in order to convert the ^{39}K to ^{39}Ar. ^{39}Ar does not occur naturally, so it can be used as a way of measuring the potassium content, from which the ^{40}K content can be calculated because the ratios between the different isotopes are so constant (Walter 1997). This may seem like a rather roundabout way of approaching the measurement, but it brings several advantages. Firstly, the measurements are made at the same time using the same equipment, so there is less variability due either to equipment or to differences between aliquots. In addition, less material is needed, and there is no need to determine absolute quantities of argon isotopes, only the ratios between the different isotopes (Renne 2000; Walter 1997).

The lower limit of applicability of K–Ar and Ar–Ar dating is one of detection of the tiny initial amounts of argon. This limit will depend in part on the potassium content of the material to be dated, and may be as low as a few thousand years, or up to around 500,000 years, if very small samples are used (Renne 2000; Walter 1997). The upper age limit of the technique is the age of the universe.

Luminescence

Luminescence dating is another radiogenic dating method. It is applied directly to sand or silt-sized crystals of quartz or feldspar, both of which are very common in sediments (Aitken 1997; 1998). It has two advantages for physical geographers:

1.   The sediment itself is dated, not the material that is included within it but which may be of a different age.

2.   The age determined is for the arrival of the sediment, not for the creation of the mineral grains themselves.

When the quartz or feldspar crystals are exposed to background ionizing radiation (as they are all the time in nature), charged particles are released within the crystals. The charge is then held at the natural faults ('traps') that occur in a crystal lattice. Over time, that trapped charge population grows, and eventually the traps fill up – the crystal becomes saturated. When the crystal is heated or exposed to light, energy is given to the trapped charge that releases it and allows it to return to the ground state. This is the zeroing event, and in most cases where physical geographers are interested in using luminescence it will occur when sediment is exposed to light during transport and deposition. That event needs to expose the sediment to enough light to reset the signal fully, so aeolian sediments are much more amenable to dating than are, for example, glacial sediments, which may not have been well exposed to light when they were deposited. Once the sediment is buried, charge once again starts to build up in the traps.

In order to date the sediment, samples need to be taken without exposing them to light, for example by hammering a black tube into an exposed section. All laboratory measurements are made in dim red light – a bit like a photographic darkroom. Subsamples (aliquots) of each sample are exposed to controlled amounts of heat (thermoluminescence – TL) or to narrow band wavelengths of light (optically stimulated luminescence – OSL – which is sometimes also known as optical dating). The heat or light gives energy to the trapped charge causing it to be released from the traps. As it recombines, light of different wavelengths is given off: this light is luminescence. The amount of luminescence that is given off is proportional to the size of the trapped charge population, and so to the amount of radiation the sample was exposed to while buried. In order to find out how old the sample is, aliquots are given controlled doses of radiation in the laboratory to build up a growth curve of increased luminescence with radiation dose. The natural signal is compared with this growth curve to find out what radiation dose it received in nature (known as the equivalent dose or ED). In order to calculate the age, measurements are also made of the background radiation dose rate which the sample received while buried.

$$\text{Age} = \frac{\text{Equivalent Dose (ED)}}{\text{Dose Rate}}$$

Generally, OSL is considered better for sediment dating than TL as the OSL signal is reduced more rapidly by exposure to light at deposition than is the TL signal (Duller 2000; Forman *et al.* 2000).

Electron spin resonance (ESR)

ESR is based on the same process of trapped charge as luminescence dating, but measurement is made of that trapped charge in a different way, and it is usually applied to different materials, in particular teeth, shells and carbonates such as stalagmites and stalactites. Unlike luminescence, ESR is usually used to date the time of formation of the material, or the time it was last heated (Grün 1997). The trapped charge population in the crystals will affect the behaviour of the sample in a magnetic field, and this behaviour is measured in order to determine an equivalent dose (the radiation dose received since the sample was formed or last heated). Just as with luminescence, this ED is then divided by the dose rate received in nature to obtain an age. The lower limit of dating is set by ability to detect the amount of trapped charge and is usually a few thousand years. The upper limit can be over a million years, depending on the material being dated (Grün 1997).

Fission track

The main way in which ^{238}U (uranium-238) undergoes radioactive decay is though the emission of an alpha particle to form ^{234}Th (thorium-234). About one in two million decays, however, is through spontaneous fission, in which the nucleus of the ^{238}U splits into two new nuclei (Dumitru 2000). These nuclei are both strongly positively charged, and so repel each other, travelling in a straight line in opposite directions. When the uranium is included in a mineral, such as zircon, the passage of the particles damages the lattice of that mineral, and a fission track is formed. This is around 17 μm long and 0.008 μm in diameter (Dumitru 2000). The number of tracks present is proportional to the age of the crystal or volcanic glass (Dumitru 2000). The fission of ^{238}U has a half-life of around 8 billion years, so this is quite a slow process.

Fission track dating is used for volcanic materials, and in rocks that have been buried at great depth. Holding crystals at high temperature for extended periods of time will anneal the tracks – that is, make them gradually heal up (Westgate *et al.* 1997). This is the way in which the dating clock is reset – in volcanic materials, for example, crystals are likely to have been totally annealed during eruption. It also, however, leads to the 'fading' of the tracks over time (Dumitru 2000). Different minerals anneal at different rates and under different temperature conditions. Zircon is considered to be a very useful mineral for fission track dating as it is stable to high temperatures and so the tracks, once created, are long lasting. Volcanic glass, on the other hand, will fade at surface temperatures over a few million years (Westgate et al. 1997).

To date a sample, it is split into two parts, or aliquots. One aliquot is irradiated in order to induce fission tracks, and the other is not. Both aliquots are held for a time at a high temperature in order to take account of fading effects, and both are etched in order to increase the size of the tracks so that they are visible using an optical microscope and can be counted (Westgate et al. 1997).

The bottom end of the age range over which fission track is useful is defined by the uranium content of the material to be dated. Since the half-life of fission of ^{238}U is so long, it takes some considerable time before there are sufficient tracks created to produce good counting statistics. This means that the relative precision available on younger samples is not as high as for older ones where there are more tracks available to count (Dumitru 2000). It is not often possible to date samples less than 100,000 years old; the upper limit is around 1 billion years (Dumitru 2000).

Cosmogenic nuclides

When cosmic rays collide with elements on Earth, new 'cosmogenic' nuclides may be created. This phenomenon was initially investigated mainly for atmospheric production (for example ^{14}C is produced by cosmic ray interactions in the atmosphere, as discussed above), but rocks and sediments on the surface of the Earth also receive a cosmic ray dose and so cosmogenic nuclides are produced there as well (Lal 1991). The penetration of cosmic rays is reduced by the atmosphere, and diminishes even more rapidly with depth into the surface of the Earth, so that, for example, there is a 20-fold decrease in the rate of cosmogenic nuclide production under 2.5 m of till (Bierman et al. 1999). The implications of this rapid drop-off in production are that if the surface erodes, then material is exposed underneath which has had little or no production of cosmogenic nuclides, but in which they now begin to develop over time (Zreda and Phillips 2000). This is the basis of the dating technique.

The use of cosmogenic nuclides offers the possibility of dating erosion surfaces, in other words, as Duller (2000, 112) puts it, managing to 'date something that is no longer there!'. Dating erosion events or rates has been a significant problem, so these very new techniques, while expensive and experimental, offer a great deal of promise for investigating landscape change

Some of the isotopes that are created in this way are radioactive (e.g. ^{10}Be, ^{26}Al, ^{36}Cl), others are stable (e.g. ^{3}He, ^{21}Ne). The radionuclides have a variety of different half-lives. This means that although one can refer to cosmogenic nuclide dating, there is in fact a suite of techniques based around different isotopes and situations. For the radionuclides, there is a limit to the dating range of around 2 or 3 half-lives. This is because over the time period during which the nuclides are created, they are also decaying. Eventually, an equilibrium situation will emerge between the two. ^{36}Cl, for example, usually has an upper limit of 200,000–500,000 years (Zreda and Phillips 2000). The upper limit for the stable ^{3}He, by contrast, is effectively infinite. All of the cosmogenic nuclide dating techniques have a lower age limit which relates to the ability to detect the presence of the nuclide.

Two main criteria must be fulfilled for this dating method to work (Lal 1991; Zreda and Phillips 2000):

1. The rate of production of the cosmogenic nuclides must be known.

2. There must be no cosmogenic nuclides in the material to begin with – i.e. the surface must have been sufficiently well stripped to have taken away the surface layers where they had formed. Having some residual cosmogenic nuclides remaining is termed inheritance.

There has been a great deal of both empirical research and modelling into the determination of the rates of production of the nuclides, but this is still an on-going process (Clark *et al.* 1995; Lal 1991). Some rates, for example, have been determined through the analysis of samples that have been dated through other methods, and any refinement of those methods may mean a consequent refinement of the production rate. The rate of production is also not constant over time or space. It varies with altitude, latitude and the behaviour of the Earth's magnetic field, since this will affect incoming cosmic rays (Lal 1991). Depth below the surface is also critical, so erosion or deposition will affect the rate of production.

The problem of a zero starting point, that is, when there were no cosmogenic nuclides present after the erosion which is to be dated, raises a number of questions. For one thing, erosion is not always a clear-cut one-off event. It is also not necessarily for all time – surfaces may be reburied. This latter issue has been tackled by the use of more than one radioactive nuclide in the dating of a surface (Bierman *et al.* 1999). Because two radioactive isotopes such as ^{10}Be and ^{26}Al will decay at different rates, if they do not give the same age, it is an indicator of a more complex history of exposure and burial. This still, however, has to assume that there were no inherited nuclides (Bierman *et al.* 1999). Studies have also used the build-up of cosmogenic nuclides to assess rates of erosion (e.g. Summerfield *et al.* 1999).

It is also possible to use cosmogenic nuclides to date deposition. When sediments are deposited which have been eroded from elsewhere, cosmogenic nuclides will begin to build up in them, and will eventually allow them to be dated. This might generally rely on their not suffering from inheritance; however, for well-mixed sediments, there may be ways in which even this can be dealt with. The way in which the cosmogenic nuclides build up is depth-dependent, as already discussed. Thus, if all the material had, on average, the same concentration of inherited cosmogenic nuclides, the degree of inheritance can be assessed by dating a depth-profile and modelling the build-up (Hancock *et al.* 1999).

CORRELATION METHODS

Fossils

Individual, or more usually assemblages of fossils, can be used to indicate the age of a unit. This may be through the occurrence of a range of fossils that are known only to exist during a certain time period. It is also possible to use palaeoecological methods to indicate the environmental conditions which prevailed and so to assign the unit to a period of time at which those conditions were known to prevail.

Oxygen isotopes

Stable isotopes of oxygen provide a way of investigating climate change, in particular the changing volumes of global ice-sheets, and therefore **eustatic sea level change** (Shackleton 1987). They can also be used as a correlation method of dating. There are three isotopes of oxygen: ^{18}O, ^{17}O and ^{16}O; of these, only ^{18}O and ^{16}O are used for environmental reconstruction (Lowe and Walker 1997). The principle behind this method relates to oxygen's presence as a component of water (H_2O). $H_2^{16}O$ is more readily evaporated than is $H_2^{18}O$, so when evaporation occurs, the vapour is richer in ^{16}O and the remaining liquid in ^{18}O – a process termed fractionation. This effect is temperature dependent, and is even more marked when temperatures are lower (Lowe and Walker 1997).

In order to make comparisons, the amount of fractionation that has happened in a sample is compared against standard reference material with known levels of the two isotopes. The proportion of ^{18}O and ^{16}O in the sample is expressed in the following standard way, as $\delta^{18}O$, in parts per thousand (Lowe and Walker 1997):

$$\delta^{18}O\text{‰} = 1000 * \frac{{}^{18}O/{}^{16}O \text{ sample} - {}^{18}O/{}^{16}O \text{ standard}}{{}^{18}O/{}^{16}O \text{ standard}}$$

In cold phases, there is a lot of ^{16}O locked in ice sheets, having been evaporated out of the ocean, then precipitated out and frozen; this means that the oceans are relatively richer in ^{18}O. The $\delta^{18}O$ of ocean water is thus relatively high. In interglacials, ice sheets melt and more of the ^{16}O is returned to the oceans, so that they have lower $\delta^{18}O$ (Lowe and Walker 1997). The isotopic composition of the water will be recorded in the shells of foraminifera (microscopic, unicellular marine organisms) living in the oceans, and when they die, the shells sink to the ocean floor and are gradually incorporated into the sediments on the bed of the ocean, leaving a long-term record which can be sampled by coring.

The record of change in ocean cores has advantages – for example the record relates in a large part to global changes and not just to local ones. However, there are local influences on $\delta^{18}O$, such as the temperature of seawater and bioturbation and compaction in cores (Bradley 1999; Shackleton 1987; Williams *et al.* 1998). This means that a variety of cores must be used in order to gain a more representative record.

The pattern of changing $\delta^{18}O$ with climate change has been a very important part of understanding long-term climate and sea level change. However, it also provides a way of dating by correlation. The oxygen isotope stratigraphy of different cores are compared with one another, or with an idealized sequence based on a large number of cores (Shackleton 1977). The sequence of changes can be 'wiggle matched' in order to determine how a partial record fits into the overall record of change. This gives relative dating, but a generalized version of oxygen isotope stratigraphy has been compared with Milankovitch periodicities in order to place ages on its different phases (Lowe and Walker 1997), so that an absolute age can be assigned.

Oxygen isotope stratigraphy has also given rise to a way of naming the glacial and interglacial periods of the Quaternary. These are traditionally known by local names (for example, the last glacial period is termed the Wisconsinan in the

eustatic sea level change: sea level changes that have occurred worldwide and are due to changes in the level of the ocean itself, rather than the relative level of the land

USA, the Devensian in the UK and the Weichselian in Europe (Ehlers 1996)). Oxygen isotope 'stages' have been used to transcend this and to take into account more complex histories of warming and cooling. These stages are divisions of the oxygen isotope record and relate to global periods of warming and cooling. They are counted down from the top, so that the Holocene is Stage 1 (Lowe and Walker 1997). Warm phases have odd numbers and cold phases even numbers. Subdivision of the stages (e.g. Stage 5 is divided into substages a–e) further reflects complex histories of climate change.

Tephrochronology

The word tephra is used to describe the pyroclastic material that is transported through the air during an explosive volcanic eruption; it includes both ash and pumice (Sarna-Wojcicki 2000). Tephra can spread over a wide area, and may be preserved in sedimentary sequences, particularly in environments that are primarily depositional, such as lakes and bogs. Geologically, a tephra deposit represents a point in time, so it is a highly useful way of dating a sequence, or correlating between different sites. It is very important, in this context, to distinguish between primary, air-fall tephra, which represents the time of the explosion, and reworked tephra, which may be older than the deposits in which it is included.

Individual tephras, relating to a particular volcanic eruption, can be identified by their physical and chemical properties, which are derived from the chemistry of the magma and the way in which the eruption occurred. This means that the ash relating to one eruption can be identified in different locations (Sarna-Wojcicki 2000). Although the ashes from different eruptions of the same volcano may be similar, they can be distinguished. For example, Mount St Helens in the state of Washington, USA, has erupted several times over at least the last 40,000 years, and these ashes are sufficiently chemically distinct from one another that they can be used as marker horizons where they are found in sedimentary sequences around the region (Mullineaux 1986).

In themselves, therefore, tephra units act as marker horizons, but if the ash can be dated, then they can be used as absolute dating for a horizon. Volcanic ash can be dated using a variety of techniques, including radiocarbon dating of organic material included in a deposit or bracketing it, fission track, K–Ar and luminescence. Since volcanic eruptions are rather noticeable to human beings, they are also likely to be noted in historic records, potentially giving a very precise age.

Tephrochronology can be applied to sequences from the modern as far back as is allowed in a specific context by weathering and other diagenesis of the ash, which will eventually mean that it cannot be identified. It is mainly applied over the Quaternary and the **Neogene**, but can extend back much further through the geological record in favourable circumstances (Sarna-Wojcicki 2000).

> **Neogene**: a geological period comprising the Pliocene and Miocene epochs. It is the period prior to the Quaternary and spans the time from around 23.5 million to around 2 million years ago

Dendrochronology

Dendrochronology is one of the best-known ways of dating – essentially by counting the annual growth rings from a living tree, or from wood that is included in a structure or a sedimentary unit (Baillie 1995). A huge amount of environmental information can be derived from the tree ring data (for a good overview on that information and how it is obtained, see Grissino-Mayer 2004).

This includes not just the age at death of a tree, but also, for example, climate information – a field known as dendroclimatology (Baillie 1995; Dean 1997; Roberts 1998; NOAA Paleoclimatology Program and World Data Center for Paleoclimatology 2003). When a good chronology has been developed, a tree ring can be pinned down to an individual calendar year, making this a very precise technique.

Trees in parts of the world which experience strong seasons (that is, those outside the tropics) usually develop one growth ring each year, on the outside of the stem. This represents the start of growth at the end of winter, the growth through the year, and the end of growth during the next winter; these phases of growth develop different kinds of pore structure in the wood, which make the rings visible (Aitken 1990).

The key breakthrough in the use of tree rings for dating was the realization that because trees will grow more in good growing seasons, and less in poor years, a distinctive pattern of wide and narrow rings will be developed by trees in a particular area. This allowed the development of a technique called cross-dating. This is the putting together of patterns of broad and narrow rings from trees so that records from different pieces of wood can be overlapped (Figure A5) (Aitken 1990; Jacoby 2000). This can only be applied within an environmentally similar region, as the pattern is so dependent on local environmental change. Modern trees will form the young end of that sequence; rings from growing trees can be sampled by taking a thin core from the tree – it is not necessary to fell it and look at the stump! Records are then extended back from that using overlaps between different pieces of wood from a region. The more individual records that are available, and the greater the overlaps between them, the better the chronology that can be developed. Chronologies covering most or all of the

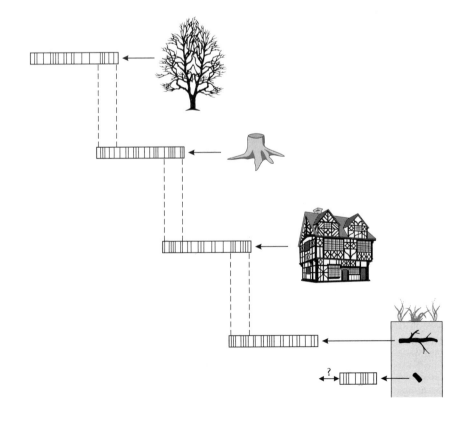

Figure A5 Cross-dating of tree ring sequences allows dating of wood that is no longer growing. Because the width of a tree ring relates to the environmental conditions in the year during which it grew, trees in a region will develop a similar growth pattern. This means that regional databases can be built up of those patterns, and tree ring sequences found in archaeological sites or sedimentary units can be fitted into that sequence, which is tied to a calendar year by being linked into growing trees at the young end

Holocene have been developed in some regions – in particular there are excellent records from bristlecone pine in California, Irish oak and European oak (Baillie, 1995; Jacoby, 2000; ORAU, 2004). Good cross-dating also allows checks to be made for very thin and missing rings in individual samples.

Varves

Varves are annual laminations in lake sediments and provide a similar form of dating to dendrochronology: both can be described as **sidereal dating methods**. Varves can occur for a variety of reasons (O'Sullivan 1983), but the most useful form for dating purposes are generally those that occur due to glacial meltwater entering a lake (Verosub 2000a). The spring and summer flush of glacial meltwater is loaded with sediment which gradually settles out of the water over the course of the year. The coarsest sediment settles first, and the finer material gradually settles out of suspension, giving an annual couplet that fines upwards, and is then overlain by the coarse material from the beginning of the next year's lamination. Years that are warmer, and so have greater amounts of meltwater and sediment, will be represented by a thicker lamination; conversely, cold years with little melting may have thin or even no lamination produced (Aitken 1990). This record of thin and thick bands can be cross-dated, in the same way as tree rings, in order to produce a long-term record for a region into which floating chronologies can be fitted to give a date (Aitken 1990). Sequences have been built up which cover the period since the last glacial maximum, one of the best known records being that from Sweden (Boygle 1993; Verosub 2000a). Even where a sequence of varves cannot be fitted into a firm chronology, and so give an absolute date, it may be useful for timing the length of events (Roberts 1998), particularly since the varves also provide environmental information such as about warming or cooling and their consequent effects on meltwater and sediment.

The varve chronology has proved useful for radiocarbon calibration, with AMS radiocarbon dates being derived for organic material and foraminifera within the varves (Bradley 1999; Lowe and Walker 1997). Correlations between varves and pollen records in peat have also been used in radiocarbon calibration (Aitken 1990).

> **sidereal dating methods:** those which give a calendar age or are based on counting annual events (Colman and Pierce 2000)

Geomagnetism

The Earth's magnetic field is often compared to a bar magnet, controlled by the activity of the Earth's core. This strong, stable field is called the dipole field, but it is not the only factor in geomagnetism. In addition, there are components in the Earth's magnetic field that are more local, less stable and not so strong. This is termed the non-dipole field (Bradley 1999). The overall magnetism at any place on Earth can be described through three components: declination (the deviation from true, geographic north); inclination (the angle from horizontal which is primarily affected by latitude, being $0°$ at the magnetic equator, and $90°$ at the magnetic pole); and intensity (the strength of the field) (Verosub 2000b; Lowe and Walker 1997). All of these components vary through time and space, and the way in which they do so forms the basis for a dating method.

Variations in the dipole field have global effects (Bradley 2000). The largest of these are called reversals. When a reversal occurs, the direction of the poles swaps, such that the magnetic north pole is at the geographic south pole. These changes have been identified over at least the last 200 million years, and the time between them may be anything from around 50,000 years to millions of years (Verosub 2000b). How and why the change in state happens is not well understood, but it appears to take a few thousand years to occur. These major changes form the basis of the palaeomagnetic timescale. A long period of time during which the Earth's magnetic field has a particular orientation is termed an epoch (or chron). These may be 'normal' – that is, with the magnetic field in its current direction – or 'reversed' – with the magnetic field at 180° to its current direction. The present epoch is called the Bruhnes, and the last major reversal occurred around 780,000 years ago (Baksi *et al.* 1992). The previous epoch was called the Matuyama and was a period of about 1.8 million years of reversed polarity. Within these epochs, however, there may be short intervals when the polarity changes; these are termed events (or subchrons). For example, during the Matuyama, with its dominantly reversed polarity, the Olduvai event (from 2.02–1.78 million years ago) had normal polarity (Verosub 200b). The boundaries between epochs have been dated by independent means, including methods such as potassium–argon and argon–argon dating of volcanic rocks and also by comparison with other long-term records such as the $\delta^{18}O$ record (Bradley 1999).

More minor changes in the magnetic field are termed secular variation (Verosub 2000b). These are by their nature less dramatic and more localized, and probably relate to variations in the non-dipole field (Bradley 1999). For example, the magnetic north pole wanders around when compared with true, geographic north, and this is recorded as a change in declination (Lowe and Walker 1997). Inclination and intensity are also variable. Master chronologies for secular variation have been built up for different regions (Lowe and Walker 1997). As with the palaeomagnetic timescale, these master chronologies have been dated using independent techniques, so that they can give absolute ages and not just act as a correlation method. Secular variation has also been directly measured at some sites (e.g. London) over the last 400 years (Lowe and Walker 1997).

The reason that these changes in the Earth's magnetic field are useful as a dating technique is that sediments and rocks contain magnetic grains, most commonly iron oxides and sulphides (Maher *et al.* 1999). These record the characteristics of the Earth's magnetic field when they are formed or when they settle out of a fluid during sedimentation (Lowe and Walker 1997). Thus, changes in magnetic field characteristics are recorded in volcanic rocks, and in lake and ocean sediments and loess, among other sediments. These individual records can be determined and matched against the palaeomagnetic timescale and the regional master chronologies for secular variation. High sedimentation rates will give the best resolution and so the most detailed records (Verosub 2000b; Lowe and Walker 1997).

Amino acid racemization

Amino acids are the building blocks of the proteins of which living organisms are made (Aitken 1990). Most common amino acids can exist in two forms, termed the L-form and the D-form. These are chemically very similar to each other, but have a slightly different structure. Living organisms have almost no D-form amino acids, but after death, there are chemical changes, and the L-form amino acids convert gradually into D-form until an equilibrium state is achieved, which is usually a roughly equal amount of each form (Wehmiller and Miller 2000). This conversion is called racemization, and the dating method that is based on it is called amino acid racemization (AAR). In order to date the materials, the D:L ratio of a sample is determined, and this will increase with increasing age, until equilibrium is reached.

The materials which amino acid racemization is used to date are generally marine shells, eggshells and bone. Different genera show different rates of conversion from L- to D-forms. The rate at which conversion occurs can be expressed as a half-life – the time taken for half the L-form amino acid to convert to D-form. The conversion, as a biological and chemical reaction, is also highly temperature-dependent (Wehmiller and Miller 2000); for example, the half-life of aspartic acid is 430,000 years at 0°C and 3000 years at 25°C (Aitken 1990).

This temperature dependence of L- to D- conversion can be modelled and be used to derive an absolute age for a sample, if the temperature history (and other chemical factors) of the site can be sufficiently constrained (Hare *et al.* 1997). This calibration of D:L ratios is termed aminochronology (Wehmiller and Miller 2000). However, this is not always possible, as sites may have complex histories (Hare *et al.* 1997). Correlation (and relative) dating is available where these strict criteria cannot be met, in the form of aminostratigraphy, when D:L ratios at sites in a single region are correlated. This may be used, for example, to correlate sites that relate to different glacial and interglacial periods and place them into a relative order (Wehmiller and Miller 2000; Lowe and Walker 1997).

Clearly, from the discussion above, the time range over which AAR can be applied is very dependent on the ambient temperature at which the fossil material has been preserved. It can be as much as 5 million years at high latitudes, using an amino acid with a long half-life, or as little as 150,000 years in the tropics (Wehmiller and Miller 2000).

GLOSSARY

abiotic: devoid of life, that is, non-biotic

accelerated erosion: soil erosion exacerbated by human activities

acid precipitation: acid rain, acid snow, acid hail, acid dew and acid fog. Scientists usually use the term 'acid rain' to mean acid precipitation

acid rain: precipitation where the raindrops have been contaminated by sulphur dioxide or nitrogen oxide in the atmosphere, causing the pH level of the rain to fall below 'normal' levels

actinomycetes: fungi-like bacteria that form long, thread-like branched filaments, which are visible in compost

active sampling: air quality monitoring method which involves passing a known volume of air through a filter or chemical collector over a specific period to determine pollutant concentrations in that volume of air

actual evaporation: the water that is actually lost given environmental constraints

adaptive radiation: the diversification of many species from a single founding species

adiabatic: occurring without a physical transfer of heat between an air parcel and its surroundings

adjacency matrix: a system represented as flows (non-zero entries) and non-flows (zero entries) in a matrix with system components as row and column headings

advection: the lateral transfer of energy

aerial photograph: a vertical or oblique photograph of the Earth's surface taken from the air

aerobic: requiring oxygen

Agenda 21: an 800-page document coming out of the Earth Summit. It was an innovative attempt to describe the policies needed for environmentally sound development

aggregates: units of many peds

agrarian: relating to ownership, cultivation and tenure of the land

air mass: a large block of air exhibiting similar characteristics over a relatively long time period and which tend to be characteristic of certain geographical areas

air pollution episodes: a term used to describe a period, usually short-lived, of very high concentrations of pollutants

albedo: reflectivity of a material measured as the amount of energy returned compared to that received

albic horizon: a light-coloured, mineral surface soil horizon from which clay and free iron oxides have been washed out

allele: a form of a gene at a locus on a chromosome

anaerobic: not requiring oxygen

analogue model: a sophisticated scale model, which includes maps and remotely sensed images

anhydrite: anhydrous calcium sulphate ($CaSO_4$) occurring as a white mineral. Usually found in association with gypsum, rock salt and other evaporite minerals

antecedent conditions: the conditions before an event that govern its effect or impact. For example, antecedent conditions of a lengthy wet period before a thunderstorm result in little infiltration and large amounts of overland flow and runoff as the saturated ground cannot accept any more water, and flooding may occur. If the antecedent conditions were dry, infiltration could occur over the unsaturated ground with less overland flow and runoff, minimizing flooding downstream

anthroposphere: the portion of the ecosphere influenced by humans

anthrosol: a soil with horizons radically altered by human activities, but excluding soils under normal agricultural use

aquifer: a saturated permeable rock

aquitard: a layer of low permeability that can store water; water can move through the layer slowly

argillic horizon: a mineral subsoil horizon in which clay washed from the overlying surface horizons accumulates

argon: one of the inert gases present in the atmosphere. Inert gases are very unreactive

artesian: where the water pressure is high enough to force the water upwards

aspect: the compass bearing or exposure of a slope

aspirator: a device attached to an instrument to provide ventilation such as a suction fan

atomic absorption spectrophotometry: a method of determining concentration by comparing the absorption of light by atoms at various wavelengths in a solution to known concentrations

attribute characteristic: thematic properties that describe what a feature is (using numbers or text)

auger: a tool for obtaining a small soil sample by boring into the ground

automated instruments: equipment that can be programmed to record information at set time intervals or over set thresholds without the operator being present

available water capacity: the amount of water available to plants between wilting point and field capacity; it varies with soil texture

azonal soils: weakly developed soils found in any climatic zone

barrow: a communal burial mound used from the Stone Age to Saxon times

baseflow: the contribution to stream flow from upwelling groundwater into the channel

benthic: a term to describe bottom dwelling organisms or sediment in aquatic systems

bias: the tendency for data to shift by some consistent amount from the 'true' value

bioaccumulation: the processes by which certain animals and plants store chemical pollutants in their cell tissues. Fish tissues, for example, concentrate low levels of DDT

bioaccumulation: the selective accumulation of nutrients and toxic substances in the cells and tissues of organisms

biocides: a poison or other substance used to kill pests

bioclimate: the climatic conditions that affect living things

biodiversity: the variety of genetic information, species, communities and ecosystems in a particular place, be it a small pond, a continent, or the entire world

bioelement: an element essential for living things to function

biogeochemical cycles: the repeated movement of elements essential to life (bioelements) through the ecosphere

bioleaching: the use of certain bacteria to remove heavy metals from contaminated soils

biological monitoring: monitoring flora and fauna to provide an indirect assessment of water quality of ecosystem 'health'

biomagnification: the increasing concentration of certain substances as they move up a food chain

biome: a community of animals and plants occupying a climatically uniform area on a continental scale

biosphere: the totality of life on Earth

biotic: pertaining to life

black-box model: a box-and-arrow diagram. The inputs and outputs to the boxes are known but process inside the boxes are not – hence black box

bocage (French): a landscape of small, enclosed fields with woodland

borehole: a well of particular depth used to tap into a strata of interest e.g. aquifers

boundary layer: the part of the atmosphere within which the effects of friction or drag of the surface can be felt

Bowen Ratio Energy Balance: an instrument to measure evaporation loss based on the Bowen ratio: the ratio of sensible heat to latent heat lost to the atmosphere by the surface under investigation

box-and-arrow diagram: a model of a system where boxes represent stores and arrows the flows or relationships between them

Brundtland Report: the final report of the World Commission on Environment and Development, chaired by Gro Harlem Brundtland, that was published in 1987 and proposed a major international conference to deal with environmental issues, which materialized as the United Nations Conference of Environment and Development (UNCED – or Earth Summit) in Rio de Janeiro in 1992

bubbler: air quality monitoring method where sampled air is literally passed through a chemical solution that retains the pollutant and which can be subsequently measured to determine its concentration in the air sample

bulk density: the mass of material in a unit volume of dry soil

bunds: low banks of earth sometimes created at the edge of human-made terraces. Grass or shrubs may be planted on them

capillary zone: a zone in the soil where water may move owing to capillary forces

carbon dioxide (CO_2): a greenhouse gas produced from a wide variety of natural (e.g. respiration) and human-made sources (e.g. combustion of carbon-based fuels)

carbon partitioning: the allocation of carbon to different chemical forms (such as sugars, structural carbohydrates and starch) within, or to different parts (such as leaves, stem, roots) of plants

carbon sequestration: the net removal of carbon dioxide in the atmosphere and its long-term storage in live or dead plant tissues (e.g. leaves, branches, stems, roots, litter, peat)

carbonic acid: a weak acid with the chemical formula H_2CO_3

carbon–nitrogen (C/N) ratio: a useful indicator of the degree of decomposition of organic matter in soils. Well-decomposed soil humus in humid temperate soils has a C/N ratio of around 12; straw has a C/N ratio of about 40

carrying capacity: the maximum number of individuals a particular environment can support without causing environmental degradation

cartography: the discipline of map-making

catena: a linked sequence of soils running from hilltop to valley bottom

catenary: pertaining to a soil catena, that is, a linked series of soils running from a hilltop to valley bottom

cation exchange: the exchange of ionic constituents between the soil solution and the clay-humus complex

cation exchange capacity (CEC): the size of the cation exchange complex

Cenozoic: the youngest geological era, lasting from 65 million years ago to the present

chemiluminescence: analysis technique that assesses the release of light that occurs as a result of certain chemical reactions in order to determine the composition of a sample

chemoautotrophs: organisms that obtain their energy from the oxidation of inorganic compounds, such as hydrogen sulphide

chromatographic methods: analysis method where the properties of an object (for example gases in air) are determined by the separation of molecules based on differences in their structure or composition – a process called chromatography

clay-humus complex: microscopic particles of silicate clays and humus (< 0.002 mm in diameter) with a large surface area for exchange of chemical within the soil solution

climate: the long-term characteristics of the atmosphere in a particular geographic zone

climatically sensitive communities: communities that respond quickly to changes in climate, and particularly to changes in temperature and water budgets

colloids: tiny (1–100 nm) particles of organic and inorganic soil materials normally in suspension and highly dispersed

colorimetric analysis: a technique that relates the concentration or strength of a solution by its colour intensity

community: an assemblage of species living together in a particular place

comprehensive model: a complex mathematical model that resolves space and time in rich detail

conceptual model: an abstract model of a system that can be expressed as pictures, box-and-arrow diagrams, matrix models, flowcharts and various symbolic languages

conduction: the process of energy transfer through the contact of vibrating molecules

cone of depression: the localized depression around a well, caused by the rate of water abstraction exceeding that of recharge

confined aquifer: an aquifer overlain by a layer of lower hydraulic conductivity

conformal: describes map projections which maximize the preservation of the real world shape characteristics of features

constructed wetland systems: artificial complexes that simulate natural wetlands, comprising saturated substrates, emergent and submergent vegetation, fauna and water for human use and benefits

consumer: an organism that gets its energy from other organisms

consumption: community respiration

contamination: the presence or introduction of unwanted materials and compounds into the water system

contour ploughing: ploughing along contours, at right angles to the steepest slope

convection: the process of energy transfer through mass movement or mixing of a substance (gas or liquid)

coordinates: numerical locational references which may or may not be expressed using a real world referencing system

corridors: strips of land that differ from the land to either side (e.g. roads, hedgerows, fencerows and rivers)

creep: the rolling and sliding of coarse sand and small pebbles under the force of jumping sand particles and down the tiny crater-slopes produced by impacting particles

cryosphere: all the frozen waters of the Earth (snow and ice)

cryoturbation: a term denoting soil mixing through the action of ice

cup anemometer: a device for measuring wind speed based on the ability of the wind to rotate a design of three or four hemispherical cups pivoting around a spindle

current meter: a device for measuring the speed or velocity of water

cyanobacteria: unicellular organisms formerly included in the plant kingdom as blue-green algae, but now classed as a phylum in their own right under the Monera Kingdom. They feed by photosynthesis with photosynthetic pigments of chlorophyll, phycoerythrin (red) and phycocyanin (blue)

Darcy's law: an equation describing the flow of water through a porous medium

data-loggers: devices for the storage of data from automated instruments

decomposer: an organism that lives by dissolving organic matter for nourishment

deep weathering profiles: thick, much-weathered, Earth-surface mantles or crusts, often of great age

desmids: mainly single-celled, freshwater algae with cell walls made of cellulose and pectic materials

deterministic model: a mathematical model with no random components

detritivore: an organism that feeds on organic detritus, breaking it up in the process

dew point: the temperature at which saturation of the air occurs

diatoms: microscopic, single-celled, marine, planktonic (drift in ocean currents or swim a little) algae with a skeleton composed of hydrous opaline silica

Differential Absorption Lidar: technique where gases and particle concentrations can be determined through remote sensing

diffuse pollution: the discharge of pollutants into the water system from a large area where the source is not readily identifiable

diffusion tube: inexpensive and easy to use equipment for air quality monitoring which works through exposing absorbent materials to ambient air for a certain period of time and then analysing the material to determine a concentration. Available for a number of different pollutants

digital data: data which are held and stored in a computer-readable format

Digital Terrain Model (DTM): digital representation of the Earth's surface using positional (x, y) and elevation (z) data. Also referred to as Digital Elevation Model (DEM). Can be used to plot pseudo three dimensional (surface) representations of any physical or human parameter

Digital Terrain Model/Digital Elevation Model (DTM/DEM): both terms refer to a computer-based representation of a surface using location (x, y) and height (z) data. Most commonly used for representing topography but can be used to visualize any continuous data

digitizing: the process of converting data from an analogue (non-digital) to a computer readable form. Creates vector data

dilution gauging: measuring stream discharge by adding a tracer or chemical solution and measuring its dilution downstream. There are two types: constant injection and gulp injection which relate to the addition of the chemical into the water

dipwell: a slotted well used for measuring the position of the water table or phreatic surface

direct measurements: measurements taken of the object under study

disjunct subpopulations: populations of the same species separated geographically from each other

dispersal: the spread of organisms into new areas

disturbance: the perturbation of an ecosystem by abiotic (e.g. fire and wind) and biotic (e.g. pathogens) agents

diversity gradients: changes in biodiversity along environmental gradients, such as the climatic gradients associated with increasing altitude and latitude

diversity hot-spots: areas or regions with exceptionally high biodiversity

diversity indices or metrics: techniques to measure species number and species evenness

DOAS: Differential Absorption Spectrometers (see **spectrometric methods** glossary entry)

dolomite: a rock containing at least 50 per cent calcium–magnesium carbonate ($CaMg(CO_3)_2$), a mineral called dolomite

drip irrigation: an irrigation term referring to a form of irrigation where the water is delivered using pipes with holes which allow water to escape at the base of the plant

dry deposition: the deposition on surfaces of dry gases and particles in the atmosphere

Dynamic Global Vegetation Model: a complex model used to simulate the global functioning of vegetation

Earth Summit, 1992: the United Nations Conference on Environment and Development (UNCED) held in Rio de Janeiro, Brazil, in June 1992. Its main aim was to discuss ways of safeguarding biodiversity

Earth System Model of Intermediate Complexity (EMIC): a model designed to combine the advantages of simple models with the benefits of complex models, without the drawbacks of either

ecocentrism: an environmentalist viewpoint that sees humans as part of the ecosphere and subject to its laws

ecological flow: the minimum flow required to maintain ecological quality and processes, and ecosystem sustainability

ecosphere: the global ecosystem; life plus life-support systems (air, water, soil)

ecosystem: a community of living things together with their life-supporting surroundings

ecosystem management: in its most extreme form, a system of resource management that tries to balance economic needs, ecological needs, and social and cultural needs

ecotones: transition zones between two plant communities

electrical conductivity: the dissolved solids content of water measured through its ability to conduct an electrical current

electrochemical sensors: analysis method using equipment broadly similar to a battery to generate a slight current in the presence of a specific gas. The amount of current generated is proportional to the concentration of the gas. Sophisticated versions of these instruments have different components that detect different types of substances and have been likened to 'electronic noses'

electromagnetic method: technique of measuring stream velocity by creating an electromagnetic force by an electric current. The emf is directly proportional to the average velocity of the river cross-section

electromagnetic radiation: energy associated with the Sun. This energy can be in many forms

electromagnetic spectrum: shows the full range of electromagnetic energy, grouped and ordered in terms of relative wavelength

ellipsoid: a three dimensional, ellipse-like shape with dimensions longer in one plane than the other

eluviation: the washing out of soluble materials from a soil horizon

emergent properties: properties, such as the number of species in a community, that can be measured only for a system as a whole, not in its component parts

emissions: the mass release of a substance (gas or particles) to the atmosphere

emissions inventory: a schedule of sources and associated mass emissions of pollutants for a particular area over a particular period of time

empirical: knowledge based on observation, experience and experiment – evidence that is seen, heard, touched or smelt

energese: a symbolic language for representing energy and material flows in systems

environmental lapse rate: the rate of temperature change with height observed in the real atmosphere

environmental perturbation: a change in any environmental factor that disrupts the 'balance' with a community

environmentalism: a movement centring around a moral concern for the environment

epimorphic processes: processes that occur as rock is exposed to near-surface conditions. They include weathering, leaching and the formation of new minerals

equivalent: describes map projections which maximize the preservation of the real world area characteristics of features

eustatic sea level change: sea level changes that have occurred worldwide and are due to changes in the level of the ocean itself, rather than the relative level of the land

eutrophication: the enrichment of water bodies with mineral and organic nutrients. This initially supports a proliferation of plant life but can lead to reduced dissolved oxygen levels and changes in the dominance of plant groups

evaporation pan: a chamber of known dimensions, usually circular but sometimes other shapes, from which the rate of evaporation can be recorded by noting the water level over time

evaporite: a water-soluble mineral (such as halite or rock salt), or a rock composed of such minerals, precipitated out of saline water bodies such as salt lakes

evapotranspiration: the loss of water from soil, water and plant surfaces (transpiration)

field capacity: the state in the soil when all the capillary pores are just full

filtration: involving the use of filters. For example collecting particulate matter in air through passing a sample of air through a filter that will not allow solid matter of a certain size to pass through

flood desynchronization: modification of the flood peak passing through the catchment by storing floodwater and releasing it slowly to decrease flood levels

flood return period: the statistical risk of flooding from events of a particular magnitude described in terms of a return period. For example, a large flood estimated to be a 1 in 100 year flood (1:100) has a one per cent chance of occurring in any one year

flood storage: the temporary store of floodwater to relieve the risk or magnitude of flooding downstream

flow chart: a model showing an assumed ordering of processes within an environmental system

flume: the artificial channel of known dimensions built within a stream to measure flow

fluvial: of, or pertaining to, flowing water (rivers)

food chain: a simple sequence of feeding relationships in a community, starting with plants and ending with top carnivores

food web: a network of feeding relationships within a community

forest gap: an opening produced by trees falling through old age or through natural disturbing agencies such as strong winds

fossil fuels: carbon based deposits (coal, oil and gas) that have been developed over geological time scales and which produce heat and energy on combustion

Geographical Information System (GIS): a computerized system for handling and analysing geospatial data

genetic differentiation: the splitting of a gene pool leading to the evolution of different subspecies or species

genetic diversity: the mix of genes or genetic variability (expressed as the number and relative abundance of alleles) in a population, or species, or other group of organisms

genetically impoverished species: species with a reduced gene pool owing to population reduction and isolation

geodesy: the science of characterizing and modelling the true shape of the Earth

geoid: a detailed approximation of the shape of the Earth

geoinformatics: a term used to denote the collective disciplines associated with the collection, collation and analysis of geospatial data

geopotential energy: energy associated with gravity

georeferencing: the process of transforming image coordinates into geospatial coordinates through the application of a real-world coordinate system

geospatial: spatial data where the locational reference is given in geographical (real world) coordinates

geospatial database: a collection of spatial data referenced to the Earth's surface using the same referencing system and usually held in thematic layers

geosphere: the non-living spheres of the Earth

glacial troughs: valleys, often U-shaped, eroded by valley glaciers or by ice flowing in ice sheets and ice caps

glaciofluvial landforms: landforms fashioned by water from melting ice

gleying: a soil process in which iron is reduced under waterlogged (and therefore anaerobic) conditions to a soluble form and washed out of a horizon, leaving a bluish or greyish matrix with or without yellowish brown or brown spots (mottles)

global amphibian decline: the somewhat mysterious loss of amphibian biodiversity that started in the 1970s

global environmental change: environmental changes that move in roughly the same direction and are observed in most regions at about the same time; enhanced global warming is a case in point. Local effects of deforestation and urbanization may accumulate to produce changes of global extent

global warming potential: the relative capacity of a substance to absorb heat energy and thereby warm the atmosphere, usually expressed relative to the warming effect of carbon dioxide over a century

gravimetric methods: these involve the weighing of the mass of particles collected by drawing a known volume of air through a filter paper

gravimetric soil moisture measurement: the measurement of soil moisture by computing the difference in mass before and after the soil has been dried. The soil moisture is expressed as the amount of water per weight of soil

gravity-flow systems: an irrigation term relating to the distribution of water in an irrigated field using a network of ditches and small channels through which the water moves by gravity

Great Transformation: the conversion of the terrestrial land-cover from natural to human-dominated types that started to grow pace in the late eighteenth century

greenhouse gases: gases which absorb outgoing heat from the Earth's atmosphere, thereby acting to warm the atmosphere

gross primary productivity (GPP): the amount of energy (in the form of organic matter) made per unit time in a community

ground reference sites: a phrase used to describe ground data collection points, usually with reference to remote sensing studies

groundwater: water stored in the pore spaces and joints within rocks

groundwater recharge: the movement of water through the ground to replenish an aquifer

groynes: low walls, usually made of wood, built out into the sea to help prevent beach erosion

gypsum: a white or colourless mineral. Hydrated calcium sulphate, $CaSO_4.2H_2O$

gypsum blocks: a block of gypsum buried in the soil connected to an electrical resistance meter. Soil moisture is measured on the relationship between moisture content and electrical current

habitat diversity: the number of different habitats in an area

habitat fragmentation: the breaking up of large areas of habitat into smaller and more isolated parcels

habitat requirements: the environmental conditions required for a particular species to thrive. Habitat generalists have broad habitat requirements; habitat specialists have very specific habitat requirements

half-life: The half-life ($t_{\frac{1}{2}}$) of a radioactive isotope is the average time it takes for half of the atoms present to undergo radioactive decay. This does not depend on the amount that was there to begin with, so if an isotope had a half-life of 5 years, the initial quantity would be reduced to half after 5 years, to a quarter after 10 years, to an eighth after 15 years, etc.

hardware model: a model of a system in which the size of the system is made bigger or smaller

heavy metals: ions of metallic elements such as copper, zinc, and iron with densities greater than $5–6 \text{ g cm}^{-3}$

HFCs (hydroflurocarbons): man-made compounds containing hydrogen, fluorine and carbon which are used for a variety of industrial processes

hinterland: the zone of influence around a settlement, for example for work, markets or services

historical distribution: a species range at some time in the past

holistic view: an approach that emphasizes the importance of whole systems, such as the global ecosystem, and the interdependence of their parts

Holocene: the most recent slice of geological time spanning 10,000 years ago to the present

hot wire anemometer: a device to measure wind speed indirectly using the cooling power of the wind and electrical current required to maintain the resistance and thus temperature of an exposed fine wire

humus: the end-product of organic decomposition

Hydra: an instrument to measure evaporation using the eddy flux approach

hydraulic conductivity: the rate of water movement through soil or rock

hydric soils: that have formed under saturated conditions and which experience saturation during the growing season for long enough to develop anaerobic conditions

hydrograph: plot of discharge against time at a particular location on the river

hydrological cycle: the repeated movement of water molecules around the ecosphere by evaporation, condensation and precipitation

hydrophytic vegetation: vegetation adapted to grow in saturated soil conditions

hydrosphere: all the water of the Earth, including atmospheric water vapour in some definitions

ice caps: bodies of ice smaller than $50,000 \text{ km}^2$

ice sheets: bodies of ice larger than $50,000 \text{ km}^2$

ice wedges: v-shaped masses of ice that run from the ground surface to the permafrost layer

IGBP: the International Biosphere–Geosphere Programme

indirect measurements: measurements of a variable that is not the feature of interest but can be used to infer information about the desired item

induration: the hardening of a soil horizon owing to the cementation of soil grains

infiltration capacity: the maximum infiltration rate at a given point under certain conditions

infiltration rate: the quantity of water that infiltrates the soil over a certain time period

input–output model: a record of inputs and output links in a box-and-arrow diagram, often represented as a matrix

interaction matrix: an adjacency matrix with non-zero entries signed – or +

interception: the interruption of the pathway of precipitation to the ground surface by vegetation. Water is either held on plant surfaces, from where it will evaporate, or will move at a slower rate towards the ground

intrazonal soils: soils occurring within a climatic zone due to the dominance of some factor other than climate, for example, substrate

isobars: graphical depictions (lines) denoting zones of equal pressure

isothermal: refers to a condition where thermal properties have a constant value

isotopes: isotopes of an element have the same chemical properties, but differ from each other in the number of neutrons in the nucleus, so have different atomic masses. They may behave differently in their nuclear properties, so one isotope of the same element could be radioactive while another is stable

isotopic composition: an isotope is a form of an element with a different number of neutrons, the number of protons being the same in all forms of the same element. Isotopic composition is the make-up of isotopes, for example the amounts of oxygen-16 and oxygen-18

kinetic energy: energy associated with movement

Landsat Thematic Mapper: a non-photographic imaging system sensing electromagnetic radiation in seven different bands carried by Landsat satellites to record information about the land surface

landscape resistance: the ease with which an animal or plants can move through a landscape

latent energy: energy associated with the change of state of a substance, for example liquid to solid (water to ice)

leachate: the cocktail of chemicals being washed through the soil

leaching: the washing out of soluble material for the entire soil profile (or sometimes a specific horizon)

leguminous: refers to plants of the family Leguminosae, which bear pods that split in two halves with the seeds attached to the lower end of one of the halves

lidar (light detection and ranging): an active remote sensing system that measures vertical distance using a laser light beam

limestone: a rock containing at least 50 per cent calcium carbonate ($CaCO_3$)

limestone pavements: extensive benches or plains of bare rock in horizontally bedded limestones and dolomites

lithosphere: the 'sphere of rocks', the outer shell of the solid Earth

litter decomposition: the decay and disintegration of litter lying on the soil surface

loess: silt transported by aeolian processes

Long Range Transport of Atmospheric Pollution (LRTAP): the transmission of atmospheric pollutants over 500 km. Some pollutants may travel worldwide

low flows: the river flow during the dry season; anthropogenically induced low flow is the reduction in discharge and other aspects of the flow regime caused by human activities such as abstraction, flow diversion and land use change e.g. afforestation, cropping and urbanization that affect interception, storage and evaporation processes

lysimeter: a device (of varying size) for measuring the rate of water movement and storage in a soil block

macroconsumers: organisms that feed on living plant tissues

macrofauna: soil animals at least 1 cm long, including vertebrates, earthworms, and slugs and snails

macronutrient: a nutrient required in moderate to large amounts by organisms

macrophyte: a large plant; a member of the macroscopic plant community

managed realignment: establishing a new line of defence inland of current defences to improve the long-term sustainability of the defence, by breaching or removing the existing defences. This also promotes the creation of inter-tidal habitat between the old and new defences which further enhances the long-term sustainability

Manning's equation: an equation relating stream velocity and channel geometry

map: a stylized representation of the real world, usually in a two-dimensional form

mathematical model: a system represented using the formal and symbolic logic of mathematics

matrix: the background ecosystem or land use type in which patches and corridors are set (e.g. forest, arable and residential)

megacities: cities with 10 million or more inhabitants

mesofauna (meiofauna): soil animals of intermediate size, 200 μm–1 cm. Includes most of the nematodes (roundworms), rotifers, springtails and mites, with various small molluscs and arthropods

Mesozoic: a geological era lasting from 248 to 65 million years ago and comprising the Triassic, Jurassic and Cretaceous periods

metadata: data about data – used as a means of describing and cataloguing spatial and aspatial datasets

meteorological satellites: a group of Earth observation satellites designed to monitor synoptic weather systems

methane (CH_4): a greenhouse gas produced largely through natural processes (the decomposition of materials) but which can be influenced by human activity

micro-erosion meter (MEM): a device to measure surface height at a number of predetermined points, relative to datum studs set into the surface under observation

microconsumers: organisms that feed on dead and decaying organic materials

microfauna: soil animals less than 200 μm long. Includes protozoans, and some nematodes (roundworms), rotifers and mites

micronutrient: a nutrient required in trace amounts by organisms

mineralization: the release of mineral components in organic matter through decomposition

mobilization: the rendering of metals in the soil into a soluble, and therefore transportable, form

model building: the process by which humans simplify the 'real world' to help understand and explain it

mollisol: a soil with a thick, dark surface horizon rich in calcium and magnesium cations (a mollic horizon)

moraines: various kinds of landforms formed by the deposition of glacial sediments

mosaic: the landscape pattern produce by a particular combination of patches, corridors and matrix

multivariate analysis: a statistical analysis considering multiple (three or more) variables simultaneously to assess their interrelationships

Munsell Soil Colour Chart System: a standard system for describing soil colour in terms of hue, lightness and chroma

natural selection: The process by which the environmental factors sort and sift genetic variability and drive microevolution

negative feedback: a self-stabilizing change in a system – the change triggers a series of subsequent changes that eventually neutralize the effects of the original change, so bringing the system back to its original state. An example is the regulation of body temperature in warm-blooded animals

neocolonialism: where one or more foreign nations exert an influence over a host nation through economic power and social influence

Neogene: a geological period comprising the Pliocene and Miocene epochs. It is the period prior to the Quaternary and spans the time from around 23.5 million to around 2 million years ago

net primary productivity (NPP): gross primary productivity less the energy used in respiration

network: a collection of connected lines (with explicit topological information)

neural network methods: techniques of analysis using pattern recognition software that is 'trained' by an input dataset and corresponding output

neutron scattering: a technique of soil moisture measurement based on the relationship between water content and speed of scattering of neutrons from a neutron source

nitrogen-fixing processes: processes involving nitrogen-fixers (a few free-living bacteria and blue-green algae, as well as symbiotic bacteria associated with the plant roots) that manufacture biologically active forms of nitrogen – ammonia, amines and amino acids

nitrous oxide (N_2O): a greenhouse gas produced through natural processes which can be modified by humans (the activity of soil microbes) and human-made sources (e.g. industry)

non-sampling error: the measurement error, e.g. writing wrong numbers

noösphere: the sphere of influence of the mind

nutrient cycle: the repeated movement of a bioelement through the ecosphere

orographic rainfall: precipitation created by the forced ascent of moist air over topographic obstacles, such as mountains

orthophoto: a mosaic of aerial photographs

ostracods: minute, chiefly freshwater crustaceans with a bivalve carapace (hard outer covering)

overland flow: the movement of water as sheet flow over land due to ground saturation or impermeability

oxic horizon: a soil horizon at least 30 cm thick, in which weathering has removed or altered a large portion of the silica and left sesquioxides or iron and aluminium and a concentration of kaolinitic clays

oxides of nitrogen: a collective term used to group two species of oxidized nitrogen, nitric oxide (NO) and nitrogen dioxide (NO_2), the former is rapidly oxidized to the latter in the atmosphere

ozone (O_3): the highly reactive and unstable tri-atomic (O_3) form of molecular oxygen formed through the action of the Sun's energy on the more common di-atomic (O_2) form of molecular oxygen

ozone layer: term used to refer to the relatively high concentration of ozone found in parts of the stratosphere. The characteristics of ozone prevent some of the Sun's harmful UV radiation reaching the Earth's surface

ozone 'hole': an area of reduced concentration of ozone in the stratosphere which varies from place to place and time of the year

pH: a measure of the concentration of hydrogen ions in the soil; technically, when measured in soil, it is the negative logarithm of the hydrogen ion activity (meaning the hydrogen ions active in the soil solution and excluding those in organic matter and mineral structure), so a low pH value corresponds to an acid soil with more hydrogen ions in the soil solution, a high pH value with an alkali soil with fewer hydrogen ions

palaeosols: old, fossil or relict soils of soil horizons, usually buried

parent material: the material in which soil profile is formed

parent rock: the bedrock in which a soil profile is formed

particle density: the mass of soil solids

particulate matter: any solid particles or liquid droplets that are present in the atmosphere

patches: fairly uniform (homogeneous) landscape areas that differ from their surroundings (e.g. woods, fields, ponds and houses)

pedogenesis: the formation of soils

pedon: the smallest volume of material that can be called 'a soil'; it extends the idea of a soil profile, which is a vertical sequence of soil horizons seen in a soil pit, to three dimensions

pedosphere: the totality of soils on the Earth

pedoturbation: a term covering all processes that tend to mix the soil by churning it

peds: individual soil aggregates

Penman–Monteith model: the most widely accepted model for computing evaporation from a vegetated surface, based on Monteith's modification of Penman's formula for open water

periglacial landforms: landforms created in the 'periglacial zone' (the region where intense frosts during winter alternate with snow-free ground during summer)

peripheral population: a population lying at the edge of a species range

permeability: the ability of liquids or gases to move through soil and rocks, which depends upon the interconnectivity of rock spaces, and the shape and packing of the grains

permeameter: device to record the soil permeability by recording the flow of water through a soil column. There are two types: falling and constant head permeameter

photoautotrophic organisms: organisms – green plants and bacteria – that obtain their energy from sunlight

photosynthesis: the manufacture of carbohydrates from carbon dioxide and water by chloroplasts, using light as an energy source. Oxygen is a by-product

physiological drought: conditions under which a plant is unable to extract water from the soil, either because the water is held to soil particles at a higher capillary force or osmotic force (or both) than the plant root can exert, or because the water is frozen

phytoplankton: the plant community in marine and freshwater systems which floats free in the water and contains many species of algae and diatoms

phytoremediation: the use of plants to decompose organic chemical wastes and pollutants

piezometer: a well slotted at the base, used to determine the groundwater head for particular strata

Pleistocene: a slice of geological time between 2 million and 10,000 years ago

plough layer: a ploughed and cultivated organo-mineral surface soil horizon

plough pans: compacted layers of soil that build up below the ground surface owing to ploughing or other cultivation activities

podzol: a soil with a spodic B horizon (a horizon showing illuvial accumulation of sesquioxides, with

or without organic matter), beneath a light-coloured (albic) A horizon

point pollution: the discharge of pollutants into the water system at a specific point in space

polder: an area of low-lying land reclaimed from sea, lake or river, and protected by dikes

pollution: the presence or introduction of damaging loads or concentrations of unwanted materials and compounds into the water system

population (in statistics): a set of observations on a particular phenomenon; populations may be finite, as in the number of fields on the Isle of Wight or infinite, as in sand grains on a beach

pore water pressure: the pressure of water in the pores and voids of soil and rock. Under saturated conditions, this pressure may force particles apart and cause slope failure

pores: small voids within rocks, unconsolidated sediments and soils

porosity: the volume of open pores compared to the volume of soil or rock

positional characteristic: locational properties that describe where a feature is located (using coordinates)

positive feedback: a self-reinforcing change in a system – the change triggers further changes that magnify the effects of the original change. Exponential population growth is an example

postdict: predict for times past (i.e. after the event)

potential evaporation: the amount of water that could be evaporated from water or soil according to the atmospheric demand, if water supply were unlimited

potentiometric approach: concentration is measured by the difference in electrical potential between a metal electrode in a salt solution and the unknown concentration of metallic ions in the solution

precautionary principle: the view that a lack of scientific data should not prevent the implementation of measures to avert environmental degradation; a 'better safe than sorry' approach

pre-Quaternary: all geological time before the start of the Quaternary subera, 2 million years ago

producer: an organism that uses energy from the physical environment to sustain itself

production: the sum of sugars created by photosynthesis in land plants and autotrophic algae

projected: geographical data converted to a two dimensional form following a certain set of rules and procedures

projection plane: the characteristics of the two dimensional surface upon which the map will be represented

puck: a piece of equipment rather like a computer mouse which is traced around features on a map in order to capture the features digitally

Quaternary: a geological subera spanning the past 2 million years

radiation inversion: where heat loss from the Earth causes temperatures to rise with increasing height rather than falling with height as is more usual

radiocarbon dating: a technique of determining the age of organic material based on measuring the proportion of ^{14}C isotope within the carbon

radionuclides: nuclides (nuclei of atoms) that disintegrate spontaneously, releasing radioactivity in the process. Examples are nuclei of uranium and strontium atoms

rainfall intensity: the rate at which rain falls, expressed as total rainfall over a unit of time

rain gauge: a device for measuring rainfall at a particular point on the ground. Many different designs exist with a typical design of a funnel to capture rainfall over a certain area, which then funnels water either into a storage bottle or tipping device with known volume

Ramsar Convention: established in the city of Ramsar, Iran 1971, it initiated the international cooperation in the designation, conservation and wise use of wetlands

range contraction: the shrinking of a species range

range expansion: the enlargement of a species range

range shift: a change in the distribution (range) of species

raster data model: conceptual means of representing spatial data using grid cells

reactive: refers to unstable compounds that are prone to change when coming into contact with other constituents and conditions in the atmosphere

reactive nitrogen: nitrogen in a form that is available to living things

receivers: devices that pick up transmissions from GPS satellites and interpret these signals as a spatial coordinate

recharge: the addition of water to an aquifer. Can either occur naturally through precipitation infiltrating the soil and percolating down to the underlying aquifer, or it can be done artificially by pumping water into the rock

reflectometry: an analysis method for determining concentrations of particulate matter (especially black smoke) through cross-referencing the darkness of a sample against a mass concentration value using standard calibration curves

regolith: weathered mantle; all the weathered material lying above unaltered or fresh bedrock

remote sensing: observation of an object of interest without contact. In this context the collection of data about the Earth's surface without being in direct contact with the Earth

respiration: a complex series of chemical reactions in all organisms by which energy is made available for use

reversible cementation: the formation of hard layers in soils that can subsequently be removed

riparian: along the edge of a river

rock tablets: pieces of rock of known weight used to assess the rate of weathering in a particular place or environment. The rate of weathering is determined by weighing the rock over time and noting any change caused by disintegration

salinization: the accumulation of soluble salts, such as sulphates and chlorides of calcium and magnesium, in salty soil horizons

saltating grains: sediment grains being transported by bouncing along a surface. Most commonly seen in aeolian transport

saltation: the bouncing of air-borne or water-borne sand grains on hitting the ground

saltwater intrusion: the movement of saline water into an aquifer formerly holding freshwater

sample (in statistics): a representative subset of a population

sampling error: the potential difference between the 'true' value from the population and the estimation gained from the sample

sand/dust traps: devices designed to capture sand or dust over time. Regular measurement of the build up in the trap can be used to estimate transport rates

scale model (iconic model): a smaller or larger replica of a system

scanning: automated digitizing. Creates raster data

Schmidt hammer: an instrument used for the non-destructive testing of concrete, or rock based on the relationship of the rebound of the hammer and the compressive strength of the surface under investigation

sequestration: locking compounds out of circulation by combination with other materials

sesquioxides: oxides that contain two metallic atoms to every three of oxygen; for example, Al_2O_3

shelterbelts: natural or artificial windbreaks

sidereal dating methods: those which give a calendar age or are based on counting annual events (Colman and Pierce 2000)

signed digraph: a box-and-arrow diagram with positive and negative links indicated by small arrows and circles, respectively

siltation: the settling out of silt and other fine particles in a lake, reservoir, low-gradient river or estuary

slickenside: a smoothed surface within a soil showing parallel striations and grooves

slope: an inclined surface of any part of the Earth's surface

smectite: a clay mineral requiring high ionic concentrations of silica and magnesium for its neoformation (formation in the soil); also called montmorillonite

social capital: the inherent wealth of communities of people, represented for example by sets of social values and knowledge networks

soil acidification: the loss of bases from a soil, leading to a higher acidity (lower pH)

soil aggregates: basic structural units of soil, formed of individual soil particles (mineral and humus), with planes of weakness between them. Individual aggregates are called peds

soil horizon: a layer of soil, lying roughly parallel to the land surface, which has characteristics created by soil-forming processes

soil moisture deficit: the amount by which moisture held within the soil falls below the level of field capacity

soil profile: all the soil horizons from the ground surface to the unaltered bedrock or parent material; usually divided into A ('topsoil'), B ('subsoil'), and C (weathered parent material) horizons, the A and B horizons together forming the solum

soil solution: the liquid portion of a soil, that is, soil water plus the materials dissolved in it (solutes)

solar radiation: light energy emitted by the Sun, which is by convention grouped according to wavelength. The naked human eye can detect less than half of this incoming radiation, the visible spectrum, all other wavelengths are invisible

solum: the layers of the soil lying above the weathered parent rock: the A and B horizons

sorption: the uptake of an ion or molecule by adsorption and absorption. It is often used when the exact nature of the removal is unknown, or as a collective term for both processes

spatial analysis: the intelligent combination and processing of diverse locational and attribute data to produce new information

spatial data: any data that have an explicit locational reference, whether direct (e.g. coordinates) or indirect (e.g. postcodes). Compares to aspatial data that do not have any such reference

spatial resolution: the level of spatial detail or precision given through a locational reference and its associated data. In the context of remotely sensed images, this refers to the size of individual pixels and, by extension, the area on the ground that is covered by one data value

species diversity: the number and variety of species living in an area

spectral methods: a technique that relates the spectral intensity of ions in a flame to known concentrations of the same ion to determine concentration

spectral signature: the specific electromagnetic radiation reflectance and absorption characteristics associated with a feature or set of features

spectrometry: techniques for identifying chemicals in a substance

spectroscopic methods: analysis method where the properties of an object (for example, gases and particles in air) are determined based on its interaction with light – a process called spectroscopy

speleothem: a mineral deposit formed by the precipitation of calcium carbonate from the solution passing through caves: stalactites and stalagmites are examples of speleothems

spheroid: a three dimensional, sphere-like shape

spodic horizon: a soil horizon showing illuvial accumulation of sesquioxides, with or without organic matter, and lying beneath a light-coloured (albic) A horizon

spray systems: an irrigation term referring to a non-specific mass delivery of water to a crop which mimics the effect of rainfall

stable: resistant of vertical motion so if a parcel of air is lifted it remains relatively cool compared to its surroundings and tends to fall back to its original elevation

stage–discharge relationship (rating curve): a chart plotting the river stage against discharge so that flow can be estimated from water level data

statistical model: a mathematical model dealing with inherent variability in measured phenomena

stochastic model: a mathematical model that includes a random component

Stokes's law: an expression expressing the rate of settling of spherical particles in a fluid of known density and viscosity

stratosphere: the part of the atmosphere above the troposphere and which is characterized by increasing temperature with height

sulphur dioxide (SO_2): an acidifying gas most commonly produced by the anthropogenic combustion of materials containing trace amounts of sulphur or as a result of volcanic eruptions

surface: a continuous, pseudo three dimensional representation of a feature comprising both locational data (x,y coordinates) and some form of elevation attribute (z coordinate)

surface roughness: quantification of the amount of friction or drag exerted by a particular type of land cover

surface water gley: a gley soil is one which exhibits prolonged soil saturation resulting in the presence of blue/green colour caused by the reduction of iron in the soil to its ferrous state in the anaerobic conditions. Surface water gleys are soils in which poor drainage or the presence of lenses or pans causes restricted water movement down the profile causing water logging in the upper horizons

suspension: the carriage of silt and clay particles by the air

sustainable development: using natural resources to meet current human needs without harming the biosphere and with an eye to the possible needs of future human generations

synoptic weather conditions: weather conditions over a large area of the atmosphere

system: a meaningful arrangement of components, as in the solar system and ecological system

technocentrism: an environmentalist viewpoint that recognizes environmental problems and believes that humans can surmount them

temperature inversion: where temperatures increase with height rather than reduce with height, a characteristic which inhibits any further upward vertical movement of air

temporal: relating to the attribute of time. In this context, the time period a dataset refers to

tensiometer: piece of equipment comprising a porous cup sealed to a tube which is filled with water and inserted into the ground to record the soil wetness by measuring the suction or water tension

terracing: the construction of a series of flat platforms on slopes

Terrestrial Biosphere Models: models designed to simulate global patterns and dynamics of vegetation

thematic: relating to subject matter. In this context, the subject or theme of a dataset

thermal: wavelengths associated with emitted heat energy. Can be derived from natural sources (e.g. after being absorbed by materials on the surface, from animals or natural fires) or from artificial energy sources (e.g. central heating or cars)

thermokarst: irregular terrain produced by the thawing of ground ice in periglacial environments

thermosphere: the outermost layer of the atmosphere, separated from the mesosphere by the mesopause. It sits at about 80 to 550 km above the Earth's surface and extends to around 10,000 km, where it merges into space. Temperatures rise with increasing altitude in the thermosphere to over 1000 °C

till plains: almost flat, slightly rolling and gently sloping plains formed of a thick blanket of till (a mainly unsorted and unlayered deposit laid down beneath a glacier)

time domain reflectometry: an accurate method to determine moisture content and electric conductivity by dielectric permittivity and signal attenuation respectively

titrimetric techniques: balancing of reactions using coloured indicators

topological characteristic: relational properties that describe how features are related in space

toposphere: the surface features of the globe, natural and human-made

total sampling error: the combined potential error of the sample being unrepresentative and human errors taking the sample

traditional ecological knowledge: the huge mental database of knowledge about local environments and their human use held by indigenous peoples around the world

tree-line: the altitudinal and latitudinal limits of tree growth

tropopause: the point of the troposphere at which temperatures no longer reduce with increasing height, the 'top' of the troposphere

troposphere: the part of the atmosphere which is in contact with the Earth's surface and which is characterized by a general fall off in temperature with height

turbidity: the reduced transparency of the atmosphere or water caused by dust or particles in suspension

turbulence: irregular movement of the atmosphere, perhaps as a result of a physical disturbance of the flow of air

turbulent transfer processes: the movement of water or air in an irregular, eddy flow, in contrast to smooth laminar flow

tutorial models: simple, non-spatial models of environmental systems

ultrasonic method: measuring stream velocity at a certain depth by recording the time taken for sound pulses to pass through the water

ultraviolet radiation: shortwave radiation, around 200 to 400 nanometers wavelength, originating from the Sun

unconfined aquifer: an aquifer with no confining layers between the aquifer and the surface

UNESCO: United Nations Educational, Scientific and Cultural Organization, established 16 November 1945

unit hydrograph: the conceptual process where precipitation is converted into runoff by assuming rain events of equal duration produce the same hydrograph. The unit hydrograph is produced from rainfall falling in a unit of time uniformly over the catchment to a certain depth. The area under the curve of the hydrograph gives the volume of surface runoff and is equivalent to the effective depth of rainfall over the catchment area

unsaturated zone: the area between the ground surface and the water table

unstable: not resistant to vertical motion so if a parcel of air is lifted it remains relatively warm compared to its surroundings and tends to continue to rise itself as a result of its own buoyancy

urban heat island: the phenomenon of elevated temperatures within some large built-up areas relative to adjacent non-built-up areas caused by differences in surface cover and heat sources

vadose (unsaturated) zone: a subsurface zone lying between the ground surface and the water table

varves: annually laminated sediments that were deposited through water. They are usually deposited in lakes, but may also be marine

vector data model: conceptual means of representing spatial data using coordinate pairs

visible spectrum: the part of the electromagnetic spectrum that can be detected with the naked eye

volatilize: become volatile, that is, capable of being readily vaporized

water stress: a lack of soil water during at least one season

water table: the upper limit of groundwater saturation in permeable rocks; the level at which groundwater pressure is equal to atmospheric pressure

wavelengths: a measure of the distance covered by one 'wave' of electromagnetic energy, for example crest to crest or trough to trough. Measured in micrometres (10^{-6} metres) (μm) or nanometres (10^{-9} metres) (nm)

weather: short-term characteristics (temperature, pressure, humidity, clouds, precipitation, visibility and wind) of the atmosphere

weathering: the chemical, mechanical and biological breakdown of rocks on exposure to the atmosphere, hydrosphere and biosphere

weir: a dam across a stream over which the water flows. It is used for flow measurements and water level control

wet deposition: the deposition on surfaces of dissolved substances and particles by any form of precipitation

wetlands: terrain in which the soil is saturated with water, and soil saturation is the chief determinant of soil development and plant communities. Marshes, swamps, bogs and fens are examples

wilting point: the soil water storage at which plants are unable to overcome hygroscopic forces holding water in the soil matrix, fail to take up water through their roots, and therefore lose turgor and wilt

wind-tunnel: a structure shaped like a tunnel through which air moves and its speed recorded. Obstacles can be placed within the tunnel to measure their effects on wind speed

wise use: the sustainable use of wetlands for the benefit of humankind in a way compatible with the maintenance of the natural properties of the ecosystem

xerophyte: a plant adapted to dry conditions

zonal soils: soils that occur over large areas roughly corresponding to major climatic zones

zooplankton: aquatic invertebrates living in sunlit waters of lakes, rivers, seas and oceans

BIBLIOGRAPHY

ABRAHAMS, P. W. (2002) 'Soils: their implications to human health'. *The Science of the Total Environment* 29, 1–32.

ADVISORY CENTRE FOR EDUCATION (1971) *Clean Water: The Water of Rivers and Streams, the Life it Supports and the Effects of Pollution* (Things of Science Kit). Cambridge: Advisory Centre for Education.

AHRENS, C. D. (1998) *Essentials of Meteorology: An Invitation to the Atmosphere*. Belmont, USA: Wadsworth Publishing.

AIR QUALITY EXPERT GROUP (AQEG) (2004) *Nitrogen Dioxide in the UK, First Report*. London: DEFRA.

AITKEN, M. J. (1990) *Science-based Dating in Archaeology*. London: Longman.

AITKEN, M. J. (1997) Luminescence dating. In R. E. Taylor and M. J. Aitken (eds) *Chronometric Dating in Archaeology*, 183–216. New York: Plenum Press.

AITKEN, M. J. (1998) *An Introduction to Optical Dating: The Dating of Quaternary Sediments by the Use of Photon-stimulated Luminescence*. Oxford: Oxford University Press.

ALEXANDER, L. M. (1993) 'Large marine ecosystem: a new focus for marine resources management'. *Marine Policy* 17, 186–98.

ALLOWAY, B. J. (1990) *Heavy Metals in Soils*. Glasgow: Blackie.

AMANO, H., MATSUNAGAA, T., NAGAOA, S., HANZAWAA, Y., WATANABEA, M., UENOA, T. AND ONUMAB, Y. (1999) 'The transfer capability of long-lived Chernobyl radionuclides from surface soil to river water in dissolved forms'. *Organic Geochemistry* 30, 437–42.

ANAND, M., MA, K., OKONSKI, A., LEVIN, S. AND MCCREATH, D. (2003) 'Characterising biocomplexity and soil microbial dynamics along a smelter-damaged landscape gradient'. *The Science of the Total Environment* 311, 247–59.

ANDERSON, S. H., MANN, K. AND SHUGART, H. H. (1977) 'The effect of transmission-line corridors on bird populations'. *American Midland Naturalist* 97, 216–21.

ANDRES, R. J. AND KASGNOC, A. D. (1998) 'Time averaged inventory of subaerial volcanic sulphur emissions'. *Journal of Geophysical Research* 103, 251–61.

ANTONIĆ, O., BUKOVEC, D., KRIŽAN, J., MARKI, A. AND HATIĆ, D. (2000) 'Spatial distribution of major forest types in Croatia as a function of macroclimate'. *Natura Croatica* 9, 1–13.

APLIN, P. (2003) 'Remote Sensing: base mapping'. *Progress in Physical Geography* 27, 275–83.

ARMESTO, J. J., ROZZI, R. AND CASPERSEN, J. (2001) Temperate forests of North and South America. In F. S. Chapin III, O. E. Sala and E. Huber-Sannwald (eds) *Global Biodiversity in a Changing Environment: Scenarios for the 21st Century* (Ecological Studies 152), 223–49. Springer: New York.

ARNFIELD, A. J. (2003) 'Two decades of urban climate research: a review of turbulence exchanges of energy and water and the urban heat island effect'. *International Journal of Climatology* 23, 1–26.

ARNOLD, G. W., WEELDENBURG, J. R. AND STEVEN, D. E. (1991) Distribution and abundance of two species of kangaroo in remnants of native vegetation in the central wheatbelt of Western Australia and the role of native vegetation along road verges and fencelines as linkages. In D. A. Saunders and R. J. Hobbs (eds) *Nature Conservation 2: The Role of Corridors*, 273–80. Chipping Norton, Australia: Surrey Beatty & Sons.

ATTEWELL, P. B. AND TAYLOR, D. (1988) Time-dependent atmospheric degradation of building stone in a polluting environment. In P. G. Marinos and G. C. Koukis (eds) *Engineering Geology of Ancient Works, Monuments and Historical Sites*, 739–53. Rotterdam: Balkema.

BAEA, M., WATANABE, C., INAOKAB, T., SEKIYAMAA, M., SUDOC, N., BOKULD, M. H. AND OHTSUKAA, R. (2002) 'Arsenic in cooked rice in Bangladesh'. *The Lancet* 360 (9348), 1839–40.

BAIED, C. A. AND WHEELER, J. C. (1993) 'Evolution of Andean puna ecosystems: environment, climate, and culture change over the last 12,000 years in the central Andes'. *Mountain Research and Development* 13, 145–56.

BAILLIE, M. G. L. (1995) *A Slice Through Time: Dendrochronology and Precision Dating*. London: B. T. Batsford Ltd.

BAKER, A. AND GENTRY, D. (1998) 'Environmental pressures on conserving cave speleothems: effects of changing surface land use and increased cave tourism'. *Journal of Environmental Management* 53, 165–75.

BAKHSI, A. K., HSU, V., MCWILLIAMS, M. O. AND FARRAR, E. (1992) '^{40}Ar/^{39}Ar dating of the Bruhnes–Matuyama geomagnetic field reversal'. *Science* 256, 356–57.

BALDESANO, J. M., VALERA, E. AND JIMENEZ, P. (2003) 'Air quality data from large cities'. *The Science of the Total Environment* 307, 141–65.

BARRY, R. G. AND CHORLEY, R. J. (2003) *Atmosphere, Weather and Climate*, 8th edn. London: Routledge.

BARTLEIN, P. J., EDWARDS, M. E., SHAFER, S. L. AND BARKER, E. D., JR. (1995) 'Calibration of radiocarbon ages and the interpretation of palaeoenvironmental records'. *Quaternary Research* 44, 417–24.

BENNETT, A. F. (1991) Roads, roadsides and wildlife conservation: a review. In D. A. Saunders and R. J. Hobbs (eds) *Nature Conservation 2: The Role of Corridors*, 99–118. Chipping Norton, Australia: Surrey Beatty & Sons.

BENNETT, A. F. (1999) *Linkages in the Landscape: The Role of Corridors and Connectivity in Wildlife Conservation*. Gland, Switzerland and Cambridge, UK: IUCN.

BENNETT, A. F., HENEIN, K. AND MERRIAM, G. (1994) 'Corridor use and the elements of corridor quality: chipmunks and fencerows in a farmland mosaic'. *Biological Conservation* 68, 155–65.

BERNER, R. A. (1994) '3GEOCARB II: a revised model of atmospheric CO_2 over Phanerozoic time'. *American Journal of Science* 294, 56–91.

BEVEN, K. AND GERMANN, P. (1982) 'Macropores and water flow in soils'. *Water Resources Research* 18, 1311–25.

BIERMAN, P. R., MARSELLA, K. A., PATTERSON, C., DAVIS, P. T. AND CAFFEE, M. (1997) 'Mid-Pleistocene cosmogenic minimum-age limits for pre-Wisconsinan glacial surfaces in southwestern Minnesota and southern Baffin Island: a multiple nuclide approach'. *Geomorphology* 27, 25–39.

BINDÉ, J. (1998) 'Cities and environment in the 21st century: a future oriented synthesis after Habitat II'. *Futures* 30, 499–518.

BIRD, E. C. F. (2000) *Coastal Geomorphology: An Introduction*. Chichester: John Wiley & Sons.

BLAND, W. AND ROLLS, D. (1998) *Weathering: An Introduction to the Scientific Principles*. London: Arnold.

BLISS, L. C. (1990) Arctic ecosystems: patterns of change in response to disturbance. In G. M. Woodwell (ed.) *The Earth in Transition: Patterns and Process of Biotic Impoverishment*, 347–66. Cambridge: Cambridge University Press.

BLUME, H.-P. (1968) 'Die pedogenetische Deutung einer Catena durch die Untersuchung der Bodendynamik'. *Transactions of the Ninth International Congress of Soil Science, Adelaide* 4, 441–49.

BÖHNER, J., SCHÄFER, W., CONRAD, O., GROSS, J. AND RINGELER, A. (2003) 'The WEELS model: methods, results and limitations'. *Catena* 52, 289–308.

BOORMAN, D. B., HOLLIS, J. M. AND LILY, A. (1995) *Hydrology of Soil Types: a Hydrologically Based Classification of the Soils of the United Kingdom* (Institute of Hydrology Report No. 126). Wallingford: Institute of Hydrology.

BORDER, P. (1994) *Breathing in Our Cities: Urban Air Pollution and Respiratory Health*. London: Parliament Office of Science and Technology (POST).

BOTKIN, D. B. (1990) *Discordant Harmonies: A New Ecology for the Twenty-First Century*. New York: Oxford University Press.

BOTKIN, D. B., JANAK, J. F. AND WALLIS, J. R. (1972) 'Some ecological consequences of a computer model of forest growth'. *Journal of Ecology* 60, 849–73.

BOTKIN, D. B., WOODBY, D. A. AND NISBET, R. A. (1991) 'Kirtland's warbler habitats: a possible early indicator of climatic warming'. *Biological Conservation* 56, 63–78.

BOUBEL, R. W., FOX, D. L., TURNER, D. B. AND STERN, A. C. (1994) *Fundamentals of Air Pollution*, 3rd edn. San Diego: Academic Press.

BOUMA, J. (1977) *Soil Survey and the Study of Water in the Unsaturated Soil* (Soil Survey Paper 13). Wageningen: Netherlands Soil Survey Institute.

BOYGLE, J. (1993) The Swedish varve chronology – a review. *Progress in Physical Geography* 17, 1–19.

BRADLEY, R. S. (1999) *Paleoclimatology: Reconstructing Climates of the Quaternary*, 2nd edn. San Diego: Academic Press.

BRADY, N. C. (1990) *The Nature and Properties of Soils*, 10th edn. London: Collier Macmillan.

BRASSINGTON, R. (1988) *Field Hydrogeology*. Chichester: John Wiley & Sons.

BRAUER, J. (2002) The effect of war on the natural environment. In 'Arms and the Developing World', *Encyclopedia of Life Support Systems* (EOLSS). Oxford, UK: Developed under the Auspices of the UNESCO, EOLSS Publishers.

BRIDGES, E. M. (1997) *World Soils*, 3rd edn. Cambridge: Cambridge University Press.

BRIDGMAN, H., WARNER, R. AND DODSON, J. (1995) *Urban Biophysical Environments* (Meridian Australian Geographical Perspectives). Melbourne: Oxford University Press.

BRIGGS, D. J. (1997) 'Mapping urban air pollution using GIS: a regression-based approach'. *International Journal of Geographical Information Science* 11, 699–718.

BRIGHT, P. W. (1998) 'Behaviour of specialist species in habitat corridors: arboreal dormice avoid corridor gaps'. *Animal Behaviour* 56, 1485–90.

BRIMBLECOMBE, P. (1987) *The Big Smoke*. London: Methuen.

BRIMBLECOMBE, P. (2001) 'Acid rain 2000 ± 1000'. *Water Air and Soil Pollution* 130, 25–30.

BRITISH LIBRARY (2003) *Map Library Overview*. Internet site at http://www.bl.uk/collections/map_overview_history. html. Last accessed Sept 2003.

BROECKER, W. (1995) 'Chaotic climate'. *Scientific American* 273 (May), 44–50.

BROOKES, A. J. (1985) 'Downstream morphological consequences of river channelisation in England and Wales'. *The Geographical Journal* 151, 57–65.

BROWN, K. (2003) 'Three challenges for a real people-centred conservation'. *Global Ecology and Biogeography* 12, 89–92.

BROWN, L. R. and WOLF, E. C. (1984) *Soil Erosion: Quiet Crisis in the World Economy* (Worldwatch Paper 60). Washington, DC: Worldwatch Institute.

BRUTSAERT, W. (1982) *Evaporation into the Atmosphere. Theory, History and Applications*. Dordrecht: D. Reidel Publishing Co.

BULLOCK, A. AND ACREMAN, M. (2003) 'The role of wetlands in the hydrological cycle'. *Hydrology and Earth System Sciences* 7, 359–89.

BUOL, S. W., HOLE, F. D. AND MCCRACKEN, R. J. (1980) *Soil Genesis and Classification*, 2nd edn. Ames, Iowa: The Iowa State University Press.

BUREL, F. AND BAUDRY, J. (1990) Hedgerow network patterns and processes in France. In I. S. Zonneveld and R. T. T. Forman (eds) *Changing Landscapes: An Ecological Perspective*, 99–120. New York: Springer.

BURKE L., KURA, Y., KASSEM, K., REVENGA, C., SPALDING, M. AND MCALLISTER, D. (2000) 'Pilot analysis of global ecosystems: Coastal ecosystems'. World Resources Institute. http://marine.wri.org/pubs_description.cfm?PubID=3054. Last accessed April 2004.

BURNHAM, C. P. (1980) 'The soils of England and Wales'. *Field Studies* 5, 349–63.

BURRELL, R. J. W., ROACH, W. A. AND SHADWELL, A. (1966) 'Esophageal cancer in the Bantu of the Transkei associated with mineral deficiency in garden plants'. *Journal of the National Cancer Institute* 36, 201–209.

BURROUGH, P. A. (1986) *Principles of Geographical Information Systems*. Oxford: Oxford University Press.

BURROUGH, P. A. (2001) 'GIS and geostatistics: essential partners for spatial analysis'. *Environmental and Ecological Statistics* 8, 361–77.

BURROUGH, P. A. AND MCDONNELL, R. (1998) *Principles of Geographical Information Systems*. Oxford: Oxford University Press.

BURTON, L. R. (2003) 'The Mersey Basin: a

historical assessment of water quality from an anecdotal perspective'. *The Science of the Total Environment* 314–316, 53–66.

BUTLIN, R. N., COOTE, A. T., DEVENISH, M., HUGHES, I. S. C., HUTCHENS, C. M., IRWIN, J. G., LLOYD, G. O., MASSEY, S. W., WEBB, A. H. AND YATES, T. J. S. (1992a) 'Preliminary results from the analysis of stone tablets from the National Materials Exposure Programme (NMEP)'. *Atmospheric Environment* 26, 189–98.

BUTLIN, R. N., COOTE, A. T., DEVENISH, M., HUGHES, I. S. C., HUTCHENS, C. M., IRWIN, J. G., LLOYD, G. O., MASSEY, S. W., WEBB, A. H. AND YATES, T. J. S. (1992b) 'Preliminary results from the analysis of metal samples from the National Materials Exposure Programme (NMEP)'. *Atmospheric Environment* 26, 199–206.

CAIA, X., MCKINNEY, D. C. AND ROSEGRANT, M. W. (2003) 'Sustainability analysis for irrigation water management in the Aral Sea region'. *Agricultural Systems* 76, 1043–66.

CAPE, J. N., FOWLER, D. AND DAVIDSON, A. (2003) 'Ecological effects of sulfur dioxide, fluorides, and minor air pollutants recent trends and research needs'. *Environment International* 29, 201–11.

CARLEY, M. (1994) *Policy Management Systems and Methods of Analysis for Sustainable Agriculture and Rural Development.* Rome: IIED/FAO.

CARLSON, T. (2003) 'Application of remote sensing to urban problems'. *Remote Sensing of Environment* 86, 273–74.

CARPENTER, G. A., GOPAL, S., MACOMBER, S., MARTENS, S., WOODCOCK, C. E. AND FRANKLIN, J. (1999) 'A neural network method for efficient vegetation mapping'. *Remote Sensing of Environment* 70, 326–38.

CARSON, R. (1962) *Silent Spring.* Boston, Massachusetts: Houghton Mifflin.

CENTRE FOR SCIENCE AND ENVIRONMENT (1986) *The State of India's Environment 1984–5* (The Second Citizen's Report). New Delhi: Centre for Science and the Environment.

CHAKRABORTI, D., RAHMANA, M. M., PAULA, K., CHOWDHURYA, U. K., SENGUPTAA, M. K., LODHA, D., CHANDAA,

C. R., SAHAB, K. C. AND MUKHERJEEC, S. C. (2002) 'Arsenic calamity in the Indian subcontinent. What lessons have been learned?'. *Talanta* 58, 3–22.

CHAKRABORTI, D., MUKHERJEE, S. C., PATI, S., SENGUPTA, M. K., RAHMAN, M. M., CHOWDHURY, U. K., LODH, D., CHANDA, C. R., CHAKRABORTI, A. K. AND BASU, G. K. (2003) 'Arsenic groundwater contamination in Middle Ganga Plain, Bihar, India: a future danger?'. *Environmental Health Perspectives* 111, 1194–201.

CHANNELL, R. AND LOMOLINO, M. V. (2000) 'Dynamic biogeography and conservation of endangered species'. *Nature* 403, 84–86.

CHAPIN, F. S., III AND DANELL, K. (2001) Boreal forest. In F. S. Chapin III, O. E. Sala and E. Huber-Sannwald (eds) *Global Biodiversity in a Changing Environment: Scenarios for the 21st Century* (Ecological Studies 152), 101–20. Springer: New York.

CHAPMAN, D. (ed.) (1996) *Water Quality Assessments: A Guide to the Use of Biota, Sediments and Water in Environmental Monitoring,* 2nd edn. London: E. & F. N. Spon.

CHAPMAN, L. AND THORNES, J. E. (2003) 'The use of Geographical Information Systems in climatology and meteorology'. *Progress in Physical Geography* 27, 313–30.

CHAPMAN, P. (1993) *Caves and Cave Life* (New Naturalist Series No. 79). London: Harper Collins.

CHORLEY, R. J. AND KENNEDY, B. A. (1971) *Physical Geography: A Systems Approach.* London: Prentice Hall.

CLARK, D. H., BIERMAN, P. R. AND LARSEN, P. (1997) 'Improving *in situ* cosmogenic chronometers'. *Quaternary Research* 44, 367–77.

CLARKE, G., HADIWINOTO, S. AND LEITMAN, J. (1991) *Environmental Profile of Jakarta* (Draft paper). Washington DC: World Bank.

CLARKE, K. C. (2001) *Getting Started with Geographical Information Systems,* 3rd edn (Prentice Hall Series in Geographic Information Science). Upper Saddle River, New Jersey: Prentice Hall.

CLAUSSEN, M., BROVKIN. V., GANOPOLSI, A., KUBATZKI, C., PETOUKHOV, V. AND RAHMSTORF, S. (1999) 'A new model for climate system analysis: outline of the model and application to palaeoclimate simulations'. *Environmental Modeling and Assessment* 4, 209–16.

CLAYTON, K. M. (1989) 'Sediment input from the Norfolk cliffs, eastern England – a century of coast protection and its effects'. *Journal of Coastal Research* 5, 433–42.

CLERGEAU, P. AND BUREL, F. (1997) 'The role of spatio-temporal patch connectivity at the landscape level: an example in a bird distribution'. *Landscape and Urban Planning* 38, 37–43.

COHEN, J. B. AND RUSTON, A. G. (1912) *Smoke: A Study of Town Air.* London: Edward Arnold.

COLEMAN, P. (2001) Toxic Organic Micropollutant Monitoring 1999 to 2000. AEAT/R/ ENV/00301. Report available online at http://www. aeat.co.uk/netcen/airqual/reports/strategicpolicy/TO MPS_2000_report.pdf. Last accessed December 2003.

COLINVAUX, P. A. (1980) *Why Big Fierce Animals are Rare.* Harmondsworth, Middlesex: Pelican Books.

COLLS, J. (1997) *Air Pollution – An Introduction.* London: E. & F. N. Spon.

COLMAN, S. M. AND PIERCE, K. L. (2000) Classification of Quaternary dating methods. In J. S. Noller, J. M. Sowers and W. R. Lettis (eds) *Quaternary Geochronology: Methods and Applications* (AGU Reference Shelf 4), 2–5. Washington DC: American Geophysical Union.

COLVILLE, R. N., HUTCHINSON, E. J., MINDELL, J. S. AND WARREN, R. F. (2001) 'The transport sector as a source of pollution'. *Atmospheric Environment* 35, 1537–65.

COMMITTEE ON FORESTRY RESEARCH, NATIONAL RESEARCH COUNCIL (1990) *Forestry Research: A Mandate for Change.* Washington DC: The National Academy Press.

COMMITTEE ON THE MEDICAL EFFECTS OF AIR POLLUTANTS (COMEAP) (1995) *Non-biological Particles and Health* (Department of Health). London: HMSO.

CONNOR, E. F. AND MCCOY, E. D. (1979) 'The statistics and biology of the species–area relationship'. *The American Naturalist* 113, 791–833.

CONOVER, M. (2002) *Resolving Human–Wildlife Conflicts: The Science of Wildlife Damage Management.* Boca Raton: Lewis Publishers (A CRC Press Company).

COOKE, R. U. (1994) Salt weathering and the urban water table in deserts. In D. A. Robinson and R. B. G. Williams (eds) *Rock Weathering and Landform Evolution*, 193–205. Chichester: John Wiley & Sons.

COOKE, R. U., INKPEN, R. J. AND WIGGS, G. F. S. (1995) 'Using gravestones to assess changing rates of weathering in the United Kingdom'. *Earth Surface Processes and Landforms* 20, 531–46.

COOPER, P. F. AND FINDLATER, B. C. (eds) (1990) *Constructed Wetlands in Water Pollution Control.* Oxford: Pergamon Press.

CORBETT, J. AND FISCHBECK, P. (1997) 'Emissions from ships'. *Science* 278, 823–24.

CORBIT, M., MARKS, P. L. AND GARDESCU, S. (1999) 'Hedgerows as habitat corridors for forest herbs in central New York, USA'. *Journal of Ecology* 87, 220–32.

CORTNER, H. J. AND MOOTE, M. A. (1994) 'Trends and issues in land and water resources management: setting the agenda for change'. *Environmental Management* 18, 167–73.

CORWIN D. L., KAFFKA, S. R., HOPMANS, J. W., MORI, Y., VAN GROENIGEN, J. W., VAN KESSEL, C., LESCH, S. M. AND OSTER, J. D. (2003) 'Assessment and field-scale mapping of soil quality properties of a saline–sodic soil'. *Geoderma* 114, 231–59.

COUCLELIS, H. (1998) Computation in context. In P. A. Longley, S. Brooks, R. McDonnell and B. MacMillan (eds) *Geocomputation: A Primer*, 17–29. Chichester: John Wiley & Sons.

COX, G. W. (1999) *Alien Species in North America and Hawaii: Impacts on Natural Ecosystems.* Washington DC: Island Press.

CRABBE, H. (2003) Antipodean emissions sources and characteristics. Personal Communication August 2003.

CRAIGHEAD, F. (1979) *Track of the Grizzly.* San Francisco, California: Sierra Club Books.

CRESCIMANNO, G., IOVINO, M. AND PROVENZANO, G. (1995) 'Influence of salinity and sodicity on soil structural and hydraulic characteristics'. *Soil Science Society of America Journal* 59, 1701–708.

CRIGHTON, E. J., ELLIOTT, S. J., VAN DER MEERC, J., SMALLD, I. AND UPSHURA, R. (2003) 'Impacts of an environmental disaster on psychosocial health and well-being in Karakalpakstan'. *Social Science & Medicine* 56, 551–67.

CRITICAL LOADS ADVISORY GROUP (CLAG) (1994) *Critical Loads of Acidity in the United Kingdom.* London: DoE.

CROME, F. H. J. AND RICHARDS, G. C. (1988) 'Bats and gaps: microchiropteran community structure in a Queensland rainforest'. *Ecology* 69, 1960–69.

CROWLEY, T. J. (1991) Utilization of palaeoclimate results to validate projections of future greenhouse warming. In M. E. Schlesinger (ed.) *Greenhouse-gas-induced Climatic Change: A Critical Appraisal of Simulations and Observations*, 35–45. Amsterdam: Elsevier.

CRUTZEN, P. J. AND RAMANATHAN, V. (2000) 'The ascent of the atmospheric sciences'. *Science* 290, 299–304.

CURRAN, P. J. (2001) 'Remote sensing: using the spatial domain'. *Environmental and Ecological Statistics* 8, 321–44.

CURTIS, L. F. AND TRUDGILL, S. (1975) *The Measurement of Soil Moisture* (British Geomorphological Research Group, Technical Bulletin No. 13). Norwich, England: GeoAbstracts.

DANIEL, H. (2001) 'Replenishment versus retreat: the cost of maintaining Delaware's beaches'. *Ocean and Coastal Management* 44, 87–104.

DAVIDS, C. AND TYLER, A. N. (2002) 'Detecting contamination-induced tree stress within the Chernobyl exclusion zone'. *Remote Sensing of Environment* 85, 30–38.

DAVIDSON, D. A. (1992) *The Evaluation of Land Resources.* Harlow: Longman Scientific and Technical.

DAVIES, B. E. AND ANDERSON, R. J. (1987) 'The epidemiology of dental caries in relation to environmental trace elements'. *Experientia* 43, 87–92.

DAVIES, B., EAGLE, D. AND FINNEY, B. (1993) *Soil Management*, 5th edn. Ipswich: Farming Press Books.

DAY, M. AND URICH, P. (2000) 'An assessment of protected karst landscapes in Southeast Asia'. *Cave and Karst Science* 27, 61–70.

DE KLUIJVER H. AND STOTER, J. (2003) 'Noise mapping and GIS: Optimising quality, accuracy and efficiency of noise studies'. Internoise2000, Nice (France), 27-30 August 2000. Internet reference http://home.tiscali.nl/~felbers/publicaties/internoise2000_hk1.htm. Last accessed April 2004.

DEAN, J. S. (1997) Dendrochronology. In R. E. Taylor and M. J. Aitken (eds) *Chronometric Dating in Archaeology*, 31–64. New York: Plenum Press.

DENT, F. J. (1980) Major production systems and soil related constraints in Southeast Asia. In M. Drosdoff, H. Zandstra and W. G. Rockwood (eds) *Priorities for Alleviating Soil-related Constraints to Food Production in the Tropics*, 79–106. Manila, The Philippines: IRRI (International Rice Research Institute) and Ithaca: Cornell University.

DEPARTMENT OF TRANSPORT (DoT) (1996) *Digest of Transport Statistics.* London: HMSO.

DERWENT, R. G. (1995a) 'Improving air quality in the United Kingdom'. *Clean Air* 25, 70–94.

DERWENT, R. G. (1995b) Sources, distributions and fates of VOCs in the atmosphere. In R. E. Hester and R. M. Harrison (eds) *Volatile Organic Compounds in the Atmosphere* (Issues in Science and Technology), 1–16. Cambridge: Royal Society of Chemistry.

DIJAK, W. D. AND THOMPSON III, F. R. (2000) 'Landscape and edge effects on the distribution of mammalian predators in Missouri'. *Journal of Wildlife Management* 64, 209–16.

DIRZO, R. (2001) Tropical forests. In F. S. Chapin III, O. E. Sala and E. Huber-Sannwald (eds) *Global Biodiversity in a Changing Environment: Scenarios for the 21st Century* (Ecological Studies 152), 251–76. Springer: New York.

DIXON, L. F. J., BARKER, R., BRAY, M., FARRES, P., HOOKE, J., INKPEN, R., MEREL, A., PAYNE, D. AND SHELFORD, A. (1998) Analytical photogrammetry for geomorphological research. In S. Lane, K. Richards and J. Chandler (eds) *Landform Monitoring, Modelling and Analysis*, 63–94. Chichester: John Wiley & Sons.

DIXON, M. D. AND JOHNSON, W. C. (1999) 'Riparian vegetation along the Middle Snake River,

Idaho: zonation, geographical trends, and historical changes'. *Great Basin Naturalist* 59, 18–34.

DoE (Department of the Environment) (1987) *Report of the Committee of Enquiry into the Handling of Geographic Information* (the 'Chorley Report'). London: HMSO.

DORAN, J. W. AND PARKIN, T. B. (1994) Defining and assessing soil quality. In J. W. Doran, D. C. Coleman, D. F. Bezdicek and B. A. Stewart (eds) *Defining Soil Quality for a Sustainable Environment* (Soil Science Society of America Special Publication No. 35), 3–21. Madison, Wisconsin: Soil Science Society of America.

DOUGLAS, I. (1971) 'Dynamic equilibrium in applied geomorphology – two case studies'. *Earth Science Journal* 5, 29–35.

DOUGLAS, I. (1983) *The Urban Environment*. London: Arnold.

DOUGLAS, I. (1986) 'The unity of geography is obvious'. *Transactions of the Institute of British Geographers*, New Series 11, 517–40.

DOUGLAS, I. (2000) 'Fluvial geomorphology and river management'. *Australian Geographical Studies* 38, 253–62.

DOUGLAS, I. AND LAWSON, N. (2001) 'The human dimensions of geomorphological work in Britain'. *Journal of Industrial Ecology* 4, 9–33.

DOUGLASS, M. (1989) 'The environmental sustainability of development; coordination incentives and political will in land use planning for the Jakarta meteropolis'. *Third World Planning Review* 11, 2.

DUGGINS, D. O. (1980) 'Kelp beds and sea otters: an experimental approach'. *Ecology* 61, 447–53.

DUGUÉ, P., ROOSE, É. AND RODRIGUEZ, L. (1993) 'L'aménagement de terroirs villageois et l'amélioration de la production agricole au Yatenga (Bukina Faso)'. *Cahier Orstrom, Série Pédologie* 28, 385–402.

DULLER, G. A. T. (2000) 'Dating methods: geochronology and landscape evolution'. *Progress in Physical Geography* 24, 111–16.

DUMITRU, T. A. (2000) Fission-track geochronology. In J. S. Noller, J. M. Sowers and W. R. Lettis (eds) *Quaternary Geochronology: Methods and Applications* (AGU Reference Shelf 4), 131–55. Washington DC: American Geophysical Union.

DUNCAN, B. W., BOYLE, S., BREININGER, D. R. AND SCHMALZER, P. A. (1999) 'Coupling past management practice and historic landscape change on John F. Kennedy Space Center, Florida'. *Landscape Ecology* 14, 291–309.

DUNN, S. M., STALHAMB, M., CHALMERSA, N. AND CRABTREE, B. (2003) 'Adjusting irrigation abstraction to minimise the impact on stream flow in the East of Scotland'. *Journal of Environmental Management* 68, 95–107.

DUNNE, T. AND LEOPOLD, L. B. (1978) *Water in Environmental Planning*. San Francisco: W. H. Freeman.

EEA (1999) *Nutrients in European Ecosystems* (Environmental Assessment Report No. 4). Copenhagen: European Environment Agency.

EEA (2001) *Sustainable Water Use in Europe Part 3: Extreme Hydrological Events: Floods and Droughts* (Environmental Issue Report No. 21). Copenhagen: European Environment Agency.

EHLERS, J. (1996) *Quaternary and Glacial Geology* (English version by P. L. Gibbard). Chichester: John Wiley & Sons.

ELLIS, J. B., REVITT, D. M., SHUTES, J. M. AND LANGLEY, J. M. (1994) 'The performance of vegetated biofilters for highway run off control'. *The Science of the Total Environment* 146/147, 543–50.

ELLIS, S. AND MELLOR, A. (1995) *Soils and Environment*. London: Routledge.

ELSOM, D. (1992) *Atmospheric Pollution: A Global Perspective*. Oxford: Blackwell.

ELSOM, D. (1996) *Smog Alert: Managing Urban Air Quality*. London: Earthscan.

EMBERSON, L. D., ASHMORE, M. R., MURRAY, F., KUYLENSTRIENA, J. C. I., PERCY, K. E., IZUTA, T., ZHENG, Y., SHIMIZU, H., SHEU, B. H., LIU, C. P., AGRAWAL, M., WAHID, A., ABDEL-LATIF, N. M., VAN TIENHOVEN, M., DE BAUER, L. I. AND DOMINGOS, M. (2001) 'Impacts of air pollution on vegetation in developing countries'. *Water Air and Soil Pollution* 130, 107–88.

ENVIRONMENT AGENCY (2003a) *Pollution Inventory Substances Factsheet – Carbon Dioxide*. Web address: http:216.31.193.171/asp/1_search_pisubstancehep.asp?id=31. Last accessed 28 July 2003.

ENVIRONMENT AGENCY (2003b) *Pollution Inventory Substances Factsheet – Methane*. Web address: http://216.31.193.171/asp/1_search_pisubstancehep.asp?id=????. Last accessed 28 July 2003.

ENVIRONMENT AGENCY (2003c) *Pollution Inventory Substances Factsheet – Nitrous Oxide*. Web address: http://216.31.193.171/asp/1_search_pisubstancehep.asp?id=440. Last accessed 28 July 2003.

ENVIRONMENT AGENCY (2003d) *Pollution Inventory Substances Factsheet – Hydroflurocarbons (HFCs)*. Web address: http://216.31.193.171/asp/1_search_pisubstancehep.asp?id=556. Last accessed 28 July 2003.

ENVIRONMENT AGENCY (2003e) *Pollution Inventory Substances Factsheet – Perflurocarbons (PFCs)*. Web address: http://216.31.193.171/asp/1_search_pisubstancehep.asp?id=1041. Last accessed 28 July 2003.

ENVIRONMENT AGENCY (2003f) *Pollution Inventory Substances Factsheet – Sulphur Hexaflouride*. Web address: http://216.31.193.171/asp/1_search_pisubstancehep.asp?id=1044. Last accessed 28 July 2003.

ERISMAN, J. W., GRENNFELT, P. AND SUTTON, M. (2003) 'The European perspective on nitrogen deposition and emission'. *Environment International* 29, 311–25.

EUROPEAN ENVIRONMENT AGENCY (2003) *Europe's Environment: The Third Assessment*. Copenhagen: EEA.

FAO (1995) *Irrigation in Africa in Figures* (Water Report No. 7). Rome: Food and Agriculture Organization of The United Nations.

FAURE, G. (1986) *Principles of Isotope Geology*, 2nd edn. New York: John Wiley & Sons.

FEDDES, R. A., KABAT, P., VAN BAKAL, P. J. T., BRONSWIJK, J. J. B. AND HABERTSMA, J. (1988) 'Modelling soil water dynamics in the unsaturated zone – state of the art'. *Journal of Hydrology* 100, 69–111.

FEDRA, K. (1993) 'Clean air – air quality modelling and management'. *Mapping Awareness and GIS Europe* 7, 24–27.

FENGER, J. (1999) 'Urban air quality'. *Atmospheric Environment* 33, 4877–900.

FERRERUELA A., WADSWORTH, R., FRANCE, G. AND PLUMMER, S. (2002) Mapping 'historical' fire scars in the Canadian boreal forests. In Wise, S., Brindley, P., Young, H. K. and Openshaw, C. (eds) *10th Annual Conference in the UK Conference Proceedings (GISRUK)*, 3rd–5th April 2002, Sheffield: University of Sheffield.

FETTER, C. W. (1994) *Applied Hydrogeology*, 3rd edn. Englewood Cliffs, New Jersey: Prentice Hall.

FETTER, C. W., JR (1999) *Applied Hydrogeology*, 4th edn. Upper Saddle River, NJ: Prentice Hall.

FISCHER, G. AND SCHRATTENHOLZER, L. (2001) 'Global bioenergy potentials through 2050'. *Biomass and Bioenergy* 20, 151–59.

FITZPATRICK, E. A. (1974) *An Introduction to Soil Science*. Edinburgh: Oliver & Boyd.

FLORINSKY, I. V. (1998) 'Combined analysis of digital terrain models and remotely sensed data in landscape investigations'. *Progress in Physical Geography* 22, 33–60.

FOLEY, G. (1992) *The Energy Question*, 4th edn. London: Penguin.

FORMAN, R. T. T. (1995) *Land Mosaics: The Ecology of Landscapes and Regions*. Cambridge: Cambridge University Press.

FORMAN, S. L., PIERSON, J. AND LEPPER, K. (2000) Luminescence geochronology. In J. S. Noller, J. M. Sowers and W. R. Lettis (eds) *Quaternary Geochronology: Methods and Applications* (AGU Reference Shelf 4), 157–76. Washington DC: American Geophysical Union.

FORRESTER, J. W. (1971) *World Dynamics*. Cambridge, Massachusetts: Wright-Allen.

FRANKLIN, J., WOODCOCK, C. E. AND WARBINGTON, R. (2000) 'Multi-attribute vegetation maps of Forest Service lands in California supporting resource management decisions'. *Photogrammetric Engineering and Remote Sensing* 66, 1209–17.

FRENCH, H. M. (1996) *The Periglacial Environment*, 2nd edn. Harlow: Addison-Wesley Longman.

GALLOWAY, J. N. AND COWLING, E. B. (2002) 'Reactive nitrogen and the world: 200 years of change'. *Ambio* 31, 64–71.

GAO, J. AND LIU, Y. (2001) 'Applications of

remote, sensing, GIS and GPS in glaciology: a review'. *Progress in Physical Geography* 25, 520–40.

GARDINER, V. AND DACKOMBE, R. (1983) *Geomorphological Field Manual*. London: Allen and Unwin.

GARDNER, R. (1996) Developments in physical geography. In E. M. Rawling and R. A. Daugherty (eds) *Geography into the Twenty-first Century*, 95–112. Chichester: John Wiley & Sons.

GAUDIOSI, G. (1999) 'Offshore energy prospects'. *Renewable Energy* 16, 828–34.

GELDOF, G. D. AND VAN DE VEN, F. H. M. (1998) Urban water management. In P. Huisman, W. Cramer, G. van Ee, J. C. Hooghart, H. Salz and F. C. Zuidema (ed. Committee) (1998) *Water in the Netherlands*, 128–33. Delft: Netherlands Hydrological Society (NHV).

GILLHAM, C. A., COULING, S., LEECH, P. K., EGGLESTON, H. S. AND IRWIN, J. G. (1994) *UK Emissions of Atmospheric Pollutants 1970–1991 (including methodology update)* (LR 961). Stevenage: WSL.

GILLIESON, D. (1996) *Caves: Processes, Development and Management*. Oxford: Blackwell.

GINSBERG, E. O. AND COWLING, E. B. (2003) 'Future directions in air-quality science, policy, and education'. *Environment International* 29, 125–35.

GLASSER, N. F. AND BARBER, G. (1995) 'Cave conservations plans: the role of English Nature'. *Cave and Karst Science* 21, 33–36.

GLEICK, P. H. (1998) 'The human right to water'. *Water Policy* 1, 487–503.

GLEICK, P. H. (2001) 'Making every drop count'. *Scientific American* 284 (February), 28–33, 40–45.

GONZÁLEZ, O. M. (2001) 'Assessing vegetation and land cover changes in northeastern Puerto Rico: 1978–1995'. *Caribbean Journal of Science* 37, 95–106.

GOODKIND, D. AND WEST, L. A. (2002) 'China's floating population: definitions, data and recent findings'. *Urban Studies* 39, 2237–50.

GOUDIE, A. S. (1986) 'The integration of human and physical geography'. *Transactions of the Institute of British Geographers*, New Series 11, 454–58.

GOUDIE, A. S. (ed.) (1990) *Geomorphological Techniques*. London: Routledge.

GRAEDEL, T. E. AND KEENE, W. C. (1999) 'Overview: Reactive Chlorine Emissions Inventory'. *Journal of Geophysical Research* 104, 8331–33.

GRAY, J. M. (1993) 'Quaternary geology ands waste disposal in south Norfolk, England'. *Quaternary Science Reviews* 12, 899–912.

GREGORY, K. J. (2000) *The Changing Nature of Physical Geography*. London: Arnold.

GRIGG, D. (1995) *An Introduction to Agricultural Geography*, 2nd edn. London: Routledge.

GRIMMOND, C. S. B. AND OKE, T. R. (1986) 'Urban water balance. 2. Results from a suburb of Vancouver, British Columbia'. *Water Resources Research* 22, 1404–12.

GRISSINO-MAYER, H. D. (2004) Ultimate Tree-Ring Web Pages [online]. Available at: http://web.utk.edu/~grissino/principles.htm. Last accessed 21 January 2004.

GRUMBINE, R. E. (1994) 'What is ecosystem management?'. *Conservation Biology* 8, 27–38.

GRÜN, R. (1997) Electron spin resonance dating. In R. E. Taylor and M. J. Aitken (eds) *Chronometric Dating in Archaeology*, 217–60. New York: Plenum Press.

GUNN, J. AND SANDOY, S. (2001) 'Northern Lakes Recovery Study (NLRS) Biomonitoring at the ecosystem level'. *Water, Air and Soil Pollution* 130, 131–40.

GUPTA, S. C. AND LARSON, W. E. (1979) 'Estimating soil water retention characteristics from particle size distribution, organic matter percent, and bulk density'. *Water Resources Research* 15, 1633–35.

HAKLAY, M. E. (2000) 'Public access to environmental information – challenges and research directions: a review of current issues with environmental information provision'. Presented at the *Fourth International Conference on Integrating GIS and Environmental Modelling GIS/EM4 Problems, Prospects and Research Needs*. Banff, Ontario, Canada 2–4 September 2000. Available on-line at http://www.casa.ucl.ac.uk/muki/envinfo.htm. Last accessed 2 January 2004.

HAMZA, M. AND ZETTER, R. (1998) 'Structural adjustment, urban systems and disaster vulnerability in developing countries'. *Cities* 15, 291–99.

HANCOCK, G. S., ANDERSON, R. S., CHADWICK, O. A. AND FINKEL, R. C. (1997)

'Dating fluvial terraces with [10]Be and [26]Al profiles: application to the Wind River, Wyoming'. *Geomorphology* 27, 41–60.

HANNAH, L., LOHSE, D., HUTCHINSON, C., CARR, J. L. AND LANKERANI, A. (1994) 'A preliminary inventory of human disturbance of world ecosystems'. *Ambio* 23, 246–50.

HARDOY, J. E., MITLIN, D. AND SATTERTHWAITE, D. (1999) The rural, regional and global impacts of cities in Africa, Asia and Latin America. In D. Satterthwaite (ed.) *Sustainable Cities*, 426–60. London: Earthscan.

HARDY, J. T. (2003) *Climate Change – Causes, Effects and Controls.* Chichester: John Wiley and Son.

HARE, P .E., VON ENDT, D. W. AND KOKIS, J. E. (1997) Protein and amino acid diagenesis dating. In R. E. Taylor and M. J. Aitken (eds) *Chronometric Dating in Archaeology*, 261–96. New York: Plenum Press.

HARMAN, J. R., HARRINGTON, J. A. AND CERVENY, R. S. (1998) 'Balancing scientific and ethical values in environmental science'. *Annals of the Association of American Geographers* 88, 277–86.

HARMS, W. B. AND OPDAM, P. (1990) Woods as habitat patches for birds: application in landscape planning in The Netherlands. In I. S. Zonneveld and R. T. T. Forman (eds) *Changing Landscapes: An Ecological Perspective*, 73–97. New York: Springer.

HARRIS, L. D. (1984) *The Fragmented Forest: Island Biogeography Theory and the Preservation of Biotic Diversity.* With a Foreword by Kenton R. Miller. Chicago and London: Chicago University Press.

HARRISON, P. A., BERRY, P. M. AND DAWSON, T. P. (2001) *Climate Change and Nature Conservation in Britain and Ireland: Modelling Natural Resource Response to Climate Change* (The MONARCH Report). Oxford: UKCIP Technical Report.

HAUGHTON, G. (1999) 'Environmental justice and the sustainable city'. *Journal of Planning Education and Research* 18 (3), 233–43. Association of Collegiate Schools of Planning. Reprinted in Satterthwaite, D. (ed.) (1999) *The Earthscan Reader in Sustainable Cities.* London: Earthscan, 62-79.

HAYMAN, G. (2000) Acid Deposition Monitoring in the UK: 1986–1998 AEAT/EEQC-0143. Report available online at http://www.aeat.co.uk/netcen/airqual/reports/acidrain/acid98_v3.pdf. Last accessed December 2003.

HELLAWELL, J. M. (1989) *Biological Indicators of Freshwater Pollution and Environmental Management.* London: Elsevier Applied Science.

HENEIN, K., WEGNER, J. AND MERRIAM, G. (1998) 'Population effects of landscape model manipulation on two behaviourally different woodland small mammals'. *Oikos* 81, 168–86.

HEYWOOD, I., CORNELIUS, S. AND CARVER, S. (2002) *An Introduction to Geographical Information Systems*, 2nd edn. Harlow: Prentice Hall.

HILLEL, D. (1998) *Environmental Soil Physics.* San Diego, California: Academic Press.

HINSLEY, S. A., BELLAMY, P. E., ENOKSSON, B., FRY, G., GABRIELSEN, L., MCCOLLIN, D. AND SCHOTMAN, A. (1998) 'Geographical and land-use influences on bird species richness in small woods in agricultural landscapes'. *Global Ecology and Biogeography Letters* 7, 125–35.

HOUGH, M., PALMER, S., WEIR, A., LEE, M. AND BARRIE, I. (1997) *The Meteorological Office Rainfall and Evaporation Calculation System: MORECS Version 2.0 1995.* London: HMSO.

HOUGHTON, J. T., DING, Y., GRIGGS, D. J., NOQUET, M., VAN DER LINDEN, J. P., DAI, X., MASKELL, K. AND JOHNSON, C. A. (eds) (2001) *Climate Change 2001: The Scientific Basis: Contribution of Working Group I to the Third Assessment Report of the Intergovernmental Panel on Climate Change.* Cambridge: Cambridge University Press and the Intergovernmental Panel on Climate Change.

HOULAHAN, J. E., FINDLAY, C. S., SCHMIDT, B. R., MEYER, A. H. AND KUZMIN, S. L. (2000) 'Quantitative evidence for global amphibian population declines'. *Nature* 404, 752–55.

HUENNEKE, L. F. (2001) Deserts. In F. S. Chapin III, O. E. Sala and E. Huber-Sannwald (eds) *Global Biodiversity in a Changing Environment: Scenarios for the 21st Century* (Ecological Studies 152), 201–22. Springer: New York.

HUGGETT, R. J. (1980) *System Analysis in Geography.* Oxford: Clarendon Press.

HUGGETT, R. J. (1985) *Earth Surface Systems* (Springer Series in the Environment, Volume 1). Heidelberg: Springer.

HUGGETT, R. J. (1993) *Modelling the Human Impact on Nature: Systems Analysis of Environmental Problems.* Oxford: Oxford University Press.

HUGGETT, R. J. (1995) *Geoecology: An Evolutionary Approach.* London: Routledge.

HUGGETT, R. J. (1997) *Environmental Change: The Evolving Ecosphere.* London: Routledge.

HUGGETT, R. J. (1998) *Fundamentals of Biogeography.* London: Routledge.

HUGGETT, R. J. (2003) *Fundamentals of Geomorphology.* London: Routledge.

HUGGETT, R. J. AND CHEESMAN, J. E. (2002) *Topography and the Environment.* Harlow: Prentice Hall.

HULME, D. AND MURPHREE, M. (1999) 'Communities, wildlife and "new conservation" in Africa'. *Journal of International Development* 11, 277–85.

HUNTINGTON, H. P. AND MYMRIN, N. I. (2002) *Beluga Whale TEK: Traditional Ecological Knowledge of the Beluga Whale* (www.si.edu/arctic/html/tek.html). Last accessed July 2003.

HUSAIN, T. (1998) 'Terrestrial and atmospheric environment during and after the Gulf War'. *Environment International* 24, 189–96.

HUSZAR, P.C. AND PIPER, S.L. (1986) 'Estimating the off-site costs of wind erosion in New Mexico'. *Journal of Soil and Water Conservation*, 41, 414–16.

HYLANDER, L. D., SILVA, E. C., OLIVEIRA, L. J., SILVA, S. A., KUNTZE, E. K. AND SILVA, D. X. (1994) 'Mercury levels in Alto Pantanal: a screening study'. *Ambio* 23, 478–84.

INKPEN, R. J., COOKE, R. U. AND VILES, H. A. (1994) Processes and rates of urban limestone weathering. In D. A. Robinson and R. B. G. Williams (eds) *Rock Weathering and Landform Evolution*, 119–30. Chichester: John Wiley & Sons.

INNES, J. L. (1985) 'Lichenometry'. *Progress in Physical Geography* 9, 187–254.

INSTITUTE OF HYDROLOGY (1999) *Flood Estimation Handbook – Procedures for Flood Frequency Estimation*, 5 vols. Wallingford: Institute of Hydrology.

IPCC (1996) *Climate Change 1995: The Science of Climate Change.* Cambridge: IPCC and Cambridge University Press.

IPCC (2001) *IPCC Climate Change 2001: Synthesis Report Summary for Policymakers.* Geneva: IPCC and World Meteorological Society.

ISO (1980) *Water Flow Measurement in Open Channels Using Weirs and Venturi Flumes – Part 1: Thin Plate Weirs.* ISO. 1438-1 International Organization of Standards.

IUCN (International Union for Conservation of Nature and Natural Resources) (2003) *Red List of Threatened Species 2003.* Internet reference at http://www.redlist.org. Last accessed November 2003.

IYENGAR, G. V. AND NAIR, P. P. (2000) 'Global outlook on nutrition and the environment: meeting the challenges of the next millennium'. *The Science of the Total Environment* 249, 331–46.

JACOBY, G. C. (2000) Dendrochronology. In J. S. Noller, J. M. Sowers and W. R. Lettis (eds) *Quaternary Geochronology: Methods and Applications* (AGU Reference Shelf 4), 11–20. Washington DC: American Geophysical Union.

JENNINGS, J. N. (1971) *Karst* (An Introduction to Systematic Geomorphology, Vol. 7). Cambridge, Massachusetts and London, England: The MIT Press.

JENSEN, S., MAZHITOVA, Z. AND ZETTERSTROM, R. (1997) 'Environmental pollution and child health in the Aral Sea region in Kazakhstan'. *The Science of the Total Environment* 206, 187–93.

JOHANSSON, R. C., TSURB, Y., ROEC, T. L., DOUKKALID, R. AND DINARE, A. (2002) 'Pricing irrigation water: a review of theory and practice'. *Water Policy* 4, 173–99.

JOHNSON, M. P. AND SIMBERLOFF, D. S. (1974) 'Environmental determinants of island species numbers in the British Isles'. *Journal of Biogeography* 1, 149–54.

JOHNSON, R. H. AND VAUGHAN R. D. (1983) 'The Alport Castles, Derbyshire: a South Pennine slope and its geomorphic history'. *The East Midland Geographer* 8, 79–88.

JOHNSTON, R. J. (1986) 'Four fixations and the

quest for unity in geography'. *Transactions of the Institute of British Geographers*, New Series 11, 449–53.

JOHNSTON, R. J. (1997) *Geography and Geographers: Anglo-American Human Geography Since 1945*, 5th edn. London: Arnold.

JONES, J. A. A. (1997) *Global Hydrology: Processes, Resources and Environmental Management*. Harlow: Longman.

KARL, T. R. AND KNIGHT, R. W. (1998) 'Secular trends of precipitation amount and frequency and intensity in the United States'. *Bulletin of the American Meteorological Society* 79, 231–41.

KAUFMANN, M. R., GRAHAM, R. T., BOYCE, D. A. JR, MOIR, W. H., PERRY, L., REYNOLDS, R. T., BASSETT, R. L., MEHLHOP, P., EDMINSTER, C. B., BLOCK, W. M. AND CORN, P. S. (1994) *An Ecological Basis for Ecosystem Management* (General Technical Report RM–246, United States Department of Agriculture, Forest Service, Rocky Mountain Forest and Range Experiment Station, Fort Collins, Colorado). Washington, DC: United States Government Printing Office.

KEELING, C. D. (2003) Mauna Loa carbon dioxide data. Internet site http://ingrid.ldgo.coluimbroia.edu/sources/.keeling/.mauna_loa.cdf. Last accessed September 2003.

KENNEDY, M. (2002) *The Global Positioning System and GIS: An Introduction*, 2nd edn. New York: Taylor & Francis.

KENWORTHY, J. AND LAUBE, F. (1999) 'A global review of energy use in transport systems and its implication for urban transport and land-use policy'. *Transportation Quarterly* 53, 23–48.

KESSLER, W. B., SALWASSER, H., CARTWRIGHT, C., JR AND CAPLAN, J. (1992) 'New perspective for sustainable natural resources management'. *Ecological Applications* 2, 221–25.

KEYZER, M. A. AND SONNEVELD, B. G. J. S. (1998) 'Using the mollifier method to characterize datasets and models: the case of the Universal Soil Loss Equation'. *ITC Journal* 3–4, 263–72.

KIRKBY, A. V. T. AND KIRKBY, M. J. (1974) 'Surface wash at the semi-arid break in

slope'. *Zeitschrift für Geomorphologie Supplementband* NF 21, 151–76.

KIVAISI, A. K. (2001) 'The potential for constructed wetlands for wastewater treatment and reuse in developing countries: a review'. *Ecological Engineering* 16, 545–60.

KNAAPEN, J. P., SCHEFFER, M. AND HARMS, W. B. (1992) 'Estimating habitat isolation in landscape planning'. *Landscape and Urban Planning* 23, 1–16.

KNIGHT, R. L. AND KAWASHIMA, J. Y. (1993) 'Responses of raven and red-tailed hawk populations to linear right-of-ways'. *Journal of Wildlife Management* 57, 266–71.

KNUTSON, R. (1987) *Flattened Fauna: A Field Guide to Common Animals of Roads, Streets and Highways*. Berkeley, California: Ten Speed Press.

KOHL, B. (2002) 'Stabilizing neoliberalism in Bolivia: popular participation and privatization'. *Political Geography* 21, 449–72.

KOHUT, R. (2003) 'The long-term effects of carbon dioxide on natural systems: issues and research needs'. *Environment International* 29, 171–80.

KOMIVES, K. (2001) 'Designing pro-poor water and sewer concessions: early lessons from Bolivia'. *Water Policy* 3, 61–79.

KOUSSOULAKOU, A. (1994) Spatial-temporal analysis of urban air pollution. In A. M. MacEachren and D. R. Fraser-Taylor (eds) *Visualization in Modern Cartography* (Modern Cartography, Vol. 2), 243–55. London: Pergamon Press,

KOZOL, A. J., TRANIELLO, J. F. A. AND WILLIAMS, F. M. (1994) 'Genetic variation in the endangered burying beetle *Nicrophorus americanus* (Coleoptera: Silphidae)'. *Annals of the Entomological Society of America* 87, 928–35.

KRAAK, M-J. AND ORMELING, F. (2003) *Cartography Visualisation of Geospatial Data*. Harlow: Pearson Education.

KRAUSS, J., STEFFAN-DEWENTER, I. AND TSCHARNTKE, T. (2003) 'How does landscape context contribute to effects of habitat fragmentation on diversity and population density of butterflies?'. *Journal of Biogeography* 30, 889–900.

KRINGS, T. (2002) 'A critical review of the Sahel Syndrome concept from the view of political ecology'. *Geographische Zeitschrift* 90, 129–41.

KROEZE, C. (2003) Review of information currently on the GEIA files concerning N_2O. Global Atmospheric Emissions Inventory project. http://www.Geiacenter.org. Last accessed September 2003.

KU, T.-L. (2000) Uranium-series methods. In J. S. Noller, J. M. Sowers and W. R. Lettis (eds) *Quaternary Geochronology: Methods and Applications* (AGU Reference Shelf 4), 101–14. Washington DC: American Geophysical Union.

KUENY, J. A. AND DAY, M. J. (1998) 'An assessment of protected karst landscapes in the Caribbean'. *Caribbean Geography* 9, 87–100.

KUTZBACH, J. E. AND STREET-PERROTT, F. A. (1985) 'Milankovitch forcing in fluctuations in the level of tropical lakes from 18 to 0 kyr BP'. *Nature* 317, 130–34.

KUYLENSTRIENA, J. C. I., HICKS, W. K. CINDERBY, S., VALLACK, H. W. AND ENGARDT, M. (2001) 'Acidification in developing countries: ecosystem sensitivity and the critical load approach on a global scale'. *Ambio* 30, 20–28.

LAL, D. (1991) 'Cosmic ray labeling of erosion surfaces: *in situ* nuclide production rates and erosion models'. *Earth and Planetary Science Letters* 104, 424–39.

LANE, S., RICHARDS, K. AND CHANDLER, J. (1998) Landform monitoring, modelling and analysis: land *form* in geomorphological research. In Lane, S., Richards, K. and Chandler, J. (eds) *Landform Monitoring, Modelling and Analysis*, 1–17. Chichester: John Wiley & Sons.

LANIER-GRAHAM, S. D. (1993) *The Ecology of War: Environmental Impacts of Weaponry and Warfare*. New York: Walker and Company.

LAWRENCE, P. L., NELSON, J. G. AND PEACH, G. (1993) 'Great Lakes shoreline management plan for the Saugeen Valley Conservation Authority'. *Operational Geographer* 11, 26–33.

LEFOHN, A. S., JANJA, D. H. AND HUSAR, R. B. (1999) 'Estimating historical anthropogenic global sulphur emission patterns for the period 1850 to 1990'. *Atmospheric Environment* 33, 3435–44.

LEFSKY, M. A., COHEN, W. B., PARKER, G. G. AND HARDING, D. J. (2002) 'Lidar remote sensing for ecosystem studies'. *BioScience* 52, 19–30.

LERER, L. B. AND SCUDDERB, T. (1999) 'Health impacts of large dams'. *Environmental Impact Assessment Review* 19, 113–23.

LI, L.-A., WANG, C.-H., LIAW, S.-H., YUNG, Y.-L., LI, Y.-H., CHEN, Y.-C. AND LI, P.-C. (1997) 'The impact of worldwide volcanic activities on local precipitation'. *Journal of the Geological Society of China* 40, 299–311.

LICZNAR, P. AND NEARING, M. A. (2003) 'Artificial neural networks of soil erosion and runoff prediction at the plot scale'. *Catena* 51, 89–114.

LILLESAND, T. M. AND KIEFER, R. W. (2000) *Remote Sensing and Image Interpretation*, 4th edn. Chichester: John Wiley & Sons.

LIN, C. Y., LIN, W. T. AND CHOU, W.C. (2002) 'Soil erosion prediction and sediment yield estimation: the Taiwan experience'. *Soil & Tillage Research* 68, 143–52.

LIVERNASH, R. AND RODENBURG, E. (1998) 'Population change, resources and the environment'. *Population Bulletin* 53, 12–42.

LIVINGSTONE, I. AND WARREN, A. (1996) *Aeolian Geomorphology: An Introduction*. Harlow: Longman.

LLOBET, J. M., DOMINGO, J. L., BOCIO, A., CASAS, C., TEIXIDO, A. AND MULLER, L. (2003) 'Human exposure to dioxins through the diet in Catalonia, Spain: carcinogenic and non carcinogenic risk'. *Chemosphere* 50, 1193–200.

LOADER, A. (2001) UK Nitrogen Dioxide Network. AEAT/ENV/R/0669. Report available online at http://www.aeat.co.uk/netcen/airqual/reports/nonauto/no2_report_2000.pdf. Last accessed December 2003.

LODÉ, T. (2000) 'Effect of a motorway on mortality and isolation of wildlife populations'. *Ambio* 29, 163–66.

LONGLEY, P., GOODCHILD, M. AND RHIND, D. (2001) *Geographic Information Systems and Science*. Chichester: John Wiley and Sons.

LOWE, J. J. AND WALKER, M. J. C. (1997) *Reconstructing Quaternary Environments*, 2nd edn. Harlow: Longman.

LOWMAN, P., YATES, J., O'LEARY, D., SALISBURY, D., MASUOKA, P. AND MONTGOMERY, B. (1999) 'A digital tectonic activity map of the Earth'. *Journal of Geoscience Education* 47, 5.

LÜDEKE, M. K. B., PETSCHEL, G. AND SCHELLNHUBER, H.-J. (2002) Syndromes of global change: the first panoramic view – Frontiers in Ecology and the Environment WGBU. Internet reference at http://www.tkuentzel.de/pik/9syndrom.html. Last accessed December 2003.

LUFAFA, A., TENYWA, M. M., ISABIRYE, M., MAJALIWA, M. J. G. AND WOOMER, P. L. (2002) 'Prediction of soil erosion in Lake Victoria basin catchment using a GIS-based Universal Soil Loss model'. *Agricultural Systems* 76, 883–94.

LUNDARDINI, V. J. (1996) 'Climatic warming and the degradation of warm permafrost'. *Permafrost and Periglacial Processes* 7, 311–20.

LUTGENS, F. K. AND TARBUCK, E. J. (1998) *The Atmosphere: An Introduction to Meteorology.* Upper Saddle River, New Jersey: Prentice Hall International Inc.

LUTZ, W., SANDERSON, W. AND SCHERBOV, S. (2001) 'The end of world population growth'. *Nature* 412, 543–45.

LYON, J. G. (ed.) (2003) *GIS for Water Resources and Watershed Management* London: Taylor & Francis.

MABRY, K. E., DREELIN, E. A. AND BARRETT, G. W. (2003) 'Influence of landscape elements on population densities and habitat use of three small-mammal species'. *Journal of Mammalogy* 84, 20–25.

MACKENZIE, S. H. (1993) 'Ecosystem management in the Great Lakes: some observations from three RAP sites'. *Journal of Great Lakes Management* 19, 136–44.

MACKEY, E. C. AND TUDOR, G. J. (2000) Land cover changes in Scotland over the past 50 years. In R. W. Alexander and A. C. Millington (eds) *Vegetation Mapping: From Patch to Planet*, 235–58. Chichester: John Wiley & Sons.

MAFF (2000) *Climate Change and Agriculture in the United Kingdom*. London: MAFF Publications.

MAHER, B. A., THOMPSON, R. AND HOUNSLOW, M.W. (1999) Introduction. In B. A. Maher and R. Thompson (eds) *Quaternary Climates, Environments and Magnetism*, 1–48. Cambridge: Cambridge University Press: Cambridge.

MAIDMENT, D. R. (ed.) (2002) *ArcHydro GIS for Water Resources*. Redlands: ESRI Press.

MALALLAHA, G., AFZALB, M., KURIANB, M., GULSHANB, S. AND DHAMIC, M. S. I. (1998) 'Impact of oil pollution on some desert plants'. *Environment International* 24, 919–24.

MALTBY, E. (1986) *Waterlogged Wealth: Why Waste the World's Wet Places?* London: International Institute for Environment and Development.

MANCHESTER AREA COUNCIL FOR CLEAN AIR AND NOISE CONTROL (MACCANC) (1982) *25-Year Review*. Manchester: MACCANC.

MANIES, K. L. AND MLADENOFF, D. J. (2000) 'Testing methods to produce landscape-scale presettlement vegetation maps from the U.S. public land survey records'. *Landscape Ecology* 15, 741–54.

MANSFIELD, T. A. AND LUCAS P. W. (1996) 'Effects of gaseous pollutants on crops and trees'. In R. M. Harrison (ed.) *Pollution: Causes, Effects and Control*, 3rd edn, 296–326. Cambridge: Royal Society of Chemistry.

MARGULIS, L. AND SCHWARTZ, K. V. (1998) *Five Kingdoms: An Illustrated Guide to the Phyla of Life on Earth*, 3rd edn. San Francisco: W. H. Freeman.

MARLAND, G., BODEN, T. A., ANDRES, R. J. (2002) Global Regional and national fossil fuel carbon dioxide emissions. In *Trends, A Compendium of Data on Global Change*. Oak Ridge National Laboratory, USA: Carbon Dioxide Information Analysis Center.

MARLOWE, I. T., BONE, C. M., BYFIELD, M. A., EMMOT, M. A., FROST, R., GIBSON, N., HAGAN, J. P., HARMAN, N., HAYMAN, G. D., JENKIN, M. E., LINSELL, P. H. R., ROSE, C. L., RUDD, H. AND STACEY, A. (1995) *The Categorisation of Volatile Organic Compounds* (DoE report no. DoE/HMIP/RR/95/009). London: DoE.

MARTIN, A. A. AND TYLER, M. J. (1978) 'The introduction into western Australia of the frog *Limnodynastes tasmaniensis*'. *Australian Zoologist* 19, 320–44.

MASSOUD, M. S., AI-ABDALI, F. AND AI-GHADBAN, A. N. (1998) 'The status of oil pollution in the Arabian Gulf by the end of 1993'. *Environment International*, 24, 11–22.

MATHER, J. D., SPENCE, I. M., LAWRENCE, A. R. AND BROWN, M. J. (1996) Man-made hazards. In G. J. H. McCall, E. F. J. de Mulder and

B. R. Marker (eds) (1996) *Urban Geoscience*, 127–62. Rotterdam: A. A. Balkema.

MATTHEWS, E., PAYNE, R., ROWEDER, M. AND MURRAY, S. (2000) *PAGE Technical Report – Forest Ecosystems*. World Resources Institute. Internet reference at http://www.wri.org. Last accessed November 2003.

MATTSON, P., LOHSE, K. A. AND HALL, S. J. (2002) 'The globalization of nitrogen deposition: consequences for terrestrial ecosystems'. *Ambio* 31, 113–19.

McBETH, J. (1995) 'Water peril: Indonesia's urbanization may precipitate a water crisis'. *Far Eastern Economic Review* 158 (22), 61.

McCOLLIN, D. (1993) 'Avian distribution patterns in a fragmented wooded landscape (North Humberside, U.K.): the role of between-patch and within-patch structure'. *Global Ecology and Biogeography Letters* 3, 48–62.

McCOLLIN, D. (1998) 'Forest edges and habitat selection in birds: a functional approach'. *Ecography* 21, 247–60.

McGOWAN, J. G. AND CONNORS, S. R. (2000) 'Windpower: a turn of the century review'. *Annual Review of Energy and the Environment* 25, 147–97.

McHARG, I. L. (1969) *Design with Nature*. Garden City NY: Natural History Press for the American Museum of Natura.

McNEAL, B. L. (1974) Soil salts and their effect on water movement. In van Schilfgaarde, J. (ed.) *Drainage for Agriculture* (Agronomy 17), 408–68. Madison, Wisconsin: American Society of Agronomy.

McTAINSH, G. (1999) Dust transport and deposition. In Goudie, A. S., Livingstone, I. and Stokes, S. (eds) *Aeolian Environments, Sediments and Landforms*, 181–211. Chichester: John Wiley & Sons Ltd.

McVICAR, T. R. AND JUPP, D. L. B. (1998) 'The current and potential operational uses of remote sensing to aid decisions on drought exceptional circumstances in Australia: a review'. *Agricultural Systems* 57, 399–468.

MERSEY BASIN CAMPAIGN (2000) *Progress Report: 1985–2000*. Manchester: Mersey Basin Campaign.

MESSING, I. AND JARVIS, N. J. (1990) 'Seasonal variation in field saturated hydraulic conductivity in two swelling clay soils in Sweden'. *Journal of Soil Science* 41, 229–37.

METEOROLOGICAL OFFICE (2002) The website of the UK Meteorological Office. Available at: http://www.meto.gov.uk/. Last accessed October 2002.

MILLINGTON, A. C., MUTSIO, S. K., KIRKBY, S. J. AND O'KEEFE, P. (1989) 'African soil erosion – nature undone and the limitations of technology'. *Land Degradation and Rehabilitation* 1, 279–90.

MILLWARD, A. A. AND MERSEY, J. E. (2001) 'Conservation strategies for effective land management of protected areas using an erosion prediction information system, (EPIS)'. *Journal of Environmental Management* 61, 329–43.

MIMURA, N. (1999) 'Vulnerability of island countries in the South Pacific to sea level rise and climate change'. *Climate Research* 12, 137–43.

MITSCH, W. J. AND GOSSELINK, J. G. (2000) *Wetlands*, 3rd edn. New York: John Wiley & Sons.

MONMONIER, M. (1991) *How to Lie with Maps*. Chicago: University of Chicago Press.

MOONEY, H. A., KALIN ARROYO, M. T., BOND, W. J., CANADELL, J., HOBBS, R. J., LAVORAL, S. AND NEILSON, R. P. (2001) Mediterranean-climate ecosystems. In F. S. Chapin III, O. E. Sala and E. Huber-Sannwald (eds) *Global Biodiversity in a Changing Environment: Scenarios for the 21st Century* (Ecological Studies 152), 157–99. Springer: New York.

MOORE, J. W., SCHINDLER, D. E., SCHEUERELL, M. D., SMITH, D. AND FRODGE, J. (2003) 'Lake eutrophication at the urban fringe, Seattle Region, USA'. *Ambio* 32, 13–18.

MOORE, M., GOULD, P. AND KEARY, B. S. (2003) 'Global urbanization and impact on health'. *Internal Journal of Hygiene and Environmental Health* 206, 269–78.

MORAGUES, A. AND ALAIDE, T. (1996) 'The use of a geographical information system to assess the effect of traffic pollution'. *Science of the Total Environment* 189/190, 267–73.

MORGAN, M. A. (1967) Hardware models in

geography. In R. J. Chorley and P. Haggett (eds) *Models in Geography*, 727–74. London: Methuen.

MORKILL, A. E. AND ANDERSON, S. H. (1991) 'Effectiveness of marking powerlines to reduce sandhill crane collisions'. *Wildlife Society Bulletin* 19, 442–49.

MORSE, J. K., LESTER, J. N. AND PERRY, R. (1993) *The Economic and Environmental Impact of Phosphorus Removal from Wastewater in the European Community*. London: Selper Publications.

MULLIGAN, C. N., YONG, R. N. AND GIBBS, B. F. (2001) 'Remediation technologies for metal-contaminated soils and groundwater: an evaluation'. *Engineering Geology* 60, 193–207.

MULLIN, B. H. (1998) 'The biology and management of purple loosestrife (*Lythrum salicaria*)'. *Weed Technology* 12, 397–401.

MULLINEAUX, D. R. (1986) 'Summary of pre-1980 tephra-fall deposits erupted from Mount St Helens, Washington State, USA'. *Bulletin of Volcanology* 48, 17–26.

MYATT, L. B., SCRIMSHAW, M. D. AND LESTER, J. N. (2003) 'Public perceptions and attitudes towards a forthcoming managed realignment scheme: Freiston Shore, Lincolnshire, UK'. *Ocean & Coastal Management* 46, 565–82.

NAKIĆENOVIĆ, N. AND SWART, R. (eds) (2000) *Emissions Scenarios 2000: Summary for Policymakers: A Special Report of Working Group III of the IPCC*. Cambridge: Cambridge University Press and the Intergovernmental Panel on Climate Change.

NAQVI, S. M., HOWELL, R. D. AND SHOLAS, M. (1993) 'Cadmium and lead residues in field-collected red swamp crayfish (*Procambarus clarkii*) and uptake by alligator weed, *Alternanthera philoxiroides*'. *Journal of Environmental Science and Health* B28, 473–85.

NASA (2003a) *Hurricane Fabian*. Internet Reference at http://earthobservation.basa.gov/Natural_Hazards. Last accessed Sept 2003.

NASA (2003b) *European Heat Wave*. Internet Reference at http://earthobservation.basa.gov/Natural_Hazards. Last accessed Sept 2003.

NATIONAL SOCIETY FOR CLEAN AIR (NSCA) (1986) *1986 Pollution Handbook*. Brighton: NCSA.

NEARING, M. A. (1998) 'Why soil erosion models over-predict small soil losses and under-predict large soil losses'. *Catena* 32, 15–22.

NEILSON, R. P. (1993) 'Transient ecotone response to climatic change: some conceptual and modelling approaches'. *Ecological Applications* 3, 385–95.

NETCEN (2002) The National Air Quality Information Archive. Website at http:www.airquality.co.uk. Last accessed December 2002.

NEW, M., TODD, M., HULME, M. AND JONES, P. (2001) 'Precipitation measurements and trends in the twentieth century'. *International Journal of Climatology* 21, 1899–922.

NEWSON, M. (1994) *Hydrology and the River Environment*. Oxford: Clarendon Press.

NEWSON, M. (1997) *Land, Water and Development. Sustainable Management of River Basin Systems*, 2nd edn. London: Routledge.

NEWSON, M. D. (1992) 'Twenty years of systematic physical geography: issues for a "New Environmental Age"'. *Progress in Physical Geography* 16, 209–21.

NICKLING, W. G. AND MCKENNA NEUMAN, C. (1999) Recent investigations of airflow and sediment transport over desert dunes. In A. S. Goudie, I. Livingstone and S. Stokes (eds) *Aeolian Environments, Sediments and Landforms*, 15–47. Chichester: John Wiley & Sons.

NIXON, S. C., LACK, T. J., HUNT, D. T. E, LALLANA, C. AND BOSCHET, A. F. (2000) *Sustainable Use of Europe's Water? State, Prospects and Issues* (Environmental Assessment Series No. 7). Copenhagen: European Environment Agency.

NOAA PALEOCLIMATOLOGY PROGRAM AND WORLD DATA CENTER FOR PALEOCLIMATOLOGY (2003) Tree Ring. [online]. Available at: http://www.ngdc.noaa.gov/paleo/treering.html. Last accessed: 21 January 2004.

NOLLER, J. S. (2000) Lead-210 geochronology. In J. S. Noller, J. M. Sowers and W. R. Lettis (eds) *Quaternary Geochronology: Methods and Applications* (AGU Reference Shelf 4), 115–210. Washington DC: American Geophysical Union.

NOLLER, J. S. AND LOCKE, W. W. (2000) Lichenometry. In J. S. Noller, J. M. Sowers and W.

R. Lettis (eds) *Quaternary Geochronology: Methods and Applications* (AGU Reference Shelf 4), 261–72. Washington DC: American Geophysical Union.

O'HARE, G. (1988) *Soils, Vegetation, Ecosystems* (Conceptual frameworks in Geography). Edinburgh: Oliver & Boyd.

O'RIORDAN, T. (1988) Future directions for environmental policy. In D. C. Pitt (ed.) *The Future of the Environment: The Social Dimensions of Conservation and Ecological Alternatives*, 168–98. London: Routledge.

O'RIORDAN, T. (1996) Environmentalism on the move. In I. Douglas, R. J. Huggett and M. E. Robinson (eds) *Companion Encyclopedia of Geography: The Environment and Humankind*, 449–76. London: Routledge.

O'SULLIVAN, P. E. (1983) 'Annually-laminated lake sediments and the study of Quaternary environmental change – a review'. *Quaternary Science Reviews* 1, 245–313.

ODUM, H. T. (1971) *Environmental, Power, and Society*. New York: Wiley Interscience.

ODUM, H. T. (1983) *Systems Ecology: An Introduction*. New York: Wiley.

OKE, T. R. (1982) 'The energetic basis of the urban heat island'. *Quarterly Journal of the Royal Meteorological Society* 108, 1–24.

OKE, T. R. (1988) 'The urban energy balance'. *Progress in Physical Geography* 12, 471–508.

OLD, G. H., LEEKS, G. J. L., PACKMAN, J. C., SMITH, B. P. G., LEWIS, S., HEWITT, E. J., HOLMES, M. AND YOUNG, A. (2003) 'The impact of a convectional summer rainfall event on river flow and fine sediment load in a highly urbanized catchment: Bradford, West Yorkshire'. *Science of the Total Environment* 314–316, 495–512.

OLDEMAN, L. R., HAKKELING, R. T. A. AND SOMBROEK, W. G. (1991) *World Map of the Status of Human-induced Soil Degradation*. Nairobi/Wageningen: UNEP/ISRIC GLASOD Project.

OMAR, S. A. S., MISAK, R., KING, P., SHAHID, S. A., ABO-RIZQ, H., GREALISH, G. AND TORY, W. (2001) 'Mapping the vegetation of Kuwait through reconnaissance soil survey'. *Journal of Arid Environments* 48, 341–55.

ORAU (2004) Oxford Radiocarbon Accelerator Unit. [online]. Available at: http://www.rlaha.ox.ac.uk/orau/index.html. Last accessed 21 January 2004.

OSBORN, T. J., HULME, M., JONES, P. D. AND BASNETT, T. A. (2000) 'Observed trends in the daily intensity of United Kingdom precipitation'. *International Journal of Climatology* 20, 347–64.

OSTERKAMP, T. E., VIERECK, L., SHUR, Y., JORGENSON, M. T., RACINE, C., DOYLE, A. AND BOONE, R. D. (2000) 'Observations of thermokarst and its impact on boreal forests in Alaska, USA'. *Arctic, Antarctic, and Alpine Research* 32, 303–15.

OSTERKAMP, W. R. AND HUPP, C. R. (1996) The evolution of geomorphology, ecology and other composite sciences. In B. L. Rhoads and C. E. Thorn (eds) *The Scientific Nature of Geomorphology*, 415–41. Chichester: John Wiley & Sons.

OUIN, A., PAILLAT, G., BUTET, A. AND BUREL, F. (2000) 'Spatial dynamics of wood mouse (*Apodemus sylvaticus*) in an agricultural landscape under intensive use in the Mont Saint Michel Bay (France)'. *Agriculture, Ecosystems and Environment* 78, 159–65.

PARKER, D. J. (1995) 'Floodplain development policy in England and Wales'. *Applied Geography* 15, 341–63.

PARMESAN, C. (1996) 'Climate and species' range'. *Nature* 382, 765–66.

PARMESAN, C., RYRHOLM, N., STEFANESCU, C., HILL, J. K., THOMAS, C. D., DESCIMON, H., HUNTLEY, B., KAILA, L., KULLBERG, J., TAMMARU, T., TENNENT, T. W., THOMAS, J. A. AND WARREN, M. (1999) 'Poleward shifts in geographical ranges of butterfly species associated with regional warming'. *Nature* 399, 579–83.

PARSONS, G. R. AND POWELL, M. (2001) 'Measuring the cost of beach retreat'. *Coastal Management* 29, 91–103.

PASSANT, N. R. (1993) *Emissions of Volatile Organic Compounds from Stationary Sources in the United Kingdom*. Stevenage: Warren Spring Laboratory, LR 990.

PATTEN, B. C. (1985) 'Energy cycling in ecosystems'. *Ecological Modelling* 28, 1–71.

PAULEIT, S., SLINN, P., HANDLEY, J. AND LINDLEY, S. (2003) 'Promoting greenstructure of towns and cities: English Nature's Accessible Greenspace Natural Greenspace Standards model'. *Built Environment* 29, 2.

PEARSON, S. M., SMITH, A. B. AND TURNER, M. G. (1998) 'Forest patch size, land use, and mesic forest herbs in the French Broad River Basin, North Carolina'. *Castanea* 63, 382–95.

PECK, A. J. AND HATTON, T. (2003) 'Salinity and the discharge of salts from catchments in Australia'. *Journal of Hydrology* 272, 191–202.

PELKEY, N. W., STONER, C. J. AND CARO, T. M. (2000) 'Vegetation in Tanzania: assessing long term trends and effects of protection using satellite imagery'. *Biological Conservation* 94, 297–309.

PEREIRA, L. S., OWEIS, T. AND ZAIRI, A. (2002) 'Irrigation management under water scarcity'. *Agricultural Water Management* 57, 175–206.

PETSCHEL-HELD, G., BLOCK, A., CASSEL-GINTZ, M., KROPP, J., LÜDEKE, M. K. B., MOLDENHAUER, O., REUSSWIG, F. AND SCHELLNHUBER, H-J. (1999) 'Syndromes of global change: a qualitative modelling approach to assist global environmental management'. *Environmental Modeling and Assessment* 4, 295–314.

PETTIGREW, W. W. (1928) *The Influence of Air Pollution on Vegetation.* Harrogate: Lecture to the Smoke Abatement League of Great Britain.

PFAUL, S., CARIUS, A., MARCH, A., BAECHLER, G., ALCAMO, J., SIMON, K.-H. AND HYVERINEN, J. (2000) *The Use of Global Monitoring in Support of Environment and Security.* Berlin: Ecologic Centre for International and European Environmental Research, EC Contract 15520-1999-12 F1ED DG DE.

PHILLIPS, J. D. (1991) 'The human role in Earth surface systems: some theoretical considerations'. *Geographical Analysis* 23, 316–31.

PIPER, S. AND HUSZAR, P. C. (1989) 'Re-examination of the off-site costs of wind erosion in New Mexico'. *Journal of Soil and Water Conservation* 44, 332–34.

PRESTON, C. D., SHEAIL, J., ARMITAGE, P. AND DAVY-BOWKER, J. (2003) 'The long-term impact of urbanization on aquatic plants: Cambridge and the River Cam'. *The Science of the Total Environment* 314–316, 67–87.

PULFORD, I. D. AND WATSON, C. (2002) 'Phytoremediation of heavy metal-contaminated land by trees – a review'. *Environment International* 29, 529–40.

PYE, K. (1987) *Aeolian Dust and Dust Deposits.* London: Academic Press.

QUARG (1993a) *Urban Air Quality in the United Kingdom* (First Report of the Quality of Urban Air Review Group (QUARG)). London: DoE.

QUARG (1993b) *Diesel Vehicle Emissions and Urban Air Quality* (Second Report of the Quality of Urban Air Review Group (QUARG)). London: DoE.

QUARG (1996) *Airborne Particulate Matter in the United Kingdom* (Third Report of the Quality of Urban Air Review Group (QUARG)). London: DoE.

RADCLIFFE, J. E. (1982) 'Effects of aspect and topography on pasture production in hill country'. *New Zealand Journal of Agricultural Research* 25, 485–96.

RAHU, M. (2002) 'Health effects of the Chernobyl accident: fears, rumours and the truth'. *European Journal of Cancer* 39, 295–99.

RAMAKRISHNAN, U., COSS, R. G. AND PELKEY, N. W. (1999) 'Tiger decline caused by the reduction of large ungulate prey: evidence from a study of leopard diets in southern India'. *Biological Conservation* 89, 113–20.

RANDOLPH, R. C., HARDY, J. T., FOWLER, S. W., PRICE, A. R. G. AND PEARSON, W. H. (1998) 'Toxicity and persistence of nearshore sediment contamination following the 1991 Gulf War'. *Environment International* 24, 33–42. See also: www.wgbu.de/wbgu_jg1996_engl.html. Last accessed April 2004.

REACTIVE CHLORINE EMISSIONS INVENTORY (2003) *RCEI: The Reactive Chlorine Emissions Inventory.* Subgroup of the Global Atmospheric Emissions Inventory group. Internet reference at http://www.Geiacenter.org. Last accessed August 2003.

REES, W. E. (1995) 'Achieving sustainability: reform or transformation'. *Journal of Planning Literature* 9 (4), 343–61, Sage Publications. Reprinted in Satterthwaite, D. (ed.) (1999) *The Earthscan Reader in Sustainable Cities.* London: Earthscan, 22–52.

REH, W. (1989) Investigations into the influence of roads on the genetic structure of populations of the common frog *Rana temporaria*. In T. E. S. Langton (ed.) *Amphibians and Roads*, 101–103. Shefford, Bedfordshire: ACO Polymer Products.

REID, D. (1995) *Sustainable Development: An Introductory Guide*. London: Earthscan.

REN, W., ZHONG, Y., MELIGRANA, J., ANDERSON, B., WATT, E., CHEN , J. AND LEUNG, H.-L. (2003) 'Urbanization, land use and water quality in Shanghai 1947–1996'. *Environment International* 29, 649–59.

RENARD, K. G., FOSTER, G. R., WEESIES, G. A. AND PORTER, J. P. (1991) 'RUSLE: revised Universal Soil Loss Equation'. *Journal of Soil and Water Conservation* 46, 30–33.

RENARD, K. G., FOSTER, G. R., WEESIES, G. A., McCOOL, D. K. AND YODER, D. C. (1998) *Predicting Soil Erosion by Water: A Guide to Conservation Planning with Revised Universal Soil Loss Equation (RUSLE)* (Agriculture Handbook No. 703). Washington DC: USDA–ARS.

RENNE, P. R. (2000) K–Ar and ⁴⁰Ar/³⁹Ar dating. In J. S. Noller, J. M. Sowers and W. R. Lettis (eds) *Quaternary Geochronology: Methods and Applications* (AGU Reference Shelf 4), 77–100. Washington DC: American Geophysical Union.

RENTON, D. (2003) *Forsaken by History: The Miners of Katanga*. Internet reference at http://www.labournet.net/workd/0306/drc02.html. Last accessed April 2004.

REVENGA, C, BRUNNER, J., HENNINGER, N., KASSEM, K. AND PAYNE, R. (2000) *PAGE Technical Report – Freshwater Ecosystems*. World Resources Institute. Internet reference at http://www.wri.org. Last accessed November 2003.

RIEHL, H. (1965) *An Introduction to the Atmosphere*, 1st edn. New York: McGraw-Hill.

RIVETT, M. O., PETTS, J., BUTLER, B. AND MARTIN, I. (2002) 'Remediation of contaminated land and groundwater: experience in England and Wales'. *Journal of Environmental Management* 65, 251–68.

ROBERTSON, D. AND ROBINSON, J. (1998) 'Darwinian Daisyworld'. *Journal of Theoretical Biology* 195, 129–34.

ROBERTS, N. (1998) *The Holocene*, 2nd edn. Oxford: Blackwell.

ROCHELEAU, D. E., STEINBERG, P. E. AND BENJAMIN, P. A. (1995) 'Environment, development, crisis, and crusade: Ukambani, Kenya, 1890–1990'. *World Development* 23, 1037–51.

ROGNERUD, S. AND FJELD, E. (2001) 'Trace element contamination in Norwegian lake sediments'. *Ambio* 30, 11–19.

ROOSE, É., DUGUÉ, P. AND RODRIGUEZ, L. (1992) 'La G.C.E.S. [Gestion Conservatoire de l'Eau, de la biomasse et de la fertilité des Sols]. Une nouvelle stratégie de lutte anti-érosive appliquée à l'aménagement de terroirs en zone soudono-sahélienne du Burkina Faso'. *Revue des Bois et Forêts des Tropiques* 233, 49–63.

ROSEGRANT, M. W., CAI, X. AND CLINE, S. A. (2002) *World Water and Food to 2025. Dealing with Scarcity*. International Food Policy Research Institute (IFPRI) & International Water Management Institute (IWMI). Available for download at: http://www.ifpri.org/pubs/books/water2025book.htm Last accessed December 2003.

ROTMANS, J. (1990) *IMAGE: An Integrated Model to Assess the Greenhouse Effect*. Dordrecht: Kluwer Academic Publishers.

ROWELL, D. L. (1994) *Soil Science: Methods and Applications*. Harlow: Longman.

RUSTAD, L. E. (2001) 'Matter of time on the prairie'. *Nature* 413, 578–79.

SACHS, W., LOSKE, R. AND LINZ, M. (1998) *Greening the North – A Post-Industrial Blueprint for Ecology and Equity*. London and New York: Zed Books.

SALA, O. E. (2001) 'Temperate grasslands'. In F. S. Chapin III, O. E. Sala and E. Huber-Sannwald (eds) *Global Biodiversity in a Changing Environment: Scenarios for the 21st Century* (Ecological Studies 152), 121–37. Springer: New York.

SALA, O. E., CHAPIN III, F. S., GARDNER, R. H., LAUENROTH, W. K., MOONEY, H. A. AND RAMAKRISHNAN, P. S. (1999) Global change, biodiversity and ecological complexity. In B. H. Walker, W. L. Steffen, J. Canadell and J. S. I. Ingram (eds) *The Terrestrial Biosphere and Global Change: Implications for Natural and Managed Ecosystems*, 304–28. Cambridge: Cambridge University Press.

SALA, O. E., CHAPIN III, F. S. AND HUBER-SANNWALD, E. (2001) Potential biodiversity change: global patterns and biome comparisons. In F. S. Chapin III, O. E. Sala and E. Huber-Sannwald (eds) *Global Biodiversity in a Changing Environment: Scenarios for the 21st Century* (Ecological Studies 152), 351–67. Springer: New York.

SALA, O. S., CHAPIN III, F. S., ARMESTO, J. J., BERLOW, E., BLOOMFIELD, J., DIRZO, R., HUBER-SANNWALD, E., HUENNEKE, L. F., JACKSON, R. B., KINZIG, A., LEEMANS, R., LODGE, D. M., MOONEY, H. A., OESTERHELD, M., POFF, N. L., SYKES, M. T., WALKER, B. H., WALKER, M. AND WALL, D. H. (2000) 'Global biodiversity scenarios for the year 2100'. *Science* 287, 1770–74.

SALWAY, A. G., EGGLESTON, H. S., GOODWIN, J. W. L. AND MURRELLS, T. P. (1996) *UK Emissions of Air Pollutants 1970–1994* (AEA/RAMP/2009001/R/003/Issue 1). Culham, Oxon: National Environment Technology Centre, AEA Technology.

SALWAY, A.G., MURRELLS T. P., MILNE, R. AND HIDRI, S. (2003) *Greenhouse Gas Inventories for England, Scotland, Wales and Northern Ireland: 1990–2000 Report* Reference AEAT/R/ENV/1182, National Environmental Technology Centre, Culham, Abingdon, UK Internet reference available at http://www.airquality.co.uk/archive/reports/cat07/aeat-r-env-1182.pdf. Last accessed April 2004.

SARACCO, J. F. AND COLLAZO, J. A. (1999) 'Carolina bottomland hardwood forest'. *Wilson Bulletin* 111, 541–49.

SARNA-WOJCICKI, A. (2000) Tephrochronology. In J. S. Noller, J. M. Sowers and W. R. Lettis (eds) *Quaternary Geochronology: Methods and Applications* (AGU Reference Shelf 4), 357–77. Washington DC: American Geophysical Union.

SATTERTHWAITE, D. (ed.) (1999) *The Earthscan Reader in Sustainable Cities*. London: Earthscan.

SAVARD, J.-P., CLERGEAU, P. AND MENNECHEZ, G. (2000) 'Biodiversity concepts and urban ecosystems'. *Landscape and Urban Planning* 48, 131–42.

SAXENA, V. K., YU, S. AND ANDERSON, J. (1997) 'Impact of stratospheric volcanic aerosols on climate, evidence for aerosol shortwave and longwave forcing in the southeastern U.S.'. *Atmospheric Environment* 31, 4211–21.

SCHADLICH, S., GOTTSCHE, F. M. AND OLESEN, F. S. (2001) 'Influence of land surface parameters and atmosphere on METEOSAT brightness temperatures and generation of land surface temperature maps by temporally and spatially interpolating atmospheric correction'. *Remote Sensing of Environment* 75, 39–46.

SCHEIDLEDER, A. (2003) *Indicator Fact Sheet (WEU01) Nitrate in Groundwater.* European Environment Agency. Available for download at: http://themes.eea.eu.int/Specific_media/water/indicators/WEU01%2C2003.1001/Fact_sheet_15_Nitrate_in_Groundwater_final.pdf. Last accessed December 2003.

SCHEIDLEDER, A., WINKLER, G., GRATH, J. AND VOGEL, W. R. (1996) *Human Interventions in the Hydrological Cycle*. European Topic Centre on Inland Waters, Copenhagen: European Environment Agency.

SCHELLNHUBER, H.-J. (1998) *Syndromes of Global Change: An Integrated Analysis of Environmental Development Issues*. Villa Borsig Workshop Series – A policy dialogue on the World Development Report 1999: Development Issues in the 21st Century. Internet reference at http://www.dse.de/ef/vbws98/contents.htm. Last accessed November 2003.

SCHIMEL, D., ALVES, D., ENTING, I., HEIMANN, M. AND JOOS, F. (1996) Radiative forcing of climate change. In J. T Houghton, L. G. Meira Filho, B. A. Callender, N. Harris, A. Kattenberg and K. Maskell (eds) *Climate Change 1995: The Science of Climate Change*, 77. Contribution of Working Group I to the Second Assessment of the Intergovernmental Panel on Climate Change, Cambridge: Cambridge University Press, UK.

SCHNITZLER, A. (1994) 'European alluvial hardwood forests of large floodplains'. *Journal of Biogeography* 21, 605–23.

SCHULZE, E.-D., BECK, E. AND MÜLLER-HOHENSTEIN, K. (2002) *Pflanzenökologie*. Heidelberg: Spektrum Akademischer Verlag.

SCHUMM, S. A. (1956) 'The evolution of drainage basin systems and slopes in badlands at Perth Amboy, New Jersey'. *Bulletin of the Geological Society of America* 67, 597–646.

SCHUMM, S. A. (1985) 'Explanation and extrapolation in geomorphology; seven reasons for geologic uncertainty'. *Transactions of the Japanese Geomorphological Union* 6, 1–18.

SCHUMM, S. A. (1991) *To Interpret the Earth: Ten Ways to be Wrong.* Cambridge: Cambridge University Press.

SCHWARCZ, H. P. (1997) Uranium series dating. In R. E. Taylor and M. J. Aitken (eds) *Chronometric Dating in Archaeology*, 159–82. New York: Plenum Press.

SCULL, P., FRANKLIN, J., CHADWICK, O. A. AND MCARTHUR, D. (2003) 'Predictive soil mapping: a review'. *Progress in Physical Geography* 27, 171–97.

SEIDENSTICKER, J., CHRISTIE, S. AND JACKSON, P. (1999) Preface. In J. Seidensticker, S. Christie and P. Jackson (eds) *Riding the Tiger: Tiger Conservation in Human-dominated Landscapes*, xv–xix. London: The Zoological Society of London; Cambridge: Cambridge University Press.

SEIXAS, J. (2000) 'Assessing heterogeneity from remote sensing images: the case of desertification in southern Portugal'. *International Journal of Remote Sensing* 21, 2645–63.

SELBY, M. J. (1993) *Hillslope Materials and Processes*, 2nd edn. Oxford: Oxford University Press.

SERREZE, M. C., WALSH, J. E., CHAPIN III, F. S., OSTERKAMP, T., DYURGEROV, M., ROMANOVSKY, V., OECHEL, W. C., MORISON, J., ZHANG, T. AND BARRY, R. G. (2000) 'Observational evidence of recent change in the northern high-latitude environment'. *Climatic Change* 46, 159–207.

SEXTON, K. AND RYAN, P. B. (1988) Assessment of human exposure to air pollution methods measurements and models. In A. Y. Watson, R. R. Bates and D. Kennedy (eds) *Air Pollution, the Automobile and Public Health*, 207–38. Washington DC: National Academy Press.

SHACKLETON, N. J. (1977) 'The oxygen isotope stratigraphic record of the Late Pleistocene'. *Philosophical Transactions of the Royal Society of London* 280B, 169–82.

SHACKLETON, N. J. (1987) 'Oxygen isotopes, ice volume and sea level'. *Quaternary Science Reviews* 6, 183–90.

SHARP, A. D., TRUDGILL, S. T., COOKE, R. U., PRICE, C. A., CRABTREE, R. W., PICKLE, A. M. AND SMITH, D. I. (1982) 'Weathering of the balustrade on St Paul's Cathedral, London'. *Earth Surface Processes and Landforms* 7, 387–89.

SHAW, E. M. (1994) *Hydrology in Practice*, 3rd edn. London: Chapman and Hall.

SHERLOCK, R. L. (1922) *Man as a Geological Agent: An Account of His Action on Inanimate Nature*. With a foreword by A. S. Woodward. London: H. F. & G. Witherby.

SHUGART, H. H., JR AND WEST, D. C. (1977) 'Development of an Appalachian deciduous forest succession model and its application to assessment of the impact of the chestnut blight'. *Journal of Environmental Management* 5, 161–79.

SHUTES, R. B. E. (2001) 'Artificial wetlands and water quality improvement'. *Environment International* 26, 441–47.

SIMAS, T., NUNES, J. P. AND FERREIRA, J. G. (2001) 'Effects of global climate change on coastal salt marshes'. *Ecological Modelling* 139, 1–15.

SIMPSON, T. (1996) Urban soils. In McCall, G. J. H., de Mulder, E. F. J. and Marker, B. R. (1996) *Urban Geoscience*. Rotterdam: A. A. Balkema.

SKEI, J., LARSSON, P., ROSENBERG, R., JONSSON, P., OLSSON, M. AND BROMAN, D. (2000) 'Eutrophication and contaminants in aquatic systems'. *Ambio* 29, 184–94.

SKLAR, F. H., COSTANZA, R. AND DAY, J. W. (1990) Model conceptualization. In B. C. Patten (ed.) *Wetlands and Shallow Continental Water Bodies*, Volume 1, 625–58. The Hague: SPB Academic Publishing

SKWARZEC, B., STRUMINSKAA, D. I. AND BORYLOA, A. (2001) 'Bioaccumulation and distribution of plutonium in fish from Gdansk Bay'. *Journal of Environmental Radioactivity* 55, 167–78.

SLAYMAKER, O. AND SPENCER, T. (1998) *Physical Geography and Global Environmental Change*. Harlow, Essex: Addison Wesley Longman.

SMAKHTIN, V. U. (2001) 'Low flow hydrology: a review'. *Journal of Hydrology* 240, 147–86.

SMIL, V. (2001) *Enriching the Earth: Fritz Haber, Carl Bosch and the Transformation of World Food Production*. Cambridge, Massachusetts: The MIT Press.

SMITH, M., ALLEN, R. G., MONTEITH, J. L., PERRIER, A., PEREIRA, L. AND SEGEREN, A. (1992) *Report of the Expert Consultation on Procedures for Revision of FAO Guidelines for Prediction of Crop Water Requirements*. Rome: UN–FAO.

SNYDER, N. F. R. AND SNYDER, H. (2000) *The California Condor: A Saga of Natural History and Conservation*. San Diego: Academic Press.

SOIL SURVEY STAFF (1975) *Soil Taxonomy: A Basic System of Soil Classification for Making and Interpreting Soil Surveys*. Washington DC: USDA Soil Conservation Series.

SOIL SURVEY STAFF (1992) *Keys to Soil Taxonomy* (Soil Management Support Services, Technical Monograph No. 19). Blacksberg, Virginia: Pocahontas Press.

SOIL SURVEY STAFF (1999) *Soil Taxonomy: A Basic System of Soil Classification for Making and Interpreting Soil Surveys*, 2nd edn (Agricultural Handbook No. 436). US Government Printing Office, Pittsburgh, Virginia: USDA, Natural Resource Conservation Service.

SOLOMON, A. M. (1986) 'Transient response of forests to CO_2-induced climatic change: simulation modeling experiments in eastern North America'. *Oecologia* 68, 567–79.

SONNEVELD, B. G. J. S. AND NEARING, M. A. (2003) 'A nonparametric/parametric analysis of the Universal Soil Loss Equation'. *Catena* 52, 9–21.

SPENCER, T. (1995) 'Potentialities, uncertainties and complexities in the response of coral reefs to future sea-level rise'. *Earth Surface Processes and Landforms* 20, 49–64.

SPENCER, J. W. AND KIRBY, K. J. (1992) 'An inventory of ancient woodland for England and Wales'. *Biological Conservation* 62, 77–93.

SPERLING, C. H. B. AND COOKE, R. U. (1985) 'Laboratory simulation of rock weathering by salt crystallization and hydration processes in hot, arid environments'. *Earth Surface Processes and Landforms* 10, 541–55.

STANGER SCIENCE (2002) http://www.stanger.co.uk/siteinfo/SiteClassification.asp. Last accessed October 2002.

STANHILL, G. L. AND KALMA, J. D. (1995) 'Solar dimming and urban heating at Hong Kong'. *International Journal of Climatology* 15, 933–41.

STODDART, D. M. (1987) 'To claim the high ground: geography for the end of the century'. *Transactions of the Institute of British Geographers*, New Series 12, 327–36.

STRAHLER, A. H. (1980) 'Systems theory in physical geography'. *Physical Geographer* 1, 1–27.

STRAHLER, A. N. (2001) *Physical Geography. Science and Systems of the Human Environment*, 2nd edn. New York: John Wiley & Sons.

STRAHLER, A. N. AND STRAHLER, A. H. (1973) *Environmental Geoscience*. Santa Barbara, California: Hamilton.

STRAHLER, A. N. AND STRAHLER, A. H. (1974) *Introduction to Environmental Science*. Santa Barbara, California: Hamilton.

STRUTHERS, W. A. K. (1997) 'From Manchester Docks to Salford Quays: ten years of environmental improvements in the Mersey Basin Campaign'. *Journal of the Chartered Institution of Water and Environmental Management* 11, 1–7.

STUDLEY, J. (1998) *Dominant Knowledge Systems and Local Knowledge*. Mountain-Forum On-line Library Document, http://www.mtnforum.org/resources/library/stud98a2.htm. Last accessed July 2003.

STUIVER, M., REIMER, P. J., BARD, E., BECK, J. W., BURR, G. S., HUGHEN, K. A., KROMER, B., McCORMAC, F. G., VAN DER PLICHT, J. AND SPURK, M. (1998) 'INTCAL98 Radiocarbon age calibration 24,000–0 cal BP'. *Radiocarbon* 40,1041–83.

STUIVER, M., REIMER, P. J. AND REIMER, R. (2003) *CALIB Radiocarbon Calibration* [online]. Available at: http://radiocarbon.pa.qub.ac.uk/calib/. Last accessed 3 December 2003.

SUMMERFIELD, M. A., STUART, F. M., COCKBURN, H. A. P., SUGDEN, D. E., DENTON, G. H., DUNAI, T. AND MARCHANT, D. (1999) 'Long-term rates of denudation in the Dry Valleys, Transantarctic Mountains, southern Victoria Land, Antarctica based on in-sit-produced cosmogenic ^{21}Ne'. *Geomorphology* 27, 113–29.

SUNQUIST, M., KARANTH, K. U. AND SUNQUIST, F. (1999) Ecology, behaviour and resilience of the tiger and its conservation needs. In J. Seidensticker, S. Christie and P. Jackson (eds)

Riding the Tiger: Tiger Conservation in Human-dominated Landscapes, 5–18. London: The Zoological Society of London; Cambridge: Cambridge University Press.

SURIAN, N. (1999) 'Channel changes due to river regulation: the case of the Piave River, Italy'. *Earth Surface Processes and Landforms* 24, 1135–51.

SUTTON, P. (2003) 'A scale-adjusted measure of urban sprawl using nighttime satellite imagery'. *Remote Sensing of Environment* 86, 353–69.

SYKES, M. J. AND HAXELTINE, A. (2001) Modelling the response of vegetation distribution and biodiversity to climate change. In F. S. Chapin III, O. E. Sala and E. Huber-Sannwald (eds) *Global Biodiversity in a Changing Environment: Scenarios for the 21st Century* (Ecological Studies 152), 5–22. Springer: New York.

TAYLOR, R. E. (1997) Radiocarbon dating. In R. E. Taylor and M. J. Aitken (eds) *Chronometric Dating in Archaeology*, 65–96. New York: Plenum Press.

THEAKSTONE, W. H. AND JACOBSEN, F. M. (1997) 'Digital terrain modelling of the surface and bed topography of the Glacier Austre Okstindbreen, Okstindan, Norway'. *Geografiska Annaler* 79A, 201–14.

THOMAS, D. S. G. AND GOUDIE, A. S. (eds) (2000) *The Dictionary of Physical Geography*, 3rd edn. Oxford: Blackwell.

THOMAS, R. W. AND HUGGETT, R. J. (1980) *Modelling in Geography: A Mathematical Approach*. London: Harper & Row.

THOMPSON, R. D. (1998) *Atmospheric Processes and Systems* (Introductions to Environment). London: Routledge.

THOMPSON, R. D., MANNION, A. M., MITCHELL, C. W., PARRY, M. AND TOWNSHEND, J. R. G. (1992) *Processes in Physical Geography*. London: Longman Scientific and Technical.

TICHÝ, L. (1999) 'Predictive modelling of the potential natural vegetation pattern in the Podyjí National Park, Czech Republic'. *Folia Geobotanica* 34, 243–52.

TITUS, J. G. AND RICHMAN, C. (2001) 'Maps of land vulnerable to sea level rise: modeled elevations along the US Atlantic and Gulf coasts'. *Climate Research* 18, 205–28.

TRAN, L. T., RIDGLEY, M. A., NEARING, M. A., DUCKSTEIN, L. AND SUTHERLAN, R. (2002) Using fuzzy logic-based modeling to improve the performance of the revised Universal Soil Loss Equation. In D. E. Stott, R. H. Mohtar and G. C. Steinhardt (eds), *Sustaining the Global Farm* (Selected Papers from the 10th International Soil Conservation Organization Meeting, May 24–29, 1999), 919–23. West Lafayette, Indiana: International Soil Conservation Organization. CD-ROM Available from the USDA–ARS National Soil Erosion Laboratory.

TRAVIS, J. M. J. (2003) 'Climate change and habitat destruction: a deadly anthropogenic cocktail'. *Proceedings of the Royal Society, London,* 270B, 467–73.

TRUDGILL, S. T., WALLING, D. E. AND WEBB, B. W. (eds) (1999) *Water Quality: Processes and Policy*. Chichester: John Wiley & Sons.

TRUMBORE, S. E. (2000) Radiocarbon geochronology. In J. S. Noller, J. M. Sowers and W. R. Lettis (eds) *Quarternary Geochronology: Methods and Applications* (AGU Reference Shelf 4), 41–60. Washington DC: American Geophysical Union.

UN CENTRE FOR HUMAN SETTLEMENTS (1999) Cities as solutions in an urbanizing world. In D. Satterthwaite (ed.) *Sustainable Cities*, 55–61. London: Earthscan.

UN CENTRE FOR HUMAN SETTLEMENTS (2001) *The State of the World's Cities Report.* Nairobi: UNCES (Habitat).

UN/WWAP (UNITED NATIONS/WORLD WATER ASSESSMENT PROGRAMME) (2003) *UN World Water Development Report: Water for People, Water for Life.* Paris, New York and Oxford: UNESCO (United Nations Educational, Scientific and Cultural Organization) and Berghahn Books.

UNEP (1991) Urban Air Pollution GEMS Environment Library No. 04: Urban Air Pollution, UNEP: Kenya.

UNITED NATIONS (2002) *The Human Consequences of the Chernobyl Nuclear Accident: A Strategy for Recovery.* New York, UN: The United Nations Scientific Committee on the Effects of Atomic Radiation (UNSCEAR).

US BUREAU OF THE CENSUS (2003) Total midyear population for the world: 1950–2050. US Bureau of the Census International Data Base, July 2003.

VAN BEEK, E. (1998) Modelling for integrated water management. In P. Huisman, W. Cramer, G. van Ee, J. C. Hooghart, H. Salz and F. C. Zuidema (ed. Committee) (1998) *Water in the Netherlands*, 162–66. Delft: Netherlands Hydrological Society (NHV).

VAN DER LINDEN, J. AND VERBEEK, M. (2001) 'Het meetnet flora en vegetatie van de Provincie Noord-Brabant'. *Gorteria* 27, 31–39.

VANNOTE, R. L., MINSHALL, G. W., CUMMINS, K. W., SEDELL, J. R. AND CUSHING, C. E. (1980) 'The river continuum concept'. *Canadian Journal of Fisheries and Aquatic Sciences* 37, 130–37.

VARDOULAKIS, S., FISHER, B. E. A., PERICLEOUS, K. AND GONZALEZ-FLESCA, N. (2003) 'Modelling air quality in street canyons: a review'. *Atmospheric Environment* 37, 155–82.

VEROSUB, K. L. (2000a) Varve dating. In J. S. Noller, J. M. Sowers and W. R. Lettis (eds) *Quaternary Geochronology: Methods and Applications* (AGU Reference Shelf 4), 21–24. Washington DC: American Geophysical Union.

VEROSUB, K. L. (2000b) Palaeomagnetic dating. In J. S. Noller, J. M. Sowers and W. R. Lettis (eds) *Quaternary Geochronology: Methods and Applications* (AGU Reference Shelf 4), 339–56. Washington DC: American Geophysical Union

VITOUSEK, P. M., EHRLICH, P. R., EHRLICH, A. H. AND MATSON, P. A. (1986) 'Human appropriation of the products of photosynthesis'. *BioScience* 36, 368–73.

VITOUSEK, P. M., MOONEY, H. A., LUBCHENCO, J. AND MELILLO, J. M. (1997) 'Human domination of Earth's ecosystems'. *Science* 277, 494–99.

VITOUSEK, P. M., HÄTTENSCHWILER, S., OLANDER, L. AND ALLISON, S. (2002) 'Nitrogen and nature'. *Ambio* 31, 97–101

VOOGT, J. A. AND OKE, T. R. (2003) 'Thermal remote sensing of urban climates'. *Remote Sensing of Environment* 86, 370–84.

WACE, N. M. (1977) 'Assessment of dispersal of plant species – the car-borne flora in Canberra'. *Proceedings of the Ecological Society of Australia* 10, 167–86.

WALKER, B. (2001) Tropical savanna. In F. S. Chapin III, O. E. Sala and E. Huber-Sannwald (eds) *Global Biodiversity in a Changing Environment: Scenarios for the 21st Century* (Ecological Studies 152), 139–56. Springer: New York.

WALKER, D. A., WEBBER, P. J., BINNIAN, E. F., EVERETT, K. R., LEDERER, N. D., NORDSTRAND, E. A. AND WALKER, M. D. (1987) 'Cumulative impacts of oil fields on northern Alaskan landscapes'. *Science* 238, 757–61.

WALKER, J. (1999) 'The application of geomorphology to the management of river-bank erosion'. *Journal of the Chartered Institution of Water and Environmental Management* 13, 297–300.

WALKER, M. D., GOULD, W. A. AND CHAPIN III, F. S. (2001) Scenarios of biodiversity changes in Arctic and alpine tundra. In F. S. Chapin III, O. E. Sala and E. Huber-Sannwald (eds) *Global Biodiversity in a Changing Environment: Scenarios for the 21st Century* (Ecological Studies 152), 83–100. Springer: New York.

WALTER, R. C. (1997) Potassium–argon/argon–argon dating methods. In R. E. Taylor and M. J. Aitken (eds) *Chronometric Dating in Archaeology*, 97–126. New York: Plenum Press.

WARBURTON, J. AND DANKS, M. (1998) Historical and contemporary channel change, Swinhope Burn. In J. Warburton (ed.) *Geomorphological Studies in the North Pennines: Field Guide*, 77–91. Durham: Department of Geography, University of Durham, British Geomorphological Research Group.

WARBURTON, J., DANKS, M. AND WISAHART, D. (2002) 'Stability of an upland gravel-bed stream, Swinhope Burn, Northern England'. *Catena* 49, 309–29.

WARD, J. K., ANTONOVICS, J., THOMAS, R. B. AND STRAIN, B. R. (2000) 'Is atmospheric CO_2 a selective agent on model C_3 annuals?'. *Oecologia* 123, 330–41.

WARREN, A. (2002) 'Land degradation is contextual'. *Land Degradation and Development* 13, 449–59.

WARREN, A., OSBAHR, H., BATTERBURY, S.

AND CHAPPELL, A. (2002) 'Indigenous views of soil erosion at Fandou Béri, southwestern Niger'. *Geoderma* 111, 439–56.

WASHINGTON, W. M. AND MEEHL, G. A. (1991) Characteristics of coupled atmosphere–ocean sensitivity experiments with different ocean formulations. In M. E. Schlesinger (ed.) *Greenhouse-gas-induced Climatic Change: A Critical Appraisal of Simulations and Observations*, 79–110. Amsterdam: Elsevier.

WATSON, A. J. (1999) Coevolution of the Earth's environment and life: Goldilocks, Gaia and the anthropic principle. In G. Y. Craig and J. H. Hull (eds) *James Hutton – Present and Future* (Geological Society, London, Special Publication 150), 75–88. London: The Geological Society.

WATSON, A. J. AND LOVELOCK, J. E. (1983) 'Biological homeostasis of the global environment'. *Tellus* 35B, 284–89.

WATSON, R. T., RODHE, H., OESCHGER, H. AND SIEGENTHALER, U. (1990) Greenhouse gases and aerosols. In J. T. Houghton, G. J. Jenkins and J. J. Ephraums (eds) *Climate Change: The IPCC Assessment*, 1–40. Cambridge: Cambridge University Press.

WATTS, S. AND HALLIWELL, L. (eds) (1996) *Essential Environmental Science: Methods and Techniques*. London: Routledge.

WBGU (WISSENSCHAFTLICHER BERIAT DER BUNDESREGIERUNG GLOBALE UMWELTVERÄNDERUNGEN; GERMAN ADVISORY COUNCIL ON GLOBAL CHANGE) (1996) *World in Transition: The Research Challenge Annual Report 1996*. Berlin: Springer. Internet reference http://www.wbgu.de/wgbu_jg1996_engl.pdf. Last accessed December 2003.

WBGU (1999) *World in Transition: Ways Towards Sustainable Management of Freshwater Resources*. German Advisory Council on Global Change (WBGU)'. Berlin: Springer Verlag. 1999 ANNUAL REPORT 1997. http://www.wbgu.de/wbgu_jg1997_engl.html. Last accessed April 2004.

WCED (1987) *Our Common Future*. Oxford: Oxford University Press for the UN World Commission on Economy and Environment.

WEBB, P. AND ISKANDARANI, M. (1998) *Water Insecurity and the Poor: Issues and Research Needs* (ZEF – Discussion Papers On Development Policy No. 2). Bonn: Center for Development Research.

WEBB, T. H. AND BURGHAM, S. J. (1997) 'Soil–landscape relationships of downland soils formed from loess, eastern South Island, New Zealand'. *Australian Journal of Soil Research* 35, 827–42.

WEBER, R. M. AND DUNNO, G. A. (2001) 'Riparian vegetation mapping and image processing techniques, Hopi Indian Reservation, Arizona'. *Photogrammetric Engineering and Remote Sensing* 67, 179–86.

WEHMILLER, J. F. AND MILLER, G. H. (2000) Aminostratigraphic dating methods. In J. S. Noller, J. M. Sowers and W. R. Lettis (eds) *Quaternary Geochronology: Methods and Applications* (AGU Reference Shelf 4), 187–222. Washington DC: American Geophysical Union.

WEIER, J. (2003) *Putting Earthquakes in their Place – A New Map of Global Tectonic Activity*. Internet reference NASA website http://earthobservatory.nasa.gov. Last accessed September 2003.

WELLS, N. (1997) *The Atmosphere and Ocean: A Physical Introduction*, 2nd edn. Chichester: John Wiley & Sons.

WENTZEL, J., STEPHENS, J. C., JOHNSON, W., MENOTTI-RAYMOND, M., PECON-SLATTERY, J., YUHKI, N., CARRINGTON, M., QUIGLEY, H. B., MIQUELLE, D. G., TISLON, R., MANANSANG, J., BRADY, G., ZHI, L., WENSHI, P., SHI-QIANG, H., JOHNSTON, L., SUNQUIST, M., KARANTH, K. U. AND O'BRIEN, S. J. (1999) Subspecies of tigers: molecular assessment using 'voucher specimens' of geographically traceable individuals. In J. Seidensticker, S. Christie and P. Jackson (eds) *Riding the Tiger: Tiger Conservation in Human-dominated Landscapes*, 409. London: The Zoological Society of London; Cambridge: Cambridge University Press.

WESTGATE, J., SANDHU, A. AND SHANE, P. (1997) Fission-track dating. In R. E. Taylor and M. J. Aitken (eds) *Chronometric Dating in Archaeology*, 127–58. New York: Plenum Press.

WHITE, K., BRYANT, R. AND DRAKE, N. (1998) 'Techniques for measuring rock weathering: application to a dated fan segment in southern Tunisia'. *Earth Surface Processes and Landforms* 23, 1031–43.

WHITE, R., MURRAY, S. AND ROHWEDER, M. (2000) *PAGE Technical Report – Grassland Ecosystems*. World Resources Institute. Internet reference at http://www.wri.org. Last accessed November 2003.

WHITTEN, T., MUSTAFA, M. AND HENDERSON, G. S. (2002) *The Ecology of Sulawesi*, 2nd edn (The Ecology of Indonesia Series, Volume IV) Hong Kong: Periplus Edition (HK) Ltd.

WIGLEY, T. M. L. AND RAPER, S. C. B. (1993) 'Thermal expansion of sea water associated with global warming'. *Nature* 330, 127–31.

WILLIAMS, M., DUNKERLEY, D., DE DECKKER, P., KERSHAW, P. AND CHAPPELL, J. (1998) *Quaternary Environments*, 2nd edn. London: Arnold.

WILLIAMS, R. B. G. AND ROBINSON, D. A. (1981) 'Weathering of sandstone by the combined action of frost and salt'. *Earth Surface Processes and Landforms* 6, 1–9.

WILLIS, I. C., ARNOLD, N. S., SHARP, M. J., BONUM, J.–M. AND HUBBARD, B. P. (1998) Mass balance and flow variations of Haut Glacier d'Arrola, Switzerland, calculated using digital terrain modelling techniques. In S. Lane, K. Richards and J. Chandler (eds) *Landform Monitoring, Modelling and Analysis*, 343–61. Chichester: John Wiley & Sons.

WILSON, J. S., CLAY, M., MARTIN, E., STUCKEY, D. AND VEDDER-RISCH, K. (1993) 'Evaluating environmental influences of zoning in urban ecosystems with remote sensing'. *Remote Sensing of Environment* 86, 303–21.

WILSON, L. (1993) 'A dumpsite guilty of the greatest crime'. *Waste Management* 1993 (October), 44–46.

WIND EROSION AND WATER CONSERVATION (WEWC), USDA Agricultural Research Service (2003) Selected WEWC Techniques and Tools. http://www.lbk.ars.usda.gov/wewc/tooltech.htm. Last accessed March 2004.

WISCHMEIER, W. H. AND SMITH, D. D. (1978) *Predicting Rainfall Erosion Losses – A Guide to Conservation Planning* (US Department of Agriculture, Handbook 537). Washington, DC: US Government Printing Office.

WISMAR, R. C. (1993) 'The need for long-term stream monitoring programs in forest ecosystems of the Pacific Northwest'. *Environmental Monitoring and Assessment* 26, 219–34.

WOOD, C. M., LEE, N, LUKER, J. A. AND SAUNDERS, P. J. W. (1974) *The Geography of Pollution – A Study of Greater Manchester*. Manchester: Manchester University Press.

WOOD, S., SEBASTIAN, K. AND SCHERR, S. J. (2000) *PAGE Technical Report – Agro-ecosystems*. A joint report of the World Resources Institute and Food Policy Research Institute (IFPRI). Internet reference at http://www.wri.org. Last accessed November 2003.

WOODRUFF, N. P. AND SIDDOWAY, F. H. (1965) 'A wind erosion equation'. *Proceedings of the Soil Science Society of America* 29 602–608.

WOODS, M. AND MORIARTY, P. V. (2001) 'Strangers in a strange land: the problem of exotic species'. *Environmental Values* 10, 163–91.

WCD (WORLD COMMISSION ON DAMS) (2000) *Dams And Development: A New Framework For Decision-Making*. Cape Town, South Africa: World Commission on Dams.

WORLD HEALTH ORGANIZATION (1996) *Trace Elements in Human Nutrition and Health*. Geneva: WHO.

WORLD METEOROLOGICAL ORGANISATION (WMO) (1994) *Guide to Hydrological Practices*, 5th edn (WMO Publication Number 168). Geneva: World Meteorological Organization.

WORLD METEOROLOGICAL ORGANISATION (WMO) (2003) *Twenty-First Status Report on Implementation of the World Weather Watch*. WMO No. 957. Internet reference available at http://www.wmo.int/web/www/Status-Reports/21st/index.html. Last accessed April 2004.

WORLD RESOURCES INSTITUTE (1999) *Urban Growth*. Washington DC: World Resources Institute.

WORLD RESOURCES INSTITUTE (2003) *Jakarta: a Booming Megacity*. Internet reference at http://www.wri.org/enved/suscom-jakarta.htm. Last accessed November 2003.

WORLD RESOURCES INSTITUTE ECOSYSTEMS AND HUMAN WELLBEING (2003) *Ecosystems and Human Well-Being: A Framework for Assessment* World Resources Institute. Internet reference at http://www.wri.org. Last accessed November 2003.

WORLD RESOURCES INSTITUTE, UNITED NATIONS ENVIRONMENT PROGRAMME, UNITED NATIONS DEVELOPMENT PROGRAMME AND THE WORLD BANK (1996a) *World Resources 1996-7: The Urban Environment*. New York: Oxford University Press.

WORLD RESOURCES INSTITUTE, UNITED NATIONS ENVIRONMENT PROGRAMME, UNITED NATIONS DEVELOPMENT PROGRAMME AND THE WORLD BANK (1996b) *World Resources 1996–7, The Urban Environment*. Oxford University Press: New York.

WORSTER, D. (1994) *Nature's Economy: A History of Ecological Ideas*, 2nd edn. Cambridge: Cambridge University Press.

WU, J., VANKAT, J. L. AND BARLAS, Y. (1993) 'Effects of patch connectivity and arrangement on animal metapopulation dynamics: a simulation study'. *Ecological Modelling* 65, 221–54.

WYATT, B. K. (2000) Vegetation mapping from ground, air and space – competitive or complementary techniques. In R. W. Alexander and A. C. Millington (eds) *Vegetation Mapping: From Patch to Planet*, 3–15. Chichester: John Wiley & Sons.

YOUNG, D. AND DENG, H. H. (1998) 'Urbanization, agriculture and industrialization in China 1952–91'. *Urban Studies* 35, 1439–55.

ZHU, Y. (1998) 'Formal and informal urbanization in China – trends in Fujan province'. *Third World Planning Review* 20, 267–84.

ZREDA, M. G. AND PHILLIPS, F. M. (2000) Cosmogenic nuclide buildup in surficial materials. In J. S. Noller, J. M. Sowers and W. R. Lettis (eds) *Quaternary Geochronology: Methods and Applications* (AGU Reference Shelf 4), 61–76. Washington DC: American Geophysical Union.

INDEX